T0220802

Communications in Computer and Information Science   **2061**

Editorial Board Members

Joaquim Filipe ⓘ, *Polytechnic Institute of Setúbal, Setúbal, Portugal*
Ashish Ghosh ⓘ, *Indian Statistical Institute, Kolkata, India*
Raquel Oliveira Prates ⓘ, *Federal University of Minas Gerais (UFMG),*
Belo Horizonte, Brazil
Lizhu Zhou, *Tsinghua University, Beijing, China*

Rationale

The CCIS series is devoted to the publication of proceedings of computer science conferences. Its aim is to efficiently disseminate original research results in informatics in printed and electronic form. While the focus is on publication of peer-reviewed full papers presenting mature work, inclusion of reviewed short papers reporting on work in progress is welcome, too. Besides globally relevant meetings with internationally representative program committees guaranteeing a strict peer-reviewing and paper selection process, conferences run by societies or of high regional or national relevance are also considered for publication.

Topics

The topical scope of CCIS spans the entire spectrum of informatics ranging from foundational topics in the theory of computing to information and communications science and technology and a broad variety of interdisciplinary application fields.

Information for Volume Editors and Authors

Publication in CCIS is free of charge. No royalties are paid, however, we offer registered conference participants temporary free access to the online version of the conference proceedings on SpringerLink (http://link.springer.com) by means of an http referrer from the conference website and/or a number of complimentary printed copies, as specified in the official acceptance email of the event.

CCIS proceedings can be published in time for distribution at conferences or as post-proceedings, and delivered in the form of printed books and/or electronically as USBs and/or e-content licenses for accessing proceedings at SpringerLink. Furthermore, CCIS proceedings are included in the CCIS electronic book series hosted in the SpringerLink digital library at http://link.springer.com/bookseries/7899. Conferences publishing in CCIS are allowed to use Online Conference Service (OCS) for managing the whole proceedings lifecycle (from submission and reviewing to preparing for publication) free of charge.

Publication process

The language of publication is exclusively English. Authors publishing in CCIS have to sign the Springer CCIS copyright transfer form, however, they are free to use their material published in CCIS for substantially changed, more elaborate subsequent publications elsewhere. For the preparation of the camera-ready papers/files, authors have to strictly adhere to the Springer CCIS Authors' Instructions and are strongly encouraged to use the CCIS LaTeX style files or templates.

Abstracting/Indexing

CCIS is abstracted/indexed in DBLP, Google Scholar, EI-Compendex, Mathematical Reviews, SCImago, Scopus. CCIS volumes are also submitted for the inclusion in ISI Proceedings.

How to start

To start the evaluation of your proposal for inclusion in the CCIS series, please send an e-mail to ccis@springer.com.

Linqiang Pan · Yong Wang · Jianqing Lin
Editors

Bio-Inspired Computing: Theories and Applications

18th International Conference, BIC-TA 2023
Changsha, China, December 15–17, 2023
Revised Selected Papers, Part I

 Springer

Editors
Linqiang Pan ⓘ
Huazhong University of Science
and Technology
Wuhan, China

Yong Wang
Central South University
Changsha, China

Jianqing Lin
Huazhong University of Science
and Technology
Wuhan, China

ISSN 1865-0929   ISSN 1865-0937 (electronic)
Communications in Computer and Information Science
ISBN 978-981-97-2271-6   ISBN 978-981-97-2272-3 (eBook)
https://doi.org/10.1007/978-981-97-2272-3

© The Editor(s) (if applicable) and The Author(s), under exclusive license
to Springer Nature Singapore Pte Ltd. 2024

This work is subject to copyright. All rights are solely and exclusively licensed by the Publisher, whether the whole or part of the material is concerned, specifically the rights of translation, reprinting, reuse of illustrations, recitation, broadcasting, reproduction on microfilms or in any other physical way, and transmission or information storage and retrieval, electronic adaptation, computer software, or by similar or dissimilar methodology now known or hereafter developed.
The use of general descriptive names, registered names, trademarks, service marks, etc. in this publication does not imply, even in the absence of a specific statement, that such names are exempt from the relevant protective laws and regulations and therefore free for general use.
The publisher, the authors and the editors are safe to assume that the advice and information in this book are believed to be true and accurate at the date of publication. Neither the publisher nor the authors or the editors give a warranty, expressed or implied, with respect to the material contained herein or for any errors or omissions that may have been made. The publisher remains neutral with regard to jurisdictional claims in published maps and institutional affiliations.

This Springer imprint is published by the registered company Springer Nature Singapore Pte Ltd.
The registered company address is: 152 Beach Road, #21-01/04 Gateway East, Singapore 189721, Singapore

Paper in this product is recyclable.

Preface

Bio-inspired computing is a field of study that abstracts computing ideas (data structures, operations with data, ways to control operations, computing models, artificial intelligence, multisource data-driven analysis, etc.) from living phenomena or biological systems such as cells, tissue, the brain, neural networks, immune systems, ant colonies, evolution, etc. The areas of bio-inspired computing include Neural Networks, Brain-Inspired Computing, Neuromorphic Computing and Architectures, Cellular Automata and Cellular Neural Networks, Evolutionary Computing, Swarm Intelligence, Fuzzy Logic and Systems, DNA and Molecular Computing, Membrane Computing, Artificial Intelligence and its Application in other disciplines such as machine learning, deep learning, image processing, computer science, cybernetics, etc. Bio-Inspired Computing: Theories and Applications (BIC-TA) is a series of conferences that aims to bring together researchers working in the main areas of bio-inspired computing, to present their recent results, exchang ideas, and cooperate in a friendly framework.

Since 2006, BIC-TA has taken place at Wuhan (2006), Zhengzhou (2007), Adelaide (2008), Beijing (2009), Liverpool and Changsha (2010), Penang (2011), Gwalior (2012), Anhui (2013), Wuhan (2014), Anhui (2015), Xi'an (2016), Harbin (2017), Beijing (2018), Zhengzhou (2019), Qingdao (2020), Taiyuan (2021), and Wuhan (2022). Following the success of the previous editions, the 18th International Conference on Bio-Inspired Computing: Theories and Applications (BIC-TA 2023) was held in Changsha, China, during December 15–17, 2023, organized by Central South University and co-organized by Xiangtan University, with the support of Operations Research Society of Hubei.

We would like to thank the keynote speakers for their excellent presentations: Weigang Chen (Tianjin University, China), Shaoliang Peng (Hunan University, China), Ke Tang (Southern University of Science and Technology, China), and Xingyi Zhang (Anhui University, China).

A special thank you is given to the general chair, Chunhua Yang, for her guidance and support of the conference. We gratefully thank Yalin Wang and Chengqing Li, for their warm welcome and inspiring speeches at the opening ceremony of the conference. We thank the local chair, Juan Zou, for her significant contribution to the organization and guidance of the conference. We thank Shouyong Jiang, Bing-chuan Wang, Zhi-zhong Liu, and Pei-qiu Huang, for their contribution in organizing the conference. We thank Zixiao Zhang for his help in collecting the final files of the papers and editing the volume and maintaining the website of BIC-TA 2023 (http://2023.bicta.org/). We also thank all the other volunteers, whose efforts ensured the smooth running of the conference.

BIC-TA 2023 received 168 submissions on various aspects of bio-inspired computing, of which 64 papers were selected for the two volumes of *Communications in Computer and Information Science*. Each paper was peer reviewed by at least three reviewers with expertise in the relevant subject area in a single-blind peer-review process. The warmest thanks should be given to the reviewers for their careful and efficient work in the reviewing process.

Special thanks are due to Springer Nature for their skilled cooperation in the timely production of these volumes.

February 2024

Linqiang Pan
Yong Wang
Jianqing Lin

Organization

Steering Committee

Xiaochun Cheng	Middlesex University London, UK
Guangzhao Cui	Zhengzhou University of Light Industry, China
Kalyanmoy Deb	Michigan State University, USA
Miki Hirabayashi	National Institute of Information and Communications Technology, Japan
Joshua Knowles	University of Manchester, UK
Thom LaBean	North Carolina State University, USA
Jiuyong Li	University of South Australia, Australia
Kenli Li	University of Hunan, China
Giancarlo Mauri	Università di Milano-Bicocca, Italy
Yongli Mi	Hong Kong University of Science and Technology, China
Atulya K. Nagar	Liverpool Hope University, UK
Linqiang Pan (Chair)	Huazhong University of Science and Technology, China
Gheorghe Păun	Romanian Academy, Romania
Mario J. Perez-Jimenez	University of Seville, Spain
K. G. Subramanian	Liverpool Hope University, UK
Robinson Thamburaj	Madras Christian College, India
Jin Xu	Peking University, China
Hao Yan	Arizona State University, USA

General Chair

Chunhua Yang	Central South University, China

Program Committee Chairs

Yong Wang	Central South University, China
Linqiang Pan	Huazhong University of Science and Technology, China

Publication Chair

Bingchuan Wang   Central South University, China

Publicity Chair

Guangwu Liu   Wuhan University of Technology, China

Local Chair

Juan Zou   Xiangtan University, China

Registration Chair

Shouyong Jiang   Central South University, China

Program Committee

Muhammad Abulaish	South Asian University, India
Andy Adamatzky	University of the West of England, UK
Guangwu Liu	Wuhan University of Technology, China
Chang Wook Ahn	Gwangju Institute of Science and Technology, South Korea
Adel Al-Jumaily	University of Technology Sydney, Australia
Bin Cao	Hebei University of Technology, China
Junfeng Chen	Hohai University, China
Wei-Neng Chen	Sun Yat-sen University, China
Shi Cheng	Shaanxi Normal University, China
Xiaochun Cheng	Middlesex University London, UK
Tsung-Che Chiang	National Taiwan Normal University, China
Sung-Bae Cho	Yonsei University, South Korea
Zhihua Cui	Taiyuan University of Science and Technology, China
Kejie Dai	Pingdingshan University, China
Ciprian Dobre	University Politehnica of Bucharest, Romania
Bei Dong	Shanxi Normal University, China
Xin Du	Fujian Normal University, China
Carlos Fernandez-Llatas	Universitat Politècnica de València, Spain

Shangce Gao	University of Toyama, Japan
Marian Gheorghe	University of Bradford, UK
Wenyin Gong	China University of Geosciences, China
Shivaprasad Gundibail	Manipal Academy of Higher Education, India
Ping Guo	Beijing Normal University, China
Yinan Guo	China University of Mining and Technology, China
Guosheng Hao	Jiangsu Normal University, China
Cheng He	Southern University of Science and Technology, China
Shan He	University of Birmingham, UK
Tzung-Pei Hong	National University of Kaohsiung, China
Pei-qiu Huang	Central South University, China
Florentin Ipate	University of Bucharest, Romania
Sunil Kumar Jha	Banaras Hindu University, India
He Jiang	Dalian University of Technology, China
Qiaoyong Jiang	Xi'an University of Technology, China
Shouyong Jiang	Central South University, China
Licheng Jiao	Xidian University, China
Liangjun Ke	Xi'an Jiaotong University, China
Ashwani Kush	Kurukshetra University, India
Hui Li	Xi'an Jiaotong University, China
Kenli Li	Hunan University, China
Lianghao Li	Huazhong University of Science and Technology, China
Yangyang Li	Xidian University, China
Zhihui Li	Zhengzhou University, China
Jing Liang	Zhengzhou University, China
Jerry Chun-Wei Lin	Western Norway University of Applied Sciences, Norway
Jianiqng Lin	Huazhong University of Science and Technology, China
Qunfeng Liu	Dongguan University of Technology, China
Xiaobo Liu	China University of Geosciences, China
Zhi-zhong Liu	Hunan University, China
Wenjian Luo	University of Science and Technology of China, China
Lianbo Ma	Northeastern University, China
Wanli Ma	University of Canberra, Australia
Xiaoliang Ma	Shenzhen University, China
Francesco Marcelloni	University of Pisa, Italy
Efrén Mezura-Montes	University of Veracruz, Mexico

Hongwei Mo	Harbin Engineering University, China
Chilukuri Mohan	Syracuse University, USA
Abdulqader Mohsen	University of Science and Technology in Yemen, Yemen
Holger Morgenstern	Albstadt-Sigmaringen University, Germany
Andres Muñoz	Universidad Católica San Antonio de Murcia, Spain
G. R. S. Murthy	Lendi Institute of Engineering and Technology, India
Akila Muthuramalingam	KPR Institute of Engineering and Technology, India
Yusuke Nojima	Osaka Prefecture University, Japan
Linqiang Pan	Huazhong University of Science and Technology, China
Andrei Paun	University of Bucharest, Romania
Gheorghe Păun	Romanian Academy, Romania
Xingguang Peng	Northwestern Polytechnical University, China
Chao Qian	University of Science and Technology of China, China
Balwinder Raj	NITTTR, India
Rawya Rizk	Port Said University, Egypt
Rajesh Sanghvi	G. H. Patel College of Engineering and Technology, India
Ronghua Shang	Xidian University, China
Zhigang Shang	Zhengzhou University, China
Ravi Shankar	Florida Atlantic University, USA
V. Ravi Sankar	GITAM University, India
Bosheng Song	Hunan University, China
Tao Song	China University of Petroleum, China
Jianyong Sun	University of Nottingham, UK
Yifei Sun	Shaanxi Normal University, China
Bing-chuan Wang	Central South University, China
Handing Wang	Xidian University, China
Yong Wang	Central South University, China
Hui Wang	Nanchang Institute of Technology, China
Hui Wang	South China Agricultural University, China
Gaige Wang	Ocean University of China, China
Sudhir Warier	IIT Bombay, India
Slawomir T. Wierzchon	Polish Academy of Sciences, Poland
Zhou Wu	Chongqing University, China
Xiuli Wu	University of Science and Technology Beijing, China
Bin Xin	Beijing Institute of Technology, China

Gang Xu	Nanchang University, China
Yingjie Yang	De Montfort University, UK
Zhile Yang	Shenzhen Institute of Advanced Technology, Chinese Academy of Sciences, China
Kunjie Yu	Zhengzhou University, China
Xiaowei Zhang	University of Science and Technology of China, China
Jie Zhang	Newcastle University, UK
Gexiang Zhang	Chengdu University of Technology, China
Defu Zhang	Xiamen University, China
Haiyu Zhang	Wuhan University of Technology, China
Peng Zhang	Beijing University of Posts and Telecommunications, China
Weiwei Zhang	Zhengzhou University of Light Industry, China
Yong Zhang	China University of Mining and Technology, China
Xinchao Zhao	Beijing University of Posts and Telecommunications, China
Yujun Zheng	Zhejiang University of Technology, China
Aimin Zhou	East China Normal University, China
Fengqun Zhou	Pingdingshan University, China
Xinjian Zhuo	Beijing University of Posts and Telecommunications, China
Shang-Ming Zhou	Swansea University, UK
Dexuan Zou	Jiangsu Normal University, China
Juan Zou	Xiangtan University, China
Xingquan Zuo	Beijing University of Posts and Telecommunications, China

Contents – Part I

Membrane Computing and DNA Computing

Contents – Part II

Machine Learning and Applications

Intelligent Control and Application

Evolutionary Computation and Swarm Intelligence

Collaborative Scheduling of Multi-cloud Distributed Multi-cloud Tasks Based on Evolutionary Multi-tasking Algorithm

Tianhao Zhao(✉), Linjie Wu, Zhihua Cui, and Xingjuan Cai

Shanxi Laboratory of Big Data Analysis and Parallel Computing, Taiyuan University
of Science and Technology, Taiyuan 030024, China
zhaotianhao1015@163.com

Abstract. With the rapid development of information technology making the scale of the Internet increasing day by day, collaborative optimization of multiple scheduling tasks in a multi-cloud environment provides users with faster scheduling options. Meanwhile, there is a certain similarity between cloud scheduling tasks, and in order not to waste the similarity between tasks, similar tasks are linked together to find an optimal scheduling solution for multiple tasks, making it possible to handle multiple scheduling tasks simultaneously. Firstly, we construct a multi-objective optimization model considering time, cost and VM resource load balance; secondly, since there are not only independent optimization problems in real scenarios, we adapt the constructed multiple similar optimization models and propose a multi-task multi-objective optimization model; finally, to be able to solve the constructed model better, we use a proposed objective function-based Finally, we propose an evolutionary multitasking algorithm based on weighted summation of the objective functions, which allows the algorithm to find the optimal solution among multiple multi-objective models. Simulation experiments show that the proposed algorithm has better performance.

Keywords: Evolutionary Multitasking algorithm · Task scheduling · Cloud Computing · Optimization

1 Introduction

With the rapid development of science and technology, cloud computing technology has brought a revolution in information technology and other fields through its efficient and powerful architecture, and computing technology has been widely used for personal and commercial purposes [1]. Cloud computing is known as the primary solution for complex computing and large-scale data manipulation [8,20]. Cloud computing has brought extreme convenience to users and businesses with its advantages of hyper-scale, virtualization, high reliability, versatility, high scalability, and pay-as-you-go. The large-scale task scheduling problem makes it a highly researchable NP-hard problem due to the complexity of its

© The Author(s), under exclusive license to Springer Nature Singapore Pte Ltd. 2024
L. Pan et al. (Eds.): BIC-TA 2023, CCIS 2061, pp. 3–13, 2024.
https://doi.org/10.1007/978-981-97-2272-3_1

tasks and the heterogeneity of its resources. Currently, many scholars have done a lot of work on the cloud computing task scheduling problem, including cost, time, resource utilization, energy consumption, load, and many other objective problems [13,21].

To meet the demand of today's cloud computing services that may need to solve multiple optimization tasks simultaneously [15], we propose to construct a multi-task cloud computing scheduling task model based on evolutionary multi-task optimization algorithms to solve the problem of multi-task scheduling requirements on the cloud, improve computational efficiency, and be able to handle different cloud computing task scheduling problems simultaneously. Multi-task learning is a branch of migration learning, where multi-task processing is dedicated to solving different optimization problems or tasks simultaneously. Currently, evolutionary multitasking optimization has become a key technique for solving problems in a wide range of scientific and engineering fields, and the idea of this class of algorithms is based on the similarity between tasks and finding a promising solution to solve each task.

The current cloud computing scheduling problem is due to the complexity of its tasks and the heterogeneity of its resources, which makes it an NP-hard problem with great research value, at the same time, with the development of information technology, the demand for multi-tasking is becoming more and more obvious, and the solution of a single task wastes away the similarity between the tasks and occupies more computational resources, and the advantages of parallel processing of multiple tasks lie in the saving of computational resources, faster convergence speed, and satisfying the needs of multiple tasks at the same time. The current evolutionary multitasking algorithm [12] has a very good research prospect as a new research direction of the current flow shape.

In this paper, we design a multi-task based multi-objective cloud computing scheduling model to meet the demand of multi-tasking on cloud computing and consider different types of scheduling tasks while designing a multi-task-based multi-objective model that preserves the individuality of each task while finding the most suitable scheduling scheme to meet the demand of each task at the same time. By analyzing the characteristics of the model, the traditional evolutionary multitasking algorithm is improved to improve the scheduling efficiency of resources, increase the occurrence of positive migration, and reduce negative migration, thus solving multiple optimization problems simultaneously. The main research objective of this paper is to obtain the best solution to solve multiple scheduling problems simultaneously by constructing multiple multi-objective cloud computing scheduling task models, then constructing multiple models into a cloud computing scheduling task model based on multitasking processing, and then improving the classical evolutionary multitasking algorithm MFEA to improve the optimal allocation strategy for cloud tasks. The main contributions of this paper are as follows:

1. The constructed multiple multi-objective cloud scheduling model is adjusted and constructed into a multi-task multi-objective cloud scheduling model.

2. An evolutionary multitasking algorithm that solves multitasking multi-objective models is proposed, which enables the algorithm to better handle the constructed models by aggregating multiple objective function values and treating them as multitasking single-objective models.

The remaining Sections of this paper are organized as follows. In Sect. 2, the existing related work and the current status of the research are presented. In Sect. 3, the flow and design of the proposed framework are presented. Section 4 gives the benchmark test suits and parameter settings used for the experiments. The experimental results and analysis are presented. Finally, a detailed summary and an outlook for future work are given in the Conclusion.

2   Related Work

Researchers have done a lot of research work on the virtual machine scheduling problem in cloud computing. Task scheduling is an important factor in cloud computing to utilize the resources efficiently and meet the user's needs. In the past few years, researchers have introduced various task-scheduling algorithms and discussed some related works. GENG et al. [6] design a multi-constrained cloud task scheduling model for the perspective that only the objectives are considered in the currently proposed cloud task scheduling process and fewer constraints cannot accurately describe the problem, and construct a multi-cloud model-based scheduling problem in [3]. Considering both users and providers and setting up clouds of different sizes to solve the task scheduling problem in a multi-cloud environment. Jialei Xu et al. [23] consider that the randomness, operation style, and unpredictability of user requirements in a cloud environment pose great challenges for task scheduling, and construct an objective function based on user satisfaction from the user's perspective. Rjoub et al. [17] proposed the trust-aware task scheduling model and reported, that it improves the scheduling performance and also improves the security characteristics of cloud computing. Three state processes calculate the virtual machine trust level, prioritize the tasks and finally schedule the tasks in a trust-aware manner. This trust model achieves better performance with the shortest completion time compared to existing techniques. In workflow applications, task scheduling length is considered to avoid SLA conflicts and deadline constraints. For parallel tasks in cloud systems, minimizing the task scheduling length is critical. To this end, Mugunthan et al. [14] proposed an efficient priority and relative distance algorithm. An efficient priority and relative distance algorithm is proposed, which prioritizes tasks in the first stage and maps tasks to virtual machines based on the relative distance to improve the resource utilization of the cloud. Huang et al. [9] describe the task offloading problem for edge environments as a two-layer optimisation problem, with an optimal offloading decision problem in the upper layer and a computational resource allocation problem in the lower layer, which sufficiently tightly combines the dependencies between the two problems and represents the offloading problem in a new problem paradigm.

The computing application virtualization technology divides the huge amount of physical resources into various virtual resources [22]. A large number of users can use these virtual resources on the cloud platform anytime and anywhere [2]. With the application and development of cloud computing, big data, and artificial intelligence, more and more applications are deployed on the cloud platform and the scale of resources is increasing. The demand for cloud resources proposed to them also shows diversity, suddenness, and emergence. Most of the existing cloud resource allocation methods do not support contingency mode and cannot guarantee the timeliness and optimization of resource allocation. However, subscribers are more focused on the timeliness and optimization of emergent resource requirements, and cloud service providers are highly concerned about managing massive resources and improving resource utilization. To achieve these goals, an effective resource allocation methodology is critical. The resource allocation process is a virtual machine placement problem, i.e., finding the right physical servers to place virtual machines (VMs). This process not only meets the resource requirements of VMs but also improves the resource utilization of the cloud platform. Finding the optimal solution is also a problem. Some simple heuristic algorithms such as RR (Round Robin) [16], BF (Best Fit) [18], and Min-Max [10] are used in the resource allocation process for small cloud platforms. These algorithms are simple and easy to implement, but they also tend to waste resources, especially in large cloud platforms.

Meanwhile, the multi-task scheduling problem exists in many application areas, such as the Internet service field [11] and the medical field [5]. In response to the above research characteristics, it can be found that the current scholars are studying the cloud task scheduling problem by optimizing for a single task and proposing innovations by considering the objective function of the user or provider; and by solving the cloud task scheduling problem for scheduling tasks that may appear unexpectedly and by using hybrid algorithms with multiple policies. However, with the growth of information technology and the increasing number of scheduling tasks, there will be a need to solve multiple optimization tasks simultaneously in current cloud computing services. To meet this need, a cloud computing scheduling model based on multitasking needs to be proposed, which establishes a multitasking model and uses an evolutionary multitasking algorithm to solve the problem, using a single population to optimize multiple cloud computing scheduling problems simultaneously. Multitasking has both explicit and implicit processing, and a better-targeted knowledge migration method is selected according to different tasks. Considering the multiple task scheduling problems that exist on the cloud, the next scheduling problem can only be executed after the completion of one problem by optimization. Although finding synergistic relationships between multiple tasks is considered, how to optimize multiple scheduling problems at the same time, consider the similarity between tasks and tasks, and control the strength or weakness of the amount of knowledge migration or the appropriate occurrence of knowledge migration according to the similarity degree of tasks, and by migrating knowledge Finding an optimal scheduling scheme that can satisfy these multiple scheduling problems simultaneously is the main problem addressed in this paper.

3   The Proposed Model and Algorithm

3.1   Building a Multi-objective Cloud Computing Scheduling Model

The scheduling problems of all scales in this paper address the same objective function and solve the same problem. This section describes the three objective functions of the constructed model:

(1) Task execution time

$$Total\_Time = \max(Time\_VM_i) \tag{1}$$

$$Time\_VM_i = \sum_{i=1}^{N_i} task_{k_i} \tag{2}$$

Each VM may be assigned multiple tasks, and all VMs execute in parallel in each task, i.e., the time with the maximum execution time of all VMs in each task is used as the total execution time of that task. $Time\_VM_i$ is the execution time of the $i - th$ VM. N_i is the number of tasks assigned on the $i - th$ VM, and $task_{ki}$ is the execution time of the task k on the $i - th$ VM.

(2) Task execution cost Design the billing method of cloud computing resources according to the bandwidth utilization. The billing part refers to the Ali-Cloud price setting. The bandwidth utilization rate is below ten percent according to 0.08 Yuan/G billing, bandwidth utilization rate is higher than ten percent according to the first 5Mdbs according to 0.063 \$/Mdbs billing, greater than part according to 0.25 \$/Mdbs billing.

$$Cost = \begin{cases} 0.08 \times (filesize + outputsize) & PerBW < 10\% \\ 5 \times 0.063 + (cloudletbw - 5) \times 0.25 & PerBW < 10\% \end{cases} \tag{3}$$

(3) Virtual Machine Load Balancing

To avoid assigning multiple tasks to the same VM, load balancing of VMs is considered by calculating the ratio of the demand for tasks on each VM to the computing power, the smaller the ratio represents the more balanced the energy consumption of processing tasks on the VM, and the VM load balancing objective function is expressed as:

$$LB_{vm} = \frac{\sqrt{\dfrac{\sum\limits_{i=1}^{M}\sum\limits_{j=1}^{N}(t_{ij} * \dfrac{T_{j\_cpu} + T_{j\_ram} + T_{j\_bw}}{VM_{i\_cpu} + VM_{i\_ram} + VM_{i\_bw}})}{N}}}{M} \tag{4}$$

where T_{j_cpu}, T_{j_ram} and T_{j_bw} denotes the computational power required by the $j-th$ independent task that would require the size of the allocated CPU, ram and bw resources, respectively; denotes the computational power of the $i - th$ VM for the three attributes of CPU, ram and bw.

In this paper, we consider the multitask scheduling problem on a multiple cloud, assuming the existing scenario of 2 scheduling tasks on the current cloud, and the number of independent tasks of all scheduling tasks is 300. The task length of each independent task in scheduling task 1 and task 2, the file size, the output size, and other attributes are controlled in different intervals, indicating that each task is not identical to the other. And the most suitable scheduling scheme is found to solve all scheduling tasks simultaneously. Figure 1 depicts the model design.

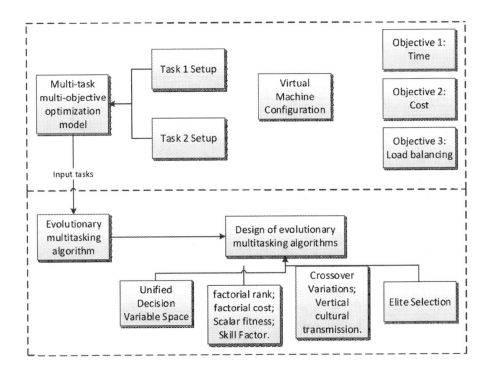

Fig. 1. Multi-tasking and multi-objective cloud computing scheduling model.

3.2   MFEA

In MFEA [7], all tasks evolve simultaneously by a population, and each individual is potentially responsible for each task. In this way, all the individuals need to be encoded in a uniform search space Y. Generally normalize Y to $[0, 1]^D$, where D is the dimension of the unified space so that tasks from different domains can co-evolve. In general, if there are K tasks to be optimized, the dimensionality of all tasks is $D_1, D_2, ..., D_K$. Then the dimensionality of the uniform search space (i.e., chromosome length) is $D = \max\{Dj\}$, where $j = 12...K$. Note that the solution encoded in space Y needs to be evaluated by decoding $x = Lj + (Uj - Lj) * y$ to the domain associated with the problem for its fitness value. Lj and Uj

correspond to the upper and lower bounds of task j, respectively, and x is the feasible solution for the corresponding task in the particular problem space X_j.

MFEA uses two strategies to generate offspring: 1. Selective mating: crossover of individuals with the same skill factor; 2. Vertical cultural propagation: direct inheritance from the parent when the skill factor of the offspring cannot be evaluated.

For the characteristics of the MFEA algorithm, a population is used to solve multiple problems simultaneously and optimally to find the most suitable set of solutions. First, unify the dimension of the search space. And calculate 4 attributes for each individual.

(1) Factorial cost: The factorial cost is defined as the value of the objective function of an individual task or the value obtained by some evaluation index, while the factor cost on other tasks is set to an infinite number (inf). For the deficiency of calculating the factorial cost only by a single objective function value in MFEA, the result of factorial cost is changed to the average value obtained by adding multiple objective function values as the calculated value of the solution cost.

(2) Factorial rank: Assuming that an individual belongs to the first task, the rank is defined as the index position of the individual in ascending order of the factor cost of all individuals evaluated for the first task.

(3) Skill factor: The skill factor determines which task the individual corresponds to and has the lowest factor rank in the corresponding task:

$$\tau = \mathrm{argmin}\{r_{ik}\} \tag{5}$$

(4) Scalar fitness: the scalar fitness of an individual is defined as the inverse of the factor ranking, as shown in Eq. 6. In the environmental selection operator, the individuals with larger scalar fitness survive to the next generation.

$$\phi = 1/r_{ik} \tag{6}$$

4   Experiment

This paper uses the Cloudsim platform for simulation and scheduling experiments. CloudSim is a function library developed on the discrete event simulation package SimJava, which can create a variety of entities in the cloud computing environment, including cloud data centers, hosts, services, agents, and virtual machines, and supports the processing of event queues, message passing in components and the management of simulation clocks. The experimental equipment is a processor Inter(R) Core(TM) i5-4200M, CPU @2.50 GHz, with 8G of memory it can meet the basic hardware requirements for simulation experiments.

The experiment sets the number of multitasking to 2, the number of independent tasks for each task is 300, and the task length, file size, and output size of scheduling task 1 are 500 1500, 300 900, and 300 900 respectively; the three attributes of task 2 are set to 600 1800, 100 300, 100 300; and 12 virtual machines of different configurations are set. Since the selected comparison algorithms, MOEAD [24], NSGA-III [4], GrEA [23], RVEA [3], KnEA [25], VaEA [19] are traditional multi-objective optimization algorithms, the results in Tables 1 to 4 show the experimental results for the scheduling task 1 model. From the experimental results, it can be seen that the performance on individual tasks has good results after the introduction of the multitasking concept.

1) PD: Pure diversity evaluation index, which is used to reflect the diversity of the solution set.
2) DM: Metric for diversity evaluation index.

Table 1. Performance metrics of different algorithms in the cloud task scheduling problem

Metrics	PD	DM
GrEA	3.2112e+3 (3.17e+3)	3.9297e−1 (7.85e−2)
MOEAD	3.2593e+4 (1.16e+4)	2.2372e−1 (4.89e−2)
NSGA-III	5.4990e+3 (4.31e+3)	4.4238e−1 (8.95e−2)
RVEA	3.8879e+3 (4.98e+3)	4.5290e−1 (1.06e−1)
KnEA	4.7189e+3 (3.24e+3)	3.8332e−1 (1.26e−1)
VaEA	4.2301e+3 (3.12e+3)	4.1494e−1 (1.64e−1)
MFEA	**4.7642e+4 (1.53e+4)**	**5.2494e−1 (1.18e−1)**

Table 1 shows the comparison of the evaluation metrics results of all the compared algorithms under the problem proposed in this paper. The experiments show that all the MFEA algorithms used in this paper obtain the best results.

In Table 2 we compare the experimental results of the three objective functions on the optimal, the worst, and the mean values, respectively, using the constructed multi-objective problem as an experimental comparison.

Table 2. Objective function values for different algorithms in cloud task scheduling problems.

	Algorithm	Task execution time	Task execution cost	VM load balancing
Best	GrEA	0.000513332	0.000613333	0.001323138
	MOEAD	0.046822487	0.014002542	0.120589401
	NSGA-III	0.000592789	0.000526764	0.000310926
	RVEA	0.000212332	0.00068655	**0.000120358**
	KnEA	**0.000152247**	0.000534836	0.000689693
	VaEA	0.000531305	**0.000522492**	0.000115574
	MFEA	0.001507207	0.031326227	0.00578593
Worst	GrEA	0.999999992	0.999821579	0.998232536
	MOEAD	0.983804354	0.964753327	0.989257331
	NSGA-III	0.999782095	0.99989639	0.99999997
	RVEA	0.999462296	0.997675656	0.998230828
	KnEA	0.996000857	0.999217738	0.999717749
	VaEA	0.998155433	0.999583803	0.99898225
	MFEA	0.998615337	**0.903476633**	0.993583871
Average	GrEA	0.279847479	0.334569544	0.379217445
	MOEAD	**0.983804354**	0.964753327	**0.989257331**
	NSGA-III	0.349325183	0.388216146	0.445261858
	RVEA	0.266792273	0.270037787	0.293097795
	KnEA	0.254749123	0.3060632	0.34417317
	VaEA	0.323209582	0.311636366	0.358110951
	MFEA	**0.115475791**	**0.284026783**	**0.215366863**

5   Conclusions

The cloud computing scheduling problem has been studied for many years, and the optimization model based on evolutionary algorithms has been recognized by scholars. The evolutionary multitasking algorithm has the advantages of saving computational resources and being able to process multiple tasks in parallel, while the optimization model based on the evolutionary multitasking algorithm has been applied to various fields. Meanwhile, to meet the demand for cloud task scheduling in cloud computing, it is necessary to design a multi-objective cloud computing scheduling model based on the evolutionary multitasking algorithm. In this paper, we analyze the similarity of multiple cloud computing scheduling task models on a multiple cloud and construct a multi-task multi-objective cloud computing task scheduling model by unifying the objective functions of multiple tasks into a unified representation. To address the above multi-task model and the shortcomings of the used evolutionary multi-task algorithm, the algorithm is endowed with the ability to re-process the multi-objective multi-task optimization algorithm.

   In future work, to address the shortcomings of the constructed multitasking model, and the evolutionary multitasking algorithm, the explicit and implicit knowledge migration approaches in the algorithm are combined, and the inclusion of an adaptive strategy is considered so that the model can adaptively select the appropriate solution and reduce the occurrence of negative transfer.

Acknowledgements. This work is supported by the Science and Technology Development Foundation of the Central Guiding Local under Grant No. YDZJSX2021A038; National Natural Science Foundation of China under Grant No.61806138; Postgraduate Joint Training Demonstration Base of Taiyuan University of Science and Technology Fund (Grant NO. JD2022003).

References

1. Addya, S.K., Satpathy, A., Ghosh, B.C., Chakraborty, S., Ghosh, S.K., Das, S.K.: CoMCLOUD: virtual machine coalition for multi-tier applications over multi-cloud environments. IEEE Trans. Cloud Comput. **11**(1), 956–970 (2021)
2. Armbrust, M., et al.: Above the clouds: a berkeley view of cloud computing. Technical report UCB/EECS-2009-28, EECS Department, University of California (2009)
3. Cai, X., Geng, S., Wu, D., Cai, J., Chen, J.: A multicloud-model-based many-objective intelligent algorithm for efficient task scheduling in Internet of Things. IEEE Internet Things J. **8**(12), 9645–9653 (2020)
4. Deb, K., Jain, H.: An evolutionary many-objective optimization algorithm using reference-point-based nondominated sorting approach, part I: solving problems with box constraints. IEEE Trans. Evol. Comput. **18**(4), 577–601 (2013)
5. Gao, L., Zhan, H., Sheng, V.S.: Mitigate gender bias using negative multi-task learning. Neural Process. Lett. **55**(8), 11131–11146 (2023)
6. Geng, S., Wu, D., Wang, P., Cai, X.: Many-objective cloud task scheduling. IEEE Access **8**, 79079–79088 (2020)
7. Gupta, A., Ong, Y.S., Feng, L.: Multifactorial evolution: toward evolutionary multitasking. IEEE Trans. Evol. Comput. **20**(3), 343–357 (2015)
8. He, X., Tu, Z., Wagner, M., Xu, X., Wang, Z.: Online deployment algorithms for microservice systems with complex dependencies. IEEE Trans. Cloud Comput. **11**(2), 1746–1763 (2023)
9. Huang, P.Q., Wang, Y., Wang, K., Liu, Z.Z.: A bilevel optimization approach for joint offloading decision and resource allocation in cooperative mobile edge computing. IEEE Trans. Cybern. **50**(10), 4228–4241 (2019)
10. Katyal, M., Mishra, A.: Application of selective algorithm for effective resource provisioning in cloud computing environment. arXiv preprint arXiv:1403.2914 (2014)
11. Liu, Y., Xu, X., Zhang, L., Wang, L., Zhong, R.Y.: Workload-based multi-task scheduling in cloud manufacturing. Robot. Comput.-Integr. Manuf. **45**, 3–20 (2017)
12. Lu, J., et al.: A multi-task oriented framework for mobile computation offloading. IEEE Trans. Cloud Comput. **10**(1), 187–201 (2019)
13. Marahatta, A., Pirbhulal, S., Zhang, F., Parizi, R.M., Choo, K.K.R., Liu, Z.: Classification-based and energy-efficient dynamic task scheduling scheme for virtualized cloud data center. IEEE Trans. Cloud Comput. **9**(4), 1376–1390 (2019)
14. Mugunthan, D.S.: Novel cluster rotating and routing strategy for software defined wireless sensor networks. J. IoT Soc. Mob. Anal. Cloud **2**(3), 140–146 (2020)
15. Pan, L., Liu, X., Jia, Z., Xu, J., Li, X.: A multi-objective clustering evolutionary algorithm for multi-workflow computation offloading in mobile edge computing. IEEE Trans. Cloud Comput. **11**(2), 1334–1351 (2021)

16. Pradhan, P., Behera, P.K., Ray, B.: Modified round robin algorithm for resource allocation in cloud computing. Procedia Comput. Sci. **85**, 878–890 (2016)
17. Rjoub, G., Bentahar, J., Wahab, O.A.: BigTrustScheduling: trust-aware big data task scheduling approach in cloud computing environments. Future Gener. Comput. Syst. **110**, 1079–1097 (2020)
18. Shirvastava, S., Dubey, R., Shrivastava, M.: Best fit based VM allocation for cloud resource allocation. Int. J. Comput. Appl. **158**(9), 25–27 (2017)
19. Sutcliffe, A., Vaea, K., Poulivaati, J., Evans, A.M.: Fast casts': evidence based and clinical considerations for rapid Ponseti method. Foot Ankle Online J. **6**(9), 2 (2013)
20. Wang, B., Hou, Y., Li, M.: QuickN: practical and secure nearest neighbor search on encrypted large-scale data. IEEE Trans. Cloud Comput. **10**(3), 2066–2078 (2020)
21. Xiong, Y., Huang, S., Wu, M., She, J., Jiang, K.: A Johnson's-rule-based genetic algorithm for two-stage-task scheduling problem in data-centers of cloud computing. IEEE Trans. Cloud Comput. **7**(3), 597–610 (2017)
22. Xu, H., Liu, Y., Wei, W., Zhang, W.: Incentive-aware virtual machine scheduling in cloud computing. J. Supercomput. **74**, 3016–3038 (2018)
23. Xu, J., Zhang, Z., Hu, Z., Du, L., Cai, X.: A many-objective optimized task allocation scheduling model in cloud computing. Appl. Intell. **51**, 3293–3310 (2021)
24. Zhang, Q., Li, H.: MOEA/D: a multiobjective evolutionary algorithm based on decomposition. IEEE Trans. Evol. Comput. **11**(6), 712–731 (2007)
25. Zhang, X., Tian, Y., Jin, Y.: A knee point-driven evolutionary algorithm for many-objective optimization. IEEE Trans. Evol. Comput. **19**(6), 761–776 (2014)

Transfer Learning-Based Evolutionary Multi-task Optimization

Shuai Li, Xiaobing Zhu, and Xi Li[✉]

School of Information Engineering, Hebei GEO University, Shijiazhuang 050031, China
lixi_sjz@foxmail.com

Abstract. Multi-task optimization (MTO) is an emerging research topic to optimize multiple related tasks simultaneously. It aims to enhance task interrelationships by leveraging shared information and features, thereby improving model performance. Evolutionary transfer optimization (ETO), applied to address multi-task problems using evolutionary algorithms, incorporates the principles of transfer learning. It utilizes knowledge and experience from source tasks to expedite the optimization process of target tasks. We introduce a transfer learning-based strategy where valuable information from one task is transferred as comprehensively as possible to another task. This article proposes an idea that is based on joint distribution adaptation (JDA) and employs population individual replacement methods as knowledge transfer, differential evolution as the underlying optimizer, called transfer learning-based evolutionary multi-task optimization algorithm (TLEMTO). To validate the effectiveness of the proposed algorithm, the experiment is conducted on CEC17 multi-task optimization problem benchmarks, the results show that TLEMTO is superior to the compared state-of-the-art algorithms.

Keywords: Evolutionary transfer optimization · Multi-task optimization · Differential evolution · Transfer learning · Joint distribution adaptation

1 Introduction

With the rapid development of machine learning and optimization fields, multi-task learning (MTL) [1] has gradually emerged as a research hotspot. MTL aims to enhance overall learning performance by simultaneously addressing multiple related tasks.

Multi-task optimization (MTO) [2] stands out as a significant research direction in machine learning and optimization. MTO focuses on concurrently tackling multiple related tasks, enhancing overall performance and efficiency by leveraging correlations and information sharing among tasks. Typically, a task with marked samples or domain knowledge is defined as a source task, while a task to be optimized or learned is designated as a target task. The source task typically possesses a large number of labeled samples, while the target task lacks sufficient labeled samples. In multi-task optimization, tasks can share underlying feature representations or model parameters, enabling mutual learning among different tasks. By exchanging information and experiences, models can better adapt to diverse task requirements, providing more comprehensive and accurate

© The Author(s), under exclusive license to Springer Nature Singapore Pte Ltd. 2024
L. Pan et al. (Eds.): BIC-TA 2023, CCIS 2061, pp. 14–28, 2024.
https://doi.org/10.1007/978-981-97-2272-3_2

solutions. MTO not only strengthens synergies among tasks but also reduces computational and storage costs for training and inference. By concurrently considering multiple related tasks, it avoids the need for training and deploying separate models for each task, thereby enhancing system efficiency and scalability.

Evolutionary transfer optimization (ETO) [3] is an optimization approach grounded in evolutionary computation and transfer learning. It seamlessly integrates evolutionary algorithms and transfer learning principles to tackle the challenge of insufficient labeled samples in target tasks. At its core, ETO aims to expedite the optimization process of target tasks by leveraging knowledge and experience gained from source tasks. The fundamental idea behind ETO is the transfer of population information, facilitating the migration of optimization results and experience from source tasks to target tasks. This shared information guides the search direction and space of target tasks, accelerating their convergence speed and enhancing the quality of optimization outcomes. The utilization of shared population information not only expedites convergence but also contributes to the diversification of the population, leading to an overall improvement in optimization efficiency.

In the realm of solving multi-task optimization problems, notable progress has been achieved, yet certain limitations and challenges persist. A key concern lies in the prevalent use of a single population in most existing multi-task learning methods. This approach may inadvertently allocate excessive population resources to one task, potentially compromising the optimization performance of other tasks due to competitive relationships among them. This can result in performance instability or unreasonably slow optimization speeds. Conversely, independent training of models for each task may not fully harness shared information. Limited to the sharing of underlying feature representations, this approach might fail to optimally utilize the pertinent knowledge between tasks.

To address the aforementioned challenges, this paper presents a transfer learning-based evolutionary multi-task optimization algorithm (TLEMTO). In the algorithm, each population faces a task to evolve independently, DE is used as the optimizer and joint distribution adaptation (JDA) [4] is incorporated after a specified number of evolutionary iterations. Additionally, individual evaluation and population information exchange occur at regular intervals, the worst solutions in the population are replaced by elite solutions from the source task. This process fosters effective information sharing between the correlative tasks. Concurrently, each task maintains a certain level of independence during the solving process, this ensures that exceptional individuals continue their search for optimal solutions within their respective search spaces. Furthermore, it facilitates knowledge transfer, enabling outstanding individuals to expedite the search process or navigate away from local optima, thereby enhancing the overall performance of the optimization algorithm.

2   Related Work

To integrate the ETO method into multi-task optimization problems, Gupta et al. [5] introduced multifactorial optimization (MFO) as a novel paradigm in evolutionary computation. The authors contended that while traditional population-based evolutionary algorithms typically address one problem at a time, the inherent parallelism of these algorithms could be harnessed to develop effective multitasking engines capable of simultaneously tackling multiple problems. They proposed an evolutionary multi-task optimization algorithm named multifactorial evolutionary algorithm (MFEA). The MFEA model emulates phenomena such as selection mating and vertical cultural transmission. By capitalizing on the potential gene complementarity between different tasks, valuable information discovered in one task is seamlessly transferred, facilitating efficient knowledge sharing across tasks. Additionally, Gupta et al. [6] transformed two-layer optimization problems into multi-task optimization problems and applied the multifactorial optimization algorithm to resolve them.

Since its inception, MFEA has attracted considerable attention from scholars, leading to the development of numerous improved algorithms. Chen et al. [7] identified shortcomings in traditional MFEA and introduced an innovative evolutionary multi-task optimization algorithm named MaTEA. MaTEA utilizes multiple subpopulations to handle various tasks, with its most notable innovation being real-time awareness of inter-task relationships. It dynamically adjusts the knowledge transfer process based on these relationships. Bali et al. [8] incorporated linear domain adaptation (LDA) methods from the realms of machine learning and multi-task learning into the MFEA framework. This integration resulted in a new evolutionary multi-task algorithm framework called LDA-MFEA. The algorithm transforms the search space of a simple task into a highly similar space for complex tasks that need simultaneous addressing, along with their solution representations. Wen et al. [9] introduced the concept of "parting ways", signifying the point when knowledge transfer becomes ineffective and starts hindering information utilization. They proposed a method to detect this separation point using the accumulative survival rate discrepancy (ASRD) metric. Expanding on traditional MFEA and incorporating a resource reallocation mechanism, they developed an evolutionary multi-task optimization algorithm with resource reallocation capabilities, known as multifactorial evolutionary algorithm with resource reallocation (MFEARR). This approach successfully facilitates the exploration and utilization of transfer information throughout the entire evolutionary process, thereby enhancing the quality of the acquired knowledge.

In addition to multi-factorial optimization-based multi-task algorithms, researchers both domestically and internationally have introduced various multi-task optimization algorithms by enhancing traditional single-task evolutionary algorithms. Feng et al. [10] pioneered the integration of the multi-factorial optimization concept with particle swarm optimization (PSO), leading to the innovative multifactorial particle swarm optimization (MFPSO) algorithm. This algorithm effectively leverages the PSO algorithm's capability to simultaneously optimize multiple tasks in a multi-task environment. In the same study, Feng et al. [10] combined the concept of multi-factorial optimization with the differential evolution (DE) algorithm, giving rise to the multifactorial differential evolution (MFDE) algorithm. The DE algorithm's distinctive evolutionary search mechanism is employed, and a novel selection-crossover strategy is designed based on the DE algorithm, while retaining other methodological strategies within the MFEA framework to facilitate individual updates. Li et al. [11] introduced a comprehensive multi-population evolution framework (MPEF). In this framework, distinct populations can employ unique genetic material transfer strategies tailored to address their specific tasks independently. Concurrently, the authors integrated the SHADE algorithm into MPEF, creating a multi-population differential evolution algorithm (MPEF-SHADE) to tackle multifactorial optimization (MFO) problems. Shang et al. [12] proposed an innovative multi-task optimization algorithm based on the autoencoder with noise reduction. By deriving autoencoders with closed-form solutions, task mappings were established between multiple tasks. This enabled the proposed MTO algorithm to effectively utilize the search preferences of different single-task optimization algorithms.

3 Transfer Learning Based Evolutionary Multi-task Optimization

In this section, we present an algorithm named transfer learning-based evolutionary multi-task optimization (TLMTO). This algorithm optimizes each tasks using independent populations and incorporates knowledge transfer strategies between tasks through joint distribution adaptation and population individuals' exchange. TLMTO facilitates the exchange of information across tasks during the optimization process by integrating transfer learning mechanisms. It maintains independent evolutionary search processes for each task, for the goal of maximizing the effectiveness of multi-task optimization.

3.1 Maximum Mean Discrepancy

Maximum mean discrepancy (MMD) [13] stands as a frequently employed nonparametric statistical method for quantifying the dissimilarity between two probability distributions. At its core, MMD gauges the similarity between these distributions by evaluating the disparity in their means within the feature space.

In this context, it is assumed that there exists a source domain represented by $X_s = [x_{s1}, x_{s2}, \ldots, x_{sn}]$, following the distribution P, and there exists a target domain represented by $X_t = [x_{t1}, x_{t2}, \ldots, x_{tm}]$. It is further assumed that H represents the reproducing kernel Hilbert space (RKHS). The function $X_s = [x_{s1}, x_{s2}, \ldots, x_{sn}]$ denotes the mapping from the original feature space to the RKHS. When $n, m \to \infty$ holds, the MMD between X_s and X_t in the reproducing kernel Hilbert space can be represented by the following equation:

$$dist(X_s, X_t) = \left\| \frac{1}{n} \sum_{i=1}^{n} f(x_{s_i}) - \frac{1}{m} \sum_{j=1}^{m} f(x_{t_j}) \right\|_{\mathcal{H}} \tag{1}$$

MMD quantifies the similarity between two distributions in the feature space through the application of a kernel function. Minimizing the MMD value allows us to narrow the gap between distributions across various tasks or domains, fostering effective information sharing and transfer.

3.2 JDA

Joint distribution adaptation (JDA) is a widely employed method for tackling distribution discrepancies and transfer issues in multi-task optimization problems. Grounded in the principles of Maximum Mean Discrepancy and kernel methods, this approach seeks optimization by minimizing differences between task feature distributions, thereby improving the interrelatedness and transfer performance among tasks.

To minimize the marginal probability distance between source domain data and target domain data, we employ formula (1) to measure the disparity in marginal probability distributions between the source and target domains. The specific expression is provided below:

$$\left\| \frac{1}{n}\sum_{i=1}^{n} A^{\mathrm{T}} x_{s_i} - \frac{1}{m}\sum_{j=1}^{m} A^{\mathrm{T}} x_{t_i} \right\|_{\mathcal{H}}^{2} = \mathrm{tr}(A^{\mathrm{T}} X M_0 X^{\mathrm{T}} A). \tag{2}$$

where X represents the integrated data from the source and target domains; $X_s = [x_{s1}, x_{s2}, \ldots, x_{sn}]$ is an MMD matrix, defined as:

$$(M_0)_{ij} = \begin{cases} \frac{1}{n^2}, & x_i, x_j \in \mathcal{D}_s \\ \frac{1}{m^2}, & x_i, x_j \in \mathcal{D}_t \\ -\frac{1}{mn}, & \text{else} \end{cases} \tag{3}$$

where $X_s = [x_{s1}, x_{s2}, \ldots, x_{sn}]$ is the labeled source domain; D_t represents the unlabeled target domain; n and m denote the sample sizes of the source and target domains, respectively.

Similarly, it is necessary to adapt the conditional probability distributions of the source and target domains. Based on Bayes' theorem $X_s = [x_{s1}, x_{s2}, \ldots, x_{sn}]$ and the definition of sufficient statistics, when there are numerous unknowns in the samples, and the samples are of sufficient quality, one can select certain statistics from the samples to approximate the distributions that need to be estimated in practice, here, we can use $P(X_t|y_t)$ to approximate $P(y_t|X_t)$. Employ $X_s = [x_{s1}, x_{s2}, \ldots, x_{sn}]$ to train a K-nearest neighbor classifier. Utilize this classifier to make predictions on the known X_t, obtaining predicted labels $X_s = [x_{s1}, x_{s2}, \ldots, x_{sn}]$ for calculation as the target domain labels. At this point, the MMD distance between classes can be expressed as:

$$\sum_{c=1}^{C} \left\| \frac{1}{n_c}\sum_{X_{s_i} \in D_s^{(c)}} A^{\mathrm{T}} X_{s_i} - \frac{1}{m_c}\sum_{X_{t_i} \in D_t^{(c)}} A^{\mathrm{T}} X_{t_i} \right\|_{\mathcal{H}}^{2} = \mathrm{tr}(A^{\mathrm{T}} X M_C X^{\mathrm{T}} A) \tag{4}$$

where $X_s = [x_{s1}, x_{s2}, \ldots, x_{sn}]$ and m_c represent the number of samples from class c in the source and target domains, respectively. m_c is defined as:

$$(\boldsymbol{M}_c)_{ij} = \begin{cases} \frac{1}{n_c^2}, & \boldsymbol{x}_i, \boldsymbol{x}_j \in \mathcal{D}_s^{(c)} \\ \frac{1}{m_c^2}, & \boldsymbol{x}_i, \boldsymbol{x}_j \in \mathcal{D}_t^{(c)} \\ -\frac{1}{m_c n_c}, & \begin{cases} \boldsymbol{x}_i \in \mathcal{D}_s^{(c)}, \boldsymbol{x}_j \in \mathcal{D}_t^{(c)} \\ \boldsymbol{x}_i \in \mathcal{D}_t^{(c)}, \boldsymbol{x}_j \in \mathcal{D}_s^{(c)} \end{cases} \\ 0, & \text{else} \end{cases} \tag{5}$$

Joint distribution adaptation requires that both the adaptation of marginal distributions between the source and target domains and the adaptation of conditional distributions should be minimized as much as possible. Therefore, by combining the contents of formulas (2) and (4), the objective function for joint distribution adaptation is obtained:

$$\min \sum_{c=1}^{C} \operatorname{tr}(\boldsymbol{A}^T \boldsymbol{X} \boldsymbol{M}_c \boldsymbol{X}^T \boldsymbol{A}) + \lambda \|\boldsymbol{A}\|_F^2 \tag{6}$$

where $X_s = [x_{s1}, x_{s2}, \ldots, x_{sn}]$ represents the regularization term. Simultaneously, there is a constraint: the variance of the data before and after transformation remains unchanged. An additional optimization objective $X_s = [x_{s1}, x_{s2}, \ldots, x_{sn}]$ is introduced, where H is the centroid matrix. Combining this with the original optimization objective and applying the Lagrange method, we obtain:

$$(\boldsymbol{X} \sum_{c=0}^{C} \boldsymbol{M}_C \boldsymbol{X}^T + \lambda \boldsymbol{I})\boldsymbol{A} = \boldsymbol{X} \boldsymbol{H} \boldsymbol{X}^T \boldsymbol{A} \boldsymbol{\Phi} \tag{7}$$

where $X_s = [x_{s1}, x_{s2}, \ldots, x_{sn}]$ is the Lagrange multiplier, and $X_s = [x_{s1}, x_{s2}, \ldots, x_{sn}]$. Through iterative processes, the predicted labels $\hat{y}_t$ approach the true labels with increasing accuracy. Additionally, the transformation matrix A can be computed, aiming to continuously reduce the joint distribution adaptation between the source and target domains. The iteration continues until the desired result is achieved or the iteration process concludes.

3.3 Transfer Learning Based Multi-task Optimization

Based on the above methods, we integrate joint distribution adaptation techniques with DE to formulate a multi-task evolutionary algorithm. The pseudocode for TLEMTO is presented in Algorithm 1.

As demonstrated in [13], the effectiveness of joint distribution adaptation techniques in one-way migration has been established. Given that multi-task problems necessitate information exchange between tasks, the TLEMTO incorporates joint distribution adaptation techniques for information exchange after a specified number of iterations. Additionally, to ensure multiple interactions of information and enhance population diversity, TLEMTO introduces individual exchange techniques. This entails transferring more competent individuals from the population of one task to the population of another task, thereby expediting population evolution and aiding in escaping from local optima.

Algorithm 1 TLEMTO

1: Set the necessary algorithm parameters and generate the initial population;
2: Apply a fitness evaluation function to each individual, calculating its fitness values on various tasks;
3: **while** stopping conditions are not satisfied **do**:
4: Apply the mutation operation to the population individuals to increase the diversity of the population;
5: Generate new individuals through the crossover operation;
6: **if** $g = g\_$ **then**
7: Utilize formula (6) to reduce the joint distribution difference between the source and target domains, integrating the data distributions of the two tasks;
8: Adjust the relationship between populations of different tasks through a transfer mechanism to obtain data after knowledge information interaction between tasks;
9: forming a new population;
10: **end if**
11: **if** $g \% 50 = 0$ **then**
12: Sort the individuals in the population of the source task on fitness value;
13: Take 3 individuals with the smallest fitness values and 7 individuals by the roulette method, replace 10 individuals with the largest fitness values in the target task;
14: **end if**
15: Compute the fitness values of newly generated individuals;
16: Use the individuals selected by fitness values or obtained through the transfer learning method to form a new next population for producing the next generation of individuals;
17: **end while**

4 Experiments and Results Analysis

4.1 Comparing with MFEA and DE Algorithms

In this work, we assess the performance of the proposed MTO algorithm using the same benchmarks employed in the existing literature. These benchmarks encompass nine commonly used multi-task optimization benchmark problems (MTOP), each featuring two minimization problems with distinct characteristics. These characteristics include various function types, the degree of inter-task crossover, and the similarity between tasks. Inter-task similarity is quantified using Pearson correlation [14], where higher values denote increased similarity between tasks. Table 1 provides a summary of the attributes of the MTOP.

To validate the efficacy of the algorithm proposed in this paper, we employed the DE as the fundamental single-objective single-task optimizer for addressing each task within every MTOP benchmark. The classical multi-task algorithm, MFEA, was introduced for comparison purposes. For a fair assessment, the MFEA algorithm utilized the SBX crossover operator and polynomial mutation operator, while the DE employed the DE/rand/1 operator. All algorithms were independently executed 20 times. The specific parameter settings were as follows:

Table 1. Test problem attributes

Number	Optimization Function	Landscape	Search Space	Degree of Overlap	Inter-task similarity
1	Griewank (T1)	multimodal, nonseparable	$x \in [-100, 100]^d$	Complete intersection	1.000 0
	Rastrigin (T2)	multimodal, nonseparable	$x \in [-50, 50]^d$		
2	Ackley (T1)	multimodal, nonseparable	$x \in [-50, 50]^d$	Complete intersection	0.226 1
	Rastrigin (T2)	multimodal, nonseparable	$x \in [-50, 50]^d$		
3	Ackley (T1)	multimodal, nonseparable	$x \in [-50, 50]^d$	Complete intersection	0.000 2
	Schwefel (T2)	multimodal, separable	$x \in [-500, 500]^d$		
4	Rastrigin (T1)	multimodal, nonseparable	$x \in [-500, 500]^d$	Partial intersection	0.867 0
	Sphere (T2)	unimodal, separable	$x \in [-50, 50]^d$		
5	Ackley (T1)	multimodal, nonseparable	$x \in [-50, 50]^d$	Partial intersection	0.215 4
	Rosenbrock (T2)	multimodal, nonseparable	$x \in [-50, 50]^d$		
6	Ackley (T1)	multimodal, nonseparable	$x \in [-50, 50]^d$	Partial intersection	0.072 5
	Weierstrass (T2)	multimodal, nonseparable	$x \in [-0.5, 0.5]^d$		
7	Rosenbrock (T1)	multimodal, nonseparable	$x \in [-50, 50]^d$	No intersection	0.943 4
	Rastrigin (T2)	multimodal, nonseparable	$x \in [-50, 50]^d$		
8	Griewank (T1)	multimodal, nonseparable	$x \in [-100, 100]^d$	No intersection	0.366 9
	Weierstrass (T2)	multimodal, nonseparable	$x \in [-0.5, 0.5]^d$		
9	Rastrigin (T1)	multimodal, nonseparable	$x \in [-50, 50]^d$	No intersection	0.001 6
	Schwefel (T2)	multimodal, separable	$x \in [-500, 500]^d$		

1) Parameter settings in TLEMTO:

 – population size of each task n: 50
 – maximum number of iterations g: 1000
 – the number of iterations completed when executing JDA $g\_$: 100

- JDA kernel type: primal
- the number of JDA iterations T: 3

2) Parameter settings in DE:

- population size n: 100
- maximum number of iterations g: 1000
- crossover rate CR: 0.6
- scaling factor F: 0.5

3) Parameter settings in MFEA:

- population size n: 100
- maximum number of iterations g: 2000
- random mating probability rmp: 0.3
- distribution index of SBX: 2
- distribution index of PM: 5

The average objective values obtained by various algorithms for each test problem set are documented in Table 2. Across the 9 sets with a total of 18 test problems, TLEMTO, utilizing cross-task knowledge transfer, notably outperforms MFEA. Our proposed algorithm, employing DE as the foundational optimizer with parameters identical to those in TLEMTO, outperforms both MFEA and DE in 15 out of the 18 test problems, confirming its efficacy in enhancing optimization outcomes.

4.2   Comparing with the Improved Multi-task Algorithms Proposed in Other Literature

In this section, we compare the experimental results of TLEMTO on the average objective values from other multi-task optimization algorithms proposed in the literature, these are MFEARR, MFPSO, MFDE, MTO, and MPEF-SHADE. The parameters for the introduced comparative algorithms in the original paper, along with the parameter settings for TLEMTO, are shown in Table 3. The comparison of results between TLEMTO and other optimization algorithms is presented in Table 4.

To ensure a fair comparison, all algorithms were assessed with 100,000 evaluations per group of tasks. The TLEMTO algorithm excels among other multi-task optimization algorithms, achieving the best average objective values for 11 out of the 18 tasks. Notably, in tasks T1 of PI+HS, T2 of NI+HS, and T1 of NI+LS, TLEMTO approaches the global optimum value of 0.

4.3   Analysis of Convergence Process and Convergence Speed

To further affirm the effectiveness of our algorithm, this section analyzes the convergence traces of different optimizers while solving tasks T1 and T2. Figure 1 illustrates the convergence traces of TLEMTO, MFEA, and DE in task T1. Meanwhile, Fig. 2 depicts the convergence traces of the same algorithms in task T2. In these figures, the horizontal axis represents the number of iterations, and the vertical axis represents the minimum objective value at the current iteration.

Table 2. Average Objective Values and Standard Deviations of the Algorithms on Test Problems.

Number	Category	Task	The average objective values and standard deviations		
			DE	MFEA	TLEMTO
1	CI+HS	T_1	7.49E−03 (1.64E−03)	1.51E−01 (1.74E−02)	**1.22E−12 (5.94E−13)**
		T_2	3.04E+02 (1.29E+01)	1.34E+02 (4.61E+01)	**1.28E−09 (5.71E−10)**
2	CI+MS	T_1	3.57E−03 (4.39E−04)	4.62E+00 (8.05E−01)	**1.03E−06 (2.33E−07)**
		T_2	3.12E+02 (1.55E+01)	1.91E+02 (5.24E+01)	**6.81E−10 (2.49E−10)**
3	CI+LS	T_1	2.12E+01 (3.26E−02)	2.01E+01 (7.35E−02)	**2.00E+01 (7.25E−02)**
		T_2	2.05E+04 (3.37E+02)	**3.54E+03 (4.48E+02)**	2.07E+04 (2.84E+01)
4	PI+HS	T_1	3.10E+02 (1.36E+01)	5.52E+02 (6.79E+01)	**2.01E−09 (1.12E−09)**
		T_2	5.60E−03 (1.52E−03)	5.17E−01 (1.14E−01)	**2.30E−06 (1.20E−06)**
5	PI+MS	T_1	**3.81E−03 (6.18E−04)**	3.03E+00 (6.32E−01)	2.41E+00 (7.17E−02)
		T_2	4.57E+01 (3.65E−01)	2.56E+02 (8.82E+01)	**4.56E+01 (3.43E−01)**
6	PI+LS	T_1	3.76E−03 (6.13E−04)	1.99E+01 (1.46E−01)	**1.60E−06 (5.11E-07)**
		T_2	3.19E+01 (8.36E-01)	2.36E+01 (2.52E+00)	**7.11E−03 (1.82E−03)**
7	NI+HS	T_1	4.58E+01 (2.48E−01)	2.37E+02 (6.97E+01)	**4.57E+01 (4.44E−01)**
		T_2	3.18E+02 (1.01E+01)	1.94E+02 (6.58E+01)	**1.82E−09 (8.36E−10)**
8	NI+MS	T_1	1.63E−01 (2.79E−02)	9.47E−02 (2.27E−02)	**3.09E−04 (1.79E−04)**

(continued)

From Fig. 1, it is evident that, except for CI+LS and PI+MS, TLEMTO demonstrates superior convergence accuracy and faster convergence speed compared to MFEA and DE with the same optimizer on the remaining seven problems. In the case of PI+MS, although MFEA exhibits faster convergence speed in the early iterations, TLEMTO significantly accelerates its convergence in the later iterations and converges to better solutions. Additionally, TLEMTO requires fewer iterations to reach the optimal solution compared to

Table 2. (*continued*)

Number	Category	Task	The average objective values and standard deviations		
			DE	MFEA	TLEMTO
		T_2	7.46E+01 (1.54E+00)	2.64E+01 (3.00E+00)	**1.15E+00 (4.54E−01)**
9	NI+LS	T_1	3.10E+02 (1.08E+01)	5.94E+02 (1.25E+02)	**2.11E−09 (1.19E−09)**
		T_2	2.06E+04 (8.10E+01)	**3.35E+03 (3.54E+02)**	2.07E+04 (2.52E+01)

Table 3. Parameter Settings of Different Algorithms

Algorithm	Parameter settings	Algorithm	Parameter settings
MFEARR	population size n: 100 random mating probability rmp: 0.3 the small number ϵ: 0.01 distribution index of SBX: 2 distribution index of PM: 5	MFPSO	population size n: 100 random mating probability rmp: 0.3 ω: linearly decreases from 0.9 to 0.4 $c_1 = c_2 = c_3 = 0.2$
MFDE	population size n: 100 random crossover probability rmp: 0.3 scaling factor F: 0.5 crossover rate CR: 0.9	MPEF-SHADE	population size n: 100 learning parameter c: 0.3
MTO	population size n: 100 cross-task solution migration interval g: 10 cross-task migrated solution count q: 10	TLEMTO	population size for each task n: 50 the number of iterations completed when introducing the JDA algorithm t: 100 cross-task solution migration interval g: 50 cross-task migrated solution count q: 3

MFEA. Moreover, owing to the joint distribution adaptation technique, Fig. 1 illustrates that at the 100th iteration, the current optimal solutions for all nine problems have improved to some extent, expediting the overall convergence process. Moving to Fig. 2, it can be observed that on CI+HS, CI+MS, PI+MS, PI+LS, and NI+HS, TLEMTO demonstrates a similar convergence speed to MFEA in the early stages but exhibits noticeably faster convergence in the later stages, achieving superior quality in converging to the optimal solution. On PI+HS and NI+MS, despite the initial convergence not being ideal, and PI+HS even experiencing the negative effects of transfer, both tasks achieve faster

Table 4. Comparison Results of the Algorithm with Other Multi-Task Algorithms.

Category	Task	The average objective values and standard deviations of the test problems					
		MFEARR	MFPSO	MFDE	MTO	MPEF-SHADE	TLEMTO
CI+HS	T_1	2.37E−01	2.10E−01	1.00E−03	1.20E−05	1.33E−09	**1.22E−12**
	T_2	2.14E+02	8.11E+00	2.61E+00	2.16E−02	3.36E−06	**1.28E−09**
CI+MS	T_1	4.36E+00	6.00E−02	1.00E−03	2.77E−01	3.21E−06	**1.03E−06**
	T_2	2.06E+02	6.26E+00	3.00E−03	1.37E+01	3.19E−08	**6.81E−10**
CI+LS	T_1	2.02E+01	5.60E+00	2.12E+01	2.10E+01	2.11E+01	**2.00E+01**
	T_2	3.81E+03	**2.23E+03**	1.18E+04	6.73E+03	5.65E+03	2.07E+04
PI+HS	T_1	5.12E+02	2.06E+02	7.83E+01	3.15E+01	2.57E+02	**2.01E−09**
	T_2	4.58E+00	3.84E+03	2.20E−05	1.74E+02	**1.18E−13**	2.30E−06
PI+MS	T_1	3.32E+00	3.58E+00	1.00E−03	1.77E+00	**3.63E−05**	2.41E+00
	T_2	3.73E+02	1.24E+02	6.03E+01	1.57E+02	4.86E+01	**4.56E+01**
PI+LS	T_1	1.61E+01	1.00E−02	4.60E−01	2.30E−05	5.92E−06	**1.60E−06**
	T_2	1.62E+01	5.00E−02	2.20E−01	2.35E−03	**5.17E−04**	7.11E−03
NI+HS	T_1	5.94E+02	4.39E+01	8.93E+01	6.00E+01	**4.27E+01**	4.57E+01
	T_2	2.51E+02	3.96E+01	2.05E+01	2.84E+01	3.77E+01	**1.82E−09**
NI+MS	T_1	3.04E−01	4.80E−01	2.00E−03	8.07E−01	**1.68E−09**	3.09E−04
	T_2	2.69E+01	1.21E+01	2.97E+00	2.63E+00	1.64E+00	**1.15E+00**
NI+LS	T_1	5.49E+02	3.33E+02	9.62E+01	3.43E+01	2.66E+02	**2.11E−09**
	T_2	**3.75E+03**	9.26E+03	3.94E+03	7.08E+03	5.98E+03	2.07E+04

convergence in the later iterations and converge to superior solutions. By comparing the two figures, we further validate the advantages of TLEMTO in terms of convergence accuracy and speed, significantly enhancing its optimization performance.

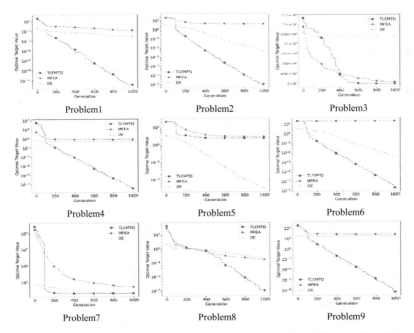

Fig. 1. Minimum Value Trajectories during Iterative Solving of Test Problem T1

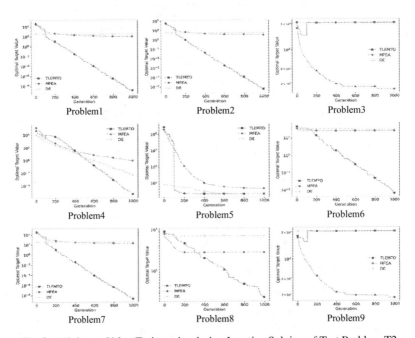

Fig. 2. Minimum Value Trajectories during Iterative Solving of Test Problem T2

5 Conclusion

This paper enables knowledge transfer among related tasks in multi-task optimization problem by introducing a novel transfer learning algorithm named the joint distribution adaptation method. Comparative analysis with traditional multi-task optimization algorithms revealed that the proposed algorithm facilitates more frequent and comprehensive information exchange among different tasks. Consequently, the experimental results on common multi-task benchmark test functions demonstrated the effectiveness of the proposed approach. In the future work, we will consider improving the aforementioned inter-task mapping approach to avoid negative knowledge transfer and applying stronger search engines.

Acknowledgements. This research was funded by the National Natural Science Foundation of China (No. 61806069), the Science and Technology Project of Hebei Education Department (No. ZD2022083) and the Postgraduate Educational and Teaching Reform Project of Hebei GEO University (No. YJGX2023020).

References

1. Zhang, Y., Yang, Q.: A survey on multi-task learning. IEEE Trans. Knowl. Data Eng. **34**(12), 5586–5609 (2022)
2. Gupta, A., Ong, Y.S.: Genetic transfer or population diversification? deciphering the secret ingredients of evolutionary multitask optimization. In: 2016 IEEE Symposium Series on Computational Intelligence (SSCI), pp. 1–7. IEEE (2016)
3. Tan, K.C., Feng, L., Jiang, M.: Evolutionary transfer optimization - a new frontier in evolutionary computation research. IEEE Comput. Intell. Mag. **16**(1), 22–33 (2021)
4. Long, M., Wang, J., Ding, G., Sun, J., Yu, P.S.: Transfer feature learning with joint distribution adaptation. In: 2013 IEEE International Conference on Computer Vision (ICCV), Sydney, Australia, pp. 2200–2207 (2013)
5. Gupta, A., Ong, Y.S., Feng, L.: Multifactorial evolution: toward evolutionary multitasking. IEEE Trans. Evol. Comput. **20**(3), 343–357 (2016)
6. Gupta, A., Mańdziuk, J., Ong, Y.S.: Evolutionary multitasking in bi-level optimization. Complex Intell. Syst. **1**(1), 83–95 (2015)
7. Chen, Y., Zhong, J., Feng, L., Zhang, J.: An adaptive archive-based evolutionary framework for many-task optimization. IEEE Trans. Emerg. Top. Comput. Intell. **4**(3), 369–384 (2019)
8. Bali, K.K., Gupta, A., Feng, L., Ong, Y.S., Siew, T.P.: Linearized domain adaptation in evolutionary multitasking. In: 2017 IEEE Congress on Evolutionary Computation (CEC), pp. 1295–1302. IEEE (2017)
9. Wen, Y.W., Ting, C.K.: Parting ways and reallocating resources in evolutionary multitasking. In: 2017 IEEE Congress on Evolutionary Computation (CEC), pp. 2404–2411. IEEE (2017)
10. Feng, L., et al.: An empirical study of multifactorial PSO and multifactorial DE. In: IEEE Congress on Evolutionary Computation (CEC), San Sebastian, Spain, pp. 921–928 (2017)
11. Li, G., Zhang, Q., Gao, W.: Multipopulation evolution framework for multifactorial optimization. In: Proceedings of the Genetic and Evolutionary Computation Conference Companion, pp. 215–216 (2018)
12. Shang, Q., Zhou, L., Feng, L.: Multi-task optimization algorithm based on denoising auto-encoder. J. Dalian Univ. Tech. **59**(4), 417–426 (2019)

13. Li, X., Li, S., Feng, Y.H., et al.: Unidirectional transfer differential evolution algorithm based on joint distribution adaptation. J. Zhengzhou Univ. (Eng. Sci.) **44**(5), 24–31 (2023)
14. Da, B., et al.: Evolutionary multitasking for single-objective continuous optimization: benchmark problems, performance metric, and baseline results (2017). https://doi.org/10.48550/arXiv.1706.03470

A Surrogate-Based Optimization Method for Solving Economic Emission Dispatch Problems with Green Certificate Trading and Wind Power

Chen Lu[1] , Huijun Liang[1]([✉]) , Heng Xie[2] , Chenhao Lin[1] ,
and Shuxin Lu[1]

[1] College of Intelligent Systems Science and Engineering, Hubei Minzu University,
Enshi 445000, China
lhj@hbmzu.edu.cn
[2] Guoneng Changyuan Enshi Hydropower Development Co., Ltd.,
Enshi 445099, China

Abstract. To reduce the impact of greenhouse effect, the deployment and utilization of renewable energy sources such as wind power has become an inevitable trend. Therefore, the decision of economic emission dispatch (EED) problems is particularly important. In this paper, to incentivize renewable energy generation, the green certificate trading mechanism is introduced to solve from the economic level. However, as the dimensions of the EED problems are increased, the current methods cannot make proper scheduling decisions in a short time. Therefore, the EED problems are categorized as computationally expensive EED problems. To solve the above problems, a surrogate-based multi-objective optimization method is proposed. On one hand, the artificial neural network (ANN) surrogate models are proposed to replace the traditional objective function, which greatly reduces the time to obtain feasible decisions. On the other hand, a modified multi-objective gray wolf optimizer (MOGWO) is proposed to execute EED optimization accurately and quickly. This algorithm improves the search ability and convergence of the original MOGWO algorithm through improving the position update strategy and introducing the difference algorithm. The effectiveness of the surrogate-based MOGWO is testified through simulations of benchmark functions and computationally expensive EED optimization problems within the actual Taipower 40-unit test system.

Keywords: Economic emissions dispatch · Random wind power · Grey wolf optimization algorithm · Artificial neural network

1 Introduction

Due to population expansion, economic growth, and rising living standards, the demand for energy is increasing alarmingly over the past few decades. With the massive exploitation of fossil energy, greenhouse effect and global warming have

Supported by the National Natural Science Foundation of China under Grant 62163013.

© The Author(s), under exclusive license to Springer Nature Singapore Pte Ltd. 2024
L. Pan et al. (Eds.): BIC-TA 2023, CCIS 2061, pp. 29–43, 2024.
https://doi.org/10.1007/978-981-97-2272-3_3

an impact on agricultural production and human health [1]. Thus, the clean energy source, such as wind energy, is encouraged to grow. The installed wind power capacity in China has gone up dramatically in recent years. Furthermore, to incentivize wind power and reduce power generation costs and emissions of polluting gases, the green certificate trading mechanism has been introduced into the electricity market. Naturally, the question of increasing wind power utilization and reduce pollution emissions as well as the cost of power generation must be considered. In other words, EED [2] problems with green certificate trading and wind power deserve further study.

The objectives of a typical multi-objective EED optimization problem are minimize cost and emissions, heuristic algorithm is an effective way to solve these problems [3]. For example, improved sailfish optimization (ISFO) algorithm [4], enhanced multi-objective exchange market algorithm [5], application of grasshopper optimization algorithm with the binary approach [6], and etc. had been applied to EED problems solving. It is worth mentioning that the power system presents high-dimensional large-scale characteristics, and the addition of wind power further increase the dimension, EED problems are modified into computationally expensive problems [7]. Although the effectiveness of heuristic algorithm can be demonstrated, the computation time is excessively long for the computationally expensive EED problems [8]. In summary, finding a time-saving method is an urgent problem.

Surrogate-based optimization methods are effective ways for solving computationally expensive optimization problems [9]. Specifically, surrogate models replace the complex high-order mathematical functions in the original models to save time [10]. A support vector regression surrogate model based on feature engineering was suggested in [11] to replace the objective functions in multi-area EED problems. In [12], a computationally efficient near-field shaping method for non-destructive thermal therapy applications was proposed. The traditional coordinate optimization process was replaced by ANN-based regression algorithm. In summary, surrogate models are used in a wide range of industries.

Based on the above reviews, the computationally expensive EED problems can be solved effectively in two steps. First, using a time-saving surrogate model instead of the originally computationally expensive objective function [13,14]. Second, the application of heuristic algorithms is worth investigating.

This paper proposes the surrogate methods for solving the computationally expensive EED problems quickly. Based on the improved MOGWO algorithm, ANN surrogate-based models are suggested to replace the traditional mathematical functions. There are two main contributions in this paper:

1) ANN-based surrogate model is proposed to replace the objective functions of computationally expensive EED problems. First of all, the historical operational data is used as the ANN model 's training and testing data sets. Secondly, the above model learns from the environment through the continuous adjustment of neuronal weights and thresholds. Then, the output results are computed through the hidden layer. Finally, the weight is corrected repeatedly according to the error between the actual output and the expected output, until the actual output reaches the expected result. The

model obtained through training is then tested with the testing data set. The error of the model is computed through the test results to measure the accuracy of the proposed method. The error of the ANN-surrogate model is less than 20%, which has good accuracy and saves computing time.

2) A chaotic selection mechanism-based multi-objective grey wolf algorithm (CS-MOGWO) is proposed which has better performance compared with MOGWO to solve the computationally expensive EED problem. First, the position update strategy of the original algorithm is changed. Further, the sine chaotic map is introduced to replace the selection probability operator used to select the different update mode in the strategy. Above improvements help to improve the search ability of the original algorithm. Second, the introduction of differential algorithm makes the result of the current iteration of the algorithm compare with the optimal solution of the previous iteration. It avoids algorithms falling into local optimality. In summary, CS-MOGWO has stronger performance compared with MOGWO.

The rest of this paper is described below. Section 2 introduces traditional and computationally expensive EED problems. Section 3 introduces the surrogate models based on ANN and an improved MOGWO algorithm are proposed. Section 4 provides the simulation results of the actual Taipower 40-unit test system, and the simulation results are compared. At last, Sect. 5 summarizes this paper and puts forward the future research direction.

2   Mathematical Modeling

This section formulates the EED problems and the computationally expensive EED problems. The objective functions and constraints are described below.

2.1   Original EED Problem Formulation

The construction of the original EED problem includes the objective functions and constraints involved.

Objective Functions. Two objective functions are expressed as follows:

The Minimum Total Cost

$$C(\cdot) = \sum_{k=1}^{N_{\mathrm{gt}}} f_k(P_k) + \sum_{l=1}^{N_{\mathrm{wt}}} g_l(W_l) - h(W_{\mathrm{wt}}), \tag{1}$$

where the total cost function is described by $C(\cdot)$, the total number of thermal units is denoted by N_{gt}, the fuel cost function for the kth thermal unit is described by $f_k(P_k)$, the total number of wind farms is denoted by N_{wt}, the cost function for the lth wind farm is $g_l(W_l)$, and the green certificate trading mechanism function is $h(W_{\mathrm{wt}})$. Then, the fuel cost function for the kth thermal unit is described by:

$$f_k(P_k) = a_k P_k^2 + b_k P_k + c_k, \tag{2}$$

where the output of the kth thermal unit is denoted by P_k, the fuel cost factors of the kth thermal unit are denoted by a_k, b_k, and c_k. Then, the mathematical formula for calculating costs for the l-th wind farm is described by [15]:

$$g_l(W_l) = q_l W_l + C_{\mathrm{rwt},l} E(Y_{\mathrm{oe},l}) + C_{\mathrm{pwt},l} E(Y_{\mathrm{ue},l}), \tag{3}$$

where the output of the lth wind farm is denoted by W_l, the cost factor of the lth wind farm is denoted by q_l, the under-estimation cost factor of residual energies in l^{th} wind farm is denoted by $C_{\mathrm{rwt},l}$, the over-estimation cost factor of purchasing energies through other means is denoted by $C_{\mathrm{pwt},l}$, the under-estimation cost of the lth wind farm is denoted by $C_{\mathrm{rwt},l} E(Y_{\mathrm{oe},l})$, the over-estimation cost of the lth wind farm is denoted by $C_{\mathrm{pwt},l} E(Y_{\mathrm{ue},l})$. Then, the green certificate trading mechanism function is described by:

$$h(W_{\mathrm{wt}}) = \begin{cases} 2/3(W_{\mathrm{wt}} - L_{\mathrm{green}})R_{\mathrm{g}}, & W_{\mathrm{wt}} \geq L_{\mathrm{green}}, \\ (W_{\mathrm{wt}} - L_{\mathrm{green}})R_{\mathrm{g}}, & W_{\mathrm{wt}} < L_{\mathrm{green}}, \end{cases} \tag{4}$$

where the total outputs of wind farms are denoted by W_{wt}, the renewable energy power quota is denoted by L_{green}, the green certificate transaction unit price is denoted by R_{g}. If the total output of the wind farms exceeds the renewable energy quota, the excess power is converted into the green certificate sold at two-thirds of the unit price. Conversely, the green certificate needs to be purchased to convert the equivalent power to supplement the renewable energy output.

The Minimum Emissions

$$E(P) = \sum_{k=1}^{N_{\mathrm{gt}}} [10^{-2}(\alpha_k P_k^2 + \beta_k P_k + \gamma_k) + \varepsilon_k \exp(\lambda_k P_k)], \tag{5}$$

where the emission is described by $E(P)$, emission factors of the kth thermal unit are denoted by α_k, β_k, γ_k, ε_k, and λ_k.

Constraints. The solution of the original EED problem must meet different constraints, which are described in detail below:

Capacity Constraints

$$P_k^{\min} \leq P_k \leq P_k^{\max}, \tag{6}$$

where the minimum and maximum output limits of the kth thermal unit are denoted by $P_k^{\min}$ and $P_k^{\max}$, respectively.

$$W_l^{\min} \leq W_l \leq W_l^{\max}, \tag{7}$$

where the minimum and maximum output limits of the l-th wind farm are denoted by $W_l^{\min}$ and $W_l^{\max}$, respectively.

Power Balance Constraint

$$\sum_{k=1}^{N_{\mathrm{gt}}} P_k + \sum_{l=1}^{N_{\mathrm{wt}}} W_l = P_{\mathrm{d}} + P_{\mathrm{loss}}, \tag{8}$$

where load demand and line loss are P_{d} and P_{loss}, respectively.

2.2 Computationally Expensive EED Problems Formulation

The scale and dimension of a power system are large. Therefore, solving the EED problems will spend much time on computing the objective function [16]. As shown in Fig. 1, using surrogate-based models to replace original objective functions is an effective solution to avoid the drawback of long time.

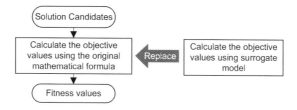

Fig. 1. The traditional objective functions are replaced by surrogate models.

3 Construction of the Proposed Methods

The section describes the process of the computationally expensive EED problems solving through the surrogate-based models. First, historical operational data is taken to construct the ANN-based surrogate models. Second, an improved MOGWO is used to solve the EED optimization problems.

3.1 Building ANN Surrogate-Based Models to Replace the Original Objective Functions

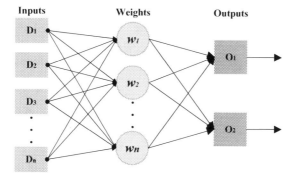

Fig. 2. Structure of ANN surrogate-based model.

In large-scale power systems, solving high-dimensional EED problems using the original objective functions take much computational time and cannot meet the time bound of real-time grid dispatch [17]. To overcome the drawback,

ANN surrogate-based models are constructed to replace the complex high-order mathematical functions. The process of ANN surrogate-based model is shown as Fig. 2.

To verify the effectiveness of proposed ANN surrogate models, the comparison results of operation time with original functions are shown as Table 1. In the actual Taipower 40-unit test system with two wind farms, the operating time of ANN surrogate models is reduced by **46.17%**. In the test system with wind power and green certificate trading, the operating time is reduced by **48.01%**. In summary, the proposed models substantially reduce the time to compute the objective functions. To further illustrate the effectiveness of ANN surrogate models, the results of the accuracy of ANN surrogate models are shown as Table 2.

Table 1. Comparison results of operation time.

The actual Taipower 40-unit test system	Original function	ANN surrogate model
With two wind farms	0.004871	**0.002622**
With wind power and green certificate trading mechanism	0.005120	**0.002662**

Table 2. Accuracy of ANN surrogate models.

The actual Taipower 40-unit test system	Objective functions	Accuracy of ANN surrogate models
With two wind farms	Cost	97.37%
	Emission	83.19%
With wind power and green certificate trading	Cost	98.66%
	Emission	83.18%

3.2   An Improved MOGWO Algorithm

GWO. GWO algorithm simulates the hierarchy and hunting strategy of wolves in nature. The basic idea of the algorithm is to build a model that simulates Wolf hunting. When the algorithm is solving the optimization problem, the current optimal, optimal, and suboptimal solutions are labeled α Wolf, β Wolf, and δ Wolf. The rest of the population is localized in three heads under the guidance of constantly adjust the position towards the prey's orientation. The guiding equation of the ω position of the head Wolf to the individual Wolf is:

$$D_p = CX_p^t - X_{\text{position}}^t, \tag{9}$$

$$X_{\text{position}}^{t+1} = 1/3 \sum_{p=\alpha,\beta} (X_p^t - AD_p) \tag{10}$$

$$\begin{cases} A = 2\alpha w_1 - \alpha, \\ C = 2w_2, \\ \alpha = 2 - 2(I/I_{\max}), \end{cases} \tag{11}$$

where D_p is the leading bit, X^t_{position} is the position of the individual ωwolf, I and $I_{\max}$ are the number of iterations and the maximum number of iterations, A and C are the wobble factor, α is the control parameter, and its value decreases with the increase of the number of I in the range greater than 1 and less than 2, w_1 and w_2 are the random number in the range of 0 to 1.

MOGWO. The MOGWO algorithm is proposed based on the original GWO algorithm. MOGWO algorithm is improved in the following two aspects:

The External Population Archive Mechanism is Introduced. MOGWO algorithm uses external population archive to store the non-dominated optimal solution bodies. New individuals are generated after each iteration of the algorithm, and the newly generated individuals and the individuals in the current Archive are updated by means of non-support ratio.

The Mode the Head Wolf Chooses. The MOGWO algorithm selected the head wolf in the Archive population using the formula of a wheel bet. In order to improve the search ability of the algorithm, the calculation formula of the selection probability P_i is:

$$P_i = (1/N_i)^c \tag{12}$$

where c is a constant number greater than 1, which needs to be set according to reality, N_i is the total number of populations to which the individual belongs.

Improvement of MOGWO Algorithm. The shortcomings of the original MOGWO algorithm are its apt to fall into the local optimum and the lack of searching ability. The wolves always update the position based on the average of the position that needs to be adjusted due to the relative position of the three leader wolves during each iteration. This strategy limits the search randomness of the algorithm. To solve the above-mentioned problems, CS-MOGWO improves the position update strategy, in which wolves have the probability to choose the average of the positions adjusted due to the relative position of one two or three leader wolves as its new position. This strategy enhances the randomness in the particle optimization process, enlarges the particle search range, and makes the algorithm jump out of the local optimal more easily. Second, the sine map is introduced to replace the selection probability operator. In addition, the introduction of differential algorithm makes the result of the current iteration of the algorithm compare with the optimal solution of the previous iteration. The above improved methods enhance the searching ability of the algorithm and make it easier for the algorithm to obtain the optimal solution. The sine chaotic map is described as (13), The flow tree of CS-MOGWO and original MOGWO are shown as Fig. 3.

$$\begin{cases} M_i = 0.99\mathrm{sin}(\pi M_{i-1}) \\ M_0 = \mathrm{rand}(1) \end{cases} \tag{13}$$

In order to prove the effectiveness of CS-MOGWO, the results of running benchmark functions of UF1, UF2 and UF3 with different algorithms are shown as Table 3. Compared with the original MOGWO [18] algorithm and multi-objective particle swarm optimization (MOPSO) [18], CS-MOGWO algorithm shows better searching ability and convergence.

4   Comparison of Simulation Results

To demonstrate the effectiveness of the methods, the actual Taipower 40-unit test system with different objective functions is simulated. The active power unit is MW, the cost unit is \$/h, the emission unit is tons/h, the operation time unit

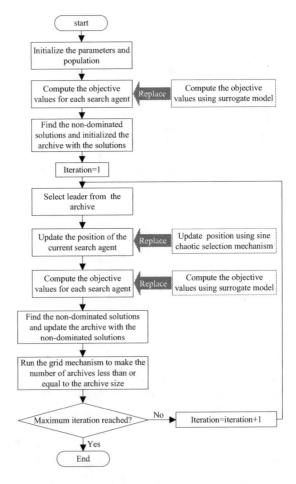

Fig. 3. The flow tree of CS-MOGWO and original MOGWO.

Table 3. Comparison results of different benchmark test functions.

Problem	Indicator	CS-MOGWO	MOGWO	MOPSO
UF1	Average	**0.06319**	0.11442	0.13701
	Best	**0.04362**	0.08023	0.08990
	Worst	**0.09712**	0.15774	0.22786
	Media	**0.05774**	0.13300	0.13175
	STD.DEV	**0.01857**	0.01954	0.04407
UF2	Average	**0.04120**	0.05825	0.06041
	Best	**0.02881**	0.04980	0.03699
	Worst	**0.05430**	0.07322	0.13051
	Media	**0.04084**	0.05778	0.04835
	STD.DEV	0.00874	**0.00739**	0.02763
UF6	Average	**0.19918**	0.27938	0.64752
	Best	**0.12828**	0.19338	0.37933
	Worst	**0.24356**	0.55036	1.24281
	Media	**0.21249**	0.24435	0.55037
	STD.DEV	**0.04152**	0.10449	0.26612

is s, the unit price of green certificate is 21.73 \$/h, the renewable energy quota is 52.5 MW, and the thermal unit parameters as well as the wind farms parameters are referred to [19].

4.1 Case 1: Simulation for the Actual Taipower 40-Unit Test System with Two Wind Farms

For Case 1, to prove the effectiveness of the proposed methods, they are tested on the actual Taipower 40-unit test system with two wind farms. The total load demand is 10500 MW. The Pareto non-dominated solutions set of the simulation results are presented in Fig. 4. Outputs of 40 thermal units and two wind farms are shown as Table 4. The minimum total cost is 144820.21 \$/h and the minimum emissions is **155435.56 tons/h**, respectively. The operation time is **24.13 s**. The symbiotic organisms search algorithms (SOS) [20], the modified and highly efficient species of the original SOS algorithm (NSOS) [20], and the multi-objective competitive swarm optimizer (MOCSO) [21] methods are used to compare with the proposed methods. SOS, NSOS and MOCSO are tested on the actual Taipower 40-unit test system without wind power. The operation time of the above methods are 408.8 s, 425.2 s, and 180.0 s. Obviously, the test system tested by the proposed methods is more complex than that tested by the above algorithms. However, compared with the above algorithms, the operation time of the proposed methods has been reduced by **94.10%**, **94.33%**, and **86.59%**, respectively. In summary, the proposed methods substantially reduce the operation time of the computationally expensive EED problems.

Table 4. Unit output results for Case 1 (Power: MW).

Items	Cost	Emission	Item	Cost	Emission	Item	Cost	Emission	Item	Cost	Emission
P_1	103.25	112.79	P_{12}	328.41	286.65	P_{23}	498.71	442.19	P_{34}	196.98	174.17
P_2	103.25	112.79	P_{13}	305.42	391.14	P_{24}	498.71	442.19	P_{35}	196.98	174.17
P_3	99.54	108.96	P_{14}	305.42	391.14	P_{25}	498.71	442.19	P_{36}	196.98	174.17
P_4	188.74	179.97	P_{15}	305.42	391.14	P_{26}	498.71	442.19	P_{37}	95.33	107.07
P_5	90.89	91.64	P_{16}	305.42	391.14	P_{27}	24.63	90.72	P_{38}	95.33	107.07
P_6	126.21	136.96	P_{17}	464.09	408.81	P_{28}	24.63	90.72	P_{39}	95.33	107.07
P_7	220.77	284.14	P_{18}	464.09	408.81	P_{29}	24.63	90.72	P_{40}	508.85	432.11
P_8	220.35	276.86	P_{19}	483.41	428.79	P_{30}	94.79	91.60	W_1	59.05	77.22
P_9	220.35	276.86	P_{20}	483.41	428.79	P_{31}	182.95	173.46	W_2	72.90	78.33
P_{10}	200.26	159.15	P_{21}	498.71	442.19	P_{32}	182.95	173.46	Total cost: 144820.21 $/h		
P_{11}	328.41	286.65	P_{22}	498.71	442.19	P_{33}	182.95	173.46	Emission: 155435.56 tons/h		

Table 5. Comparison of test solutions for Case 1.

Method	Minimum cost ($/h)	Minimum emission (tons/h)	Time (s)
ANN-CS-MOGWO	144820.21	**155435.56**	**24.13**
CS-MOGWO	144321.43	157937.78	28.51
MOGWO	145632.55	168825.62	30.21
EMA	144356.27	172595.59	/
GAEPSO	146035.32	172268.13	/

To further illustrate the effectiveness of the proposed methods, the comparison with the CS-MOGWO, MOGWO [18], exchange market algorithm (EMA) [22], gravitational acceleration enhanced particle swarm optimization (GAEPSO) [23] methods are shown as Table 5. Since the simulation environment of the proposed methods is the same as that of the first part of Case 1, the simulation results are based on the data in Case 1. CS-MOGWO, MOGWO, EMA, GAEPSO methods are tested on the same test system as the proposed methods. Compared with above methods, the proposed methods reduce the emissions by **1.58%, 7.93%, 11.30%, 11.13%**, respectively. Specially, compared with CS-MOGWO and MOGWO, the operation time has been reduced by **15.36%** and **20.12%**. Moreover, compared with MOGWO, CS-MOGWO reduce the cost and emission by **0.90%** and **6.45%**. In summary, the improved algorithm can get better results. Moreover, the proposed methods can obtain better solutions and save time for solving the computationally expensive EED problems.

4.2   Case 2: Simulation for the Actual Taipower 40-Unit Test System with Wind Power and Green Certificate Trading

For Case 2, to illustrate the scalability of the proposed methods, they are tested on the actual Taipower 40-unit test system with wind power and green certificate trading. The total load demand is 10500 MW. The Pareto non-dominated solutions set of the simulation results are provided in Fig. 5. Outputs of 40 thermal units and two wind farms are shown as Table 6. The minimum cost is **144070.42 \$/h** and the minimum emissions is **153086.22 tons/h**, respectively. The operation time is **24.86 s**. The comparison with CS-MOGWO and MOGWO is shown as Table 7. It is tested on the same test system as the proposed methods. Compared with CS-MOGWO and MOGWO, the minimum emissions are reduced by **1.17%** and **7.64%**, the operation time are reduced by **12.90%** and **23.44%**, respectively. Even compared with Case 1, the operation times are **93.92%**, **94.15%** and **86.19%** less than SOS, NSOS, and MOCSO, respectively. Moreover, compared with MOGWO, CS-MOGWO reduce the cost and emission by **0.91%** and **6.54%**. First, the proposed methods reduce the emissions and operation time significantly on the basis of minimal cost increase. Second, ANN-based surrogate models produce the lowest operation time. It proves the scalability of surrogate models to solve computationally expensive EED problems. In addition, the effect of the improved algorithm is further demonstrated.

Table 6. Unit output results for Case 2 (Power: MW).

Items	Cost	Emission	Item	Cost	Emission	Item	Cost	Emission	Item	Cost	Emission
P_1	103.36	108.31	P_{12}	290.57	298.12	P_{23}	465.01	443.83	P_{34}	193.94	194.54
P_2	103.36	108.31	P_{13}	371.67	382.67	P_{24}	465.01	443.83	P_{35}	193.94	194.54
P_3	113.79	115.70	P_{14}	371.67	382.67	P_{25}	465.01	443.83	P_{36}	193.94	194.54
P_4	182.89	176.22	P_{15}	371.67	382.67	P_{26}	465.01	443.83	P_{37}	97.68	105.13
P_5	62.21	85.60	P_{16}	371.67	382.67	P_{27}	32.50	68.40	P_{38}	97.68	105.13
P_6	121.57	124.56	P_{17}	421.02	418.61	P_{28}	32.50	68.40	P_{39}	97.68	105.13
P_7	288.11	293.19	P_{18}	421.02	418.61	P_{29}	32.50	68.40	P_{40}	473.01	438.50
P_8	260.72	263.62	P_{19}	461.69	438.68	P_{30}	63.92	88.95	W_1	75.69	70.19
P_9	260.72	263.62	P_{20}	461.69	438.68	P_{31}	176.03	172.51	W_2	77.60	75.90
P_{10}	285.11	241.96	P_{21}	465.01	443.83	P_{32}	176.03	172.51	Total cost: 144070.42 \$/h		
P_{11}	290.57	298.12	P_{22}	465.01	443.83	P_{33}	176.03	172.51	Emission: 153086.22 tons/h		

Table 7. Comparison of test solutions for Case 2.

Method	Minimum cost (\$/h)	Minimum emission (tons/h)	Time (s)
ANN-CS-MOGWO	144070.42	**153086.22**	**24.86**
CS-MOGWO	143999.47	154904.65	28.53
MOGWO	145329.68	165754.41	32.47

Furthermore, to verify the effectiveness of green certificate trading mechanism, the comparison with the proposed methods tested for Case 1 is shown
as Table 8. Since the simulation environment of the proposed methods is the
same as that of the first part of case 2, the simulation results are based on the
data in Table 7. Obviously, after the introduction of green certificate trading, the
minimum cost and emissions are reduced by **0.52%**, and **1.51%**, respectively. In
addition, compared with the unit output at minimum cost data results in Table 4
and 6, the wind farms of Case 2 produce **16.40%** more output than Case 1. In
summary, the green certificate trading mechanism reduce the cost and emissions
also incentivizes wind turbines output.

Table 8. Effectiveness analysis of green certificate trading mechanism.

Case	Minimum cost ($/h)	Minimum emission (tons/h)	Wind power output (MW)
2	**144070.42**	**153086.22**	**159.29**
1	144820.21	155435.56	131.95

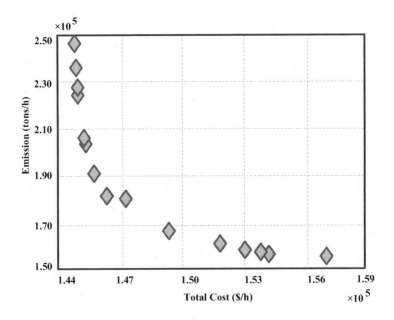

Fig. 4. Simulation results for Case 1.

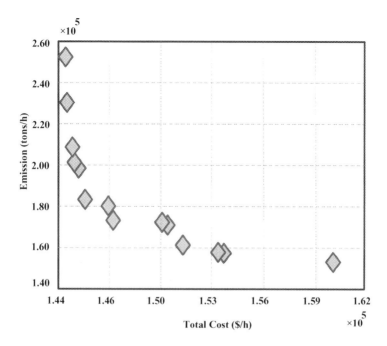

Fig. 5. Simulation results for Case 2.

5   Concluding Remarks

This paper describes the surrogate-based methods to solve the computationally expensive EED problems. ANN-based surrogate models greatly reduce the computation time by replacing the original mathematical functions in the problems. Based on this, an improved MOGWO algorithm, named CS-MOGWO, is proposed for solving the Pareto front of the computationally expensive EED problems. The convergence and searching ability of the algorithm are accelerated by introducing the differential algorithm and improving the position update strategy. On one hand, the simulation comparison results show that the methods can actually solve the computationally expensive problems. On the other hand, green certificate trading mechanism can incentivize renewable energy generation, thereby reducing power generation costs and polluting gas emissions. Other complex EED problems with renewable energy sources can also be overcame by surrogate modeling, although these problems are up for future studies.

Acknowledgements. This work was supported by the National Natural Science Foundation of China under Grant 62163013.

References

1. Yu, X., Dong, Z., Zhou, D.: Integration of tradable green certificates trading and carbon emissions trading: how will Chinese power industry do? J. Clean. Prod. **279**, 123485 (2021). https://doi.org/10.1016/j.jclepro.2020.123485

2. Liang, H., Liu, Y., Li, F.: A multi-objective hybrid bat algorithm for combined economic/emission dispatch. Int. J. Electr. Power Energy Syst. **101**, 103–115 (2018). https://doi.org/10.1016/j.ijepes.2018.03.019

3. Leena, V.A., Ajeena, B.A.S., Rajasree, M.S.: Genetic algorithm based bi-objective task scheduling in hybrid cloud platform. Int. J. Comput. Appl. **8**, 7–13 (2016)

4. Li, L.L., Shen, Q., Tseng, M.L.: Power system hybrid dynamic economic emission dispatch with wind energy based on improved sailfish algorithm. J. Clean. Prod. **316**, 128318 (2021). https://doi.org/10.1016/j.jclepro.2021.128318

5. Nourianfar, H., Abdi, H.: Economic emission dispatch considering electric vehicles and wind power using enhanced multi-objective exchange market algorithm. J. Clean. Prod. **415**, 137805 (2023). https://doi.org/10.1016/j.jclepro.2023.137805

6. Sharifian, Y., Abdi, H.: Solving multi-zone combined heat and power economic emission dispatch problem considering wind uncertainty by applying grasshopper optimization algorithm. Sustain. Energy Technol. Assess. **53**, 102512 (2022). https://doi.org/10.1016/j.seta.2022.102512

7. Younes, M., Khodja, F., Kherfane, R.L.: Multi-objective economic emission dispatch solution using hybrid FFA (firefly algorithm) and considering wind power penetration. Energy **67**, 595–606 (2014). https://doi.org/10.1016/j.energy.2013.12.043

8. Hazra, S., Roy, P.K.: Quasi-oppositional chemical reaction optimization for combined economic emission dispatch in power system considering wind power uncertainties. Renew. Energy Focus. **53**, 45–62 (2019). https://doi.org/10.1016/j.ref.2019.10.005

9. Wang, H., Jin, Y., Jansen, J.O.: Data-driven surrogate-assisted multiobjective evolutionary optimization of a trauma system. IEEE Trans. Evol. Comput. **20**, 939–952 (2016). https://doi.org/10.1109/TEVC.2016.2555315

10. He, C., Tian, Y., Wang, H.: A repository of real-world datasets for data-driven evolutionary multiobjective optimization. Complex Intell. Syst. **6**, 189–197 (2019). https://doi.org/10.1007/s40747-019-00126-2

11. Lin, C., Liang, H., Pang, A.: A fast data-driven optimization method of multi-area combined economic emission dispatch. Appl. Energy **337**, 120884 (2023). https://doi.org/10.1016/j.apenergy.2023.120884

12. Unal, M., Mahouti, P., Turk, A.S.: A novel near field radiation shaping technique by using data driven surrogate based optimization for nondestructive hyperthermia. Int. J. Numer. Model. Electron. Netw. Dev. Fields **36**, 3061 (2023). https://doi.org/10.1002/jnm.3061

13. Mittal, S., Sharma, P.: An optimized and efficient multi parametric scheduling approach for multi-core systems. Int. J. Comput. Theory Eng. **56**, 391–395 (2013)

14. Zhang, G., Shao, X., Li, P.: An effective hybrid particle swarm optimization algorithm for multi-objective flexible job-shop scheduling problem. Comput. Ind. Eng. **56**, 1309–1318 (2009). https://doi.org/10.1016/j.cie.2008.07.021

15. Hetzer, J., David, C.Y., Bhattarai, K.: An economic dispatch model incorporating wind power. IEEE Trans. Energy Convers. **23**, 603–611 (2008). https://doi.org/10.1109/TEC.2007.914171

16. Pang, A., Liang, H., Lin, C.: A surrogate-assisted adaptive bat algorithm for large-scale economic dispatch. Energies **16**, 1011 (2023). https://doi.org/10.3390/en16021011
17. Zare, M., Narimani, M.R., Terzija, V.: Reserve constrained dynamic economic dispatch in multi-area power systems: an improved fireworks algorithm. Int. J. Electr. Power Energy Syst. **126**, 106579 (2021). https://doi.org/10.1016/j.ijepes.2020.106579
18. Mirjalili, S., Saremi, S., Mirjalili, S.M.: Multi-objective grey wolf optimizer: a novel algorithm for multi-criterion optimization. Expert Syst. Appl. **47**, 106–119 (2016). https://doi.org/10.1016/j.eswa.2015.10.039
19. Liang, H., Liu, Y., Shen, Y.: A hybrid bat algorithm for economic dispatch with random wind power. IEEE Trans. Power Syst. **33**, 5052–5061 (2018). https://doi.org/10.1109/TPWRS.2018.2812711
20. Secui, D.C.: Large-scale multi-area economic/emission dispatch based on a new symbiotic organisms search algorithm. Energy Convers. Manag. **154**, 203–223 (2017). https://doi.org/10.1016/j.enconman.2017.09.075
21. Wang, J., Chen, C., Ye, J.: Multi-objective optimal control of bioconversion process considering system sensitivity and control variation. J. Process Control **119**, 13–24 (2022). https://doi.org/10.1016/j.jprocont.2022.09.006
22. Hagh, M.T., Kalajahi, S.M., Ghorbani, N.: Solution to economic emission dispatch problem including wind farms using Exchange Market Algorithm Method. Appl. Soft Comput. **88**, 106044 (2020). https://doi.org/10.1016/j.asoc.2019.106044
23. Jiang, S., Ji, Z., Wang, Y.: A novel gravitational acceleration enhanced particle swarm optimization algorithm for wind-thermal economic emission dispatch problem considering wind power availability. Int. J. Electr. Power Energy Syst. **73**, 1035–1050 (2015). https://doi.org/10.1016/j.ijepes.2015.06.014

MODMOA: A Novel Multi-objective Optimization Algorithm for Unmanned Aerial Vehicle Path Planning

Qian Wang[✉], Xiaobo Li, Peng Su, Yuxin Zhao, and Qiyong Fu

School of Computer Science and Technology, Zhejiang Normal University, Jinhua 321004, China
19548810051@zjnu.edu.cn

Abstract. Multi-Objective UAV path planning problem is to find an optimal path, which can satisfy multiple objectives at the same time and optimize other performance indicators in the case of considering multiple conflicting objectives, constraints and trade-offs. In order to solve the challenge of considering multiple constraints and responding to environmental changes in real time, we propose a UAV path planning method based on multi-objective dwarf mongoose optimization algorithm. Considering the general search efficiency and insufficient global convergence of the original algorithms, nonlinear factorial augmented search strategies, chaotic mapping, and on-the-fly search strategies have been proposed to address the above problems and to enhance the algorithmic population diversity and local optimization capability. Experimental results show that, compared with other algorithms, the proposed method has better performance and robustness in multi-target UAV path planning, and can effectively find high-quality non-inferior solution sets, which provides an effective solution for UAV path planning.

Keywords: Dwarf Mongoose · Meta-heuristic · Pareto Solutions · Multi-objective Optimization · UAV path planning

1 Introduction

Metaheuristic Algorithms [1] are a class of efficient and adaptive optimization algorithms used to solve various complex optimization problems. Compared to traditional deterministic algorithms, metaheuristic algorithms can flexibly explore the solution space during the search process and find solutions that approach the optimal solution within an acceptable time frame. However, traditional single-objective metaheuristic algorithms can only handle a single objective and are difficult to solve problems with multiple objectives and constraints. To address this challenge, researchers have improved the algorithm to simultaneously consider multiple objectives and constraints. By finding non-inferior solutions in the solution set (i.e. the Pareto Front), multiple choices are provided to meet the needs of different stakeholders. This improved algorithm is known as multi-objective optimization algorithm.

© The Author(s), under exclusive license to Springer Nature Singapore Pte Ltd. 2024
L. Pan et al. (Eds.): BIC-TA 2023, CCIS 2061, pp. 44–58, 2024.
https://doi.org/10.1007/978-981-97-2272-3_4

Unmanned aerial vehicle (UAV) path planning [2] refers to the planning and control of UAV autonomous flight paths through algorithms and technology in a given environment. With the continuous development and application of UAV technology, UAV path planning has become an important issue in the field of UAV applications. For example, in agriculture, surveying and mapping, environmental monitoring, logistics delivery, and security monitoring, UAVs need to fly along predetermined paths to complete specific tasks. In complex urban environments or areas with complex terrain such as mountain and forests, UAV path planning is even more challenging. During this process, there are inevitably multiple conflicting objectives, such as the shortest path, height restrictions, minimum obstacle avoidance distance, etc. Multi-objective optimization algorithms can balance these objectives, find a set of optimal solutions, provide different path choices, and meet different task requirements.

Therefore, multi-objective optimization algorithms [3] have significant application value in UAV path planning, which can help UAVs efficiently plan paths in complex environments and balance different objectives according to task requirements. Using these algorithms can improve the effectiveness and performance of path planning, further promoting the development and application of UAV technology.

2   Related Work

2.1   The DMO Algorithm Optimization

The Dwarf Mongoose Algorithm is an optimization algorithm inspired by the foraging and migration behaviors of ferret colonies. It divides the ferret colony into three distinct groups: the Alpha Group, the Scout Group, and the Nanny Group.

The Alpha Group consists of mongooses with strong foraging abilities and is responsible for the main search task, which involves finding the optimal or non-inferior solutions in the solution space. To maintain global search capability and diversity, the algorithm periodically facilitates information exchange and knowledge sharing among the members of the Alpha Group to promote fusion and collaboration.

The Scout Group comprises mongooses with excellent exploration capabilities. They are tasked with exploring unknown regions of the solution space and discovering new solutions. Through periodic burst behaviors and a covert search strategy, the Scout Group can avoid getting trapped in local optima, thus enhancing the global search capability.

Lastly, the Nanny Group consists of stable and high-quality individuals responsible for maintaining the already discovered excellent solutions in the solution space, preventing them from being eliminated or forgotten. The Nanny Group improves the convergence speed and search efficiency of the entire colony through periodic information sharing and information sharing.

In summary, the Ferret Algorithm leverages the collaborative work of the Alpha Group, Scout Group, and Nanny Group to achieve global search and diversity maintenance, thereby enhancing the algorithm's search efficiency and its ability to solve complex optimization problems.

The DMO algorithm optimization process consists of three phases, as shown in Fig. 1.

Fig. 1. The optimization procedures of the proposed DMO.

2.2   Related Research

Due to the strong adaptability and superiority of multi-objective optimization algorithms in solving UAV path planning problems [4], more and more researchers have used some classical multi-objective optimization algorithms for solving UAV path planning problems and achieved remarkable results. However, in the process of practical application, more problems to be optimized also emerge. For example, In 2020, Xu et al. [5] proposed an improved Multi-Objective Particle Swarm Optimization (MOPSO) algorithm aimed at finding a collision-free feasible path for a rotating unmanned aerial vehicle (R-UAV) in a known static rough terrain environment with minimum height, length, and angle variability. However, the MOPSO algorithm may suffer from path instability due to insufficient control over velocity and position, which fails to meet the safety requirements of the UAV. In 2021, Wang et al. [6] proposed an improved NSGA-II algorithm for multi-objective optimization of UAV path planning. The algorithm adjusts the crossover probability and mutation probability adaptively and introduces an improved directional mutation strategy. The algorithm is able to search the optimal path for UAVs under the consideration of multiple objectives such as path length, threat and concealment, however, the NSGA-II algorithm may not be able to fully take into account the influence of environmental factors on UAVs, resulting in the generated paths not meeting the actual conditions. In addition, the algorithm may fall into the local optimal solution and fail to find the global optimal solution due to the elite strategy. In 2021, Tong et al. [7] proposed a UAV path planning and autonomous formation improvement method based on pigeon optimization and differential evolution. Firstly, the mathematical model of UAV trajectory planning was designed as a multi-objective optimization model with three metrics: path length, path sinuosity, and path risk, and then, a variation strategy based on pigeon optimization and differential evolution was developed to optimize the feasible paths. However, the pigeonholing algorithm is more sensitive to the modeling of the problem and parameter setting, and may have lower robustness to the variation and noise of the problem. In UAV path planning, environmental conditions and constraints may change, and the robustness of the algorithm is an important factor in ensuring the reliability of path planning. In 2022, Peng et al. [8] proposed a decomposition based constrained multi-objective evolutionary algorithm (M2M-DW), which has a local infeasibility utilization mechanism for unmanned aerial vehicle trajectory planning. Among them, M2M-DW is used as a solution optimizer because it can utilize infeasible individuals. In addition, an improved mutation scheme was designed to further explore promising areas. But the algorithm relies on many parameter settings, such as population size, mutation factor, etc. For unmanned aerial vehicle path planning problems, excessive or inappropriate parameter selection may lead to slow or difficult optimization.There are also many multi-objective optimization algorithms applied to unmanned aerial vehicle path planning problems, such as SPEA2 [9], which is a commonly used multi-objective optimization algorithm that improves the quality and diversity of non-dominated solution sets by dividing individuals into external archives and internal archives. However, in unmanned aerial vehicle path planning, due to the need to consider many constraint conditions and the impact of dynamic environment, SPEA2 may find it difficult to find a global optimal solution that satisfies all constraints.

3   Methodology

Considering the problems in the above research, such as insufficient convergence, easy to fall into local optimal, low efficiency, and risk caused by algorithm instability, we proposed an improved multi-objective algorithm MODMOA [10] with strong robustness and stability. By adding nonlinear factorial enhanced search strategy, the algorithm convergence and local search ability were enhanced. PWLCM chaotic mapping to enhance the algorithm's initial population diversity. And LEVY flight strategy enhanced the local optimal level of jumping out.

3.1   Improvement of MODMOA

We constructed the MOO [11] version of DMOA using three components: archive controller [12], grid [13] and elite strategy [14], which retains the advantages of the original algorithm such as simple parameters, high efficiency, adaptivity, robustness and easy implementation. In addition, we propose several innovative improvements: introducing the original nonlinear factors a and p, applying PWLCM chaos to the initial population of mongoose, and applying the LEVY flight strategy to the exploration phase of the algorithm.

1) **Nonlinear Factorial Enhanced Search Strategy**

Due to the weak local search capability of the initial formula (1) in simulating the reconnaissance behavior of the scout team, and the slowing down of convergence speed of linear optimization algorithms when approaching the optimal solution, it is possible to get trapped in a stable region. To overcome these issues, we introduce non-linear factors.

By incorporating non-linear factors, we can increase the diversity and exploratory nature within the search space, thus accelerating the convergence speed of the algorithm. This is particularly advantageous when dealing with non-convex, multimodal, and complex problems.

Our strategy introduces three non-linear factors, namely a, p, and w. The purpose of incorporating these factors is to enhance the local search capability of the algorithm and accelerate the convergence speed. They help the algorithm escape from local optima and explore towards the global optimum. Additionally, these factors contribute to solving non-smooth and non-convex problems, thereby improving the robustness and adaptability of the algorithm.

By appropriately adjusting and optimizing these non-linear factors, we can achieve a better balance between global search and local search, resulting in faster and more accurate optimization results. The above formula is improved as follows:

$$
X_{i+1} = \begin{cases} Xi - SF * ph * ran * \left[Xi - \vec{N}\right] & if \ \varphi i + 1 > \varphi i \\ Xi + SF * ph * ran * \left[Xi - \vec{N}\right] & else \end{cases} \tag{1}
$$

$$
X_{i+1} = \begin{cases} X_g - cos(a) * \exp(-CF * ph * rand) * \\ \quad \left[tan(w * p) * X_i - p * \vec{N}\right], \\ \qquad if \ \varphi_{i+1} > \varphi_i \\ X_g + cos(a) * \exp(-CF * ph * rand) * \\ \quad \left[tan(w * p) * X_i - p * \vec{N}\right], \\ \qquad else \end{cases} \tag{2}
$$

Among them, t is the current iteration count, T is the maximal iteration count, and the calculation formulas of nonlinear factors a, p and w are as follows:

$$
a = \pi - \pi \left(\frac{t}{T}\right)^2 \tag{3}
$$

$$
p = 1 - 0.8 * \sqrt{\frac{t}{T}} \tag{4}
$$

$$
w = a * rand - a \tag{5}
$$

2) PWLCM Chaos Mapping

PWLCM Chaos mapping [15] is a method for generating chaotic sequences. Compared to traditional chaotic mapping algorithms, the PWLCM algorithm exhibits better randomness in control parameters and initial conditions, allowing it to generate more complex and diverse initial populations. Additionally, the PWLCM algorithm has a simpler computational process and lower computational complexity, which enhances its efficiency and practicality in real-time applications and large-scale data processing. By applying it to the MODMOA algorithm, the initial mongoose population can be fully optimized, and the computational complexity after fusion is also reduced. The mathematical formula of the PWLCM algorithm is as follows:

When $0 \leq f(t) < p$:

$$
f(t + 1) = \frac{f(t)}{p} \tag{6}
$$

When $p \leq f(t) < 0.5$:

$$f(t+1) = \frac{f(t) - p}{0.5 - p} \tag{7}$$

When $0.5 \leq f(t) < 1 - p$:

$$f(t+1) = \frac{1 - p - f(t)}{0.5 - p} \tag{8}$$

When $1 - p \leq f(t) < 1$:

$$f(t+1) = \frac{1 - f(t)}{p} \tag{9}$$

among them, $p = 0.4$. The overall flowchart of this algorithm refers to Fig. 2 below:

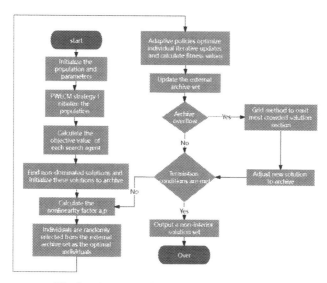

Fig. 2. Flowchart of the proposed MODMOA.

3) **The Levy strategy**

Levy Flight [16] is a type of random walk mechanism. By introducing long-distance random step sizes, it may jump out of local optima during the search process, expanding the search space and increasing breadth and diversity in the search. This helps to avoid the algorithm getting trapped in local optima, discover global optima more effectively, and enables the search to traverse the search space quickly and approach the optimal solution. Incorporating this strategy into MODMAO can significantly enhance the algorithm's ability to escape local optima and accelerate convergence towards the optimal solution.

$$\text{Le}(y)^\alpha \mu - 1 - \varphi \tag{10}$$

$$\mu = \frac{D}{\frac{|G|}{\varphi}} \tag{11}$$

$$\sigma^2 = \left\{ \frac{\gamma(1+\varphi)}{\varphi^\gamma((1+\varphi)/2)} \frac{\sin\left(\frac{\pi\varphi}{2}\right)}{\frac{2(1+\varphi)}{2}} \right\}^{\frac{2}{\varphi}} \tag{12}$$

Among them, $0 < \varphi \leq 2$, D, $G \sim N(0, \sigma^2)$, $\gamma(x)$ is the number of Gamma graphs, μ represents the step size, $\varphi = 2/3$.

The formula for generating candidate food positions in the original algorithm is (Eq. (10)):

$$X_{t+1} = X_t + ph * hoot \tag{13}$$

Since formula (4) relies on the position update at the previous moment, it is inefficient and prone to local optimization, the formula is improved to:

$$X_{t+1} = X_g + ph * hoot * levy(D) \tag{14}$$

where X_g is the global optimal position, $levy(*)$ means Levy flight, D is the dimension of the problem.

3.2 Application of MODMOA in UAV Path Planning

In drone trajectory planning, the role of the objective function is to evaluate the effectiveness of trajectory planning from multiple perspectives, usually including a series of constraints and optimal conditions, with a smaller objective function as the goal to achieve higher trajectory quality. On this basis, this article proposes an algorithm model based on multiple objectives such as path cost, threat cost, flight altitude cost, and flight angle cost. The path cost is set as F1, and the other three cost components are set as F2, F3, and F4, so the two objective functions are set as:

$$f_1(X_i) = F_1(X_i) \tag{15}$$

$$f_2(X_i) = F_2(X_i) + F_3(X_i) + F_4(X_i) \tag{16}$$

By representing the position of the ferret individual i in MODMOA algorithm as Ω_i and the velocity vector associated with the ferret individual i as the position change vector $\Delta\Omega_i$:

$$\Delta\Omega_i = (\Delta\rho_{i1}, \Delta\theta_{i1}, \Delta z'_{i1}, \ldots, \Delta\rho_{in}, \Delta\theta_{in}, \Delta z'_{in}) \tag{17}$$

The velocity update equation for individual i in the MODMOA algorithm, denoting the trajectory encoding vector of the $i - th$ individual's result at the $t - th$ iteration as Y^t_{ij} ($\rho_{ij}, \theta_{ij}, z'_{ij}$), and the velocity vector at the t-th iteration as $\Delta Y^t_{ij}(\Delta\rho_{ij}, \Delta\theta_{ij}, \Delta z'_{ij})$, can be expressed as:

$$\Delta Y^t_{ij} = \omega_b \Delta Y^{t-1}_{ij} + f_1 r_{1j}\left(P^t_{ij} - Y^{t-1}_{ij}\right) + f_2 r_{2j}\left(P^t_{qj} - Y^{t-1}_{ij}\right) \tag{18}$$

$$\boldsymbol{P}^t_{ij} = (P_{i1}, P_{i2}, \ldots, P_{in}) \tag{19}$$

$$\boldsymbol{P}^t_{qj} = \left(P_{q1}, P_{q2}, \ldots, P_{qn}\right) \tag{20}$$

and $\boldsymbol{P}^t_{ij}$ and $\boldsymbol{P}^t_{qj}$ is the individual and global optimal position of the $t - th$ iteration.

4 Experiments

4.1 Experiment Settings

Experiments are implemented in matlab R2020a, and run in Windows 10, 64-bit operating system, x64-based processor, 1.99 GHz, 8 GB memory. The population size was set to 100, the maximum number of iterations was set to 100, the external archive was set to 200, the grid size was set to 60, and 15 independent repeated experiments were performed for each comparison algorithm.

4.2 MODMOA Performance

To verify the performance of DODMOA, we conducted comparative experiments on six multi-objective algorithms using four evaluation metrics on two benchmark test functions[17]. Among them, WFG (1, 2, 3, 4, 6) is a set of weighted feature generation multi-objective optimization test functions. The good performance on the ZDT test function indicates that the algorithm has strong global search ability, convergence performance, and performance stability. DTLZ (6, 7) is a set of high-dimensional test functions, and if the algorithm performs well on these functions, it indicates that the algorithm has the ability to handle high-dimensional complex diversity problems.

Figure 3 illustrates the convergence of MODMOA on the ZDT test functions, demonstrating that the resulting solution set approaches the Pareto optimal set indefinitely.

Table 1 compares the effectiveness of the MODMOA method on the ZDT test function with five known algorithms. The results of calculating the evaluation indicators for ZDT (1, 2, 3, 4, 6) show that the MODMOA algorithm proposed in this paper performs better than other algorithms on almost all evaluation indicators. Among them, MODMOA performs best in terms of IGD, GD, and HV, which reflects that compared with other algorithms, it obtains a non-inferior solution set closer to the true Pareto front on ZDT1-ZDT6, with better convergence and optimization ability. From Fig. 5, it can also be seen intuitively that the non-inferior Pareto results of the MODMOA algorithm are very close to the Pareto solutions of the ZDT test function. However, the SP indicator performance of this algorithm in ZDT6 is average, which may indicate that the uniformity of the solution set on higher-dimensional complex diversity problems still needs further verification and improvement.

The experimental results fully demonstrate that MODMOA has strong global search ability, convergence performance, and stability, and can effectively handle multi-objective optimization problems without prior knowledge or link information. However, when facing overly high dimensions and complexity, the concentration of the solution set may lead to the problem of falling into local search.

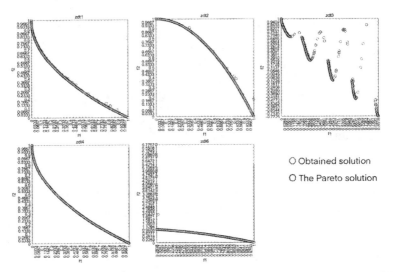

Fig. 3. The convergence of selection approaches on ZDT1, ZDT2, ZDT3, ZDT4, ZDT6.

Table 2 In experiments based on ZDT test functions, we have raised questions about whether the algorithm can be well applied to high-dimensional complex problems. To answer this question, we compared the MODMOA algorithm with five other well-known methods on the high-dimensional multi-objective test function DTLZ (6, 7). The experimental results well demonstrate the ability of MODMOA to handle high-dimensional problems. The only slight deficiency is the SP indicator performance on DTLZ7, which indicates that the distribution of the solution set is not uniform enough compared to the MODA/D algorithm.

4.3 Application of MODMOA in UAV Path Planning

In order to verify the effectiveness of the improved algorithm, simulation modeling of unmanned aerial vehicle path planning was conducted in the Matlab R2022a environment. Designed a simulation environment [18]: there are cylindrical obstacles in the environment, corresponding to a mountainous environment with complex terrain. The simulation formula for mountainous environment is as follows:

$$\begin{cases} r_1(x, y) = \sin(y + a_m) + b_m\sin x + c_m\cos\left(d_m\sqrt{x^2 + y^2}\right) \\ \qquad\quad + e_m\cos y + f_m\sin\left(f_m\sqrt{x^2 + y^2}\right) + g_m\cos y \\ r_2(x, y) = \sum_{i=1}^{n'} H_i\exp\left(-\left(\frac{x-x_i}{x_{pi}}\right)^2 - \left(\frac{y-y_i}{y_{pi}}\right)^2\right) \end{cases} \quad (21)$$

Among them, n' is the number of slopes, H_i is the control parameter related to the height of the mountain slope, (x_i, y_i) is the center coordinate of the i-th mountain slope, x_{pi}, y_{pi} corresponds to the attenuation of the slope along the x-axis and y-axis directions, used to constrain slope data, a_m~g_m represents various coefficients. $r_1(x, y)$ Simulate the diversified baseline terrain features of digital maps through the combination of different

Table 1. Results achieved with suggested and competing methods on the ZDT benchmark.

F	Index	NSGA-II	NSGA-III	MOEA/D	NSWOA	NSMFO	MODMOA
ZDT1	IGD	1.60E+00	1.70E+00	7.45E−01	1.71E−02	5.95E−02	**2.20E−03**
	GD	2.16E−01	3.23E−01	9.87E−02	1.98E−03	2.80E−03	**1.12E−04**
	HV	4.20E−04	1.11E−04	7.60E−03	6.44E−01	6.98E−01	**7.87E−01**
	SP	7.85E−02	9.20E−02	2.32E−01	1.44E−02	7.92E−02	**3.42E−03**
ZDT2	IGD	1.44E+00	2.34E+00	1.33E+00	2.51E−02	1.79E−02	**2.24E−03**
	GD	3.90E−01	3.79E−01	8.98E−02	8.79E−05	1.31E−03	**3.18E−05**
	HV	3.23E−05	2,11E−05	1.02E−05	4.36E−01	4.31E−01	**4.58E−01**
	SP	6.46E−02	6.45E−02	3.80E−03	2.94E−03	2.47E−02	**2.92E−03**
ZDT3	IGD	7.35E−01	7.31E−01	3.87E−01	4.52E−03	1.39E−02	**2.54E−03**
	GD	3.03E−01	1.98E−01	2.01E−02	1.97E−04	6.57E−04	**1.08E−04**
	HV	1.49E−01	8.89E−02	2.98E−01	6.01E−01	5.34E−01	**6.13E−01**
	SP	3.90E−02	5.03E−02	1.66E−02	6.41E−03	1.66E−02	**3.97E−03**
ZDT4	IGD	4.52E+00	2.67E+00	5.35E+00	2.22E−03	2.48E−03	**2.21E−03**
	GD	3.16E+00	4.23E−01	3.68E−01	4.56E−05	3.89E−05	**3.88E−05**
	HV	5.33E−05	4.21E−06	1.11E−05	7.11E−01	7.13E−01	**7.33E−01**
	SP	2.39E−01	1.29E−01	1.31E−02	3.56E−03	2.32E−03	**3.34E−03**
ZDT6	IGD	6.17E+00	6.33E+00	6.23E+00	1.34E−02	2.39E+00	**1.80E−03**
	GD	1.59E+00	1.41E+00	4.19E−01	2.22E−02	9.87E−01	**1.39E−02**
	HV	0.00E+00	0.00E+00	0.00E+00	3.59E−01	0.00E+00	**3.90E−01**
	SP	5.34E−02	6.21E−02	**2.73E−04**	2.69E−02	2.81E−01	2.12E−01

Table 2. Results achieved with suggested and competing methods on the DTLZ (6, 7).

F	Index	NSGA-II	NSGA-III	MOEA/D	NSWOA	NSMFO	MODMOA
DTLZ6	IGD	7.98E+00	8.14E+00	9.47E−01	2.43E−03	5.44E−02	**2.73E−03**
	GD	6.02E−01	6.18E−01	7.10E−02	3.57E−05	7.61E−02	**3.49E−05**
	HV	0.00E+00	0.00E+00	0.00E+00	2.01E−01	1.39E−01	**1.99E−01**
	SP	3.09E−01	3.19E−01	4.98E−02	4.87E−03	1.29E−01	**4.52E−03**
DTLZ7	IGD	4.32E+00	3.43E+00	4.67E+00	5.63E−02	8.78E−02	**5.55E−02**
	GD	4.68E−01	4.66E−01	2.56E−01	2.71E−03	3.82E−03	**2.31E−03**
	HV	0.00E+00	0.00E+00	0.00E+00	2.74E−01	2.61E−01	**2.77E−01**
	SP	1.11E−01	1.12E−01	**5.60E−04**	5.26E−02	7.99E−02	4.33E−02

coefficient settings, $r_2(x, y)$ is used to represent hill and mountain slope data in digital maps to increase the degree of terrain undulation, The spatial area of this terrain is 1000 m × 1000 m × 400 m, the flight environment is equipped with 8–12 threat zones with varying sizes of obstacle radiation areas; Minimum flying altitude of unmanned aerial vehicle $H_{min} = 80$ m, maximum flight altitude $H_{max} = 320$ m.

Table 3. Triple trajectory planning parameters.

Flight task	Starting point coordinates	End point coordinates	Number of obstacles
First time	[900; 900; 50]	[160; 60; 150]	8
Second time	[160; 60; 50]	[900; 900; 150]	10
Third time	[900; 60; 50]	[160; 900; 150]	12

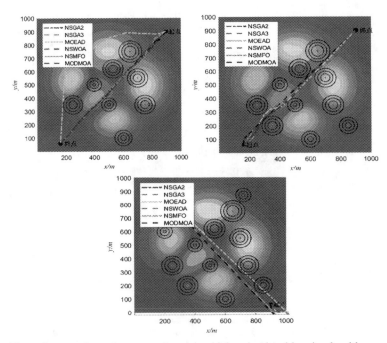

Fig. 4. The trajectory plots of unmanned aerial vehicles simulated by six algorithm models in three different environments.

To fully evaluate the performance of the MODMOA based unmanned aerial vehicle trajectory planning model [19], this article set up three flight missions, with parameter settings shown in Table 3. The number of drone flight path points for each mission was 12. In terms of algorithm testing, the comparison algorithm of MODMOA algorithm includes five classic and efficient multi-objective algorithms, each with a population size of 100 and a maximum number of iterations of 60, which are independently repeated

10 times. The flight trajectory plots for the six algorithm models simulating the UAV in three different environments are shown in Fig. 4.

Figure 5 shows the results of various algorithms solving the cost function for drone flight. The number of threat areas for the three flight tasks were 8, 10, and 12 respectively, and the complexity of the terrain in the drone flight environment gradually increased. The experimental results showed that compared with the various control algorithms, the MODMOA-UAV algorithm model had good convergence and optimization ability in each flight. The scores for the three flights also demonstrated the algorithm's strong stability, global search capability, and robustness. Specific numerical values are shown in Table 4.

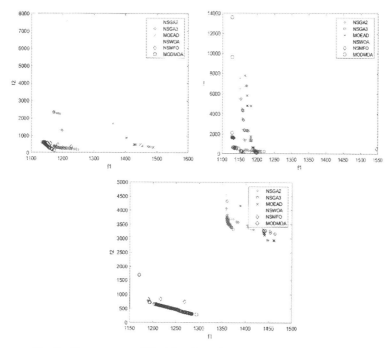

Fig. 5. Convergence of Six Algorithms in Three Flight Environments.

From Fig. 6, it is evident that the total cost of the MODMOA-UAV model in all three flight environments is lower than that of other comparative algorithm models, fully demonstrating the low cost and efficiency of the MODMOA-UAV model. As the environment becomes more complex, the cost of some previously performing algorithms significantly increased during the second and third flights. Specifically, in the second flight, the MOEAD, NSWOA, and NSMFO algorithms were unable to adapt well to the environment of increasing obstacles and changing terrain kurtosis, resulting in a significant increase in flight altitude costs. In the third flight, two large-scale obstacles and more complex paths were added, resulting in a significant increase in the corner and threat costs of the NSGA II, NSGA III, and MOEAD algorithms. In contrast, MODMOA performs stably in every environment, with a controllable increase in cost within a small

Table 4. Comparison of Algorithm Cost Fitness Values.

flight	Index	NSGA-II	NSA-III	NSWOA	MOEA/D	NSMFO	MODMOA
1-th	Max	1340.25	1355.49	1615.31	1385.21	1373.93	**1178.29**
	Min	1604.43	1762.43	2103.54	1743.38	1911.57	**1321.56**
	Avg	1440.25	1455.49	1815.31	1485.21	1573.93	**1306.29**
2-th	Max	1302.56	1377.54	1523.77	1334.53	1256.69	**1105.34**
	Min	1737.12	1698.66	1956.43	1627.95	1565.23	**1309.67**
	Avg	1465.95	1520.83	1721.86	1433.89	1439.54	**1288.35**
3-th	Max	4709.34	4209.56	3343.09	1645.67	1799.76	**1435.34**
	Min	5134.56	4907.55	4035.77	2201.56	2134.67	**1709.67**
	Avg	4849.67	4631.10	3796.35	1938.35	1991.32	**1641.28**

range, and has strong robustness and the ability to adapt to high-dimensional complex environments.

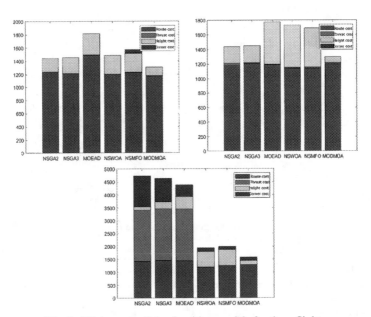

Fig. 6. Flight costs of six algorithm models for three flights.

5 Conclusions and Future Works

After conducting performance verification experiments, MODMOA has demonstrated excellent performance in terms of its global search capability, convergence, robustness, and ability to obtain optimal solutions. It exhibits stability when dealing with various complex problems. The only drawback is that compared to other algorithms, the distribution of solution sets in MODMOA is not uniform, which may affect the diversity of solutions.

To address longstanding issues in unmanned aerial vehicle (UAV) path planning, such as the inability to simultaneously satisfy multiple constraints, the challenge of balancing low-cost consumption and high performance, and the inability to adapt to real-time changes in complex environments, we have integrated MODMOA with UAV path planning, resulting in the MODMOA-UAV algorithm model. Experimental results have shown that compared to other commonly used classical UAV path planning algorithms, our model demonstrates lower cost consumption, stronger obstacle avoidance and optimization capabilities, as well as greater adaptability and stability when facing complex environments. Therefore, MODMOA is considered an excellent algorithm suitable for high-dimensional, complex, and diverse applications involving UAVs.

In the future, with further refinement and optimization, MODMOA holds the potential for widespread application in related fields, effectively addressing longstanding limitations of such problems.

Acknowledgements. This research was supported by the Key Project of the Zhejiang Provincial Natural Science Foundation of China under Grant No. LZ24F020005.

References

1. Agrawal, P., Abutarboush, H.F., Ganesh, T., Mohamed, A.W.: Metaheuristic algorithms on feature selection: a survey of one decade of research. IEEE Access **9**(3), 26766–26791 (2021)
2. Hayat, S., Yanmaz, E., Bettstetter, C., Brown, T.X.: Multi-objective drone path planning for search and rescue with quality-of-service requirements. Auton. Robot. **44**(7), 1183–1198 (2020)
3. Deb, K., Sindhya, K., Hakanen, J.: Multi-objective optimization. Decis. Sci. **27**(5), 161–200 (2016)
4. Majeed, A., Hwang, S.O.: A multi-objective coverage path planning algorithm for UAVs to cover spatially distributed regions in urban environments. Aerospace **8**(11), 343 (2021)
5. Zhen, X.U., Enze, Z., Qingwei, C.: Rotary unmanned aerial vehicles path planning in rough terrain based on multi-objective particle swarm optimization. J. Syst. Eng. Electron. **31**(1), 130–141 (2020)
6. Wang, H., Tan, L., Shi, J., Lv, X., Lian, X.: An improved NSGA-II algorithm for UAV path planning problems. J. Internet Technol. **22**(3), 583–592 (2021)
7. Tong, B., Chen, L., Duan, H.: A path planning method for UAVs based on multi-objective pigeon-inspired optimisation and differential evolution. Int. J. Bio-Inspired Comput. **17**(2), 105–112 (2021)
8. Peng, C., Qiu, S.: A decomposition-based constrained multi-objective evolutionary algorithm with a local infeasibility utilization mechanism for UAV path planning. Appl. Soft Comput. **118**, 108495 (2022)

9. Gupta, M., Varma, S.: Optimal placement of UAVs of an aerial grid network in an emergency situation. J. Ambient. Intell. Humaniz. Comput. **12**, 343–358 (2021)

10. Agushaka, J.O., Ezugwu, A.E., Abualigah, L.: Dwarf mongoose optimization algorithm. Comput. Methods Appl. Mech. Eng. **391**, 114570 (2022)

11. Gharehchopogh, F.S., Namazi, M., Ebrahimi, L., Abdollahzadeh, B.: Advances in sparrow search algorithm: a comprehensive survey. Arch. Comput. Methods Eng. **30**(1), 427–455 (2023)

12. Kuo, R.J., Gosumolo, M., Zulvia, F.E.: Multi-objective particle swarm optimization algorithm using adaptive archive grid for numerical association rule mining. Neural Comput. Appl. **31**(4), 3559–3572 (2019)

13. Dhiman, G., et al.: EMoSOA: a new evolutionary multi-objective seagull optimization algorithm for global optimization. Int. J. Mach. Learn. Cybern. **12**, 571–596 (2021)

14. Dhiman, G., Kumar, V.: Emperor penguin optimizer: a bio-inspired algorithm for engineering problems. Knowl.-Based Syst. **159**(7), 20–50 (2018)

15. Zhang, S., Liu, L.: A novel image encryption algorithm based on SPWLCM and DNA coding. Math. Comput. Simul. **190**, 723–744 (2021)

16. Kaidi, W., Khishe, M., Mohammadi, M.: Dynamic levy flight chimp optimization. Knowl.-Based Syst. **235**, 107625 (2022)

17. Ishibuchi, H., Nan, Y., Pang, L.M.: Performance evaluation of multi-objective evolutionary algorithms using artificial and real-world problems. In: Emmerich, M., et al. (eds.) EMO 2023. LNCS, vol. 13970, pp. 333–347. Springer, Cham (2023). https://doi.org/10.1007/978-3-031-27250-9_24

18. Phung, M.D., Ha, Q.P.: Safety-enhanced UAV path planning with spherical vector-based particle swarm optimization. Appl. Soft Comput. **107**, 107376 (2021)

19. Lin, J., Pan, L.: Multiobjective trajectory optimization with a cutting and padding encoding strategy for single-UAV-assisted mobile edge computing system. Swarm Evol. Comput. **75**, 101163 (2022)

Decomposed Multi-objective Method Based on Q-Learning for Solving Multi-objective Combinatorial Optimization Problem

Anju Yang[1], Yuan Liu[1(✉)] ⓘ, Juan Zou[1] ⓘ, and Shengxiang Yang[2] ⓘ

[1] Hunan Engineering Research Center of Intelligent System Optimization
and Security, Xiangtan University, Xiangtan 411105, China
`liu3yuan@xtu.edu.cn`, `202121632849@smail.xtu.edu.cn`

[2] School of Computer Science and Informatics, De Montfort University,
Leicester LE1 9BH, UK

Abstract. Neural combinatorial optimization has emerged as a promising technique for combinatorial optimization problems. However, the high representation of deep learning inevitably requires a lot of training overhead and computing resources, especially in large-scale decision-making and multi-objective scenarios. This paper first provides a simple but efficient combinatorial optimization method that uses a traditional reinforcement learning (RL) paradigm to balance the computational cost and performance. We decompose the multi-objective problem into multiple scalar subproblems and only use the improved Q-learning for the sequential optimization of these subproblems. Our method employs the Temporal-Difference (TD) update strategy and provides a shared Q-table for all subproblems. The TD update strategy speeds up the optimization by learning while making decisions. The shared Q-table devotes a high-quality starting point to generate excellent solutions quickly for each subproblem. Both strategies promote the effectiveness and efficiency of the proposed method. After new solutions are generated, a selection operator keeps the historical optimal solution for each subproblem. We apply our method to various multi-objective traveling salesman problems involving up to 10 objectives and 200 decisions. Experiments demonstrate that only simple RL achieved comparable performance to state-of-the-art approaches.

Keywords: Reinforcement Learning · Q-learning ·
Temporal-Difference · Shared Q-table · Multi-objective Traveling
Salesman Problem

1 Introduction

A multi-objective combinatorial optimization problem (MOCOP) [1] involves several conflicting objectives to be optimized simultaneously and no single solution is optimal for all objectives. Instead, a set of Pareto optimal solutions with

© The Author(s), under exclusive license to Springer Nature Singapore Pte Ltd. 2024
L. Pan et al. (Eds.): BIC-TA 2023, CCIS 2061, pp. 59–73, 2024.
https://doi.org/10.1007/978-981-97-2272-3_5

different trade-offs among the objectives is required. Without loss of generality, MOCOP can be defined as follows:

$$\min_{x \in X} \quad \mathbf{F}(x) = (f_1(x), f_2(x) \cdots f_m(x)), \tag{1}$$

where x represents an n-dimensional discrete candidate solution; X represents an n-dimensional discrete boundary decision space; m is the objective dimension, and $\mathbf{F}(x)$ represents an m-dimensional objective space.

Considering the NP-hard and multi-objective characteristics of the MOCOP, multi-objective metaheuristics, such as Ant Colony Optimization (ACO) [2] and Particle Swarm Optimization (PSO) [3], have been developed in the past two decades. However, metaheuristics also have some limitations [4–6], such as meta-heuristics often require significant configuration time to select the best parameters and operators for a specific instance. Actually, as the scale of optimization problems continues to grow in applications, neural combinatorial optimization (Neural CO) using neural network models and reinforcement learning (RL) has attracted widespread concern [7,8].

Most of the existing Neural CO methods mainly focus on single-objective COPs. However, what stands out is that decomposition-based multi-objective optimization algorithms MOEA/D [9] builds a bridge for Neural CO to solve MOCOPs. For example, DRL-MOA [10], MODGRL [11], ML-DRL [12], PMOCO [13] and MOPN [14] have confirmed that Neural MOCO performs remarkably in solving MOCOPs. Unfortunately, these Neural MOCO methods typically have high resource demands using neural networks. Therefore, it is worth pondering whether it is possible to directly use traditional RL methods without neural networks to solve MOCOPs for balancing the training overhead and performance of the model.

Here, we try to use the traditional Q-learning [15] to solve the MOCOPs and verify the method's performance on the MOTSP [16] instances is within reasonable training time on the CPU.

The main contributions of this paper are as follows:

- A fast and effective RL method based Q-learning, QL-MOA, is proposed for dealing with MOTSPs. Unlike the existing methods, QL-MOA relies on empirical learning without training sets, which is easy to implement. Combining a shared Q-table and the improved Q-learning, QL-MOA enables fast decision-making and achieves high-quality solutions.
- Experiments on MOTSPs verify that traditional RL can also effectively solve MOCOPs compared to state-of-the-art approaches, even if such instances have large decision-making variables.
- QL-MOA is highly scalable and can solve MOTSP instances from 2- to 10-D objectives. Since QL-MOA solves sub-problems based on decomposition, the optimization efficiency will not be affected by the increase in the number of objectives, which is more advantageous in solving high-dimensional MOCOPs.

2   Background

2.1   Problem Formulation

We consider a generalized M-objective MOTSP with K-city nodes and M $K*K$ cost matrices $\{\mathbf{c}_1, \mathbf{c}_2, \cdots, \mathbf{c}_M\}$ for M different costs as a benchmark. The purpose of MOTSP is to find a tour sequence $\pi = \{s_1, s_2, ..., s_K\}$ to simultaneously minimize all the costs:

$$\min \mathbf{F}(\pi) = \min \ (f_1(\pi), f_2(\pi), \cdots, f_M(\pi)),$$
$$\text{where } f_i(\pi) = \sum_{j=1}^{K-1} \mathbf{c}_i(s_j, s_{j+1}) + \mathbf{c}_i(s_K, s_1), \tag{2}$$

and $\mathbf{c}_i(s_j, s_{j+1})$ represents the cost for the i-objective when moving from city j to $j+1$.

The MOTSP can be formulated into a Multi-objective Markov Decision Process (MOMDP) [13] defined by a tuple $(\mathcal{S}, \mathcal{A}, \mathcal{R}, \mathcal{T}, \gamma)$, where $\mathcal{S}$ is the state space containing the cities that tourists need to traverse; $\mathcal{A}$ is the action space representing tourists reach from the current city to the next city; $\mathcal{R} : \mathcal{S} \times \mathcal{A} \to \mathcal{R}$ is the set of reward function $\mathcal{R} = [r_1, \cdots, r_M]^T = [-\mathbf{c}_1, \cdots, -\mathbf{c}_M]^T$ reflecting the cost of different objectives between two cities; $\mathcal{T} : \mathcal{S} \times \mathcal{A} \to \mathcal{S}$ is the dynamic transition function and $\gamma \in [0, 1]$ is a discount factor. In the MOTSP, a policy $\pi : \mathcal{S} \to \mathcal{A}$ is associated with a vector of expected returns $\mathbf{Q}(\pi) = [Q_1(\pi), \cdots, Q_M(\pi)]^T$, where

$$Q_i(\pi) = \mathbb{E}\Big[\sum_{t=0}^{T} \gamma r_i(s_t, a_t) | s_0, a_t \sim \pi(s_t)\Big]. \tag{3}$$

2.2   Decomposition-Based Scalarization

The solution to a MOCO problem is a set of policies called a Pareto optimal set that contains at least one optimal policy for each possible preference that a decision maker (DM) might have. This scalarization method maps each reward function $\mathcal{R}$ and possible policy value vector $\mathbf{Q}(\pi)$ onto a scalar value, which converts MOMDP to single-objective MDPs. Currently, decomposition-based scalarization, especially MOEA/D and its variants, has become a mainstream method for solving multi-objective optimization problems (MOPs). As in MOEA/D, a MOP is decomposed into N scalar optimization subproblems by N pre-defined weight vectors. Specifically, a set of uniformly distributed weight vectors, such as $w_1, w_2, \cdots, w_8$ in Fig. 1, are designed by the weight generation method, and each weight corresponds to a scalar optimization subproblem. In this paper, we use the Weighted Sum as the objective function of each subproblem w_i as follows:

$$\min g^{ws}(\pi|w_i) = \sum_{j=1}^{M} w_{i,j} f_j(\pi), i = i, 2, \cdots, N. \tag{4}$$

Therefore, we can use standard RL to optimize different scalar subproblems, and the optimal solution of each subproblem constitutes the Pareto optimal front of the original problem.

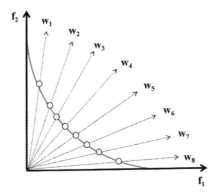

Fig. 1. Illustration of the decomposition strategy.

2.3   Tabular Q-Learning

Tabular Q-learning is a popular TD [17] algorithm capable of solving discrete single-objective MDP. In this algorithm, a table consisting of state-action pairs (s, a), $s \in \mathcal{S}$ and $a \in \mathcal{A}$, is constructed, each element of which represents an estimate of $Q^*(s, a)$. Q-values are updated according to the following:

$$Q(s_t, a_t) = (1 - \alpha)Q(s_t, a_t) + \alpha[r_{t+1} + \gamma Q^*(s_{t+1}, a_{t+1})], \qquad (5)$$

where $Q^*(s_{t+1}, a_{t+1}) = \max_{a_{t+1}} Q(s_{t+1}, a_{t+1})$, $\alpha \in [0, 1]$ is the learning rate that specifies the incremental update and $\gamma \in (0, 1)$ represents the discount factor. Q-learning is continued by updating the Q-value $Q(s_t, a_t)$ for each state using the Eq. 5 until it converges to the optimal $Q^*(s_t, a_t)$.

3   Methodology

In this section, we propose our algorithm QL-MOA that employs tabular Q-learning. First, we provide the basic framework of QL-MOA for a MOTSP instance. Then we explain how to use improved Q-learning to solve each sub-problem.

3.1   General Framework

The general framework of QL-MOA is described as Algorithm 1. We first decompose a MOTSP instance I into N subproblems by the decomposition method [9] and sort according to the similarity between weights. Then, we initialize a shared Q-table Q and Pareto optimal solutions P used for solving subproblems. For each subproblem, we use Weighted Sum function, such as Eq. 4, to construct multi-objective rewards into a single-objective reward, then maximize the cumulative reward through the Improved Q-learning (Algorithm 2). The reward r as follows:

$$r(s_t, a_t | w_i) = -\sum_{j=1}^{m} w_{i,j} c_j(s_t, s_{t+1}).$$ (6)

Algorithm 1. General framework of QL-MOA

Require: A MOTSP instance I;
Ensure: Pareto optimal solutions P;
 1: $\{w_1, w_2, \cdots, w_N\} \leftarrow$ Decomposition(I);
 2: Suproblem similarity ranking $\{w_1, w_2, \cdots, w_N\}$;
 3: $Q \leftarrow$ Initialize();
 4: $P \leftarrow \emptyset$;
 5: **for** $i \leftarrow 1, \cdots, N$ **do**
 6:      $P_i \leftarrow \emptyset$;
 7:      $P_i, Q \leftarrow$ Improved Q-learning(Q, w_i, P_i);
 8:      $p \leftarrow \min_{p \in P_i} g^{ws}(p|w_i)$;
 9:      $P \leftarrow P \cup p$;
10:      $i \leftarrow i + 1$;
11: **end for**
12: **return** P.

For a subproblem, we generate a set of diverse solutions P_i according to the shared Q-table and select an optimal solution p with their minimum travel expenses. When the current subproblem is optimized, the shared Q-table migrates to the next adjacent subproblem as a starting point for training. Ultimately, optimal solutions of these subproblems constitute the original problem's Pareto optimal solutions P. As described above, our algorithm contains two essential components: the shared Q-table strategy and the improved Q-learning algorithm.

3.2   The Shared Q-Table Strategy

Q-learning optimizes a specific problem by training a Q-table. This method brings a computational cost that cannot be ignored for solving a series of subproblems. The decomposed subproblems have approximate MDP environments, e.g., same state and action spaces as well as similar reward functions. Therefore, when the reward functions of the two subproblems are closer, the optimal decision sequence obtained by RL will also have a higher similarity. For example, in the MOTSP, the optimal decision sequence of the subproblem w_i has many sequence fragments that are identical to its adjacent subproblems w_{i-1} and w_{i+1}. Figure 2 presents the trajectories of the optimal solutions of the three adjacent subproblems in the MOTSP with 15 cities. As shown by the red and blue dashed lines in Fig. 2, most of the sequences of the optimal solutions of these three adjacent subproblems are consistent and the number of cities in these same sequences exceeds 60% of the total number of cities. Therefore, we only need to fine tune the optimal solutions to achieve transfer on adjacent subproblems.

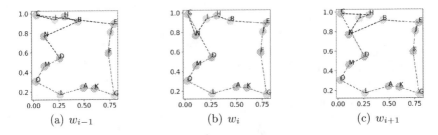

$$(a)\ w_{i-1} \qquad\qquad (b)\ w_i \qquad\qquad (c)\ w_{i+1}$$

Fig. 2. The trajectories of the optimal solutions to the three adjacent subproblems in the 15-city MOTSP, where the red and blue dashed lines represent the common sequence of w_{i-1}, w_i and w_{i+1}; the green dashed line represents the common sequence of the two subproblems w_i and w_{i+1}, and the black dashed lines represent the ordinary sequence of w_{i-1}, w_i and w_{i+1}. (Color figure online)

Based on these considerations, we provide a shared Q-table for the subproblems to speed up the training process. In the shared Q-table, each element represents the current optimal Q-value of the state-action pairs (s_t, a_t) for the subproblem. Since the Q-table converges to the optimum, the current subproblem w_i will be solved. The optimal Q-table then migrates to the next subproblem w_{i+1} as the initial table for fine-tuning. In order to make the initial Q-table most effective, we select the subproblem w_{i+1} that is most similar to the current subproblem w_i for migration, and the similarity can be reflected from the angle between the weight vectors.

3.3   Improved Q-Learning Algorithm

As shown in Algorithm 2, we first initialize an empty archive set P_i to collect the sampled solutions of the current subproblem w_i. Then we use an agent to perform E episodes, which includes the pre-training and sampling phases for the subproblem w_i. In each episode, a starting city (or state) is randomly initialized as s_0. Based on the current state s_t, the agent will select the next state from a feasible set of states s_{t+1}. To this end, the ε-greedy selection strategy is used to balance exploration and exploitation during the decision-making process, as the following formula.

$$\pi(a|s) = \begin{cases} \underset{a}{argmax}(Q^\pi(s_t, a_t)), & if\ p_t >= \varepsilon. \\ random(), & otherwise. \end{cases} \quad (7)$$

In the ε-greedy selection, a random probability p_t is first generated. If the probability p_t is greater than or equal to the threshold ε, the agent will select

Algorithm 2. Improved Q-Learning algorithm

Require: The shared Q-table Q, subproblem w_i and empty solution set P_i;
Ensure: P_i, Q;
1:   $P_i \leftarrow \emptyset$;
2: **for** $episode \leftarrow 1, 2, \cdots, E$ **do**
3:      $\pi \leftarrow \emptyset$;
4:      $s_0 \leftarrow$ Random_Initialization();
5:      $\pi \leftarrow \pi \cup s_0$;
6:      **for** $t \leftarrow 0, \cdots, T$ **do**
7:         $a_t \leftarrow$ Action_Selection(s_t, ε);
8:         $s_{t+1}, r_t \leftarrow$ Action_Execution(a_t);
9:         $\pi \leftarrow \pi \cup s_{t+1}$;
10:        Update the Q by Eq. (5);
11:        Decay ε by multiplying with d;
12:        $t \leftarrow t + 1$;
13:      **end for**
14:      **if** $episode \% r_f == 0$ **then**
15:        $\varepsilon \leftarrow 1$;
16:      **end if**
17:      **if** $E - episode <= N_s$ **then**
18:        $P_i \leftarrow P_i \cup \pi$;
19:      **end if**
20:      $episode \leftarrow episode + 1$;
21: **end for**
22: **return** P_i, Q.

---

the next state s_{t+1} with the largest Q-value $Q(s_t, a_t)$. Otherwise, the agent will randomly select a feasible city as the next city s_{t+1}. The selection of the next state is accompanied by an immediate reward r_t. After the selection of the new city, the Q-table is updated by Eq. 5. We update the value of ε until an episode is traversed.

As mentioned, threshold ε is crucial to which city the agent selects. If the value of ε is set too small, the agent will over-exploit and fall into the local optimum. In contrast, if the value is too high, the agent will continue to explore and find it difficult to converge. In the improved Q-learning, we will dynamically adjust the ε value through the decaying and resetting operations. After each episode, the ε will gradually decay by multiplying itself with a coefficient d; the d must be as close to 1 as possible to prevent incomplete exploration. The decaying of ε can make the agent gradually change from the exploration process to the exploitation process, which enables the agent to obtain a more efficacious experience to train the shared Q-table. However, persistent decaying will cause the agent to only focus on optimal local decisions, resulting in insufficient exploration in the later episodes. The resetting of ε can mitigate this dilemma to a certain extent. In the improved Q-learning, we reset the value of ε to 1 after r_f episodes. The r_f can be set according to the maximum number of episodes. By periodically adjusting the ε parameter, we can prevent the agent from prematurely converging to a locally optimal decision.

Along with the training process, the improved Q-learning will sample E solutions for each subproblem. To do this, we keep the generated solutions in the archive set P_i for the last N_s episodes. When the training is over, we return the latest Q and solution set P_i.

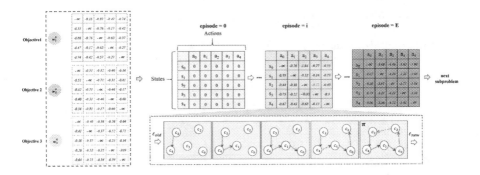

Fig. 3. Illustration of the main execution process of Algorithm 2. The black dashed line represents the instance. The green dashed line represents the generating a solution π. (Color figure online)

To facilitate understanding of the Algorithm 2, we provide the update process of the shared Q-table Q under the three-objective distance MOTSP instance with 5-city in Fig. 3. Taking the first subproblem w_0 as an example, the shared Q-table of all zeros is initialized to indicate that no experience has been obtained. In the Q-table, the rows represent the state space $\mathcal{S} = \{s_0, s_1, \cdots, s_4\}$ including all possible cities, and the columns represent the action space $\mathcal{A} = \{a_0, a_1, \cdots, a_4\}$ that can be performed in a certain state. Addressing specific subproblem w_0, the agent randomly selects the city c_3 as the initial state s_0. Then the search strategy performs an action a_0 to visit the new city c_4 as the next state s_1. Simultaneously, the instance provides an immediate reward r_0 whose composition is as Eq. 6. After a decision, the $Q(s_0, a_0)$ is computed according to Eq. 5. The agent will repeat the operation on the new state until all cities have been traversed. As shown in Fig. 3, the trajectory of the agent in an episode is such as $\pi = \{c_3, c_4, c_1, c_0, c_2, c_3\}$, and the corresponding position in the Q-table is updated. After each episode, the value of ε updates for the next trajectory. This process will continue to iterate until the maximum episode is completed. When the subproblem w_0 is optimized, the shared Q-table will be migrated to the next adjacent subproblem w_1 as a starting point for training.

4   Experiment Settings

4.1   Test Instances

Distance Objective. This objective which reflects the total Euclidean distance between cities traveled is commonly considered in MOTSPs. Undoubtedly, a traveling salesman wishes to traverse all cities by the shortest distance. Therefore, in this objective, the reward of each decision is defined by the distance between the real coordinates of two cities s_i and s_{i+1}.

Altitude Objective. This objective is another that should be considered in the MOTSPs, which is used to reflect the smoothness between cities traveled. That is, the smaller the objective, more comfortable the journey will be. Therefore, for this objective, the reward of each decision is defined by the altitude variances between two cities s_i and s_{i+1}.

Table 1. Four different MOTSP instances and their introduction.

Instances	Objective Dimension (D)	Instance Description
Type I	2-D distance objectives	It consists of two distance objectives
Type II	3-D mixed objectives	It consists of an altitude objective and two distance objectives
Type III	5-D mixed objectives	It consists of three distance objectives and two altitude objectives
Type IV	10-D distance objectives	It consists of ten distance objectives

According to these two objectives, we constructed four different instances, as shown in Table 1.

4.2   Parameter Settings

Number of Subproblems N. To balance the performance and computational overhead of the algorithm, we set 100, 105, 105, and 110 uniform weight vectors for instances of 2, 3, 5 and 10 objectives, respectively.

Learning Rate α. The learning rate [18] mainly controls the update step of the Q-values. In our paper, we set the learning rate as 0.5.

Discount Factor γ. The discount factor is a constant that reflects how much the RL agent cares about rewards in the distant future relative to those in the immediate future. A higher discount factor (such as 0.95) means that almost all the rewards of the last step will be applied to the previous step.

Parameters Related to the ε-Greedy Search. The search strategy is a method to balance exploration and exploitation by randomly choosing actions, and the ε value refers to the probability of choosing to explore. In our algorithm, the ε value is initialized to 1 and decayed by being multiplied with 0.95 at each decision and being reset to 1 after 1000 episodes, which indicates that the d is 0.95 and r_f is 100.

Episode Size E. The episode size E consists of the number of pre-training and the number of solutions collected for a subproblem. The shared Q-table will be pre-trained for 20,000 episodes on the first subproblem and 10,000 episodes on other subproblems. For each subproblem, a solution set P_i with size 100 is generated to accompany the subsequent training process, which indicates that the N_s is 100.

Parameter Settings for Compared Algorithms. The parameter settings of the DRL-MOA and MOPN and MOEA/D and NSGA-II [19] are consistent with their original papers. Additionally, we set the number of iterations of the metaheuristic methods to 4000, and the population size is consistent with N.

4.3   Metrics

Hypervolume. Hypervolume (HV) [20] which mainly reflects the performance of the algorithm is the most common metric in multi-objective optimization. Let the set P be an approximation to the Pareto front of the problem. The HV value of P is the volume of the region formed by the set P and the reference point. The larger the HV value of the set, the better the performance of the set P. In this paper, we use the maximum value of all solutions on different objectives to form the reference point.

5   Experimental Results

5.1   Performance Analysis

Performance on Type I Instances. Table 2 provides the HV values of compared method in the Type I instances. As shown in the table, our method achieves excellent performances at different city sizes, showing that our method has good scalability. Although our method is slightly inferior to the MOPN method on the 40-city instance, the difference between the two methods is insignificant. Furthermore, the results show that the learning-based methods outperform the metaheuristic methods overall, especially in large-scale decision spaces. For the metaheuristic methods, the traditional strategy based on the random search is difficult to solve in a large-scale decision space, resulting in the inability of them to find the optimal solution. For the learning-based methods, the difficulty of neural network training cost is much more significant than that of traditional Q-table training, so our method can show better advantages within an acceptable city scale. In addition, the reason MOPN could perform similar to our method is that it uses prior samples for training, while the samples of our algorithm mainly rely on interactive generation.

Figure 4 shows the solution sets obtained by the five compared algorithms on the Type I instances. Consistent with the results in Table 2, the solution set obtained by our algorithm has shown the best performance, not only in diversity but also in convergence. It is followed by the MOPN method, which shows advantages in terms of convergence, but is slightly inferior to our algorithm

Table 2. Comparison of HV on Type I instance. The best HV is marked bold.

Methods	2-D distance objectives				
	40-city	70-city	100-city	150-city	200-city
MOEA/D	312.84	793.52	1776.21	3961.19	5895.78
NSGA-II	328.76	817.36	1783.16	3683.80	5239.45
DRL-MOA	315.66	853.79	2122.73	5140.59	8405.34
MOPN	**331.50**	921.19	2303.68	5584.71	9310.16
QL-MOA	330.89	**928.78**	**2317.07**	**5610.22**	**9384.85**

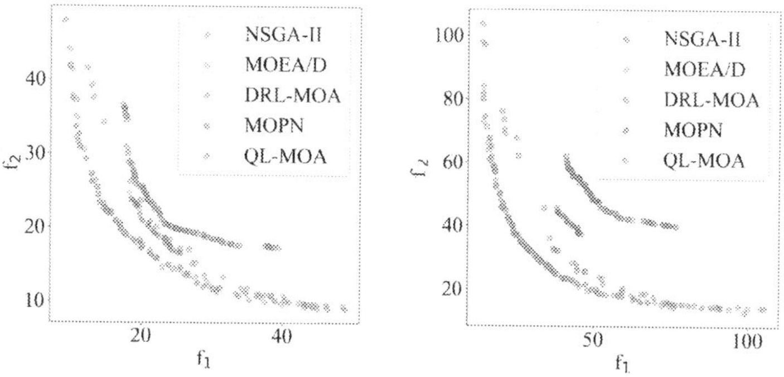

(a) 100-city Type I instance   (b) 200-city Type I instance

Fig. 4. Solution sets obtained by compared methods.

in terms of distribution. For DRL-MOA, the solution set is widely distributed in the objective space, but its uniformity and convergence are not as good as MOPN and our method. The convergence and diversity of MOEA/D and NSGA-II are not as good as the previous three methods, which further verifies that traditional meta-heuristic methods have difficulty competing with learning-based methods in large-scale decision spaces.

Performance on Type II Instances. Table 3 provides the HV values of each method in the Type II instances. The table shows that our method has a significant advantage overall, only worse than the metaheuristic methods in the instance of 40 cities. As the city size increases, the performance of the meta-heuristic methods degrades significantly, evidenced by their solutions being concentrated in the middle regions and far away from the Pareto front, as shown Fig. 5. The distribution of DRL-MOA is consistent with the situation in Type I instances, and the solution sets are basically gathered in the boundary regions. Our method has advantages in convergence compared to MOPN.

Performance on Type III and Type IV Instances. To further evaluate the scalability of the methods in the objective spaces, we examine their performance in 5-D and 10-D spaces as shown in Table 4 and Table 5. Here we only provide the HV performance of DRL-MOA in the 5-D instances, since the training cost of

Table 3. Comparison of HV on Type II instance. The best HV is marked bold.

Methods	3-D mixed objectives				
	40-city	70-city	100-city	150-city	200-city
MOEA/D	**6188.20**	10988.86	39387.87	1.01E+05	2.01E+05
NSGA-II	5298.29	6032.93	45735.60	1.01E+05	1.17E+05
DRL-MOA	4459.83	11252.09	52541.08	1.79E+05	4.31E+05
MOPN	4456.61	14366.70	63297.77	2.18E+05	5.41E+05
QL-MOA	5214.16	**16820.24**	**71933.63**	**2.48E+05**	**6.19E+05**

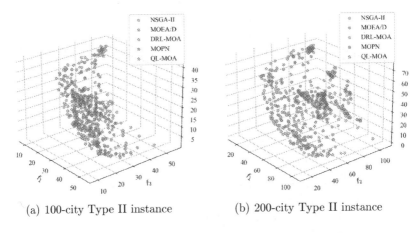

(a) 100-city Type II instance   (b) 200-city Type II instance

Fig. 5. Solution sets obtained by compared methods.

DRL-MOA in more high-dimensional objective problems is too expensive to be acceptable. As shown in two tables, our method is less affected by the increase of dimensionality, and still shows remarkable advantages in high-dimensional objective spaces. MOPN does not perform as well in high-dimensional objectives as it does in the low-dimensional objective spaces, and the performance is inferior to MOEA/D, DRL-MOA, and our method. The poor performance of NSGA-II is due to a lack of selection pressure when dealing with high-dimensional problems.

Table 4. Comparison of HV on Type III instance. The best HV is marked bold.

Methods	5-D mixed objectives	
	100-city	200-city
MOEA/D	2.64E+07	4.81E+08
NSGA-II	4.32E+06	5.98E+07
DRL-MOA	3.59E+07	9.36E+08
MOPN	1.64E+06	3.98E+08
QL-MOA	**8.23E+07**	**2.93E+09**

Table 5. Comparison of HV on Type IV instance. The best HV is marked bold.

Methods	10-D distance objectives	
	100-city	200-city
MOEA/D	3.13E+13	7.29E+15
NSGA-II	1.24E+11	4.33E+12
DRL-MOA	–	–
MOPN	2.01E+11	1.23E+13
QL-MOA	**6.00E+13**	**3.25E+16**

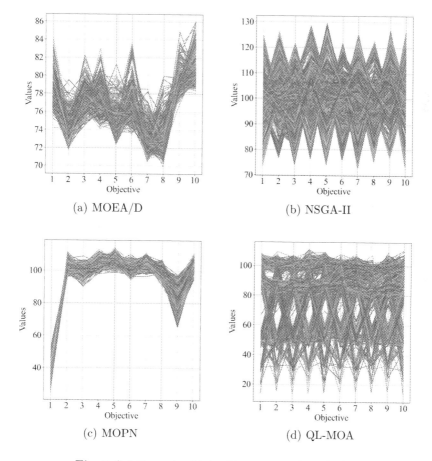

Fig. 6. Solution sets obtained by compared methods.

Figure 6 shows the solution sets obtained by MOEA/D, NSGA-II, MOPN, and QL-MOA in the Type IV MOTSP instance with 200 cities. Since the high-dimensional spaces cannot be visualized intuitively, we adapt parallel coordinate plots to describe the distribution of solution sets in the objective spaces. It can be seen from the figure that MOEA/D and MOPN cannot be widely distributed, and the solution sets are concentrated in some specific regions. The solution set obtained by NSGA-II is difficult to converge to the Pareto front. Compared with the other three algorithms, our method is compelling in terms of convergence and diversity, and the solution set is more widely distributed on the Pareto front.

With our approach, the solution of a single subproblem is solved in seconds under the CPU. Furthermore, we employ the improved Q-learning, a typical Temporal-Difference (TD) method, to update the shared Q-table during each step incrementally, which is a critical factor in our superior performance.

6   Conclusion

The MOTSPs have been widely concerned in practical applications. Metaheuristic methods have been important solutions but cannot efficiently search for instances with large-scale decision spaces. In recent years, learning-based methods, especially Neural CO, have emerged as effective optimization techniques for CO problems. Furthermore, these Neural CO have been successfully applied to MOCOPs through decomposition strategies. However, the high representation of deep learning requires a lot of training overhead to build highly reliable network models for subproblems. Currently, there is a lack of a generalizable model to face a set of subproblems. Therefore, how to balance the performance and efficiency of learning-based methods is an extremely challenging problem.

This paper has carried out primary research on the use of traditional RL, Q-learning, to solve MOTSPs. Our method abandons the training of complex models and obtains high-quality solutions for subproblems by learning Q-table. Considering the environmental similarity of adjacent subproblems, we can improve the learning ability of the algorithm for new subproblems by sharing Q-table. Based on this, we also propose an improved Q-learning method, which balances the exploitation and exploration by adaptively modifying the ϵ-greedy selection. With our approach, the solution of a single subproblem is solved in seconds under the CPU. In addition, we have verified that our method has advantages in both performance and time overhead in the five types of MOTSP instances involving 2- to 10-D objective spaces and 40- to 200-city sizes. It suggests that the performance of traditional learning-based methods is still a new field that needs further breakthroughs.

Acknowledgements. This work was supported in part by the National Natural Science Foundation of China (Grant No. 62276224, Grant No. 6230073173), in part by the Natural Science Foundation of Hunan Province, China (Grant No. 2022JJ40452).

References

1. Ehrgott, M., Gandibleux, X.: A survey and annotated bibliography of multiobjective combinatorial optimization. OR-Spektrum **22**, 425–460 (2000)
2. Wang, Y., Han, Z.: Ant colony optimization for traveling salesman problem based on parameters optimization. Appl. Soft Comput. **107**, 107439 (2021)

3. Gu, Q., Wang, Q., Li, X., Li, X.: A surrogate-assisted multi-objective particle swarm optimization of expensive constrained combinatorial optimization problems. Knowl.-Based Syst. **223**, 107049 (2021)
4. Gao, K.Z., He, Z.M., Huang, Y., Duan, P.Y., Suganthan, P.N.: A survey on meta-heuristics for solving disassembly line balancing, planning and scheduling problems in remanufacturing. Swarm Evol. Comput. **57**, 100719 (2020)
5. Lauri, J., Dutta, S., Grassia, M., Ajwani, D.: Learning fine-grained search space pruning and heuristics for combinatorial optimization. arXiv preprint arXiv:2001.01230 (2020)
6. Li, K., Zhang, T., Wang, R., Wang, Y., Han, Y., Wang, L.: Deep reinforcement learning for combinatorial optimization: covering salesman problems. IEEE Trans. Cybern. **52**(12), 13142–13155 (2021)
7. Garmendia, A.I., Ceberio, J., Mendiburu, A.: Neural combinatorial optimization: a new player in the field. arXiv preprint arXiv:2205.01356 (2022)
8. Bello, I., Pham, H., Le, Q.V., Norouzi, M., Bengio, S.: Neural combinatorial optimization with reinforcement learning. arXiv preprint arXiv:1611.09940 (2016)
9. Zhang, Q., Li, H.: MOEA/D: a multiobjective evolutionary algorithm based on decomposition. IEEE Trans. Evol. Comput. **11**(6), 712–731 (2007)
10. Li, K., Zhang, T., Wang, R.: Deep reinforcement learning for multiobjective optimization. IEEE Trans. Cybern. **51**(6), 3103–3114 (2021)
11. Perera, J., Liu, S.H., Mernik, M., Črepinšek, M., Ravber, M.: A graph pointer network-based multi-objective deep reinforcement learning algorithm for solving the traveling salesman problem. Mathematics **11**(2), 437 (2023)
12. Zhang, Z., Wu, Z., Zhang, H., Wang, J.: Meta-learning-based deep reinforcement learning for multiobjective optimization problems. IEEE Trans. Neural Netw. Learn. Syst. **34**(10), 7978–7991 (2023)
13. Lin, X., Yang, Z., Zhang, Q.: Pareto set learning for neural multi-objective combinatorial optimization. arXiv preprint arXiv:2203.15386 (2022)
14. Gao, L., Wang, R., Liu, C., Jia, Z.: Multi-objective pointer network for combinatorial optimization. arXiv preprint arXiv:2204.11860 (2022)
15. Clifton, J., Laber, E.: Q-learning: theory and applications. Ann. Rev. Stat. Appl. **7**, 279–301 (2020)
16. Lust, T., Teghem, J.: The multiobjective traveling salesman problem: a survey and a new approach. In: Coello Coello, C.A., Dhaenens, C., Jourdan, L. (eds.) Advances in Multi-Objective Nature Inspired Computing. SCI, vol. 272, pp. 119–141. Springer, Heidelberg (2010). https://doi.org/10.1007/978-3-642-11218-8_6
17. Silver, D., Sutton, R.S., Müller, M.: Temporal-difference search in computer Go. Mach. Learn. **87**, 183–219 (2012)
18. Even-Dar, E., Mansour, Y., Bartlett, P.: Learning rates for Q-learning. J. Mach. Learn. Res. **5**, 1–25 (2003)
19. Deb, K., Pratap, A., Agarwal, S., Meyarivan, T.: A fast and elitist multiobjective genetic algorithm: NSGA-II. IEEE Trans. Evol. Comput. **6**(2), 182–197 (2002)
20. Zitzler, E., Brockhoff, D., Thiele, L.: The hypervolume indicator revisited: on the design of pareto-compliant indicators via weighted integration. In: Obayashi, S., Deb, K., Poloni, C., Hiroyasu, T., Murata, T. (eds.) EMO 2007. LNCS, vol. 4403, pp. 862–876. Springer, Heidelberg (2007). https://doi.org/10.1007/978-3-540-70928-2_64

Binary Multi-objective Hybrid Equilibrium Optimizer Algorithm for Microarray Data

Peng Su, Xiaobo Li$^{(\boxtimes)}$ ⓘ, Qian Wang, and Xiaoqian Xie

School of Computer Science and Technology, Zhejiang Normal University,
Jinhua 321004, China
lxb@zjnu.edu.cn

Abstract. Feature selection aims at identifying features relevant to the target from high-dimensional data to enhance the performance of the learner. When dealing with high-dimensional data, traditional methods often exhibit lower accuracy, posing significant challenges to feature selection. In this paper, we propose a multi-objective hybrid binary balance optimizer algorithm to address this issue. This method integrates the output results of multiple filters, considering redundancy and complementarity among genes in the process. Building upon the original Equilibrium optimizer (EO), we employ an external archive to guide the population's search direction and discretize the EO using an S-shaped transfer function. In this approach, we do not consider the classification error rate as a sole optimization objective. Instead, we use the ranking of genes in the selected subset across various filters as the second optimization objective. This design allows for the selection of a smaller number of genes while capturing those most relevant to the labels. To validate the performance of the proposed method, we compare it with 5 multi-objective algorithms on 12 microarray datasets and 3 UCL datasets. The results demonstrate that the proposed method consistently achieves the optimal Pareto front on most datasets.

Keywords: Feature selection · Redundancy and complementary · Equilibrium optimizer

1 Introduction

Microarray data is a common experimental data type in genomics and biotechnology used to investigate changes in gene expression. It employs specific techniques to measure the expression levels of particular genes or RNA sequences in samples. Microarray technology allows for the simultaneous measurement of the expression levels of a large number of genes, aiding researchers in understanding which genes are activated or suppressed under specific conditions within cells, tissues, or biological samples [1]. This is of paramount importance for studying disease mechanisms, and drug development and gaining insights into

© The Author(s), under exclusive license to Springer Nature Singapore Pte Ltd. 2024
L. Pan et al. (Eds.): BIC-TA 2023, CCIS 2061, pp. 74–87, 2024.
https://doi.org/10.1007/978-981-97-2272-3_6

biological processes. Data obtained through microarray technology often exhibit high dimensionality, as the search space is typically extensive. These gene sets include a significant amount of redundancy and unnecessary features [2], making it challenging for researchers to extract valuable information from the plethora of feature sets.

Feature selection (FS), also known as gene selection (GS) in microarray data, is a crucial data dimensionality reduction technique. It aims to eliminate redundant and irrelevant features from the dataset and select those most relevant to class labels. This simplifies the learning model and improves classification accuracy [17]. Feature selection can be primarily categorized into three methods: filtering, wrapping, and embedding. The filtering method involves initially screening or ranking features before training the model, followed by selecting the most relevant features and discarding irrelevant or redundant ones. Common filtering methods include Mutual Information, Correlation-based Feature Selection, Variance Thresholding, and others [15]. The wrapping method incorporates feature selection into the performance evaluation of the model, typically involving four steps: search for feature subsets, model training and evaluation, feature subset update, and iterative search. Metaheuristic algorithms possess stochasticity and global search properties, making them suitable for solving complex optimization problems. Consequently, numerous metaheuristic algorithms have been applied to feature selection, such as Whale Optimization Algorithm (WOA) [13], Genetic Algorithm (GA) [5,11], Particle Swarm Optimization (PSO) [4], and Equilibrium Optimizer (EO) [7]. Embedded methods automatically select the most relevant features during model training, obviating the need for a separate feature selection step. These methods assess and select features during the model training process to ensure the selected features are closely related to the model's performance [16]. Common methods include L1 regularization and decision trees.

In the process of feature selection, there is a constant trade-off between achieving high classification accuracy and minimizing the number of selected features. Traditional single-objective feature selection assigns a higher weight to classification accuracy and a lower weight to the number of features, then combines them into a single objective function. However, this approach often fails to meet the requirements of most cases. In practical engineering applications, the need arises to simultaneously optimize multiple objective functions, a problem known as Multi-Objective Optimization [9]. The EO algorithm, introduced in 2020, is a novel metaheuristic algorithm that has found applications in various domains due to its strong search capabilities. While there have been studies applying EO to feature selection, research on designing it as a multi-objective approach and applying it to the medical field is still limited. This paper investigates the application of Equilibrium Optimizer (EO) algorithms for addressing large search spaces in multi-objective feature selection, specifically for dealing with high-dimensional microarray data. The approach involves the utilization of a hybrid feature selection method. Initially, a combined multi-filter strategy is employed to eliminate redundant and irrelevant features from the original feature space. Subsequently, a transformation is applied to map the original continuous

EO functions into a binary space, aiming to enhance the overall model accuracy. To assess the performance of the proposed method, a comparative analysis is conducted on 15 datasets involving 5 prominent multi-objective algorithms. In most cases, our method demonstrates superior performance.

2   Methodology

2.1   Multi-objective Optimization

In optimization problems with competing objectives, a single solution cannot simultaneously meet all optimization goals. Hence, it becomes essential to identify a set of solutions that can closely approach or attain optimal results for the respective objective functions. In the realm of multi-objective optimization, a singular globally optimal solution is absent. Instead, what exists is a collection of solutions referred to as the Pareto optimal solution set. From a mathematical perspective, multi-objective problems can be succinctly described as follows:

$$minimize\ F(x) = [f_1, f_2, f_3, \ldots, f_n] \tag{1}$$

$$subject\ g_i(x) \leq 0, i = 1, 1, 3, \ldots, k \tag{2}$$

$$h_i(x) = 0, i = 1, 2, 3, \ldots, l, \tag{3}$$

where $f_k(y)$ represents the k-th objective function to be optimized, y is the decision variable vector, and l represents the count of target functions aimed for minimization, considering constraint functions $g_k(y)$ and $h_k(y)$. In MOPs, consider two proposed solutions denoted as u and v, u and v are considered equivalent if and only if:

$$\forall_i \in \{1, 2, 3, \ldots, m\} : f_i(u) \leq f_i(v)\ and\ \exists_i \in \{1, 2, 3, \ldots, m\} : f_i(u) < f_i(v) \tag{4}$$

In this context, we use the dominating to describe the relationship where the solution u outperforms solution v. When a solution u stands unchallenged by any other feasible solutions, it is called a Pareto optimal solution.

2.2   Symmetric Uncertainty

Symmetric Uncertainty (SU), as defined in information theory, is a measure used to quantify the degree of uncertainty in a set of variables or attributes. It is commonly employed in tasks such as feature selection and data mining. The Symmetric Uncertainty between two variables, X and Y, is defined as their mutual information divided by their average entropy, calculated using the following formula:

$$SU(X, Y) = 2 * \frac{IG(X, Y)}{H(X) + H(Y)} \tag{5}$$

$$IG(X, Y) = H(X) - H(X|Y) \tag{6}$$

Here, $IG(X, Y)$ signifies the information gain related to X with respect to Y, $H(X)$ stands for the entropy of X, $H(Y)$ denotes the entropy of Y, and $H(X|Y)$ represents the conditional entropy of X given Y.

2.3   Conditional Mutual Information

Conditional Mutual Information (CMI) is employed to measure the level of dependence or correlation between two variables, taking a specific condition into account. CMI denoted as $I(X;Y|Z)$, quantifies the shared information between random variables X and Y when variable Z is provided as context. It quantifies the information we can obtain about X and Y concerning each other when we already know the value of Z. The mathematical formula for its calculation is as follows:

$$I(X,Y|Z) = H(X|Z) - H(X|Y,Z) \qquad (7)$$

3   The Proposed Method

3.1   Multi-filter Ensemble Strategy Based on Redundancy and Complementarity (RCMF)

The optimal feature subset often comprises both strongly correlated features and weakly correlated features. Weakly correlated features encompass those that are related and complementary to the strongly correlated features [10]. Since filters only consider the relationship between individual features and the target label, in this study, to avoid the occasional randomness of a single filter, we employed five mainstream methods (Neighborhood Component Analysis, Pearson correlation coefficient, Chi-Square, ReliefF, Fisher score). By setting a threshold value of N, we aggregated the top N features selected by these five filters to form a new feature subset. However, during this process, the redundancy between features is not taken into account. Therefore, to eliminate redundant features, we employ the Approximate Markov Blanket [6] to remove redundancy in the features aggregated at multiple filter stages. The features jointly selected by the five filters were considered as strongly correlated features and conditional mutual information was used to select complementary features among the removed features. Figure 1 shows the flowchart of RCMF.

3.2   Binary Multi-Objective EO Based on External Archive Updates (BMOEO)

Traditional EO algorithms were initially developed to address continuous problems and are not directly applicable to solving multi-objective feature selection problems. The original EO individuals' positions involve continuous movement in the search space, which is not suitable for the discrete space required for feature selection. Hence, it is essential to employ a transfer function that converts continuous positions into a binary discrete space to fulfill the criteria for feature selection [8].

The conventional EO algorithm employs a balance pool that selects individuals from the best individuals in the population. In multi-objective algorithms, the concept of dominance exists only among individuals. However, the external

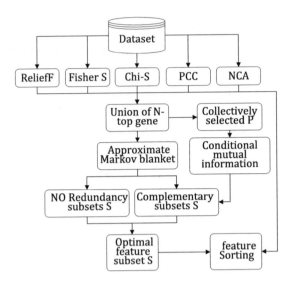

Fig. 1. The pictorial flowchart of the RCMF method.

archive stores the positions of all non-dominating solutions found since the beginning of the population iterations. In this study, we transferred the balance pool of the EO algorithm from the population to the external archive, thereby accelerating the population's search for optimal solutions. The steps of the algorithm we propose are as follows.

Initialization: Randomly generate the positions of particles within the entire search space.

$$X_i = l_b + r \left(u_b - l_b \right) \tag{8}$$

Generating non-dominated solutions: Evaluate the particles generated during initialization to produce initial non-dominated solutions and store them in the external archive.

Constructing the equilibrium pool: In traditional EO algorithms, to expedite particle exploration towards the optimal direction, the four best positions found in the population up to that point are placed in the equilibrium pool, and their average position is used to generate an average solution that is also included in the equilibrium pool. In multi-objective problems, there is no single best solution, so we use the stored external archive to guide the generation of the balance equilibrium.

$$X_{eq,POOL} = \left\{ X_{eq(1)}, X_{eq(2)}, X_{eq(3)}, X_{eq(4)}, X_{eq(ave)} \right\} \tag{9}$$

When the external archive stores fewer than four non-dominated solutions, they are all placed into the equilibrium pool. Since the positions are encoded in binary, the calculation of the average position is adjusted accordingly. Here, n represents the dimension of the entire search space, and x_i denotes the position of a randomly selected particle from the balance pool in the corresponding dimension.

$$X_{eq(ave)} = \{x_1, x_2, x_3, \ldots, x_n\} \tag{10}$$

Concentration Update: In EO, the adjustment of particle concentration involves primarily two rules, namely the concentration update rule controlled by the exponent term F, and the equilibrium state rule controlled by the generation rate G. The exponent term F determines the balance between exploration and exploitation in the algorithm and is formulated as:

$$F = a_1 sigh\,(r_1 - 0.5)\,[\exp{(-r_2 t)} - 1] \tag{11}$$

Here, where a_1 is set as a constant 2 to control the algorithm's exploration capability, r_1 and r_2 are random numbers within the range $(0,1)$, and t is an EO coefficient that receives an update in each iteration, represented as:

$$t = (1 - T/M_{iter})^{(a_2 T/M_{iter})} \tag{12}$$

where T represents the current iteration, M_{iter} represents the maximum iterations, and a_2 is a constant that controls the mining capability of EO. If a_1 is larger, the algorithm will have stronger exploration capabilities, and similarly, if a_2 is larger, the algorithm will have stronger development capabilities. Under normal circumstances, both values are commonly set to 2 and 1. The generation rate G constitutes another essential update rule for EO and is expressed as:

$$G = -P\,(X_{eq} - r_2 X(T))\,F \tag{13}$$

$$p = \begin{cases} 0.5 r_{d1} u & r_{d2} \geq GP \\ 0 & r_{d2} < GP \end{cases} \tag{14}$$

where X_{eq} is the best particle selected from the equilibrium pool in the current iteration, $X(T)$ represents the position of the current iteration's particle, r_{d1} and r_{d2} are random numbers within the interval $(0, 1)$, The vector u is normalized to a unit vector, and the parameter GP is assigned a fixed value of 0.5.

$$X\,(T+1) = X_{eq} + (X(T) - X_{eq})\,F + (1 - F)\,G/r_3 V \tag{15}$$

where $X(T+1)$ signifies the updated position of the particle, $X(T)$ denotes the particle's position from the preceding iteration. r_3 is a random number within the interval $(0, 1)$, and V is a unit vector. Transfer function: To adapt EO for feature selection, the update positions, which are continuous, need to be converted into binary representations. In this paper, an S-shaped transfer function is used for this conversion. The mathematical formula is as follows:

$$sigmoid(X) = \frac{1}{1 + e^{-10(x-0,5)}} \tag{16}$$

$$x_d^{t+1} = \begin{cases} 1 & sigmoid\,(x_d^t) < rand() \\ 0 & sigmoid\,(x_d^t) > rand() \end{cases} \tag{17}$$

where x represents the continuous value of an EO position. Using a sigmoid transfer function, we can map the continuous EO values to a binary space, when a feature is denoted as 1, it signifies that the individual has opted for that particular feature, whereas 0 indicates that the feature has not been chosen. Figure 2 shows the flowchart of the algorithm.

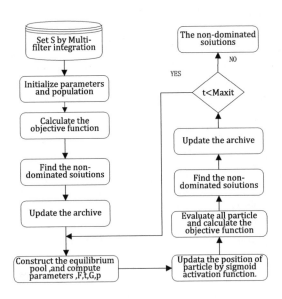

Fig. 2. The pictorial flowchart of the BMOEO.

4   Experiment Setup

4.1   Datasets

In the experiments, We comprised 12 high-dimensional biomedical datasets obtained from microarray repositories and 3 high-dimensional datasets acquired from UCL data [3,12]. The features of these datasets are summarized in Table 1 and 2.

4.2   Objective Function

In general, multi-objective feature selection aims to strike a balance by minimizing both classification accuracy and the number of features as simultaneous optimization objectives [14]. In this paper, multiple filters were employed to eliminate redundant and irrelevant features from the search space. Filters select features most relevant to the class labels. To bias the algorithm toward choosing features ranked higher in the filters, we calculated the sum of ranks for each

Table 1. Microarray datasets

NO	Dataset Name	Features	Instance	Class
1	SRBCT_4	2308	83	4
2	LUNG_5	3312	203	5
3	DLBCL	4026	47	2
4	GLIOMA	4434	50	2
5	Brain_Tumor1	5920	90	5
6	ALLAML	7129	72	2
7	CNS	7130	60	2
8	CAR	9182	174	2
9	Brain_Tumor2	10367	50	4
10	LUNG	12533	181	2
11	MLL	12583	72	3
12	BREAST	24482	97	2

Table 2. UCI datasets

NO	Dataset Name	Features	Instances	Classes
1	Gastroenterology	698	72	2
2	Color	2000	62	2
3	DBworld	4702	64	2

feature in five filters as the first objective and considered the classification error rate as the second objective.

$$minimze\ F(x) = \begin{cases} f_1(x) = sort_1 + sort_2 + \ldots + sort_n \\ f_2(x) = \frac{FP+FN}{TP+TN+FP+FN} * 100 \end{cases} \qquad (18)$$

The first objective $f_1(x)$ is defined as the sum of ranks for the feature subset, where $sort_1$ represents the sum of ranks of the first selected feature across five filters, and $sort_n$ represents the sum of ranks of the nth selected feature across the five filters. $f_2(x)$ is defined as the classification error rate.

4.3 Hypervolume

In the realm of multi-objective optimization, the Hypervolume (HV) stands out as a widely employed metric for assessing the effectiveness of multi-objective optimization algorithms and comparing diverse solutions. Specifically, it assesses the volume covered under the Pareto front. HV is utilized to assess the excellence of the Pareto front. Generally, a higher HV value suggests a larger quantity of solutions on the Pareto front and a more extensive distribution within the objective space. Its mathematical representation is as follows:

$$HV = \delta \left(\bigcup_{i=1}^{|S|} v_i \right) \qquad (19)$$

In this context, δ symbolizes the Lebesgue measure, employed for volume measurement. The variable s signifies the count of non-dominated solutions, while v_i denotes the Hypervolume formed by the reference point and the $i - th$ solution.

4.4 Parameters Setting

We used 5 multi-objective algorithms for comparison including MOMPA, MOPSO, NSGAII, MOALO, and MOSSA. To ensure the reliability of the experiments, the algorithms utilized the RCMF strategy for filtering and obtained the overall ranking of all features on the filter. The size of the population for all algorithms was established at 20, with the number of iterations set to 100. The parameters for each algorithm are listed in the Table 3.

Table 3. Parameter settings of Algorithms

Method	Parameter	Value
BMOEO	$a1$	2
	$a2$	1
	GP	0.5
MOMPA	P	0.2
		0.5
MOPSO	$c1$	rand[1.5, 2]
	$c2$	rand[1.5, 2]
	w	rand[0.1, 0.5]
NSGAII	$Cross$	1
	$Muta$	1/D
	$Dis\ index$	20
MOALO	w	2
MOSSA	$C1$	eMaxiter/(-2it)

5   Results

5.1   Comparing Pareto Optimal Sets

The Figs. 3 illustrate the Pareto-optimal results of 6 different algorithms across 15 datasets. Our proposed method is referred to as MOEO. The horizontal axis represents the sum of feature subset rankings, while the vertical axis represents the classification error rate, with the titles denoting the names of the datasets. From the figures, it is evident that among the 12 microarray datasets, our method achieves the lowest classification error rate in SRBCT-4, LUNG-5, DLBCL, Brain_Tumor1, ALLAML, CNS, MLL, and SRBCT-4. Moreover, it attains the best feature ranking in DLBCL, GLIOMA, Brain_Tumor1, ALLAML, CAR, LUNG, and BREAST. This observation demonstrates the robustness of our algorithm, both in terms of precision in feature selection and the ability to search for features most relevant to class labels. In particular, it is worth noting that in SRBCT-4, LUNG-5, DLBCL, Brain_Tumor1, ALLAML, CNS, CAR, Brain_Tumor2, and MLL datasets, our algorithm distinctly exhibits a superior Pareto front, effectively dispersing non-dominated solutions and enhancing solution diversity to adapt to various scenarios.

Within the context of 3 UCL datasets, our algorithm achieves the lowest classification error rates in Gastroenterology and Color. It is evident from the graphs that our method exhibits a notably superior Pareto front compared to other algorithms. The above analysis reveals that, in comparison to other benchmarked algorithms, our proposed method displays a high level of competitiveness.

5.2 Results of HV Metric

From Table 4, we can see that our method achieved the best results on 14 out of 15 datasets. In the case of BREAST, ALAML, CAR, GLIOMA, DLBCL, MLL, and Dbworld, it attained a peak value of 0.99, signifying its close proximity to the actual Pareto frontier. On the GLIOMA dataset, MOMPA and MOALO both obtained the same HV value, suggesting that these three algorithms performed well on this particular dataset. However, on the LUNG dataset, BMOEO achieved an HV value of 0.98, while MOMPA and MOALO achieved even better results, indicating that these two algorithms possessed stronger optimization capabilities and could escape local optima to obtain a superior Pareto front. On the remaining datasets, BMOEO still exhibited good performance, demonstrating that our method outperformed the other five methods in terms of optimization capabilities and achieved a better balance of diversity and convergence in the obtained Pareto front.

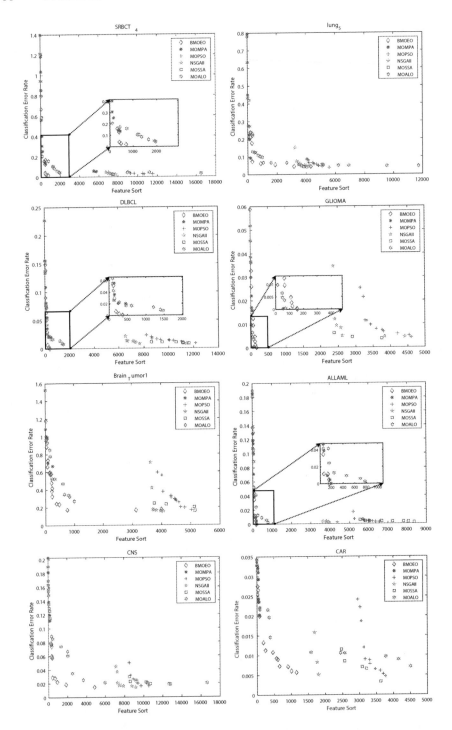

Fig. 3. The optional Pareto sets of different algorithms on all of the datasets

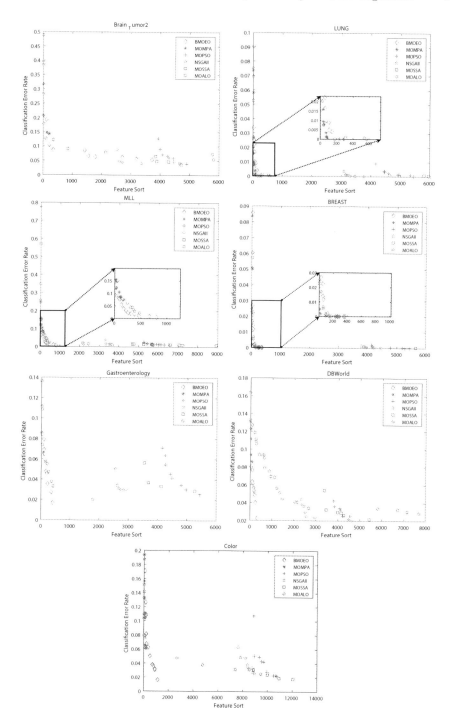

Fig. 3. (*continued*)

Table 4. BMOEO compared with other algorithms in terms of HV values

Set/Method	MOMPA	MOPSO	NSGAII	MOALO	MOSSA	BMOEO
SRBCT_4	0.96	0.33	0.39	0.97	0.09	**0.99**
ALLAML	0.97	0.43	0.60	0.98	0.39	**0.99**
Brain_Tumor1	0.60	0.51	0.55	0.85	0.53	**0.89**
Brain_Tumor2	0.91	0.56	0.71	0.93	0.57	**0.95**
CAR	0.97	0.68	0.81	0.96	0.73	**0.99**
CNS	0.93	0.53	0.61	0.96	0.53	**0.98**
GLIOMA	**0.99**	0.83	0.87	**0.99**	0.87	**0.99**
LUNG	**0.99**	0.77	0.83	**1.00**	0.75	0.98
DLBCL	0.98	0.55	0.64	0.98	0.53	**0.99**
LUNG_5	0.94	0.76	0.80	**0.96**	0.78	**0.96**
MLL	0.96	0.71	0.81	0.97	0.74	**0.99**
BREAST	0.95	0.50	0.62	**0.97**	0.59	**0.97**
Color	0.96	0.51	0.57	0.96	0.51	**0.98**
Dbworld	0.96	0.78	0.86	0.98	0.80	**0.99**
Gastroenterology	0.96	0.76	0.85	0.96	0.79	**0.98**

6   Conclusions

To overcome the limitations of traditional feature selection methods when dealing with high-dimensional microarray data, this study proposes an ensemble of multiple filter strategies (RCMF). In this strategy, five different methods are employed to mitigate the randomness associated with individual filters. During this process, we use the approximate Markov blanket to measure redundancy between features and employ conditional mutual information to identify complementary features strongly correlated with each other, thereby enhancing feature diversity. By applying this approach, we obtain a ranking of features across multiple filters, which serves as the second objective for multi-objective feature selection. Furthermore, we enhance the traditional EO method by updating it based on non-dominant solutions from an external archive, and we map continuous EO into binary space using an S-shaped transfer function.

By evaluating our approach using Pareto front and hypervolume (HV) metrics, we achieve outstanding performance on 15 different datasets. In the future, we intend to integrate this method into various scenarios to make it more versatile and apply it to imbalanced datasets to enhance the algorithm's stability.

Acknowledgements. This research was supported by the Key Project of the Zhejiang Provincial Natural Science Foundation of China under Grant No. LZ24F020005 and the National Natural Science Foundation of China under Grant No. 61373057.

References

1. Alharbi, F., Vakanski, A.: Machine learning methods for cancer classification using gene expression data: a review. Bioengineering **10**(2), 173 (2023)
2. Almugren, N., Alshamlan, H.: A survey on hybrid feature selection methods in microarray gene expression data for cancer classification. IEEE Access **7**, 78533–78548 (2019)
3. Blake, C.L.: UCI repository of machine learning databases (1998). http://www.ics.uci.edu/~mlearn/MLRepository.html
4. Chuang, L.Y., Chang, H.W., Tu, C.J., Yang, C.H.: Improved binary PSO for feature selection using gene expression data. Comput. Biol. Chem. **32**(1), 29–38 (2008)
5. Farissi, A., Dahlan, H.M., et al.: Genetic algorithm based feature selection with ensemble methods for student academic performance prediction. In: Journal of Physics: Conference Series, vol. 1500, p. 012110. IOP Publishing (2020)
6. Fu, S., Desmarais, M.C.: Markov blanket based feature selection: a review of past decade. In: Proceedings of the World Congress on Engineering, Hong Kong, China, vol. 1, pp. 321–328. Newswood Ltd. (2010)
7. Gao, Y., Zhou, Y., Luo, Q.: An efficient binary equilibrium optimizer algorithm for feature selection. IEEE Access **8**, 140936–140963 (2020)
8. Guha, R., Ghosh, K.K., Bera, S.K., Sarkar, R., Mirjalili, S.: Discrete equilibrium optimizer combined with simulated annealing for feature selection. J. Comput. Sci. **67**, 101942 (2023)
9. Gunantara, N.: A review of multi-objective optimization: methods and its applications. Cogent Eng. **5**(1), 1502242 (2018)
10. Hashemi, A., Dowlatshahi, M.B., Nezamabadi-pour, H.: Minimum redundancy maximum relevance ensemble feature selection: a bi-objective pareto-based approach. J. Soft Comput. Inf. Technol. **12**(1), 20–28 (2023)
11. Huang, C.L., Wang, C.J.: A GA-based feature selection and parameters optimizationfor support vector machines. Expert Syst. Appl. **31**(2), 231–240 (2006)
12. Li, J., Liu, H.: Kent ridge bio-medical data set repository. Institute for Infocomm Research (2002). http://sdmc.lit.org.sg/GEDatasets/Datasets.html
13. Liu, W., Guo, Z., Jiang, F., Liu, G., Wang, D., Ni, Z.: Improved WOA and its application in feature selection. PLoS ONE **17**(5), e0267041 (2022)
14. Rahimi, I., Gandomi, A.H., Chen, F., Mezura-Montes, E.: A review on constraint handling techniques for population-based algorithms: from single-objective to multi-objective optimization. Arch. Comput. Methods Eng. **30**(3), 2181–2209 (2023)
15. Remeseiro, B., Bolon-Canedo, V.: A review of feature selection methods in medical applications. Comput. Biol. Med. **112**, 103375 (2019)
16. Wei, G., Zhao, J., Feng, Y., He, A., Yu, J.: A novel hybrid feature selection method based on dynamic feature importance. Appl. Soft Comput. **93**, 106337 (2020)
17. Xue, B., Zhang, M., Browne, W.N., Yao, X.: A survey on evolutionary computation approaches to feature selection. IEEE Trans. Evol. Comput. **20**(4), 606–626 (2015)

Difference Vector Angle Dominance with an Angle Threshold for Expensive Multi-objective Optimization

Cuicui Yang[✉] and Jing Chen

Beijing University of Technology, Beijing 100124, China
yangcc@bjut.edu.cn

Abstract. For the latest two years, relation classification-based surr-ogate-assisted algorithms show good potential for solving expensive multi-objective optimization problems (EMOPs). In this category of methods, the used dominance relation that is vital for building training dataset and selecting promising solutions to reduce expensive real function evaluations (FEs). However, the existing studies are still at the initial stage and lack specific research on the dominance relation. This paper proposes a novel dominance relation called Difference Vector Angle Dominance with an angle threshold for EMOPs (called as DVAD-φ). The proposed DVAD-φ has adaptive selection pressure and considers the convergence and diversity of solutions when picking out superior solutions, which makes it beneficial to pick out promising solutions for expensive real FEs and reduce expensive real FEs. To be specific, we firstly give the definition of DVAD-φ that measures the superiority from one solution to another solution, where the angle threshold φ controls the selection pressure. Then, we propose an adaptive determination strategy of angle threshold based on bisection to set proper pressure for picking out promising solutions for expensive real FEs. Experiments have been conducted on 7 test functions from one benchmark set. The experimental results have verified the effectiveness of DVAD-φ.

Keywords: Expensive multi-objective optimization problems · Surrogate models · Surrogate assisted multi-objective evolutionary algorithms · Dominance relation

1 Introduction

Many real-world problems involve optimizing multiple objectives that require to cost lots of time and resources for evaluation [5,9], such as photonic waveguide design [15], computational electromagnetics [20,22]. These problems are called Expensive Multi-objective Optimization Problems (EMOPs). As is well-known, Evolutionary Algorithms (EAs) [8,12,21] have become the most popular methods for solving Multi-objective Optimization Problems (MOPs), which leads to a hot research area called Multi-Objective Evolutionary Algorithms (MOEAs).

© The Author(s), under exclusive license to Springer Nature Singapore Pte Ltd. 2024
L. Pan et al. (Eds.): BIC-TA 2023, CCIS 2061, pp. 88–102, 2024.
https://doi.org/10.1007/978-981-97-2272-3_7

MOEAs usually require more than tens of thousands of Function Evaluations (FEs) to solve MOPs, which is not tolerated for EMOPs that only bear a few hundreds of FEs due to the expensive feature [2,4].

To address this issue, surrogate models have been introduced into MOEAs to reduce the number of expensive real FEs. Such algorithms are named as Surrogate-Assisted Multi-objective Evolutionary Algorithms (SA-MOEAs) [11]. SA-MOEAs use surrogate models to pre-select promising solutions for expensive real FEs and thereby reduce the number of FEs needed. According to the role of the surrogate model, the existing SA-MOEAs can be roughly classified as regression-based SA-MOEAs and classification-based SA-MOEAs.

Regression-based SA-MOEAs usually build regression models to predict some indicator. The methods belonging to this category includes the Pareto-based Efficient Global Optimization (ParEGO) [10], The Kriging assisted two-archive evolutionary algorithm (KTA2) [16], The Efficient Dropout Neural Network assisted AR-MOEA (EDN-AR-MOEA) [6], etc. In general, the regression-based SA-MOEAs have good performance in the low-dimensional objective space, but usually would deteriorate for the high-dimensional objective space, since they are restricted by the small amount of training data that is made up of evaluated solutions by real objective functions [26].

Classification-based SA-MOEAs are a category of new methods emerged since 2015 [24]. They build classifiers to predict the labels of candidate solutions or relations of two solutions for picking out promising solutions for expensive real FEs and reducing the number of expensive real FEs. Since classification models are more robust for the high-dimensional objective space even in the case of the small amount of training data [14], classification-based SA-MOEAs are attracting more attention in the recent several years. According to the classification object, this category of SA-MOEAs can be further divided into two sub-categories: solution classification-based SA-MOEAs and relation classification-based SA-MOEAs.

The solution classification-based SA-MOEAs [14,24] come first and select promising solutions by predicting the labels of new candidate solutions. For example, CPS-MOEA [24] uses the Pareto non-dominated solutions as the superior solutions and the dominated solutions as the inferior solutions. CSEA [14] uses a series of reference points as boundaries to classify all the evaluated solutions into superior and inferior solutions. MCEA/D [17] uses multiple scalarization functions to decompose the multi-objective optimization problem into multiple single-objective optimization problem. Objectively speaking, this category of methods can not recognize the differences of solutions in the same class and uses a coarse-grained classification mechanism toward candidate solutions in fact. In other words, if there are multiple candidate solutions that are predicted to be superior ones, it is difficult for such methods to further distinguish the truly promising solutions within them. This may increase the consumption of expensive real FEs.

The relation classification-based SA-MOEAs [7] have appeared in the last two years and select promising solutions by predicting the relation of two solutions.

Obviously, this category of methods use a fine-grained classification mechanism and can recognize the difference of any pair of solutions, thereby they may more accurately select promising solutions. To the best of our knowledge, there are only the following two such methods. In 2022, REMO [7] firstly groups evaluated solutions into two classes uniformly based on reference points: superior solutions and inferior solutions. Then they are combined into pairs to generate the relation dataset with three labels: better than, worse than, similar to. Finally, the relations between new candidate solutions and evaluated solutions are predicted. Those solutions which are predicted to be better than inferior solutions or similar to superior solutions are considered as promising solutions for expensive real FEs. However, the behavior of this method that uniformly groups the evaluated solutions into superior solutions and inferior solutions would lose information about the dominance relations of solutions in the same class and inevitably affect the prediction performance towards promising solutions. In the same year, θ-DEA-DP [23] directly uses Pareto dominance and θ-dominance relations to label the relation of any two solutions. Firstly, pairs of evaluated solutions are labeled by Pareto dominance and θ-dominance. Then the labeled pairs of solutions are formed into two training dataset to train the classification model separately. Finally, two classifiers are used to predict the relations between new candidate solutions and the evaluated non-dominated solutions. Such candidate solutions which are predicted to dominate evaluated non-dominated solutions are selected as promising solutions. Although θ-DEA-DP focuses on learning the dominance relation of any two solutions, its prediction performance is constrained by the two used dominance relations. Specifically, Pareto dominate relation is generally considered to have low selection pressure in the high-dimensional objective space, while θ-dominance relation is susceptible to the influence of reference points.

In order to develop the potential of relation classification-based SA-MOEAs, this paper proposes a novel dominance relation called Difference Vector Angle Dominance with an angle threshold for EMOPs (called as DVAD-φ). The proposed DVAD-φ has adaptive selection pressure and considers the convergence and diversity of solutions when picking out promising solutions, which makes it beneficial to reduce expensive FEs. To be specific, we firstly give the definition of DVAD-φ that measures the superiority from one solution to another solution, where the angle threshold φ controls the selection pressure. Then, we propose an adaptive determination strategy of angle threshold based on bisection to set proper pressure for picking out promising solutions for expensive real FEs. The novelty and main contributions of this paper can be summarized as follows.

1. This paper provides a DVAD-φ for EMOPs. DVAD-φ can provide proper selection pressure by adjusting the angle threshold φ and is beneficial to meet the need of proper selection pressure for selecting truly promising solutions in EMOPs.
2. This paper proposes an adaptive determination strategy of angle threshold based on bisection, which can rapidly determine the most best angle threshold to ensure proper selection pressure for picking out the promising solutions for expensive real FEs to reduce expensive real FEs.

3. DVAD-φ is embedded into a representative algorithm and compared with seven state-of-the-art SA-MOEAs on 7 test problems from one test suites. The comparative results have verified the effectiveness of proposed DVAD-φ.

The remainder of this paper is organized as follows. Section 2 provides a brief introduction about EMOPs. Section 3 presents DVAD-φ in detail. Afterward, the experimental results and discussions are given in Sect. 4. Finally, a conclusion is given in Sect. 5.

2   Preliminaries

Expensive Multi-objective Optimization Problems: Without loss of generality, a Multi-objective Optimization Problem (MOP) can be formulated as

$$\begin{aligned} \text{minimize} \quad & F(\mathbf{x}) = (f_1(\mathbf{x}), \dots, f_m(\mathbf{x})) \\ \text{subject to} \quad & \mathbf{x} = (x_1, \dots, x_d) \end{aligned} \tag{1}$$

where $\mathbf{x} \in \mathbb{R}^d$ is a decision vector from the d-dimensional decision space $\mathbb{R}^d$, d is the number of decision variables, i.e., the dimension of decision space, $f_1, f_2, \dots, f_m$ are m (≥ 2) objective functions, $F : \mathbb{R}^d \to \mathbb{R}^m$ is a mapping function from the d-dimensional decision space $\mathbb{R}^d$ to the m-dimensional objective space $\mathbb{R}^m$. When the evaluation of the objective function F requires a large time or economic cost, such as physical experiments or numerical model simulations, this problem is called in particular an Expensive MOP (also called as EMOP). An EMOP is required to be solved with a limited number of FEs.

3   Proposed DVAD-φ

In this section, the proposed DVAD-φ will be introduced in detail. Firstly, the definition of DVAD-φ that measures the superiority from one solution to another solution is proposed. Then, an adaptive determination strategy of angle threshold based on bisection is presented. Finally a framework for embedding DVAD-φ to relation classification based-SA-MOEAs is provided.

3.1   DVAD-φ

The proposed DVAD-φ decides the dominance relation from one solution to another solution in the objective space according to the corresponding Difference Vector Angle (DVA). Some related definitions and properties are firstly given in the following.

Definition 1. *Standardized objective vector: Let $\boldsymbol{x}_i$ is a solution in population P, and $F(\boldsymbol{x}_i)$ is its corresponding solution in the objective space. The standardized objective vector of $F(\boldsymbol{x}_i)$ is defined as:*

$$V_i = F(\boldsymbol{x}_i) - Z \tag{2}$$

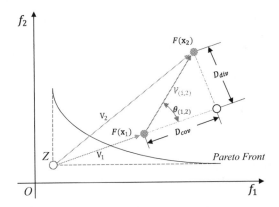

Fig. 1. Illustration of DVA. V_1, V_2 are the standardized objective vectors of $F(\mathbf{x}_1)$, $F(\mathbf{x}_2)$, respectively. $V_{(1,2)}$ is the difference vector from V_1 to V_2. $\theta_{(1,2)}$ is the DVA from V_1 to V_2. D_{cov} denotes projection length from $V_{(1,2)}$ to the direction along principle vector V_1. The larger the D_{cov}, the less convergence of $F(\mathbf{x}_2)$ with respect to $F(\mathbf{x}_1)$. D_{div} denotes projection length from $V_{(1,2)}$ to the perpendicular direction of the principle vector V_1. The larger the D_{div}, the larger the difference between $F(\mathbf{x}_1)$ and $F(\mathbf{x}_2)$.

where $Z = (z_1, z_2, \ldots, z_m)$ is the ideal solution in the objective space, each component of which is the current minimum value (i.e., the current optimal value) of the corresponding objective function and is defined as: $z_i = \min_{j=1}^{n} f_i(\mathbf{x}_j)$, n is the population size of P.

Definition 2. *Difference vector: Let $\mathbf{x}_i$, $\mathbf{x}_j$ are two solutions in population P. $F(\mathbf{x}_i)$ and $F(\mathbf{x}_j)$ are respectively their corresponding solutions in the objective space. V_i, V_j are standardized objective vectors of $F(\mathbf{x}_i)$ and $F(\mathbf{x}_j)$, respectively. The difference vector from $F(\mathbf{x}_i)$ to $F(\mathbf{x}_j)$ is defined as:*

$$V_{(i,j)} = V_j - V_i \tag{3}$$

where V_i is called the principal vector of $V_{(i,j)}$, V_j is called the auxiliary vector of $V_{(i,j)}$.

Definition 3. *Difference vector angle (DVA): Let $\mathbf{x}_i$, $\mathbf{x}_j$ are two solutions in population P, and $F(\mathbf{x}_i)$ and $F(\mathbf{x}_j)$ are their corresponding solutions in the objective space, respectively. The DVA from $F(\mathbf{x}_i)$ to $F(\mathbf{x}_j)$ is defined as:*

$$\theta_{(i,j)} = \arccos\left(\frac{V_{(i,j)}^T V_i}{\|V_{(i,j)}\| \cdot \|V_i\|} \right) \tag{4}$$

where V_i is the standardized objective vector of $F(\mathbf{x}_i)$, $V_{(i,j)}$ is the difference vector from $F(\mathbf{x}_i)$ to $F(\mathbf{x}_j)$, $\|V_i\|$ and $\|V_{(i,j)}\|$ denote the L2-norm of V_i and $V_{(i,j)}$, respectively.

According to the above formula, when comparing two solutions $\mathbf{x}_1$ and $\mathbf{x}_2$ as shown in Fig. 1, we can decompose the difference vector $V_{(1,2)}$ along the direction of the principal vector V_1 and the perpendicular direction of the principal vector V_1. The projection length of the difference vector $V_{(1,2)}$ along the direction of principal vector V_1 is called the convergence distance of two solutions $\mathbf{x}_1$ and $\mathbf{x}_2$ denoted by D_{cov}. The positive or negative of this distance tells which vector has better convergence. If this distance is positive, the solution corresponding to the principal vector has better convergence; otherwise, the solution corresponding to the auxiliary vector has better convergence. On the other hand, the projection length of the difference vector $V_{(1,2)}$ along the perpendicular direction of the principal vector V_1 is called the diversity distance of $\mathbf{x}_1$ and $\mathbf{x}_2$ denoted by D_{div}. This distance reflects how similar the two vectors are in the case of having fixed convergence distance. The larger the distance, the larger the difference between the two solutions corresponding to the V_1 and V_2, i.e., they have better diversity. Also according to the definition of trigonometric functions, $\theta_{(i,j)} = \arctan(\frac{D_{div}}{D_{cov}})$. This mathematical expression reveals a positive correlation between the $\theta_{(i,j)}$ and D_{div}, while a negative correlation exists between $\theta_{(i,j)}$ and D_{cov}. Hence, $\theta_{(i,j)}$ can be regarded as a composite indicator amalgamating the influences of both D_{div} and D_{cov}, which means DVA can be used as a comprehensive indicator of measuring the convergence and diversity of two solutions. A higher value of $\theta_{(i,j)}$ indicates a higher quality of the solution corresponding to the auxiliary vector.

From the above analysis, we can conclude that DVA reflects the degree of superiority or inferiority of the solution corresponding to the auxiliary vector with respect to the solution corresponding to the principal vector. Therefore, if an appropriate angle threshold φ is set, it is possible to evaluate whether the solution corresponding to the auxiliary vector is better than the solution corresponding to the principal vector. Thus, we define the following new dominance relation to compare two solutions:

Definition 4. *Difference vector angle dominance with angle threshold φ (DVAD-φ): Let $\mathbf{x}_i, \mathbf{x}_j \in P$, $\theta_{(i,j)}$ is the DVA from $F(\mathbf{x}_i)$ to $F(\mathbf{x}_j)$. $\mathbf{x}_i$ is said to dominate $\mathbf{x}_j$ iff $\theta_{(i,j)} \leq \varphi$, denoted as $\mathbf{x}_i \prec_{DVA}^{\varphi} \mathbf{x}_j$. And conversely, $\mathbf{x}_i$ is said to non-dominate $\mathbf{x}_j$ iff $\theta_{(i,j)} > \varphi$, denoted as $\mathbf{x}_i \not\prec_{DVA}^{\varphi} \mathbf{x}_j$.*

It should be noted that due to the mutual dominance between two solutions will occur when φ greater than $\pi/2$, we generally require $0 \leq \varphi \leq \pi/2$ for clear superiority or inferiority between each pair of solutions.

According to the above definition, the selection pressure of DVAD-φ can be changed by adjusting the angle threshold φ, which helps to meet the need of strong selection pressure for selecting truly promising solutions in EMOPs.

3.2 Adaptive Determination Strategy of Angle Threshold Based on Bisection

To ensure proper selection pressure for picking out the promising solutions, this section proposes an adaptive determination strategy of angle threshold based on bisection.

Algorithm 1. Adaptive determination of angle threshold

Input: P (population); ω (percentage of non-dominated solutions)
Output: φ^* (best angle threshold)
1: $\varphi_{\max} \leftarrow \pi/2, \varphi_{\min} \leftarrow 0$
2: $\varphi \leftarrow (\varphi_{\max} + \varphi_{\min})/2$
3: $F_S(P) \leftarrow F(P) - Z$
4: $P_{nd} \leftarrow \text{NonDominatedSolution}(F_S(P),\text{DVAD-}\varphi)$
5: **while** $\left| \frac{|P_{nd}|}{|P|} - \omega \right| > 0.01$ and $\varphi_{\max} - \varphi_{\min} > 0.01$ **do**
6:     **if** $\frac{|P_{nd}|}{|P|} > \omega$ **then**
7:         $\varphi_{\min} \leftarrow \varphi$
8:     **else**
9:         $\varphi_{\max} \leftarrow \varphi$
10:     **end if**
11:     $\varphi \leftarrow (\varphi_{\max} + \varphi_{\min})/2$
12:     $P_{nd} \leftarrow \text{NonDominatedSolution}(F_S(P),\text{DVAD-}\varphi)$
13: **end while**
14: $\varphi^* \leftarrow \varphi$

Algorithm 1 provides the implementation of this adaptive determination strategy of angle threshold based on bisection. Firstly, the relevant parameters are initialized. The maximum value of the angle threshold is set to $\pi/2$. The minimum value of the angle threshold is set to 0. The initial threshold is the mean value of maximum value and minimum value (Lines 1–2). Then all solutions in the objective space $F(P)$ are subtracted from the ideal point Z and normalized to positive values to get the set of the corresponding standardized objective vectors $F_S(P)$ (Line 3). The current non-dominated solutions P_{nd} is selected from $F_S(P)$ based on the DVAD-φ (Line 4). If the proportion of current non-dominated solution proportion is much different from ω, the angle threshold will be adjusted. Specifically, if the dominance proportion is larger than ω, the current value φ is assigned to the minimum value $\varphi_{\min}$. Conversely, the current value φ is assigned to the maximum value $\varphi_{\max}$ (Lines 5–10). Then, φ is set to be the average of $\varphi_{\max}$ and $\varphi_{\min}$, and is used as new angle threshold for DVAD-φ to determine the current non-dominated solutions P_{nd} (Lines 11–12). This loop continues until the condition is not satisfied. The final φ is used as the best angle threshold φ^* (Line 14).

3.3   Framework of DVAD-φ Based SA-MOEAs

Algorithm 2 provides the framework of embedding DVAD-φ into a relation classification-based SA-MOEA.

- **Initialization (lines 1–3):** The initial population P with $11d - 1$ initial solutions are sampled by Latin hypercube sampling [13] from the decision space, and are evaluated by expensive real FEs and these evaluated solutions P are saved to an archive Arc.

Algorithm 2. Framework of DVAD-φ relation classification based SA-MOEAs

---

Input: N(population size); FE_{max}(maximum number of FEs); ω(proportion of non-domainated solutions)

Output: P(Current optimal population)

1: $P \leftarrow$ InitializePopulation(N)
2: $Arc \leftarrow P$
3: $FEs \leftarrow N$
4: **while** $FEs < FE_{max}$ **do**
5:     $\varphi^* \leftarrow$ **AdaptiveDeterminationOfAngleThreshold**(P, ω)
6:     Generate training dataset R_{DVA} by **DVAD-φ^***
7:     $Q \leftarrow$SA-MOEA(R_{DVA}, Arc)
8:     $Q_r \leftarrow$RealFunctionEvaluation(Q)
9:     $FEs \leftarrow FEs + |Q_r|$
10:    $Arc \leftarrow Arc \cup Q_r$
11:    $P \leftarrow$EnvironmentalSelection(Arc, N)
12: **end while**

- **Loop condition (line 4):** If the number of function evaluations (FEs) does not reach the maximal function evaluations (FE_{max}), the following steps are executed.
- **Determining the best angle threshold φ^* (line 5):** The solutions in the population P are combined into solution pairs and the best angle threshold φ^* is obtained by Algorithm 1.
- **Generating training dataset (line 6):** Training dataset R_{DVA} is generated according to DVAD-φ. For each training sample, the training input is a concatenation of the decision vectors of the two solutions, and the label is the dominance relation for DVAD-φ from the first solution to the second solution.
- **SA-MOEA (line 7):** The training dataset R_{DVA} is used to train the classification model. Then the model is used to predict whether the candidate solution dominates the existing solution, in order to select the promising solution. Repeat the SA-MOEA until the maximum number of predictions is reached.
- **Updating data (lines 8–11):** The promising solutions are evaluated by expensive real FEs and used to updated with archive Arc.
- **Select optimal populations (line 12):** Finally, the Current best solutions is selected as output populations P from the archive Arc by environmental selection of original SA-MOEA.

4 Experimental Studies

In this section, the performance of DVAD-φ is thoroughly investigated. Firstly, a parameter sensitivity analysis is performed to select proper parameter setting for DVAD-φ. Secondly, two ablation experiments are conducted to verify the effectiveness of the adaptive determination strategy of angle threshold strategy based on bisection and the DVAD-φ, respectively. Finally, to further prove the performance of DVAD-φ, DVAD-φ is embedded into a representative algorithm REMO and is compared with seven state-of-the-art algorithms on 7 benchmark test problems.

4.1 Experimental Settings

Test Problems. Benchmark test suite DTLZ [3] is used to investigate the performance of DVAD-φ. It is scalable in terms of the number of objectives. DTLZ has 7 test problems, i.e., DTLZ1-DTLZ7, and covers the single modality and multi modality feature, the concave Pareto front feature, and the disconnected Pareto front feature.

Performance Metrics. Three metrics are used in the experiments: the Inverted Generational Distance (IGD) [1], normalization IGD. IGD is used for most of the experiments. Normalization IGD is used for parameter sensitivity experiments, since it can better show the results for the same test problem with different objective dimensions in one figure.

IGD [1] can comprehensively check both convergence and diversity of the obtained Pareto optimal solutions and is defined as:

$$IGD(P^*, P) = \frac{\sum_{x \in P^*} dis(x, P)}{|P^*|} \tag{5}$$

where P^* is the set of uniformly distributed solutions on the true Pareto front and P is the set of solutions obtained by a algorithm, $dis(x, P)$ is the minimum Euclidean distance between a solution x and its nearest solution in P, $|P^*|$ denotes the number of solutions in P^*. The IGD calculates the average minimum distance from each solution of P^* to P, Thus the smaller the IGD value, the closer the obtained set of solutions to the true Pareto front, and the better the corresponding algorithm.

The normalization IGD is used in the parameter sensitivity experiments. Since the magnitude order of IGD results on different objective function vary widely, the IGD results on each number of objectives are normalized. The scaling equations are as follows.

$$I_n = \frac{I - I_{\min}}{I_{\max} - I_{\min}} \tag{6}$$

where I_n is the scaled value, I is origin value of IGD, $I_{\min}$ and $I_{\max}$ are the minimum and maximum IGD obtained on the problem with same number of objectives.

Statistical Methods. The Wilcoxon rank-sum test [19] is executed to compare the results achieved by proposed DVAD-φ and other comparison algorithms. Symbols "+", "−", "≈" indicates the number of test problems for each comparison algorithm that is significantly better than, worse than, and similar to proposed DVAD-φ.

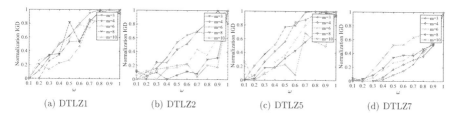

(a) DTLZ1   (b) DTLZ2   (c) DTLZ5   (d) DTLZ7

Fig. 2. The parameter sensitivity result of ω on four test problems: (a) DTLZ1, (b) DTLZ2, (c) DTLZ5, (d) DTLZ7.

Comparison Algorithms. Seven state-of-the-art comparison algorithms are selected for the experimental study, including ParEGO [10], MOEAD-EGO [25], CPS-MOEA [24], CSEA [14], REMO [7], MCEA/D [17], EDN-ARMOEA [6]. Among them, ParEGO, MOEAD-EGO, and EDN-ARMOEA are regression-based SA-MOEAs. CPS-MOEA, CSEA, and MCEA/D are solution classification-based SA-MOEAs. REMO is a relation classification-based SA-MOEA. All the algorithms are implemented in PlatEMO [18].

Parameters Setting. The common parameters in all comparison algorithms are set as follows.

- The maximum number of FEs (maxFEs) is set to 300.
- The number of objectives m is set to 3, 4, 6, 8, 10, respectively.
- The number of decision variables d is set to 10 in DTLZ.
- The population size N is set to 50.
- The number of independent runs for each algorithm on each test problem is set to 20.

All parameters specific to the comparison algorithms follow the default parameter of PlatEMO [18].

4.2   Parameter Sensitivity Analysis

In DVAD-φ, there is only one parameter, i.e. the non-dominated solution proportion (ω). To select proper value for this parameter, DVAD-φ-REMO (DVAD-φ is embedded into REMO) is tested on 3-, 4-, 6-, 8-, 10-objectives DTLZ1, DTLZ2, DTLZ5, and DTLZ7 with a angle threshold ω that is changed from 0.1 to 1 for every 0.1. These four test functions have different features and thereby can comprehensively test the algorithm performance under different parameter values of ω. Specifically, DTLZ1 has numerous local Pareto-optimal and is difficult to converge. DTLZ2 is a relatively simple test problem for testing the algorithm performance of maintaining diversity. DTLZ5 has a curved shape Pareto Front (PF) for testing the algorithm performance on irregular PF. DTLZ7 has a discontinuous PF, which tests the exploration ability of algorithms in different sub-regions.

Table 1. Results of different angle thresholds

Problem	M	D	DVAD-0-REMO	DVAD-π/4-REMO	DVAD-π/2-REMO	DVAD-φ-REMO
	3	10	8.1431e+1 (1.73e+1) −	5.2066e+1 (1.05e+1) −	4.3305e+1 (5.28e+0) ≈	4.2145e+1 (5.65e+0)
	4	10	7.1387e+1 (1.36e+1) −	4.5590e+1 (1.06e+1) −	3.1534e+1 (5.57e+0) ≈	3.3865e+1 (7.49e+0)
DTLZ1	6	10	2.6397e+1 (7.45e+0) −	2.0746e+1 (5.01e+0) −	1.5779e+1 (4.69e+0) ≈	1.7366e+1 (3.23e+0)
	8	10	9.6000e+0 (3.69e+0) −	6.8655e+0 (3.63e+0) ≈	5.4343e+0 (2.87e+0) ≈	5.0955e+0 (2.72e+0)
	10	10	4.0540e-1 (1.03e-1) −	2.5532e-1 (7.10e-2) ≈	2.2890e-1 (6.82e-2) ≈	2.2833e-1 (7.28e-2)
	3	10	3.0678e-1 (1.65e-2) −	2.9190e-1 (3.61e-2) −	2.2664e-1 (3.79e-2) ≈	2.2444e-1 (2.26e-2)
	4	10	3.5524e-1 (3.09e-2) −	3.5382e-1 (2.46e-2) −	2.8162e-1 (2.52e-2) ≈	2.7457e-1 (2.70e-2)
DTLZ2	6	10	4.9248e-1 (2.89e-2) −	4.3880e-1 (2.18e-2) −	3.6646e-1 (2.16e-2) ≈	3.7328e-1 (2.72e-2)
	8	10	5.9872e-1 (2.21e-2) −	5.4333e-1 (3.89e-2) −	4.8422e-1 (3.70e-2) ≈	4.9126e-1 (5.26e-2)
	10	10	6.6381e-1 (1.98e-2) −	6.3849e-1 (2.86e-2) −	5.9951e-1 (3.65e-2) ≈	5.8815e-1 (3.84e-2)
	3	10	2.3445e-1 (2.87e-2) −	2.0110e-1 (4.60e-2) −	1.3503e-1 (3.49e-2) ≈	1.2129e-1 (2.53e-2)
	4	10	1.9603e-1 (2.06e-2) −	1.8609e-1 (2.42e-2) −	1.0814e-1 (3.57e-2) ≈	8.9453e-2 (2.03e-2)
DTLZ5	6	10	1.1127e-1 (1.99e-2) −	8.2166e-2 (1.69e-2) −	5.3955e-2 (1.87e-2) ≈	4.6100e-2 (1.06e-2)
	8	10	5.2602e-2 (8.60e-3) −	3.2655e-2 (7.09e-3) −	2.3595e-2 (5.68e-3) ≈	2.4945e-2 (4.95e-3)
	10	10	1.4668e-2 (1.99e-3) −	1.0089e-2 (1.67e-3) ≈	9.8910e-3 (1.60e-3) ≈	1.0295e-2 (1.53e-3)
	3	10	4.8168e+0 (1.08e+0) −	3.4279e-1 (8.23e-2) +	5.4987e-1 (9.00e-2) −	4.1938e-1 (9.36e-2)
	4	10	6.4292e+0 (1.43e+0) −	4.6155e-1 (8.11e-2) +	6.4947e-1 (9.54e-2) −	5.2555e-1 (9.98e-2)
DTLZ7	6	10	9.9258e+0 (2.53e+0) −	8.1672e-1 (1.08e-1) +	8.2711e-1 (5.20e-2) ≈	8.0186e-1 (5.27e-2)
	8	10	6.7013e+0 (2.78e+0) −	1.1051e+0 (1.24e-1) −	1.1016e+0 (9.74e-2) −	1.0524e+0 (2.11e-1)
	10	10	2.0103e+0 (2.37e-1) −	1.2184e+0 (9.09e-2) ≈	1.2730e+0 (4.58e-2) −	1.2314e+0 (3.91e-2)
+/−/≈			0/20/0	2/13/5	0/4/16	

The normalized IGD results are provided in Fig. 2. We can see that on the most cases in the range of 0.2 to 1, the smaller the value of φ, the smaller the normalized IGD, i.e., the better the performance of the algorithm. But when ω = 0.1, I_n rises on DTLZ2, DTLZ5, and DTLZ7, i.e., the performance is reduced compared to $\omega = 0.2$. This may because a too small proportion of non-dominated solutions can lead to too strong convergence pressure, which would cause the population to aggregate around some local optimum. Hence, we select $\omega = 0.2$ as a general setting.

4.3   Ablation Experiments

Effect of the Adaptive Determination Strategy of Angle Threshold Based on Bisection. To investigate the effectiveness of this strategy, DVAD-φ-REMO is compared with three variant algorithms DVAD-0-REMO, DVAD-π/4-REMO, and DVAD-π/2-REMO, which respectively use fixed angle threshold 0, $\pi/4$, and $\pi/2$. The four algorithms are tested on the same problems in Sect. 4.2. The Wilcoxon ranksum test is used to determine whether there is a significant superiority relationship between the two algorithms. The experimental results are shown in Table 1, where "+", "−", "≈" indicates the number of test problems for each comparison algorithm that is significantly better than, worse than, and similar to DVAD-φ-REMO, respectively. The best IGD values are highlighted in gray background.

From this table, DVAD-φ-REMO performs best on 11 of 20 test instances. The next best performer is DVAD-π/2-REMO, which performs best in 6 out of 20 test instances. DVAD-π/4-REMO performs best in 3 out of 20 test instances. DVAD-0-REMO does not find the best result on any test instance. These results demonstrate that DVAD-φ-REMO has the best performance by adaptively adjusting the angle threshold φ. Therefore, this adaptive determination strategy of angle threshold plays a positive role of solving EMOPs.

Table 2. Comparison of DVAD-φ with other three relations

Problem	M	D	REMO	θ-REMO	Pareto-REMO	DVAD-φ-REMO
DTLZ1	3	10	5.9267e+1 (1.51e+1) −	8.7398e+1 (1.31e+1) −	7.8350e+1 (1.78e+1) −	4.7234e+1 (7.04e+0)
	4	10	4.1045e+1 (1.39e+1) ≈	5.5774e+1 (1.57e+1) −	6.2745e+1 (1.86e+1) −	3.4435e+1 (7.47e+0)
	6	10	1.6721e+1 (6.42e+0) ≈	2.1038e+1 (6.26e+0) −	3.2284e+1 (9.09e+0) −	1.8319e+1 (5.02e+0)
	8	10	4.7146e+0 (2.14e+0) ≈	6.5031e+0 (2.13e+0) ≈	9.0922e+0 (2.45e+0) −	5.5480e+0 (2.69e+0)
	10	10	2.7518e-1 (7.20e-2) −	2.4408e-1 (4.45e-2) −	3.6178e-1 (9.79e-2) −	2.1383e-1 (4.99e-2)
DTLZ2	3	10	1.9515e-1 (3.62e-2)	2.6518e-1 (2.63e-2) −	2.8517e-1 (2.88e-2) −	2.2390e-1 (3.00e-2)
	4	10	2.5254e-1 (2.07e-2)	3.0990e-1 (2.61e-2) ≈	3.6023e-1 (1.58e-2) −	2.9375e-1 (3.08e-2)
	6	10	3.9011e-1 (4.07e-2) ≈	3.9825e-1 (2.30e-2) −	4.9376e-1 (3.19e-2) −	3.7787e-1 (3.44e-2)
	8	10	5.2229e-1 (4.84e-2) ≈	4.7699e-1 (3.11e-2) ≈	6.1201e-1 (2.39e-2) −	4.6231e-1 (3.73e-2)
	10	10	5.7993e-1 (3.89e-2) ≈	5.6854e-1 (5.26e-2) ≈	6.5810e-1 (3.04e-2) −	5.9461e-1 (3.53e-2)
DTLZ5	3	10	1.0434e-1 (3.49e-2) ≈	2.1635e-1 (2.99e-2) −	1.9822e-1 (3.32e-2) −	1.1709e-1 (2.12e-2)
	4	10	1.0928e-1 (2.78e-2) ≈	1.8908e-1 (3.40e-2) −	1.9043e-1 (3.37e-2) −	1.0124e-1 (2.87e-2)
	6	10	6.5686e-2 (1.80e-2) ≈	1.1972e-1 (2.68e-2) −	1.1659e-1 (1.64e-2) −	5.0809e-2 (1.23e-2)
	8	10	3.1944e-2 (1.10e-2) −	3.8210e-2 (1.03e-2) −	5.5353e-2 (1.03e-2) −	2.5322e-2 (7.96e-3)
	10	10	1.1532e-2 (2.53e-3) −	1.1302e-2 (1.64e-3) −	1.4087e-2 (1.23e-3) −	9.4095e-3 (1.56e-3)
DTLZ7	3	10	5.3788e-1 (1.96e-1) −	2.7732e-1 (6.27e-2) +	9.6188e-1 (4.82e-1) −	4.2499e-1 (9.57e-2)
	4	10	1.2476e+0 (4.41e-1) −	1.2076e+0 (5.79e-1) −	3.3560e+0 (1.12e+0) −	4.9825e-1 (5.99e-2)
	6	10	2.3215e+0 (4.74e-1) −	1.6318e+0 (6.28e-1) −	8.6515e+0 (2.47e+0) −	8.4865e-1 (1.07e-1)
	8	10	3.4982e+0 (4.71e-1) −	1.6831e+0 (4.51e-1) −	5.9977e+0 (2.73e+0) −	1.0755e+0 (1.22e-1)
	10	10	2.0590e+0 (4.07e-1) −	1.3214e+0 (1.29e-1) −	1.9039e+0 (3.61e-1) −	1.2259e+0 (6.61e-2)
+/−/≈			2/11/7	1/13/6	0/20/0	

Comparison of DVAD-φ with Other Dominance Relations. To verify the performance of DVAD-φ, three dominance relations including the Pareto dominance relation, θ-dominance relation, and the proposed DVAD-φ are embedded into REMO respectively. The corresponding algorithms are called Pareto-REMO, θ-REMO, and DVAD-φ-REMO, respectively. The original REMO algorithm is also compared as a baseline algorithm.

The experimental results are presented in Table 2. From this table, DVAD-φ-REMO achieves 13 best results of 20 test problems. REMO gets 5 best results. θ-dominance-REMO achieves only 2 best results, and Pareto-REMO obtains 0 best results. These results show that, the proposed DVAD-φ is better than the other dominance relations for EMOPs.

4.4 Comparison with State-of-the-Art Algorithms

In this section, in order to further validate the effectiveness of DVAD-φ, DVAD-φ-REMO is selected as a representative to compare with seven state-of-the-art algorithms. "$+$", "$-$", "$\approx$" indicates the number of test problems for each comparison algorithm that is significantly better than, worse than, and similar to DVAD-φ-REMO, respectively. The best IGD values are highlighted in gray background. In the last row, "Best results" lists the number of best or statistically equivalent to best results for all each algorithm.

The results on the DTLZ test suite are summarized in Table 3. From this table, DVAD-φ-REMO obtains the best results on 20 out of 35 test instances. The six comparison algorithms including REMO, MCEA/D, ParEGO, CSEA, EDN-ARMOEA, MOEAD-EGO achieves the best results on 9, 8, 7, 6, 6, and 2 out of 35 instances, respectively. CPSMOEA does not get the best result on any test instance. Moreover, from the row of "$+/-/\approx$", DVAD-φ-REMO is significantly better or similar to the each comparison algorithm on vast majority

Table 3. Results of DVAD-φ-REMO and seven comparison algorithms on DTLZ test problems with 10 decision variables

Problem	M	D	ParEGO	MOEAD-EGO	CPSMOEA	CSEA	REMO	MCEA/D	EDN-ARMOEA	DVAD-φ-REMO
DTLZ1	3	10	6.3895e+1 (8.65e+0) −	8.4290e+1 (1.79e+1) −	7.9089e+1 (1.76e+1) −	5.8776e+1 (1.35e+1) ≈	5.9267e+1 (1.51e+1) ≈	6.4305e+1 (2.11e+1) −	9.7705e+1 (2.20e+1) −	4.7216e+1 (7.04e+0)
	4	10	5.3635e+1 (8.06e+0) −	6.5842e+1 (1.35e+1) −	5.8367e+1 (1.70e+1) −	4.4849e+1 (1.04e+1) −	4.1045e+1 (1.39e+1) −	2.6327e+1 (1.89e+1) ≈	8.0538e+1 (1.77e+1) −	3.4435e+1 (7.47e+0)
	6	10	3.5133e+1 (6.25e+0) −	3.4239e+1 (7.57e+0) −	3.4789e+1 (7.86e+0) −	3.6438e+1 (6.17e+0) ≈	3.6724e+1 (6.43e+0) ≈	1.6768e+1 (1.04e+1) ≈	3.6526e+1 (1.14e+1) −	1.8319e+1 (5.02e+0)
	8	10	1.5119e+1 (6.34e+0) −	1.1043e+1 (4.28e+0) −	1.0642e+1 (4.54e+0) −	3.3872e+0 (1.37e+0) ≈	4.7346e+0 (2.34e+0) ≈	4.0791e+0 (4.74e+0) ≈	1.1128e+1 (3.71e+0) −	5.5480e+0 (2.69e+0)
	10	10	5.4279e-1 (2.37e-1) −	4.1325e-1 (2.16e-1) −	3.9146e-1 (1.06e-1) −	2.5153e-1 (3.50e-2) −	2.7514e-1 (7.20e-2) −	4.1678e-1 (1.19e-1) −	4.4027e-1 (1.84e-1) −	2.1358e-1 (4.09e-2)
DTLZ2	3	10	3.5978e-1 (4.33e-2) ≈	3.2373e-1 (1.43e-2) −	2.9568e-1 (3.27e-2) −	2.4211e-1 (2.92e-2) −	1.9515e-1 (3.62e-2) +	1.8839e-1 (2.65e-2) +	2.9066e-1 (1.94e-2) −	2.2390e-1 (3.00e-2)
	4	10	4.2439e-1 (2.60e-2) −	3.7534e-1 (3.01e-2) −	4.4478e-1 (4.25e-2) −	3.3439e-1 (3.40e-2) −	3.3254e-1 (2.07e-2) ≈	3.6205e-1 (3.02e-2) −	3.6755e-1 (1.24e-2) −	2.9375e-1 (3.08e-2)
	6	10	5.2696e-1 (2.49e-2) −	4.7181e-1 (2.26e-2) −	6.2530e-1 (4.51e-2) −	4.9598e-1 (4.15e-2) −	3.9012e-1 (4.07e-2) ≈	4.6927e-1 (3.91e-2) −	4.2429e-1 (1.66e-2) −	3.7752e-1 (3.44e-2)
	8	10	6.3948e-1 (2.76e-2) −	5.4377e-1 (1.62e-2) −	6.6694e-1 (1.84e-2) −	6.1193e-1 (2.74e-2) −	5.2229e-1 (4.84e-2) −	6.1180e-1 (4.51e-2) −	4.8531e-1 (2.26e-2) ≈	4.6231e-1 (3.78e-2)
	10	10	6.7666e-1 (2.36e-2) −	5.2657e-1 (1.48e-2) +	6.5526e-1 (3.70e-2) −	6.7059e-1 (2.36e-2) −	5.7993e-1 (3.89e-2) −	7.4604e-1 (4.98e-2) −	4.8703e-1 (1.26e-2) ≈	5.9461e-1 (3.53e-2)
DTLZ3	3	10	1.7268e+2 (1.29e+1) −	1.9549e+2 (1.98e+1) −	2.1053e+2 (3.80e+1) −	1.6713e+2 (3.57e+1) −	1.5408e+2 (3.69e+1) −	1.0858e+2 (5.07e+1) ≈	3.0503e+2 (5.05e+1) −	1.2963e+2 (2.60e+1)
	4	10	1.4988e+2 (7.20e+0) −	1.5678e+2 (1.63e+1) −	1.6850e+2 (2.72e+1) −	1.1883e+2 (3.93e+1) ≈	1.0719e+2 (3.14e+1) ≈	7.9612e+1 (4.21e+1) ≈	2.5327e+2 (5.46e+1) −	1.1718e+2 (2.59e+1)
	6	10	1.0279e+2 (1.14e+1) −	9.1098e+1 (1.28e+1) −	1.0926e+2 (2.54e+1) −	8.2193e+1 (4.46e+1) ≈	6.2195e+1 (3.41e+1) ≈	4.8516e+1 (3.81e+1) ≈	1.1680e+2 (3.37e+1) −	5.1159e+1 (1.87e+1)
	8	10	4.8711e+1 (1.05e+1) −	2.8987e+1 (1.04e+1) −	5.1273e+1 (1.33e+1) −	3.1456e+1 (1.57e+0) ≈	1.6181e+1 (1.06e+1) ≈	3.8678e+1 (1.72e+1) ≈	3.7413e+1 (1.18e+1) −	1.4290e+1 (5.73e+0)
	10	10	1.5714e+0 (4.94e-1) −	1.2993e+0 (3.05e-1) −	6.3851e+0 (6.49e+0) −	9.3872e-1 (2.58e-1) ≈	6.0650e+0 (3.97e-1) ≈	1.2585e+0 (5.32e-1) −	1.4189e+0 (3.21e-1) −	1.0035e+0 (3.91e-1)
DTLZ4	3	10	4.7272e-1 (9.21e-2) −	6.0142e-1 (7.31e-2) −	5.6611e-1 (5.14e-2) −	3.8625e-1 (1.43e-1) −	2.9506e-1 (5.71e-2) ≈	8.4802e-1 (1.49e-1) −	2.4975e-1 (8.89e-2) ≈	2.4079e-1 (4.02e-2)
	4	10	7.2765e-1 (1.04e-1) −	7.0532e-1 (4.29e-2) −	6.2840e-1 (4.25e-2) −	3.8416e-1 (8.42e-2) −	3.4606e-1 (7.73e-2) ≈	8.1707e-1 (1.66e-1) −	3.9605e-1 (5.56e-2) −	3.2723e-1 (4.07e-2)
	6	10	5.2564e-1 (7.15e-2) −	6.8826e-1 (3.13e-2) −	6.6512e-1 (2.37e-2) −	4.9933e-1 (3.81e-2) −	4.9117e-1 (5.91e-2) −	8.0030e-1 (1.54e-1) −	5.1161e-1 (3.80e-2) −	4.4998e-1 (5.34e-2)
	8	10	6.0464e-1 (5.50e-2) −	6.5127e-1 (2.13e-2) −	6.6320e-1 (2.02e-2) −	5.8222e-1 (4.64e-2) −	5.9328e-1 (5.17e-2) −	6.9869e-1 (1.33e-1) −	5.1419e-1 (2.95e-2) ≈	5.4914e-1 (3.78e-2)
	10	10	6.4361e-1 (2.58e-2) −	6.4354e-1 (1.04e-2) −	6.7504e-1 (9.64e-3) −	6.3954e-1 (3.63e-2) −	6.6167e-1 (3.40e-2) −	6.9283e-1 (2.79e-2) −	5.6556e-1 (1.48e-2) ≈	6.0794e-1 (2.34e-2)
DTLZ5	3	10	4.3905e-1 (3.54e-2) ≈	2.6792e-1 (3.59e-2) −	2.1959e-1 (3.51e-2) −	1.2570e-1 (3.84e-2) ≈	1.0434e-1 (3.49e-2) ≈	6.8380e-2 (2.47e-2) +	1.6381e-1 (2.96e-2) −	1.1709e-1 (2.12e-2)
	4	10	3.5784e-1 (3.82e-2) ≈	2.2542e-1 (2.52e-2) −	1.8444e-1 (2.61e-2) −	1.3610e-1 (2.56e-2) −	1.0928e-1 (2.78e-2) ≈	9.1654e-2 (2.93e-2) −	1.7048e-1 (1.94e-2) −	1.0124e-1 (2.87e-2)
	6	10	8.1794e-2 (2.34e-2) −	1.4306e-1 (1.61e-2) −	1.9470e-1 (4.51e-2) −	8.1568e-2 (2.27e-2) −	6.5686e-2 (1.88e-2) −	5.6003e-2 (2.30e-2) ≈	1.2194e-1 (1.71e-2) −	5.0975e-2 (1.23e-2)
	8	10	6.7354e-2 (1.37e-2) −	3.8588e-2 (1.09e-2) −	1.2962e-1 (3.27e-2) −	4.5632e-2 (8.35e-3) −	3.1944e-2 (1.10e-2) −	4.1833e-2 (1.07e-2) −	5.6973e-2 (9.33e-3) −	2.5272e-2 (7.96e-3)
	10	10	2.0476e-2 (2.14e-3) −	2.1522e-2 (1.99e-3) −	5.6032e-2 (1.43e-2) −	1.4171e-2 (1.39e-3) −	1.1532e-2 (2.53e-3) −	2.2639e-2 (6.01e-3) −	1.4938e-2 (1.55e-3) −	5.6995e-3 (1.56e-3)
DTLZ6	3	10	1.1613e+0 (3.12e-1) −	1.9315e+0 (4.62e-1) −	3.0049e+0 (6.26e-1) −	5.1289e+0 (5.86e-1) −	3.7002e+0 (6.85e-1) −	2.1246e+0 (4.10e-1) −	5.0399e+0 (3.53e-1) −	1.7228e+0 (5.89e-1)
	4	10	1.3349e+0 (3.78e-1) −	1.5204e+0 (2.76e-1) −	2.6146e+0 (4.22e-1) −	4.2898e+0 (4.51e-1) −	3.0904e+0 (5.05e-1) −	1.6792e+0 (6.79e-1) −	4.3566e+0 (2.40e-1) −	1.9236e+0 (4.66e-1)
	6	10	1.3696e+0 (2.40e-1) −	1.5194e+0 (2.75e-1) −	2.0399e+0 (4.48e-1) −	2.6078e+0 (3.34e-1) −	1.7932e+0 (5.01e-1) −	1.3820e+0 (2.96e-1) −	2.4681e+0 (2.53e-1) −	9.3306e-1 (2.74e-1)
	8	10	1.7471e+0 (2.22e-1) −	1.8131e+0 (1.52e-1) −	1.2149e+0 (2.01e-1) +	1.4534e+0 (1.83e-1) +	1.4173e+0 (3.24e-1) −	1.6042e+0 (2.82e-1) −	1.9001e+0 (3.75e-1) −	1.5885e+0 (1.77e-1)
	10	10	2.0418e+0 (8.45e-2) ≈	2.0047e+0 (5.91e-2) −	1.9493e+0 (1.25e-1) −	1.9562e+0 (1.49e-1) −	2.0057e+0 (1.02e-1) −	2.0337e+0 (1.44e-1) −	2.9080e+0 (5.84e-1) −	1.9935e+0 (1.07e-1)
DTLZ7	3	10	1.6440e-1 (3.73e-2) ≈	2.3218e-1 (5.77e-2) −	4.1142e+0 (1.06e+0) −	1.7708e+0 (5.80e-1) −	5.3788e-1 (1.96e-1) −	1.6350e+0 (1.09e+0) −	1.3534e+0 (3.89e-1) −	4.2499e-1 (9.57e-2)
	4	10	5.2295e-1 (2.36e-2) ≈	5.1402e-1 (5.84e-2) ≈	4.3374e+0 (2.04e+0) −	3.2372e+0 (9.74e-1) −	1.2476e+0 (4.41e-1) −	2.0134e+0 (1.31e+0) −	1.6969e+0 (6.02e-1) −	4.9825e-1 (5.99e-2)
	6	10	6.6429e-1 (4.05e-2) ≈	8.8947e-1 (8.16e-2) −	3.1980e+0 (1.83e+0) −	7.9589e+0 (2.60e+0) −	2.3215e+0 (4.74e-1) −	1.1864e+0 (1.94e-1) −	1.7228e+0 (5.71e-1) −	8.4865e-1 (1.07e-1)
	8	10	1.1045e+0 (3.95e-2) ≈	1.0644e+0 (4.23e-2) ≈	2.3753e+0 (1.04e+0) −	6.3509e+0 (2.21e+0) −	3.4982e+0 (4.71e-1) −	1.2874e+0 (1.55e-1) −	1.7755e+0 (7.46e-1) −	1.0756e+0 (3.22e-1)
	10	10	1.4152e+0 (5.68e-2) −	1.2710e+0 (3.47e-2) ≈	1.7626e+0 (1.91e-1) −	1.9762e+0 (3.59e-1) −	2.0590e+0 (4.07e-1) −	1.4437e+0 (1.00e-1) −	1.4324e+0 (2.00e-1) −	1.2208e+0 (6.61e-2)
+/−/≈			6/26/3	2/28/5	1/33/1	2/24/9	3/17/15	3/22/10	5/28/2	
Best Result			7	2	0	6	9	8	6	20

of test instances. According to these results, it is reasonable to conclude that DVAD-φ-REMO performs better than its seven competitors in solving EMOPs on the whole, which indirectly reflects that the effectiveness of the proposed DVAD-φ for solving EMOPs.

5   Conclusion

EMOPs widely exist in the real-world and require to find the better solutions with a limited number of FEs. The relation classification based SA-MOEAs is a category of new emerging methods for solving EMOPs. This paper proposes a new dominance relation DVAD-φ that can help to better solve EMOPs. DVAD-φ uses DVA as a comprehensive indicator to measure the convergence and diversity from one solution to another solution. DVAD-φ can control the selection pressure by adjusting the angle threshold φ. Moreover, an adaptive determination strategy of angle threshold for DVAD-φ is proposed to select a proper angle threshold, which can provide proper selection pressure to select promising solutions and reduce the needed expensive real FEs. The results on one suite of benchmark have validated the performance of the proposed DVAD-φ.

Although DVAD-φ has achieved better results, it is only a preliminary attempt to pay attention on the dominance relation of relation-based SA-MOEAs for better solving EMOPs. It is still worth studying how to develop the dominance relation with better performance for EMOPs.

Acknowledgement. This work is supported in part by the NSFC Research Program (61906010, 62276010) and R&D Program of Beijing Municipal Education Commission (KM202010005032, KZ202210005009).

References

1. Bosman, P., Thierens, D.: The balance between proximity and diversity in multiobjective evolutionary algorithms. IEEE Trans. Evol. Comput. **7**(2), 174–188 (2003). https://doi.org/10.1109/TEVC.2003.810761

2. Deb, K., Pratap, A., Agarwal, S., Meyarivan, T.: A fast and elitist multiobjective genetic algorithm: NSGA-II. IEEE Trans. Evol. Comput. **6**(2), 182–197 (2002). https://doi.org/10.1109/4235.996017

3. Deb, K., Thiele, L., Laumanns, M., Zitzler, E.: Scalable multi-objective optimization test problems. In: Proceedings of the 2002 Congress on Evolutionary Computation, CEC 2002 (Cat. No.02TH8600), vol. 1, pp. 825–830 (2002). https://doi.org/10.1109/CEC.2002.1007032

4. Deb, K., Jain, H.: An evolutionary many-objective optimization algorithm using reference-point-based nondominated sorting approach, Part I: solving problems with box constraints. IEEE Trans. Evol. Comput. **18**(4), 577–601 (2014). https://doi.org/10.1109/TEVC.2013.2281535

5. Farzaneh, M., Mahdian Toroghi, R.: Music generation using an interactive evolutionary algorithm. In: Djeddi, C., Jamil, A., Siddiqi, I. (eds.) MedPRAI 2019. CCIS, vol. 1144, pp. 207–217. Springer, Cham (2020). https://doi.org/10.1007/978-3-030-37548-5_16

6. Guo, D., Wang, X., Gao, K., Jin, Y., Ding, J., Chai, T.: Evolutionary optimization of high-dimensional multiobjective and many-objective expensive problems assisted by a dropout neural network. IEEE Trans. Syst. Man Cybern. Syst. **52**(4), 2084–2097 (2022). https://doi.org/10.1109/TSMC.2020.3044418

7. Hao, H., Zhou, A., Qian, H., Zhang, H.: Expensive multiobjective optimization by relation learning and prediction. IEEE Trans. Evol. Comput. **26**(5), 1157–1170 (2022). https://doi.org/10.1109/TEVC.2022.3152582

8. Jiang, M., Wang, Z., Qiu, L., Guo, S., Gao, X., Tan, K.C.: A fast dynamic evolutionary multiobjective algorithm via manifold transfer learning. IEEE Trans. Cybern. **51**(7), 3417–3428 (2021). https://doi.org/10.1109/TCYB.2020.2989465

9. Jin, Y., Wang, H., Chugh, T., Guo, D., Miettinen, K.: Data-driven evolutionary optimization: an overview and case studies. IEEE Trans. Evol. Comput. **23**(3), 442–458 (2019). https://doi.org/10.1109/TEVC.2018.2869001

10. Knowles, J.: Parego: a hybrid algorithm with on-line landscape approximation for expensive multiobjective optimization problems. IEEE Trans. Evol. Comput. **10**(1), 50–66 (2006). https://doi.org/10.1109/TEVC.2005.851274

11. Lin, X., Zhang, Q., Kwong, S.: A decomposition based multiobjective evolutionary algorithm with classification. In: 2016 IEEE Congress on Evolutionary Computation (CEC), pp. 3292–3299 (2016). https://doi.org/10.1109/CEC.2016.7744206

12. Liu, S., Li, J., Lin, Q., Tian, Y., Tan, K.C.: Learning to accelerate evolutionary search for large-scale multiobjective optimization. IEEE Trans. Evol. Comput. **27**(1), 67–81 (2023). https://doi.org/10.1109/TEVC.2022.3155593

13. McKay, M.D., Beckman, R.J., Conover, W.J.: A comparison of three methods for selecting values of input variables in the analysis of output from a computer code. Technometrics **42**(1), 55–61 (2000)

14. Pan, L., He, C., Tian, Y., Wang, H., Zhang, X., Jin, Y.: A classification-based surrogate-assisted evolutionary algorithm for expensive many-objective optimization. IEEE Trans. Evol. Comput. **23**(1), 74–88 (2019). https://doi.org/10.1109/TEVC.2018.2802784

15. Shiratori, R., Nakata, M., Hayashi, K., Baba, T.: Particle swarm optimization of silicon photonic crystal waveguide transition. Opt. Lett. **46**(8), 1904–1907 (2021)

16. Song, Z., Wang, H., He, C., Jin, Y.: A kriging-assisted two-archive evolutionary algorithm for expensive many-objective optimization. IEEE Trans. Evol. Comput. **25**(6), 1013–1027 (2021). https://doi.org/10.1109/TEVC.2021.3073648

17. Sonoda, T., Nakata, M.: Multiple classifiers-assisted evolutionary algorithm based on decomposition for high-dimensional multiobjective problems. IEEE Trans. Evol. Comput. **26**(6), 1581–1595 (2022). https://doi.org/10.1109/TEVC.2022.3159000

18. Tian, Y., Cheng, R., Zhang, X., Jin, Y.: PlatEMO: a MATLAB platform for evolutionary multi-objective optimization. IEEE Comput. Intell. Mag. **12**(4), 73–87 (2017)

19. Wilcoxon, F.: Individual Comparisons by Ranking Methods. Springer, Cham (1992)

20. Xiao, J., Liang, J., Zhao, K., Yang, Z., Yu, M.: Multi-parameters optimization for electromagnetic acoustic transducers using surrogate-assisted particle swarm optimizer. Mech. Syst. Signal Process. **152**, 107337 (2021). https://doi.org/10.1016/j.ymssp.2020.107337

21. Yu, G., Ma, L., Jin, Y., Du, W., Liu, Q., Zhang, H.: A survey on knee-oriented multiobjective evolutionary optimization. IEEE Trans. Evol. Comput. **26**(6), 1452–1472 (2022). https://doi.org/10.1109/TEVC.2022.3144880

22. Yu, M., Li, X., Liang, J.: A dynamic surrogate-assisted evolutionary algorithm framework for expensive structural optimization. Struct. Multidiscip. Optim. **61**, 711–729 (2020)

23. Yuan, Y., Banzhaf, W.: Expensive multiobjective evolutionary optimization assisted by dominance prediction. IEEE Trans. Evol. Comput. **26**(1), 159–173 (2022). https://doi.org/10.1109/TEVC.2021.3098257

24. Zhang, J., Zhou, A., Zhang, G.: A classification and pareto domination based multiobjective evolutionary algorithm. In: 2015 IEEE Congress on Evolutionary Computation (CEC), pp. 2883–2890 (2015). https://doi.org/10.1109/CEC.2015.7257247

25. Zhang, Q., Liu, W., Tsang, E., Virginas, B.: Expensive multiobjective optimization by MOEA/D with gaussian process model. IEEE Trans. Evol. Comput. **14**(3), 456–474 (2010). https://doi.org/10.1109/TEVC.2009.2033671

26. Zhu, S., Xu, L., Goodman, E.D., Lu, Z.: A new many-objective evolutionary algorithm based on generalized pareto dominance. IEEE Trans. Cybern. **52**(8), 7776–7790 (2022). https://doi.org/10.1109/TCYB.2021.3051078

A Non-uniform Clustering Based Evolutionary Algorithm for Solving Large-Scale Sparse Multi-objective Optimization Problems

Shuai Shao[1]($\boxtimes$), Ye Tian[2]($\boxtimes$), and Xingyi Zhang[2]($\boxtimes$)

[1] Information Materials and Intelligent Sensing Laboratory of Anhui Province,
Anhui University, Hefei 230601, China
`freshshao@gmail.com`
[2] School of Computer Science and Technology, Anhui University, Hefei 230601, China
`field910921@gmail.com`, `xyzhanghust@gmail.com`

Abstract. Evolutionary algorithms have shown their effectiveness in solving sparse multi-objective optimization problems (SMOPs). However, for most of the existing multi-objective optimization algorithms (MOEAs) for solving SMOPs, their search granularity keeps the same for all the decision variables, which leads to significant performance deterioration when dealing with SMOPs in high-dimensional decision spaces. To tackle the issue, in this paper, a non-uniform clustering based evolutionary algorithm, termed NUCEA, is proposed for solving large-scale SMOPs. The proposed algorithm divides the decision variables into multiple groups with varying sizes, so as to reduce the search space with different granularity. These clustering outcomes inspire the development of new genetic operators, which have been proven to efficiently perform dimensionality reduction when approximating sparse Pareto optimal solutions. Experimental results on both benchmark and real-world SMOPs have shown that the proposed algorithm has significant advantages in comparison with the state-of-the-art evolutionary algorithms.

Keywords: Large-scale multi-objective optimization · Sparse Pareto optimal solutions · Evolutionary algorithms · Variable clustering

1 Introduction

In recent decades, a main focus of evolutionary algorithms has been to solve large-scale multi-objective optimization problems (MOPs) [17,24,33]. However, as indicated in [28], many large-scale MOPs in real-world scenarios include sparse Pareto optimal solutions. That is, most decision variables in these optimal solutions are zero, and these problems are referred to as sparse multi-objective optimization problems (SMOPs). For example, the influence maximization problem is to maximize the influence and minimize the cost [31], the facility location

© The Author(s), under exclusive license to Springer Nature Singapore Pte Ltd. 2024
L. Pan et al. (Eds.): BIC-TA 2023, CCIS 2061, pp. 103–116, 2024.
https://doi.org/10.1007/978-981-97-2272-3_8

problem is to minimize the facility construction costs and minimize the distance to demand points [16,30], and the power grid fault diagnosis problem is to minimize the difference between actual and expected states, as well as minimize the difference between observed and actual states [35].

Although a number of multi-objective evolutionary algorithms (MOEAs) [8,10,14,27] have effectively met the challenges posed by large-scale MOPs, these MOEAs cannot be used to solve large-scale SMOPs. Existing MOEAs for general large-scale MOPs can be roughly divided into three categories. The first category uses divide-and-conquer strategies [15] to divide decision variables into different groups and optimizes the decision variables in groups separately. The second category adopts dimensionality reduction-based strategies [4] to enhance the search ability. The last category employs novel reproduction strategies [13] to effectively generate offspring solutions. However, these approaches do not consider the sparse characteristic of large-scale SMOPs, hence their performance deteriorates on large-scale SMOPs.

In order to effectively solve large-scale SMOPs, some MOEAs have been recently customized based on the sparse characteristic of SMOPs [5,11,19,22], which design new genetic operators to identify zero variables efficiently. These algorithms adopt a bi-level encoding scheme to optimize binary vectors for searching zero variables and optimize real vectors for finding the optimal values of non-zero variables [6,28]. Since the optimization of binary vectors can affect the performance of MOEAs to a greater extent, most existing MOEAs optimize the binary vector with the assistance of a guiding vector [28] or variable clustering [37]. While existing MOEAs based on variable clustering divide the binary variable into several groups with equal size, they lead to the same size of decision space explored during the evolutionary process, and are difficult to converge to the Pareto front. Therefore, this paper proposes a non-uniform clustering method for the efficient optimization of binary vectors in solving SMOPs.

To be specific, during the evolutionary process, the proposed non-uniform clustering method categorizes all decision variables into multiple groups with varying sizes, where the groups with larger size can be significantly used for dimensionality reduction and the groups with smaller size can be used for the exploitation of the current search space. Hence, the decision space can be more effectively explored and exploited by selecting these groups with varying sizes to guide offspring generation. Based on the proposed non-uniform clustering method, an evolutionary algorithm abbreviated as NUCEA, is developed in this paper. According to the experiments on both benchmark and real-world applications, the proposed NUCEA exhibits exceptional effectiveness in comparison with the state-of-the-art sparse MOEAs.

The rest of this paper is organized as follows. Section 2 introduces existing sparse MOEAs, and explains the motivation of this work. Section 3 delineates the proposed algorithm in detail. In Sect. 4, we perform comparative experiments on benchmark and real-world SMOPs. Finally, conclusions and future work are provided in Sect. 5.

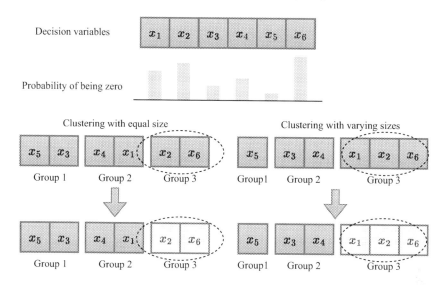

Fig. 1. Illustrative examples of different variable clustering methods, where non-uniform clustering can search the optimal solutions with varying granularity. The groups with larger sizes can considerably reduce the dimensionality of the decision space, and the groups with smaller sizes can fine-tune the values of more critical variables.

2 Related Work

Based on the idea of generating offspring solutions, existing sparse MOEAs can be roughly divided into two categories. The first category covers various guiding vectors based approaches for flipping a single binary variable. For example, SparseEA [28] and SparseEA2 [34] calculate the importance of each decision variable during initialization to estimate the probability of being zero, which flip a single binary variable via binary tournament selection based on the importance of decision variables. To dynamically update the importance of decision variables, RSMOEA [7] and MSKEA [3] suggest a dynamic guiding vector that is calculated based on the sparse distribution of the current population. The second category divides binary variables into multiple groups. For instance, SLMEA [20] proposes a fast clustering method to divide the large number of decision variables into multiple groups for solving super-large-scale SMOPs. MOEADRL divides decision variables into the all-zero group, all-one group and mixed group with the assistance of reinforcement learning [5,21]. Likewise, DSGEA [37] dynamically groups decision variables in the population that have a similar sparsity and enhance the genetic operators to optimize the decision variables in each group.

It can be concluded that the first category tends to prioritize exploitation, as only a single decision variable is changed when generating offspring solutions [28,34]. On the other hand, the second category leans towards exploration, as a group of decision variables are modified simultaneously. In order to achieve

---

Algorithm 1: Main procedure of NUCEA

Input: N (population size), FE_{max} (maximum number of evaluations)
Output: P (final population)

1   $[P, \mathbf{gv}] \leftarrow Initialization(N)$; //Algorithm 2
2   $FE \leftarrow |P|$; //Number of consumed evaluations
3   **while** $FE \leq FE_{max}$ **do**
4     $P' \leftarrow$ Select $2N$ parents from P via the mating selection strategy of SPEA2;
5     $Groups \leftarrow NUClustering(P)$; //Algorithm 3
6     $Q \leftarrow Variation(P', Groups)$;//Algorithm 4
7     $FE \leftarrow FE + |Q|$;
8     $P \leftarrow$ Select N solutions from $P \cup Q$ via the environmental selection strategy of SPEA2;
9   **return** P;

---

a better balance between exploration and exploitation, the proposed algorithm aims to divide decision variables into multiple groups of varying sizes. Unlike previous approaches that divide variables into groups with the same size [37], the proposed algorithm can make each group have an independent search granularity, as illustrated in Fig. 1. To be more specific, the critical variables are divided into smaller groups for fine-tuning, and the non-critical variables are divided into larger groups for rapid dimensional reduction. Therefore, the proposed algorithm develops a non-uniform clustering method whose group sizes can be adaptively adjusted, which can strike an excellent tradeoff between exploitation and exploration during the evolutionary process.

3 The Proposed Algorithm

In this section, we first present the main procedure of the proposed NUCEA, then elaborate on its two important components, i.e., the non-uniform clustering method and the novel crossover and mutation operators.

3.1 The Main Framework of NUCEA

Algorithm 1 presents the main framework of the proposed NUCEA, which begins with the initialization of a population and a guiding vector representing the importance of each decision variable. In the main loop, $2N$ parents are selected by binary tournament selection based on a fitness measured according to dominance relations and Euclidean distance. Then, all decision variables are divided into a number of groups with varying sizes by using the proposed non-uniform clustering. The clustering outcomes guide the generation of offspring population O. Afterwards, the N solutions are remained by using the environmental selection strategy of SPEA2 from the combined population.

In order to ensure the sparsity of the solution set, a hybrid representation of solution in [28] is adopted. Specifically, each solution $\mathbf{x} = (x_1, x_2, \ldots, x_d)$ is represented by

$$(x_1, x_2, \ldots) = (dec_1 \times mask_1, dec_2 \times mask_2, \ldots). \tag{1}$$

where dec represents the real value and $mask$ represents the binary value of each decision variable, respectively. As shown in Algorithm 2, NUCEA initializes the population based on the above presentation. In addition, a guiding vector $\mathbf{gv}$ is calculated by identifying the importance of each decision variable separately, where a larger value in $\mathbf{gv}$ indicates a higher probability that the decision variable should be set to zero. To be specific, the real vector dec of each solution is assigned with random values, while all elements in the binary vector $mask$ are set to 0 except for the i-th element in the $mask$ of the i-th solution being 1, i.e., it leads to a $D \times D$ identity matrix. The population Q is then sorted using non-dominated sorting [25], and the non-dominated front number of the i-th solution is used as the importance of the i-th decision variable.

3.2 Non-uniform Clustering

The proposed non-uniform clustering is used to divide all decision variables into multiple groups with varying sizes before generating offspring solutions at each generation. According to Algorithm 3, the mean sparsity of the current population and $GroupSize$ is first calculated. Then, all decision variables are sorted in a descending order according to $\mathbf{gv}$. Based on the rankings of decision variables, they are divided into multiple groups of varying sizes, where the size of i-th group is $\lceil i \times GroupSize \rceil$. Due to the different sizes of groups, the proposed algorithm can search for the sparse optimal solutions with varying granularity, i.e., the groups with larger sizes can reduce the dimensionality of the decision space at express speed, and the groups with smaller sizes can fine-tune the values of more critical variables. For example, the number of decision variables per group will be increased from 10 to 40 when solving the SMOPs with 100 variables and a sparsity of 0.1, which can quickly approximate the sparse Pareto front by flipping the larger groups, and focus subtly on promising decision variables by flipping the smaller groups.

3.3 Crossover and Mutation Operators

The offspring solutions are generated based on the clustering outcomes as described in Algorithm 4. Firstly, two parent solutions $\mathbf{p}$ and $\mathbf{q}$ are randomly selected from the mating pool. Then, the crossover operator sets the $mask$ of offspring $\mathbf{o}$ to the same as the parent $\mathbf{p}$, and the following two operations are conducted with the same probability: (1) Randomly selecting half the decision variables from $index \cap group$ and setting them to one; (2) randomly selecting half the decision variables from $index \cap group$ and setting them to zero. By doing so, a varying number of variables can be simultaneously flipped according to the

---

Algorithm 2: *Initialization(N)*

Input: N (population size)

Output: P (initial population), **gv** (importance of decision variables)

```
//Calculate the importance of variables
```
1 $D \leftarrow$ Number of decision variables;
2 **if** *the decision variables are real numbers* **then**
3   ⌊ $Dec \leftarrow D \times D$ random matrix;
4 **else if** *the decision variables are binary numbers* **then**
5   ⌊ $Dec \leftarrow D \times D$ matrix of ones;
6 $Mask \leftarrow D \times D$ identity matrix;
7 $Q \leftarrow$ A population whose i-th solution is generated by the i-th rows of Dec and $Mask$ according to (1);
8 $[F_1, F_2, \ldots] \leftarrow$ Do non-dominated sorting on Q;
9 **for** $i = 1$ *to* D **do**
10   ⌊ $\mathbf{gv}_i \leftarrow k$, s.t. $Q_i \in F_k$; //Q_i denotes the i-th solution in Q

```
//Generate the initial population
```
11 **if** *the decision variables are real numbers* **then**
12   ⌊ $Dec \leftarrow$ Uniformly randomly generate the decision variables of N solutions;
13 **else if** *the decision variables are binary numbers* **then**
14   ⌊ $Dec \leftarrow N \times D$ matrix of ones;
15 $Mask \leftarrow N \times D$ matrix of zeros;
16 **for** $i = 1$ *to* N **do**
17   **for** $j = 1$ *to* $rand() \times D$ **do**
18       $[m, n] \leftarrow$ Randomly select two decision varaibles;
19   **if** $\mathbf{gv}_m < \mathbf{gv}_n$ **then**
20   ⌊ Set the m-th element in the i-th binary vector in $Mask$ to 1;
21   **else**
22   ⌊ Set the n-th element in the i-th binary vector in $Mask$ to 1;

23 $P \leftarrow$ A population whose i-th solution is generated by the i-th rows of Dec and $Mask$ according to (1);
24 **return** P and **gv**;

---

gv. In order to enhance the diversity of population, the *mask* of **o** is mutated when $group < MaxGroup$. To be specific, the mutation process of *mask* of **o** involves the same probability of undergoing one of the following two operations: (1) Choosing an element from the nonzero elements in **o**.*mask* based on the values in **gv** (higher values are less importance), and setting this chosen element to zero. (2) Choosing an element from the nonzero elements in **o**.$\overline{mask}$ based on the values in **gv** (lower values are more importance), and setting this chosen element to one. Lastly, the real vector *dec* of **o** is generated as suggested in [34].

Algorithm 3: $NUClustering(P, \mathbf{gv})$

Input: P (current population), $\mathbf{gv}$ (importance of decison variables)
Output: $Groups$ (Groups of variables)

1 $Sparsity \leftarrow$ Calculate the mean sparsity of P;
2 $D \leftarrow$ Dimensions of the decision variable;
3 $GroupSize \leftarrow \lceil D \times Sparsity \rceil$;
4 $Rank \leftarrow$ Sort the decision variables in a descending order according to $\mathbf{gv}$;
5 $Groups \leftarrow$ Divide every $GroupSize \times i$ successive variables in $Rank$ into the i-th group;
6 **return** $Groups$;

---

4   Empirical Studies

4.1   Settings of Algorithms and Problems

The proposed NUCEA is compared with four state-of-the-art sparse MOEAs, namely, MOEA/PSL [23], SparseEA2 [34], TS-SparseEA [11], and DSGEA [37] on eight benchmark problems SMOP1–SMOP8 and the sparse signal reconstruction problems [12]. According to the common settings, all the algorithms have a population size of 100 on both benchmark problems and the sparse signal reconstruction problems. The maximum number of function evaluations is adopted as the termination condition, which is set to $100 \times d$ for benchmark problems and 20 000 for the sparse signal reconstruction problems.

The performance of all the algorithms is evaluated using inverted generational distance (IGD) [1,26] on benchmark problems. In addition, hypervolume (HV) [36] is employed to assess the population quality on the sparse signal reconstruction problems, with the reference point for calculating HV setting to (1,1). Statistical analysis between each compared algorithm and the proposed NUCEA is conducted using the Wilcoxon rank sum test [2] with a significance level of 0.05. This analysis is based on 30 independent runs on each test instance. The results are denoted as follows: '+' indicates that an algorithm is significantly better than NUCEA, '−' indicates that an algorithm is significantly worse than NUCEA, and '=' indicates that an algorithm is statistically similar to NUCEA. All the comparative experiments are conducted on PlatEMO [29].

4.2   Comparative Experiments

Table 1 displays the statistical results of the IGD and HV metric values of the five compared algorithms on the benchmark problems SMOP1–SMOP8 with a number of decision variables ranging from 100 to 1 000 and the sparse signal reconstruction problems with a number of decision variables ranging from 128 to 1 024. These values are averaged over 30 independent runs. It is evident that the proposed NUCEA demonstrates superior overall performance compared with the other algorithms. Specifically, out of a total of 36 test instances, NUCEA achieves the best results on 23 instances, followed by DSGEA, TS-SparseEA,

---

Algorithm 4: $Variation(P', Groups, \mathbf{gv})$

Input: P' (parent solutions), $Groups$ (Groups of variables), $\mathbf{gv}$ (importance of decision variables)

Output: O (a set of offsprings)

1 $O \leftarrow \emptyset$;
2 $MaxGroup \leftarrow |Groups|$;
3 **while** P' *is not empty* **do**
4       $[\mathbf{p}, \mathbf{q}] \leftarrow$ Randomly select two parents from P';
5       $group \leftarrow$ Randomly select a group from $Groups$;
6       $index \leftarrow xor(\mathbf{p}.mask, \mathbf{q}.mask)$;
7       $P' \leftarrow P' \setminus \{\mathbf{p}, \mathbf{q}\}$;
       //Generate the $mask$ of offspring o
8       $\mathbf{o}.mask \leftarrow \mathbf{p}.mask$; //$\mathbf{p}.mask$ denotes the binary vector $mask$ of solution $\mathbf{p}$
       //Crossover
9   **if** $rand() < 0.5$ **then**
10   Randomly select half the decision variables from $index \cap group$ and set them to one;
11   **else**
12   Randomly select half the decision variables from $index \cap group$ and set them to zero;
       //Mutation
13   **if** $group < MaxGroup$ **then**
14   **if** $rand() < 0.5$ **then**
15           $[m, n] \leftarrow$ Randomly select two decision variables from the nonzero elements in $\mathbf{o}.mask$;
16   **if** $\mathbf{gv}_m > \mathbf{gv}_n$ **then**
17   Set the m-th element in $\mathbf{o}.mask$ to 0;
18   **else**
19   Set the n-th element in $\mathbf{o}.mask$ to 0;
20   **else**
21           $[m, n] \leftarrow$ Randomly select two decision variables from the nonzero elements in $\mathbf{o}.\overline{mask}$;
22   **if** $\mathbf{gv}_m < \mathbf{gv}_n$ **then**
23   Set the m-th element in $\mathbf{o}.mask$ to 1;
24   **else**
25   Set the n-th element in $\mathbf{o}.mask$ to 1;
       //Generate the dec of offspring o
26   **if** *the decision variables are real numbers* **then**
27         $\mathbf{o}.dec \leftarrow$ Perform binary crossover and mutation as suggested in [34];
28   **else**
29         $\mathbf{o}.dec \leftarrow$ Vector of ones;
30      $O \leftarrow O \cup \{\mathbf{o}\}$;
31 **return** O;

---

and SparseEA2 achieving the best results on 9 instances, 3 instances and 1 instance, respectively. According to the statistical tests, it is found that NUCEA significantly outperforms MOEA/PSL, SparseEA2, TS-SparseEA, and DSGEA on 36, 25, 28, and 20 test instances, respectively. Consequently, the proposed NUCEA demonstrates superiority over existing sparse MOEAs.

Table 1. IGD/HV values obtained by MOEA/PSL, SparseEA2, TS-SparseEA, DSGEA, and the proposed NUCEA on SMOP1–SMOP8 with 100 to 1000 decision variables and the sparse signal reconstruction problems with 128 to 1024 decision variables.

Problem	D	MOEA/PSL	SparseEA2	TS-SparseEA	DSGEA	NUCEA
SMOP1	100	9.6963e-3 (5.14e-3) −	5.9823e-3 (1.18e-3) −	6.9292e-3 (3.00e-3) −	4.8660e-3 (1.22e-3) ≈	4.7669e-3 (5.21e-4)
	200	1.4086e-2 (3.05e-3) −	6.3097e-3 (1.17e-3) −	5.2645e-3 (5.15e-4) −	4.8845e-3 (6.37e-4) −	4.5077e-3 (5.69e-4)
	500	2.0062e-2 (3.66e-3) −	6.8701e-3 (1.04e-3) −	6.7135e-3 (1.51e-3) −	5.9810e-3 (1.17e-3) −	4.7753e-3 (3.80e-4)
	1000	2.1474e-2 (1.39e-3) −	7.1728e-3 (4.87e-4) −	8.1305e-3 (1.60e-3) −	6.3031e-3 (1.33e-3) −	4.8816e-3 (2.75e-4)
SMOP2	100	2.9550e-2 (1.37e-2) −	1.0283e-2 (4.52e-3) ≈	1.3257e-2 (4.57e-3) −	6.6907e-3 (2.94e-3) ≈	1.0531e-2 (9.24e-3)
	200	3.4108e-2 (8.84e-3) −	9.8093e-3 (4.14e-3) −	1.5450e-2 (5.06e-3) −	8.9951e-3 (3.31e-3) ≈	8.0391e-3 (3.32e-3)
	500	4.4592e-2 (5.25e-3) −	1.1121e-2 (3.07e-3) −	2.3527e-2 (6.75e-3) −	1.7185e-2 (4.87e-3) −	9.5672e-3 (4.86e-3)
	1000	5.1220e-2 (3.72e-3) −	1.3742e-2 (3.60e-3) ≈	2.8772e-2 (5.28e-3) −	1.8475e-2 (3.29e-3) −	1.2283e-2 (4.00e-3)
SMOP3	100	5.0634e-1 (6.76e-1) −	6.6877e-3 (9.23e-4) −	2.6548e-2 (5.04e-2) −	4.6205e-3 (2.06e-4) ≈	5.2058e-3 (1.88e-3)
	200	1.8136e-1 (5.09e-1) −	6.7885e-3 (1.27e-3) −	5.7814e-3 (1.61e-3) −	4.6891e-3 (1.94e-4) ≈	5.2230e-3 (1.48e-3)
	500	2.5402e-2 (2.63e-2) −	8.1916e-3 (1.19e-3) −	7.5569e-3 (2.15e-3) −	5.1805e-3 (5.22e-4) −	5.0496e-3 (1.22e-3)
	1000	2.1814e-2 (2.47e-3) −	1.0774e-2 (1.38e-3) −	7.2817e-3 (1.81e-3) −	5.4920e-3 (6.76e-4) −	4.4320e-3 (2.22e-4)
SMOP4	100	5.0470e-3 (4.63e-4) −	4.7790e-3 (2.60e-4) −	4.6166e-3 (1.70e-4) −	4.7907e-3 (2.88e-4) −	4.1585e-3 (5.53e-5)
	200	5.2321e-3 (8.29e-4) −	4.7153e-3 (2.17e-4) −	4.5925e-3 (1.41e-4) −	4.7816e-3 (2.01e-4) −	4.1645e-3 (7.99e-5)
	500	5.2319e-3 (5.07e-4) −	4.7846e-3 (3.27e-4) −	4.6098e-3 (2.14e-4) −	4.7287e-3 (2.38e-4) −	4.1405e-3 (6.90e-5)
	1000	5.1598e-3 (4.13e-4) −	4.7881e-3 (2.11e-4) −	4.6094e-3 (1.48e-4) −	4.8323e-3 (2.63e-4) −	4.1462e-3 (4.99e-5)
SMOP5	100	7.6142e-3 (2.78e-4) −	5.5786e-3 (2.97e-4) ≈	6.1004e-3 (5.55e-4) −	6.7212e-3 (4.36e-4) −	5.7956e-3 (4.51e-4)
	200	7.6676e-3 (3.24e-4) −	5.4753e-3 (2.38e-4) −	5.6642e-3 (4.84e-4) −	6.0194e-3 (4.16e-4) −	5.0125e-3 (2.26e-4)
	500	7.8949e-3 (3.73e-4) −	5.2877e-3 (2.70e-4) −	5.1763e-3 (2.52e-4) −	4.9993e-3 (1.49e-4) −	4.7941e-3 (1.76e-4)
	1000	9.2761e-3 (5.06e-4) −	5.3516e-3 (2.54e-4) −	5.1368e-3 (1.92e-4) −	5.0303e-3 (2.38e-4) −	4.7601e-3 (1.38e-4)
SMOP6	100	8.7812e-3 (7.03e-4) −	6.7738e-3 (3.46e-4) ≈	6.5977e-3 (7.70e-4) −	6.6168e-3 (7.62e-4) ≈	6.4706e-3 (4.81e-4)
	200	8.3446e-3 (4.65e-4) −	6.5353e-3 (2.89e-4) −	5.9430e-3 (3.65e-4) −	5.5077e-3 (3.46e-4) +	5.9904e-3 (4.81e-4)
	500	8.9253e-3 (6.08e-4) −	6.4563e-3 (2.48e-4) −	5.3696e-3 (1.56e-4) +	5.2339e-3 (2.13e-4) +	5.5451e-3 (2.24e-4)
	1000	1.0826e-2 (6.72e-4) −	6.7403e-3 (2.44e-4) −	5.3635e-3 (1.37e-4) +	5.4642e-3 (2.64e-4) +	5.7954e-3 (2.57e-4)
SMOP7	100	2.9982e-2 (1.48e-2) −	1.5210e-2 (6.28e-3) −	2.5017e-2 (5.88e-3) −	1.1571e-2 (7.38e-3) ≈	1.2627e-2 (7.13e-3)
	200	6.3873e-2 (6.01e-2) −	9.0896e-3 (4.01e-3) −	2.3805e-2 (5.86e-3) −	1.2900e-2 (5.83e-3) −	7.7271e-3 (4.15e-3)
	500	1.0347e-1 (1.08e-1) −	8.7614e-3 (3.30e-3) ≈	3.2916e-2 (1.03e-2) −	2.2680e-2 (4.98e-3) −	7.5264e-3 (2.98e-3)
	1000	8.5876e-2 (8.01e-2) −	1.0557e-2 (2.10e-3) −	4.1310e-2 (8.48e-3) −	2.7050e-2 (4.51e-3) −	8.7053e-3 (3.37e-3)
SMOP8	100	1.6913e-1 (4.92e-2) −	1.3253e-1 (2.17e-2) ≈	1.3789e-1 (2.42e-2) −	9.8299e-2 (2.27e-2) +	1.2730e-1 (2.67e-2)
	200	2.0440e-1 (2.61e-2) −	1.4477e-1 (1.73e-2) ≈	1.3371e-1 (1.55e-2) ≈	1.2514e-1 (1.59e-2) ≈	1.3596e-1 (1.90e-2)
	500	2.9290e-1 (3.32e-2) −	1.7162e-1 (1.13e-2) −	1.3727e-1 (1.98e-2) +	1.5413e-1 (2.44e-2) ≈	1.5812e-1 (2.25e-2)
	1000	3.7204e-1 (2.39e-2) −	2.0267e-1 (1.90e-2) −	1.3845e-1 (1.55e-2) +	1.6752e-1 (8.28e-3) ≈	1.6960e-1 (1.81e-2)
SR1	128	2.0213e-1 (1.44e-3) −	2.0428e-1 (1.30e-4) −	2.0054e-1 (1.62e-3) −	2.0436e-1 (2.22e-6) +	2.0431e-1 (1.28e-4)
SR2	256	2.0766e-1 (3.76e-3) −	2.1609e-1 (9.06e-4) −	2.0598e-1 (4.62e-3) −	2.1759e-1 (4.68e-4) ≈	2.1770e-1 (4.15e-4)
SR3	520	2.1298e-1 (8.43e-3) −	2.4942e-1 (4.82e-3) −	1.9836e-1 (7.08e-3) −	2.5629e-1 (3.51e-2) −	2.5841e-1 (4.08e-3)
SR4	1024	2.2605e-1 (1.50e-2) −	2.8795e-1 (9.29e-3) −	1.9102e-1 (9.37e-3) −	3.0892e-1 (1.38e-2) −	3.2863e-1 (1.46e-2)
+/−/≈		0/36/0	0/25/11	3/28/5	5/20/11	

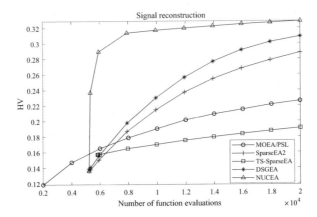

Fig. 2. Convergence profiles of HV values obtained by MOEA/PSL, SparseEA2, TS-SparseEA, DSGEA and the proposed NUCEA on the sparse signal reconstruction problem with 1024 decision variables.

Figure 2 depicts the convergence profiles of HV values obtained on the sparse signal reconstruction problem, where NUCEA has a faster convergence speed than MOEA/PSL, SparseEA2, TS-SparseEA, and DSGEA. Moreover, the decision variables obtained by the five MOEAs on SMOP7 with 1 000 decision variables are plotted in parallel coordinates. Moreover, as plotted in Fig. 3, TS-SparseEA, DSGEA, and the proposed NUCEA can obtain sparse solutions on SMOP7. It can be seen that although the solutions obtained by DSGEA are sparser than those obtained by NUCEA, its convergence speed is far slower than NUCEA, mainly due to the groups with smaller sizes in NUCEA can fine-tune the values of more critical variables. To summarize, the effectiveness of NUCEA for solving large-scale SMOPs can be evidenced.

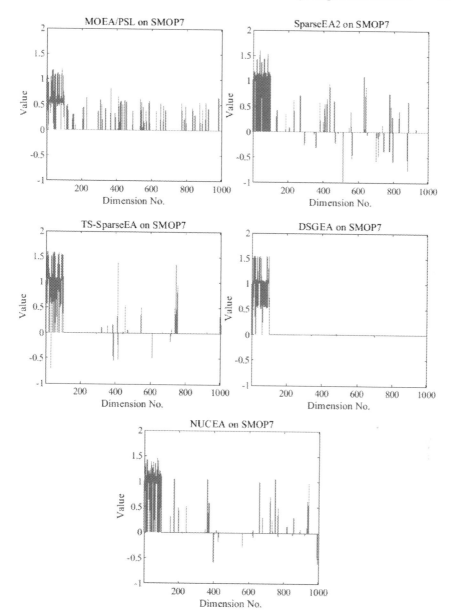

Fig. 3. Parallel coordinates of the decision variables obtained by MOEA/PSL, SparseEA2, TS-SparseEA, DSGEA and the proposed NUCEA on SMOP7 with 1000 variables.

5   Conclusions

In this paper, a non-uniform clustering based evolutionary algorithm, termed NUCEA, has proposed to solve large-scale SMOPs. In the proposed algorithm, a non-uniform clustering method has been suggested to divide all decision variables into multiple groups with varying sizes. Therefore, the proposed algorithm can use larger groups to reduce the dimensionality of the decision space and use smaller groups to fine-tune the values of more critical variables. Experimental results on both benchmark and real-world SMOPs have demonstrated the superiority of the proposed algorithm. Further investigation on this idea is still desirable. Since NUCEA divides decision variables based on a constant guiding vector, it can be enhanced by using the probability matrix [17] to analyze the interactions between decision variables. In addition, since many SMOPs are with computationally expensive objectives [19], how to combine the proposed non-uniform clustering method with surrogate models [9,18,32] for saving function evaluations remains a future work.

Acknowledgements. This work was supported in part by the National Natural Science Foundation of China (No. 62276001, No. 62136008, No. U21A20512), in part by the Anhui Provincial Natural Science Foundation (No. 2308085J03), and in part by the Excellent Youth Foundation of Anhui Provincial Colleges (No. 2022AH030013).

References

1. Bosman, P.A., Thierens, D.: The balance between proximity and diversity in multiobjective evolutionary algorithms. IEEE Trans. Evol. Comput. **7**(2), 174–188 (2003)
2. Derrac, J., García, S., Molina, D., Herrera, F.: A practical tutorial on the use of nonparametric statistical tests as a methodology for comparing evolutionary and swarm intelligence algorithms. Swarm Evol. Comput. **1**(1), 3–18 (2011)
3. Ding, Z., Chen, L., Sun, D., Zhang, X.: A multi-stage knowledge-guided evolutionary algorithm for large-scale sparse multi-objective optimization problems. Swarm Evol. Comput. **73**, 101119 (2022)
4. Feng, Y., Feng, L., Kwong, S., Tan, K.C.: A multivariation multifactorial evolutionary algorithm for large-scale multiobjective optimization. IEEE Trans. Evol. Comput. **26**(2), 248–262 (2021)
5. Gao, M., Feng, X., Yu, H., Li, X.: An efficient evolutionary algorithm based on deep reinforcement learning for large-scale sparse multiobjective optimization. Appl. Intell. 1–24 (2023)
6. Geng, H., Shen, J., Zhou, Z., Xu, K.: An improved large-scale sparse multi-objective evolutionary algorithm using unsupervised neural network. Appl. Intell. **53**(9), 10290–10309 (2023)
7. Gu, Q., Sun, Y., Wang, Q., Chen, L.: A quadratic association vector and dynamic guided operator search algorithm for large-scale sparse multi-objective optimization problem. Appl. Intell. 1–22 (2023)
8. He, C., Cheng, R., Tian, Y., Zhang, X., Tan, K.C., Jin, Y.: Paired offspring generation for constrained large-scale multiobjective optimization. IEEE Trans. Evol. Comput. **25**(3), 448–462 (2020)

9. He, C., Zhang, Y., Gong, D., Ji, X.: A review of surrogate-assisted evolutionary algorithms for expensive optimization problems. Expert Syst. Appl. 119495 (2023)
10. Hong, W.J., Yang, P., Tang, K.: Evolutionary computation for large-scale multi-objective optimization: a decade of progresses. Int. J. Autom. Comput. **18**(2), 155–169 (2021)
11. Jiang, J., Han, F., Wang, J., Ling, Q., Han, H., Wang, Y.: A two-stage evolutionary algorithm for large-scale sparse multiobjective optimization problems. Swarm Evol. Comput. **72**, 101093 (2022)
12. Liang, J., Gong, M., Li, H., Yue, C., Qu, B.: Problem definitions and evaluation criteria for the CEC special session on evolutionary algorithms for sparse optimization. Technical Report, Computational Intelligence Laboratory, Zhengzhou University, Zhengzhou, China, Report# 2018001 (2018)
13. Lin, Q., Li, J., Liu, S., Ma, L., Li, J., Chen, J.: An adaptive two-stage evolutionary algorithm for large-scale continuous multi-objective optimization. Swarm Evol. Comput. **77**, 101235 (2023)
14. Liu, S., Lin, Q., Li, J., Tan, K.C.: A survey on learnable evolutionary algorithms for scalable multiobjective optimization. IEEE Trans. Evol. Comput. (2023)
15. Liu, S., Lin, Q., Tian, Y., Tan, K.C.: A variable importance-based differential evolution for large-scale multiobjective optimization. IEEE Trans. Cybern. **52**(12), 13048–13062 (2021)
16. Rahmani, A., MirHassani, S.: A hybrid firefly-genetic algorithm for the capacitated facility location problem. Inf. Sci. **283**, 70–78 (2014)
17. Shao, S., Tian, Y., Wang, L., Yang, S., Zhang, P., Zhang, X.: A permutation group-based evolutionary algorithm for car sequencing problems in assembly lines. In: 2023 5th International Conference on Data-driven Optimization of Complex Systems (DOCS), pp. 1–8 (2023). https://doi.org/10.1109/DOCS60977.2023.10294929
18. Si, L., Zhang, X., Tian, Y., Yang, S., Zhang, L., Jin, Y.: Linear subspace surrogate modeling for large-scale expensive single/multi-objective optimization. IEEE Trans. Evol. Comput. (2023)
19. Tan, Z., Wang, H., Liu, S.: Multi-stage dimension reduction for expensive sparse multi-objective optimization problems. Neurocomputing **440**, 159–174 (2021)
20. Tian, Y., Feng, Y., Zhang, X., Sun, C.: A fast clustering based evolutionary algorithm for super-large-scale sparse multi-objective optimization. IEEE/CAA J. Automatica Sinica **10**(4), 1048–1063 (2022)
21. Tian, Y., Li, X., Ma, H., Zhang, X., Tan, K.C., Jin, Y.: Deep reinforcement learning based adaptive operator selection for evolutionary multi-objective optimization. IEEE Trans. Emerg. Top. Comput. Intell. (2022)
22. Tian, Y., Lu, C., Zhang, X., Cheng, F., Jin, Y.: A pattern mining-based evolutionary algorithm for large-scale sparse multiobjective optimization problems. IEEE Trans. Cybern. **52**(7), 6784–6797 (2020)
23. Tian, Y., Lu, C., Zhang, X., Tan, K.C., Jin, Y.: Solving large-scale multiobjective optimization problems with sparse optimal solutions via unsupervised neural networks. IEEE Trans. Cybern. **51**(6), 3115–3128 (2020)
24. Tian, Y., et al.: Evolutionary large-scale multi-objective optimization: a survey. ACM Comput. Surv. (CSUR) **54**(8), 1–34 (2021)
25. Tian, Y., Wang, H., Zhang, X., Jin, Y.: Effectiveness and efficiency of non-dominated sorting for evolutionary multi-and many-objective optimization. Complex Intell. Syst. **3**, 247–263 (2017)
26. Tian, Y., Xiang, X., Zhang, X., Cheng, R., Jin, Y.: Sampling reference points on the pareto fronts of benchmark multi-objective optimization problems. In: 2018 IEEE Congress on Evolutionary Computation (CEC), pp. 1–6. IEEE (2018)

27. Tian, Y., Yang, S., Zhang, L., Duan, F., Zhang, X.: A surrogate-assisted multiobjective evolutionary algorithm for large-scale task-oriented pattern mining. IEEE Trans. Emerg. Top. Comput. Intell. **3**(2), 106–116 (2018)
28. Tian, Y., Zhang, X., Wang, C., Jin, Y.: An evolutionary algorithm for large-scale sparse multiobjective optimization problems. IEEE Trans. Evol. Comput. **24**(2), 380–393 (2019)
29. Tian, Y., Zhu, W., Zhang, X., Jin, Y.: A practical tutorial on solving optimization problems via PlatEMO. Neurocomputing **518**, 190–205 (2023)
30. Xiang, X., Tian, Y., Xiao, J., Zhang, X.: A clustering-based surrogate-assisted multiobjective evolutionary algorithm for shelter location problem under uncertainty of road networks. IEEE Trans. Ind. Inf. **16**(12), 7544–7555 (2019)
31. Yang, J., Liu, J.: Influence maximization-cost minimization in social networks based on a multiobjective discrete particle swarm optimization algorithm. IEEE Access **6**, 2320–2329 (2017)
32. Yang, S., Tian, Y., He, C., Zhang, X., Tan, K.C., Jin, Y.: A gradient-guided evolutionary approach to training deep neural networks. IEEE Trans. Neural Netw. Learn. Syst. **33**(9), 4861–4875 (2021)
33. Yin, F., Cao, B.: A two-space-decomposition-based evolutionary algorithm for large-scale multiobjective optimization. Swarm Evol. Comput. 101397 (2023)
34. Zhang, Y., Tian, Y., Zhang, X.: Improved SparseEA for sparse large-scale multiobjective optimization problems. Complex Intell. Syst. 1–16 (2021)
35. Zhao, J., Xu, Y., Luo, F., Dong, Z., Peng, Y.: Power system fault diagnosis based on history driven differential evolution and stochastic time domain simulation. Inf. Sci. **275**, 13–29 (2014)
36. Zitzler, E., Thiele, L.: Multiobjective evolutionary algorithms: a comparative case study and the strength pareto approach. IEEE Trans. Evol. Comput. **3**(4), 257–271 (1999)
37. Zou, Y., Liu, Y., Zou, J., Yang, S., Zheng, J.: An evolutionary algorithm based on dynamic sparse grouping for sparse large scale multiobjective optimization. Inf. Sci. **631**, 449–467 (2023)

An Improved MOEA/D with Pareto Frontier Individual Selection Based on Weight Vector Angles

Qiwei Li[1,2(✉)] and Jing Guan[3]

[1] School of Computer Science, China University of Geosciences,
Wuhan 430078, China
`15090842918@163.com`
[2] Hubei Key Laboratory of Intelligent Geo-Information Processing,
China University of Geosciences, Wuhan 430078, China
[3] China Ship Development and Design Center, Wuhan 430064, China

Abstract. In this paper, we introduce an improved MOEA/D with pareto frontier individual selection based on weight vector angles (WVA-MOEA/D). This method specifically addresses premature convergence issues observed in MOEA/D when tackling high-dimensional multi-objective optimization challenges. The principal aim is to bolster the algorithm's diversity throughout its convergence journey. In this method, each weight vector is steered to select a Pareto front individual that minimizes the angle formed between the weight vector and the vector originating from the ideal point directed towards the individual. For these highlighted individuals, the replacement protocol of MOEA/D's aggregation function is only applied if a novel individual can supersede all its marked attributes comprehensively. The strategy leverages the orthogonal distance between the solution and the weight vector in the objective space, ensuring the preservation of desired diversity across the evolutionary trajectory. Such an adaptation strikes a more refined balance between convergence and diversity, especially in the realm of high-dimensional multi-objective optimization. Experimental validations suggest that our proposed algorithm consistently surpasses traditional techniques in harmonizing convergence with diversity and remains highly competitive against other prevailing algorithms in addressing many-objective optimization quandaries.

Keywords: Convergence · Pareto frontier individual · Diversity · Multi-objective optimization · Weight vector angles

1 Introduction

Multi-objective optimization problems (MOPs) [2] are pivotal across numerous disciplines, encompassing scientific inquiry, engineering domains, and practical applications. These problems predominantly involve optimizing conflicting

© The Author(s), under exclusive license to Springer Nature Singapore Pte Ltd. 2024
L. Pan et al. (Eds.): BIC-TA 2023, CCIS 2061, pp. 117–130, 2024.
https://doi.org/10.1007/978-981-97-2272-3_9

objectives, which introduces substantial computational intricacies and intricate constraints. While initial algorithms addressing MOPs borrowed foundational techniques from Single-Objective Optimization Problem (SOP) strategies-such as linear weighted methods [3], ϵ-constraint methods [3], max-min methods [10], goal programming methods [11], and goal satisfaction methods [6]-they frequently required multiple independent executions to procure an optimal solution set. Such an approach inhibited efficient inter-solution information transfer, escalating computational overheads. Moreover, contrasting the outcomes of these solution sets became a complex task, further complicating decision-making processes. Lin, He, and Cheng (2022) introduced an innovative adaptive dropout mechanism in their study, aimed at enhancing the optimization of high-dimensional expensive multiobjective optimization problems. Their approach notably emphasizes the importance of selecting key decision variables in the decision space and proposes a new infill criterion to optimize these variables, thereby assisting the decision-making process through surrogate models [8].

Zhang and Li pioneered the MOEA/D, a decomposition-centric multiobjective optimization algorithm [13]. By transmuting MOPs into SOPs, it facilitated concurrent optimization of sub-problems, tactically employing predefined weight vector distributions to guide its population's evolutionary trajectory. This ensured the solution set's distribution echoed the weight vectors' distribution. However, the heavy reliance of MOEA/D on weight vector distribution occasionally impeded the consistent distribution of its population throughout the computation [4]. Moreover, it remains a rarity for a singular multi-objective algorithm to regularly produce well-distributed solution sets across diverse frontier challenges [4]. The constraints observed in this weight vector-focused method, especially when contending with high-dimensional scenarios and complex Pareto fronts, catalyzed the emergence of various enhancement strategies. The academic community has made significant strides addressing these challenges. Beyond mere refinements to the MOEA/D via alterations in weight vector distributions, novel methodologies have emerged. Notably, the MOEA/D-DE algorithm [7] judiciously moderates population update frequencies, while the MOEA/D-M2M [9] and SPEA/R [5] algorithms implement local optimization evolution considering the distance between individuals and weight vectors. The MOEA/D-DU algorithm [12] innovatively integrates a meticulous distance constraint based on each population member's weight vector, meticulously constraining individual replication, highlighting the field's evolving dynamism.

This research offers a profound exploration into the intricacies of decomposition based MOEAs, highlighting the inherent challenges they grapple with in autonomously ensuring population diversity for advanced multi-objective tasks. Stemming from the core principles of the MOEA/D-DU algorithm and its intricate interplay between individual solutions and weight vectors, we propose an innovative strategy grounded on the angular disposition of weight vectors, denoted as WVA-MOEA/D, for Pareto front individual selection. The genesis of this approach is rooted in the observation that, in the objective space, certain individuals, despite diverging considerably from their designated weight vectors, can paradoxically achieve more favorable aggregated function values. This conundrum has the potential to skew solution selections, veering them away from true Pareto optimality. The WVA-MOEA/D algorithm seeks to reconcile this by integrating a nuanced criterion based on the angular distance between individuals and weight vectors. This metric, when juxtaposed with the traditional aggregated function value, provides a holistic measure for population update decisions. Such a dual-faceted approach ensures a balanced exploration-exploitation trade-off during the evolutionary process. Furthermore, by adopting this methodology, WVA-MOEA/D anticipates potential pitfalls in the search space and proactively strategizes to prevent premature convergence and stagnation. This avant-garde methodology aims to bolster population diversity throughout the evolutionary trajectory, offering a more resilient and adaptive mechanism against a myriad of multi-objective challenges. Our proposition was meticulously subjected to comprehensive evaluations using the renowned MaF test problems [1], with all experimental settings finetuned to align with prevailing academic norms.

The rest of the paper is arranged as follows: Sect. 2 introduce the Multi-objective evolutionary algorithm based on decomposition. Section 3 delves into the newly proposed enhancement strategy, elucidating its foundational principles and operational specifics. Section 4 showcases our experimental methods and outcomes, with Sect. 5 rounding off the discussion through a summative conclusion.

2 Multiobjective Evolutionary Algorithm Based on Decomposition

The MOEA/D algorithm, ingeniously introduced by Professors Zhang Qingfu and Li Hui, serves as a decomposition-driven approach to multi-objective optimization. At its core, MOEA/D utilizes weight vectors to systematically decompose the comprehensive Pareto front into discrete single-objective optimization sub-tasks, each corresponding to the population size. Throughout the evolutionary process, the algorithm prioritizes maintaining an optimal population for each weight vector. Notably, due to the inherent interrelation of these sub-problems, the optimal solutions of neighboring sub-problems often exhibit similarities. Within the MOEA/D framework, the rejuvenation of sub-problems is orchestrated through clusters of adjacent sub-problems. From these clusters, individuals are probabilistically selected as parents. Employing genetic algorithm techniques, these parents give rise to offspring. Subsequently, these offspring, alongside the broader population, are evaluated using a strategically chosen aggregation function, guiding the evolution of the population.

2.1 Definition of Multi-objective Optimization Problem

A multi-objective optimization problem can be defined as:

$$
\begin{cases}
\min_{\mathbf{X}} F(\mathbf{X}) = (f_1(\mathbf{X}), f_2(\mathbf{X}), \dots, f_M(\mathbf{X}))^T \\
\text{s.t. } g_i(\mathbf{X}) \leq 0, \quad i = 1, \dots, p \\
h_j(\mathbf{X}) = 0, \quad j = 1, \dots, q
\end{cases}
\tag{1}
$$

Let $\mathbf{X} = (x_1, x_2, \ldots, x_d)^T$ be a D-dimensional decision variable vector with $\mathbf{X} \in \mathbb{R}^d$. We define the objective function vector $F = (f_1, f_2, \ldots, f_m)^T$ as an M-dimensional target vector with $F \in \mathbb{R}^m$. There are p inequality constraints denoted by g_i and q equality constraints represented by h_j. All decision vectors together form the decision space Ω, and the feasible decision vectors construct the feasible region Λ.

2.2 Definition of MOEA/D

Algorithm 1 illustrates the framework of the implementation steps of the MOEA/D algorithm.

Algorithm 1. MOEA/D

Input: MOP(1); a stopping criterion; N: the number of subproblems considered in MOEA/D; a uniform spread of N weight vectors: $\lambda^1, \ldots, \lambda^N$; T: the number of weight vectors in the neighborhood of each weight vector.

Output: EP.

1: **Initialization:**
2: Set EP $= \emptyset$.
3: Compute the Euclidean distances between any two weight vectors and determine the T closest weight vectors to each weight vector. For each $i = 1, \ldots, N$, set $B(i) = \{\lambda^{i1}, \ldots, \lambda^{iT}\}$, where $\lambda^{i1}, \ldots, \lambda^{iT}$ are the T closest weight vectors to λ^i.
4: Generate an initial population $x^1, \ldots, x^N$ randomly or by a problem-specific method. Set $FV^i = F(x^i)$.
5: Initialize $z = (z^1, \ldots, z^m)^T$ by a problem-specific method.
6: **Update:**
7: **for** $i = 1, \ldots, N$ **do**
8:    Reproduction: Randomly select two indexes k, l from $B(i)$, and then generate a new solution y from x^k and x^l using genetic operators.
9:    Improvement: Apply a problem-specific repair/improvement heuristic on y to produce y'.
10:    Update of z: For each $j = 1, \ldots, m$, if $z_j < f_j(y')$, then set $z_j = f_j(y')$.
11:    Update of Neighboring Solutions: For each index $j \in B(i)$, if $g^{te}(y'|\lambda^j, z) \leq g^{te}(x^j|\lambda^j, z)$, then set $x^j = y'$ and $FV^j = F(y')$.
12:    Update of EP:
13:    **if** no vectors in EP dominate $F(y')$ **then**
14:        Add $F(y')$ to EP.
15:    **end if**
16:    Remove from EP all vectors dominated by $F(y')$.
17: **end for**
18: **Termination:** If stopping criteria is met, stop and output EP. Otherwise, go to Step 2.

The MOEA/D algorithm, while proficient for conventional multi-objective optimization problems characterized by 2 or 3 objectives-such as those in the ZDT and DTLZ test suites-was not tailored for problems that are high-dimensional or feature irregular Pareto frontiers. As such, it displays certain limitations when applied to these more nuanced multi-objective optimization scenarios. In its approach to high-dimensional multi-objective problems, MOEA/D diverges from traditional Pareto-dominance-based evolutionary algorithms by utilizing aggregation functions to discern superiority among individuals. This strategy effectively intensifies the selective pressure, pushing the population towards the optimal Pareto frontier. Yet, a major drawback is inherent: the aggregation functions often struggle to accurately balance convergence and diversity among individuals. Consequently, the population can inadvertently become ensnared in local optima throughout its evolution.

Furthermore, when tasked with problems possessing an irregular ideal Pareto frontier, the MOEA/D's capacity to generate a diverse set of solutions diminishes compared to scenarios with more regular frontiers. The solution set's distribution is critically tethered to the weight vectors' distribution. A significant disparity between the ideal Pareto frontier and the weight vector distribution can derail the algorithm's ability to guide the population's distribution aptly. To circumvent the pitfalls of premature convergence and to finetune the weight vector distribution to the problem's characteristics emerge as paramount research directions.

3   WVA-MOEA/D

In decomposition-based multi-objective algorithms, various aggregation functions are employed. The three most commonly used methods are the Weighted Sum Approach, the Tchebycheff method, and the Penalty-Based Boundary Intersection (PBI) method. Since the algorithm proposed in this paper, as well as the compared algorithms, all utilize the PBI method, this paper places a significant emphasis on introducing the PBI method.

The penalized boundary intersection method stands as a quintessential and pivotal aggregation technique. This form of aggregation ensures a more equitably dispersed population within the objective space throughout the genetic evolutionary process. Moreover, every weight vector uniquely corresponds to an optimal solution on the Pareto ideal frontier. These advantages collectively attenuate the challenges associated with maintaining population diversity.

The Penalty-Based Boundary Intersection [13]. In the Penalty-Based Boundary Intersection approach, the aggregation function for multi-objective optimization problems is given by:

$$\begin{cases} \text{Minimize } g^{bp}(x|\lambda, z^*) = d_1 + \theta d_2 \\ \text{S.T. } x \in \Omega \end{cases} \tag{2}$$

where

$$\begin{cases} d_1 = \frac{||(z^* - F(x))^T \lambda||}{||\lambda||} \\ d_2 = ||F(x) - (z^* - d_1\lambda)|| \end{cases} \tag{3}$$

The value of λ is normalized such that its L1-norm is equal to 1. The vector $F(x)$ represents the objective function values of solution x and its Euclidean norm is denoted by L_2. d_1 measures the alignment of the solution with respect to the weight vector, while d_2 measures the distance from the solution to the ideal point in the direction of the weight vector. When θ is set to 0.5, the two distances are given equal importance.

The aggregated objective function based on Penalty-Based Boundary Intersection (PBI) method is mainly used for multi-objective optimization problems where the Pareto front of the problem to be optimized is convex. When the Pareto front exhibits a concave shape, the PBI method might lead to convergence issues.

Among the decomposition approaches, the Tchebycheff method has been the most widely used due to its simplicity and efficiency, while the PBI method offers better performance in handling more complex problem landscapes.

Nevertheless, practical applications have exposed limitations, even within 2. In an ideal setting, if each multi-objective subproblem could attain its optimal solution, then the most ideal diversity distribution could be realized. However, this is not typically the case for MOEA/D. Often, the algorithm only converges to near-optimal solutions, which fall short of implicitly determining the final population's biodiversity. As depicted in Fig. 1, within a two-dimensional objective space, given weight vectors, and where individuals A and B represent the parental generation, while individual C represents the offspring. If the slope of the ideal Pareto frontier is sufficiently steep, it results in the offspring C supplanting the parent individual A. It is visually evident that the distribution of the individual set $\{B, C\}$ is less diverse than the set $\{A, B\}$. Furthermore, once the population converges close enough to the ideal Pareto frontier, individuals in the vicinity of A cease to appear, leading to a permanent decline in population diversity distribution. This predicament arises due to MOEA/D's reliance solely on aggregated values during evolution, causing potential misguidance in solution selection. Notably, during the initial stages of evolution, solutions tend to deviate significantly from the Pareto frontier, accentuating the susceptibility to this misguidance, thereby inducing a search bias towards local regions of the Pareto frontier.

The aforementioned scenario is attributed to the fact that individuals deviating considerably from their corresponding weight vectors in the objective space can still yield superior aggregated function values. This challenge exacerbates in high-dimensional multi-objective optimization problems, given the sparsely distributed solutions and the exponential growth of hypervolume. This paper integrates existing exemplary improvement strategies to address the misleading defects that occur during the evolutionary process of decomposition-based algorithms, MOEAs, solely through the aggregation function. WAV-MOEA/D is proposed and implemented in the MOEA/D algorithm.

WVA-MOEA/D is grounded on the MOEA/D algorithm. It borrows concepts from the MOEA/D-DU algorithm and introduces a new function to help

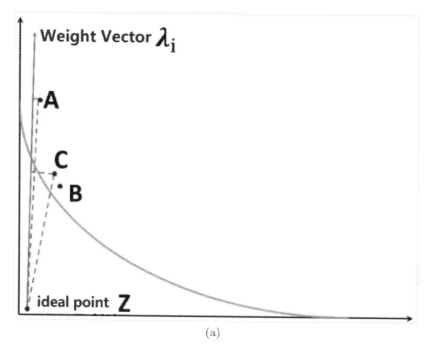

Weight Vector λ_i

A

C

B

ideal point Z

(a)

Fig. 1. Schematic representation of solution distribution in 2-M

maintain diversity during population evolution. Using the MOEA/D algorithm, Pareto individuals closest to the weight vectors are selected and marked.

The improved strategy for marking the frontier individuals of the population is illustrated in Fig. 2(a). Through a fast non-dominated sorting method, the current frontier individual set A of the population is identified. The angles between each weight vector and the lines connecting the ideal point to the frontier individuals are computed. The individual corresponding to the smallest angle is selected and marked. The marked individual is then recognized as the frontier individual chosen by that weight vector.

The improved population update method, illustrated in Fig. 2(b), involves a set of individuals corresponding to a group of weight vectors $\{\lambda_1, \lambda_2, \lambda_3\}$ with the neighboring solution set $T = \{i_1, i_2, i_3\}$. The population individuals P_{i_1} and P_{i_2} are the frontier individuals marked by weight vectors λ_1 and λ_2, respectively, while P_{i_3} is a non-marked dominated individual. For any given population individual P_{i_x}, if the offspring $OffSpring_i$ fails to replace P_{i_x} marked by weight vector λ_x, then P_{i_x} remains unchanged. Conversely, if $OffSpring_i$ replaces P_{i_x} marked by weight vector λ_x, indicating no marked individual by the weight vector, the aggregated function values of $OffSpring_i$ and P_{i_x} are compared to decide whether to update P_{i_x}. If P_{i_x} is a non-marked individual, the replacement decision is made by directly comparing the aggregated function values of $OffSpring_i$ and P_{i_x}.

In the context of the MOEA/D framework, such a selection process provides a mechanism for maintaining the diversity of the population while also ensuring convergence to the Pareto optimal solutions.

The improved process flowchart of the WVA-MOEA/D algorithm is depicted in Fig. 3.

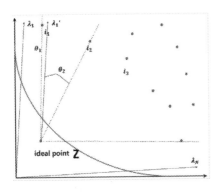 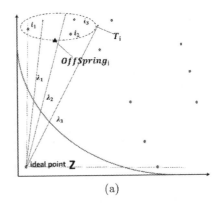

(a)

Fig. 2. Method for marking the population frontier individuals based on weight vector selection strategy in 2-M

Algorithm 2 illustrates the framework of the implementation steps of the WVA-MOEA/D algorithm.

4   Experimental Studies

In order to verify the validity of the proposed algorithm WVA-MOEA/D, comparative experiments on a set of benchmarks are conducted in this paper.

4.1   Benchmark Test Functions

To rigorously evaluate the efficacy of the improved MOEA/D strategies both qualitatively and quantitatively, we evaluate three different decomposition-based multi-objective algorithms using the MaF1-13 test suite. The MaF function test set, comprising 15 benchmark problems, captures various attributes representative of myriad real-world scenarios. These include characteristics like multimodality, disconnectedness, degeneracy, and non-separability, or situations with irregular Pareto frontiers or large-scale multi-objective optimization challenges. The term "linear features" refers to situations where the derivative of the distribution function represented by the Pareto frontier is constant. Among these, MaF1-5 problems exhibit relatively simpler Pareto Optimal Fronts (POFs), whereas MaF6-13 present more intricate frontiers. Moreover, MaF8, MaF9, and

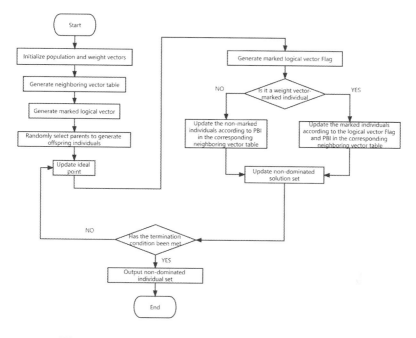

Fig. 3. The flowchart of the WVA-MOEA/D algorithm

MaF13 are unsuitable for test cases with an objective dimension of 2. Consequently, for scenarios with a 2-dimensional objective, we only experiment with the remaining 12 test functions from the MaF test set.

In this paper, Inverted Generational Distance (IGD) [14] is employed as metrics to assess algorithm performance. The computation of IGD further requires a uniformly distributed reference set derived from the ideal Pareto frontier of the multi-objective optimization problem. For the test functions, 10,000 data points are selected to serve as the reference set for IGD. A smaller IGD value suggests a higher overall algorithmic performance.

4.2   Parameter Settings

All algorithms are equipped with the same population size corresponding to the weight numbers. Specifically, the initial population for the objective dimensions of 2, 5, and 10 are set to 240, 210, and 275 respectively. Furthermore, the neighborhood size T is set to $T = \left[\frac{N}{10}\right]$. The PBI method is consistently employed as the aggregation function. Each algorithm is independently run 20 times on each test case, and the average values of the performance indicators are recorded. For objective dimensions of 2 and 5, the algorithms are iterated for more than 200 generations, with the number of evaluations set to 50,000. For an objective dimension of 10, the algorithms are iterated for more than 800 generations, with the number of evaluations set to 240,000.

Algorithm 2. WVA-MOEA/D

Input: MOP(1); a stopping criterion; N: the number of subproblems considered in
WVA-MOEA/D; a uniform spread of N weight vectors: $\lambda^1, \ldots, \lambda^N$; T: the number
of weight vectors in the neighborhood of each weight vector.

Output: EP.

1: **Initialization:**

2: Set EP to an empty set.

3: Calculate the Euclidean distances between any two weight vectors and rank the
distances for each weight vector, resulting in $B(i) = \{i_1, i_2, ..., i_T\}$.

4: Randomly generate an initial population $Pop = \{x_1, x_2, ..., x_N\}$.

5: Initialize solutions using a certain method, resulting in solutions set $Z = \{z_1, z_2, ..., z_M\}$.

6: Set a flag array $Flag = \{flag_1, ..., flag_N\}$ to maintain diversity during evolution.

7: **Update:**

8: **for** $i = 1$ to N **do**

9:     Reproduce two parent solutions from $B(i)$ to generate a new solution y.

10:     Improve y to produce an improved solution y'.

11:     Update reference point: for each dimension j from 1 to M, if $f_j(y') > Z_j$, then
set $Z_j = f_j(y')$.

12:     Update solution in $Flag$: If y' dominates any solution in $Flag$, replace that
solution with y'.

13:     Update solution in $B(i)$: If the scalar function value of y' is better, replace x_j
with y' for each j in $B(i)$.

14:     Update EP: If y' is non-dominated, add y' to EP. Remove solutions from EP
that are dominated by y'.

15: **end for**

16: **Termination:** If the stopping criterion is met, return EP. Otherwise, go back to
the Update step.

4.3   Comparisons of WVA-MOEA/D with Other Algorithm

In order to analyze the performance of WVA-MOEA/D, two different algorithms-
are compared, including MOEA/D, MOEA/D-DU. As shown in Table 1, the
comparison results of MOEA/D, MOEA/D-DU, and WVA-MOEA/D algorithms
are presented. The results in the table represent the average evaluation results
of the three algorithms running 20 times on the MaF series of test problems, as
well as their standard deviations. The best results are shown in bold. In Table 1,
it is evident that the convergence ability of WVA-MOEA/D is the best in most
test problems.

To provide a more intuitive display of the differences among the three
algorithms, Fig. 3 illustrates the true PF and PF of MOEA/D, MOEA/D-
DU, and WVA-MOEA/D in MaF1-MaF7 and MaF10-MaF11. From the provided information, it appears that the WVA-MOEA/D algorithm demonstrates
superior performance in most cases, especially when compared to MOEA/D

and MOEA/D-DU. The MOEA/D-DU algorithm excels in the MaF1, MaF5, and MaF6 test problems, but doesn't perform as well in other tests. Meanwhile, MOEA/D is consistent in converging to the optimal PF (Pareto Front). But notably, whenever MOEA/D converges to the optimal PF, so does WVA-MOEA/D, and WVA-MOEA/D even outperforms MOEA/D in the MaF4, MaF7, MaF10, and MaF11 tests. This analysis solidly indicates that among the three algorithms, WVA-MOEA/D is the most reliable and efficient. It not only consistently reaches the optimal PF but also excels in certain test problems where the other two algorithms might falter (Fig. 4).

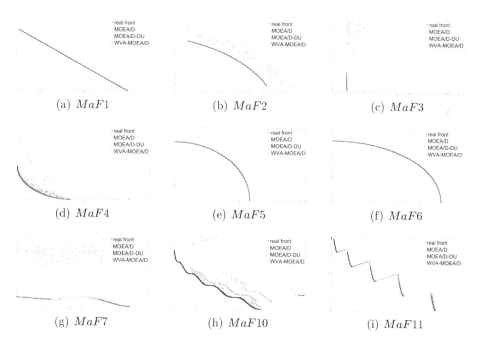

Fig. 4. Convergence curves of three multi-objective algorithms on MaF1-MaF7 and MaF10-MaF11 in 2-M

Table 1. Average IGD $\pm$ standard deviations on the MaF test problems over 20 independent runs.

Fun	m	MOEA/D	MOEA/D-DU	WVA-MOEA/D
MaF1	2	1.48E−03 ± 1.87E−07	**1.48E−03 ± 3.83E−06**	1.5274E−03 ± 2.60E−05
	5	1.46E−01 ± 4.07E−02	3.30E−01 ± 5.09E−02	**1.26E−01 ± 6.76E−04**
	10	4.83E−01 ± 2.34E−02	5.75E−01 ± 2.39E−02	**4.90E−01 ± 2.83E−02**
MaF2	2	1.00E−03 ± 2.26E−05	2.04E−03 ± 8.70E−05	**9.04E−04 ± 1.68E−05**
	5	1.11E−01 ± 4.27E−04	**1.03E−01 ± 1.73E−03**	1.13E−01 ± 7.71E−04
	10	2.62E−01 ± 1.65E−04	2.73E−01 ± 2.33E−02	**2.24E−01 ± 4.58E−03**
MaF3	2	2.31E−02 ± 1.05E−02	9.62E−01 ± 1.36E+00	**2.06E−02 ± 1.07E−02**
	5	1.04E−01 ± 7.17E−03	3.92E+00 ± 5.36E+00	**8.27E−02 ± 8.32E−03**
	10	1.41E−01 ± 6.23E−04	9.53E−02 ± 2.10E−03	**1.29E−01 ± 3.99E−03**
MaF4	2	1.93E−01 ± 3.30E−02	2.77E−01 ± 3.34E−01	**8.19E−02 ± 1.02E−02**
	5	1.24E+01 ± 1.05E+00	1.62E+02 ± 9.07E+01	**7.48E+00 ± 1.74E+00**
	10	4.56E+02 ± 2.74E+01	**3.77E+03 ± 1.68E+03**	4.35E+02 ± 1.27E+02
MaF5	2	1.06E−01 ± 4.48E−01	**7.02E−03 ± 8.73E−04**	2.06E−01 ± 6.16E−01
	5	8.20E+00 ± 1.68E+00	2.21E+00 ± 5.74E−02	**4.37E+00 ± 7.12E−01**
	10	3.02E+02 ± 9.26E−01	**8.60E+01 ± 1.18E+00**	1.58E+02 ± 2.38E+01
MaF6	2	1.77E−03 ± 5.50E−05	**1.69E−03 ± 2.39E−05**	1.82E−03 ± 9.83E−05
	5	1.24E−01 ± 1.90E−01	6.68E−02 ± 4.63E−03	**5.41E−02 ± 1.42E−01**
	10	1.25E−01 ± 2.22E−01	5.42E−02 ± 3.29E−03	**1.91E−02 ± 1.28E−03**
MaF7	2	2.91E−01 ± 2.16E−01	**4.93E−02 ± 1.48E−02**	2.46E−01 ± 2.25E−01
	5	**5.00E−01 ± 1.39E−02**	1.26E+01 ± 9.28E−01	5.42E−01 ± 3.56E−02
	10	2.97E+00 ± 6.03E−01	1.47E+01 ± 3.51E+00	**2.41E+00 ± 3.57E−01**
MaF8	5	3.02E−01 ± 9.63E−02	2.19E+02 ± 3.69E+02	**2.57E−01 ± 8.22E−02**
	10	9.15E−01 ± 7.97E−03	8.95E−01 ± 1.74E−01	**7.54E−01 ± 8.04E−02**
MaF9	5	1.57E−01 ± 5.32E−02	1.55E+00 ± 9.12E−01	**1.31E−01 ± 2.56E−02**
	10	6.83E−01 ± 1.14E+00	**6.66E−01 ± 1.66E−01**	6.73E−01 ± 1.08E+00
MaF10	2	2.32E−01 ± 3.41E−02	8.97E−01 ± 4.30E−01	**9.10E−02 ± 3.16E−02**
	5	8.77E−01 ± 4.52E−02	9.51E−01 ± 1.73E−01	**8.25E−01 ± 5.71E−02**
	10	1.73E+00 ± 7.99E−02	**1.05E+00 ± 3.56E−02**	1.68E+00 ± 4.52E−02
MaF11	2	4.41E−02 ± 3.52E−03	**1.13E−02 ± 8.64E−04**	3.55E−02 ± 6.00E−03
	5	8.21E−01 ± 4.55E−02	5.95E−01 ± 2.35E−02	**5.59E−01 ± 5.31E−02**
	10	1.85E+00 ± 1.49E−02	**1.15E+00 ± 5.81E−02**	1.66E+00 ± 1.18E−01
MaF12	2	6.62E−02 ± 7.33E−02	**1.66E−02 ± 1.50E−03**	4.90E−02 ± 6.35E−02
	5	1.64E+00 ± 7.55E−02	1.18E+00 ± 2.54E−02	**1.16E+00 ± 2.53E−02**
	10	8.78E+00 ± 2.42E−01	**4.87E+00 ± 4.70E−02**	4.89E+00 ± 1.17E−01
MaF13	5	1.92E−01 ± 9.57E−02	2.00E−01 ± 1.19E−02	**1.45E−01 ± 1.28E−02**
	10	9.81E−01 ± 9.09E−02	**3.11E−01 ± 9.87E−03**	4.80E−01 ± 6.36E−02

5   Conclusion

While the MOEA/D algorithm has indeed streamlined the solution process for multi-objective problems and has exerted strong evolutionary pressure promoting population convergence to optimal solutions, it manifests evident limitations when confronting high-dimensional multi-objective tasks or those characterized by intricate Pareto frontier properties. Recognizing the limitations of decomposition-based MOEAs in navigating high-dimensional multi-objective challenges and those with complex frontiers, this paper unveils the WVA-MOEA/D algorithm. This innovation emphasizes the imperative to evaluate solutions not merely based on their aggregation function values, but also by their status as delineated by weight vectors, thereby enriching solution diversity throughout evolutionary iterations. Additionally, we integrate a refined strategy-inspired by the MOEA/D framework-to identify Pareto frontier individuals within the population via weight vectors. Empirical assessments suggest that, relative to the original MOEA/D, our enhanced technique exhibits a more harmonious equilibrium between convergence and diversity in objective space for multi-objective optimization scenarios with fewer objectives in the majority of tested functions. Nonetheless, in isolated test instances, the augmented strategy's performance slightly trailed that of the MOEA/D. For high-dimensional multi-objective optimization scenarios, our advanced method either surpassed or paralleled the MOEA/D in performance across nearly all test functions.

Yet, the weight vector-centric enhancement proposed herein occasionally falters in realizing optimal Pareto frontier individual distributions for multi-objective optimization challenges characterized by non-standard Pareto frontiers, given its susceptibility to the weight vector distribution. Hence, tailoring the weight vector distribution to accommodate diverse irregular Pareto frontiers remains both a formidable challenge and an essential research frontier.

Acknowledgements. This work is supported by the Open Research Project of the Hubei Key Laboratory of Intelligent Geo-Information Processing (Grant No. KLIGIP-2021B04).

References

1. Cheng, R., Li, M., Tian, Y., et al.: A benchmark test suite for evolutionary many-objective optimization. Complex Intell. Syst. **3**(1), 67–81 (2017)
2. Coello, C.A.C., Lamont, G.B., Veldhuizen, D.A.V.: Evolutionary Algorithms for Solving Multi-objective Problems. Springer, New York (2007). https://doi.org/10.1007/978-0-387-36797-2
3. Cohon, J.L.: Multi-objective Programming and Planning. Courier Corporation (2013)
4. Ishibuchi, H., Setoguchi, Y., Masuda, H., et al.: Performance of decomposition-based many-objective algorithms strongly depends on pareto front shapes. IEEE Trans. Evol. Comput. **21**(2), 169–190 (2017)

5. Jiang, S., Yang, S.: A strength pareto evolutionary algorithm based on reference direction for multiobjective and many-objective optimization. IEEE Trans. Evol. Comput. **21**(3), 329–346 (2017)
6. Koski, J.: Multicriterion optimization in structural design. Technical report, DTIC Document (1981)
7. Li, H., Zhang, Q.: Multiobjective optimization problems with complicated pareto sets, MOEA/D and NSGA-II. IEEE Trans. Evol. Comput. **13**(2), 284–302 (2009)
8. Lin, J., He, C., Cheng, R.: Adaptive dropout for high-dimensional expensive multiobjective optimization. Complex Intell. Syst. **8**(1), 271–285 (2022)
9. Liu, H.L., Gu, F., Zhang, Q.: Decomposition of a multiobjective optimization problem into a number of simple multiobjective subproblems. IEEE Trans. Evol. Comput. **18**(3), 450–455 (2014)
10. Osyczka, A.: An approach to multicriterion optimization problems for engineering design. Comput. Methods Appl. Mech. Eng. **15**(3), 309–333 (1978)
11. Steuer, R.E.: Multiple Criteria Optimization: Theory, Computation, and Application. Krieger Pub. Co. (1989)
12. Yuan, Y., Xu, H., Wang, B., et al.: Balancing convergence and diversity in decomposition-based many-objective optimizers. IEEE Trans. Evol. Comput. **20**(2), 180–198 (2016)
13. Zhang, Q., Li, H.: MOEA/D: a multiobjective evolutionary algorithm based on decomposition. IEEE Trans. Evol. Comput. **11**(6), 712–731 (2007)
14. Zitzler, E., Thiele, L., Laumanns, M., et al.: Performance assessment of multiobjective optimizers: an analysis and review. IEEE Trans. Evol. Comput. **7**(2), 117–132 (2003)

A Two-Operator Hybrid DE for Global Numerical Optimization

Xiangping Li[(✉)] and Yingqi Huang

College of Information Engineering, Guizhou University of Traditional Chinese Medicine,
Guiyang 550025, China
49033548@qq.com

Abstract. Solving single objective real-parameter problem is still a challenging task. In this paper, an effective and efficient self-adaptation framework is proposed, called ToHDE, which is hybrid with CMA-ES to improve the performance. The algorithm uses two mutation strategies with linear weighted parameter to balance the exploration and exploitation. Moreover, a two-stage population size reduction and a local research are used to increase the capability of ToHDE. We evaluated the performance of ToHDE on the IEEE CEC2014 benchmark suite and compared it with six state-of-the-art peer DE variants. The statistical results show that ToHDE is competitive with the compared methods.

Keywords: Differential Evolution · Parameter adaptation · Global numerical optimization

1 Introduction

Differential evolution (DE), firstly proposed by Storn and Price in 1995 [1, 2], is a very popular evolutionary algorithm (EA) and has exhibited remarkable performance in a wide variety of problems with different characteristics. Similar to other evolutionary algorithms [3], DE is a population-based algorithm which has demonstrated effective stochastic search mechanism [4]. In DE, the population space is represented as a set of individuals being called target vectors. During the evolutionary processing, a trial vector is generated through the mutant and crossover operators. Afterward, the trial vector competes with its corresponding target vector for survival according to their fitness.

Although DE has proved its outperforming ability, many theoretical analyses and experimental studies proved that it is still quite dependent on the settings of the trial vector generation strategy and control parameters. Moreover, the premature convergence occurs due to loss of diversity when the population converges to local optimum. On the other hand, stagnation occurs when the DE algorithm is unable to generate a better offspring and the population has not yet converged. That is to say, the algorithm has not found the desired solution, and the diversity is not poor.

Based on the above considerations, in this research, a novel DE, referred as ToHDE, is presented, which includes a two-operator mixed mutation strategy and is hybrid with

© The Author(s), under exclusive license to Springer Nature Singapore Pte Ltd. 2024
L. Pan et al. (Eds.): BIC-TA 2023, CCIS 2061, pp. 131–141, 2024.
https://doi.org/10.1007/978-981-97-2272-3_10

CMA-ES [5]. Additionally, a linear weighted strategy, a two-stage population size reduction and a local search are applied to improve the search capability of ToHDE. In the proposed algorithm, the linear weighted strategy and the two-stage population size reduction are to balance the exploration, and exploitation, and the local search is to increase the exploitation capability of the proposed algorithm at the later stages. In particular, inspired by Ref. [6], a cosine perturbation related to the number of function evaluations is used to determine the probability parameter. ToHDE has been evaluated on 30 benchmark test functions of IEEE CEC2014 and is compared with six other state-of-the-art DE variants.

The rest of the paper is organized as follows. Section 2 briefly describes the classical DE paradigm and the related work on DE. ToHDE is presented in Sect. 3. The experimental results and analyses of those results are presented in Sect. 4 and Sect. 5 concludes the paper.

2 Related Work

DE is a population-based heuristic global search algorithm. A typical DE consists of four basic operators: initialization, mutation, crossover and selection. Among the four steps, initialization is performed only once at the start of a DE running while the other three are repeated in every generation. The generations are continued until a termination criterion is met, which in practice can be the exhaustion of a maximum number of Function Evaluations. Although the rand/1 mutation was proposed in the original work, most recent DE approaches often use current-to-pbest/1 strategy, which was originally proposed in JADE [7]. The current-to-pbest/1 strategy works as follows:

$$V_I^G = X_i^G + F_i^G \cdot \left(X_{pbest}^G - X_i^G\right) + F_i^G \cdot \left(X_{r_1}^G - X_{r_2}^G\right). \tag{1}$$

where pbest is an index of one of the $p*100\%$ best individuals, r1, r2 are randomly chosen indexes from the population and scaling factor F is usually in the range [0, 1]. Indexes pbest, r1 and r2 are generated different from i and each other.

On the next step, the crossover operator is applied to each pair of X_i^G and V_i^G to create a trial vector. The binomial crossover can be defined as follows:

$$u_{j,i}^G = \begin{cases} v_{j,i}^G, & if\,(rand \le CR)\,or\,j = j_{rand} \\ x_{j,i}^G, & otherwise, \end{cases} \tag{2}$$

where CR is crossover rate, which is a user-specified constant within the range [0, 1], rand is a randomly selected from the range [0,1], and j_{rand} is a uniformly distributed random integer in [1, D].

Finally, a one-to-one selection operator is used to determine the survival of the fittest among the target individual and its corresponding trial individual as follows:

$$X_i^{G+1} = \begin{cases} U_i^G, & if\,f(U_i^G \le f(X_i^G)) \\ X_i^G, & otherwise \end{cases} \tag{3}$$

There are many lots of studies which tried to improve DE via designing new mutation operators or integrating multiple operators. Yong Wang et al. [8] presented a surrogate-assisted DE with region division to solve expensive optimization problems. Li et al. [9] used the keenness to characterize the sharpness of the fitness landscape and predict the performance of DE. Wang et al. [10] presented a feature clustering-assisted feature selection method with niching-based DE to search for multiple optimal feature subsets. Sun et al. [11] introduced a novel Gaussian mutation operator and a modified "DE/rand/1" to collaboratively produce new mutant vectors. Liang et al. [5] proposed a novel DE algorithm based on local fitness landscape, where fitness landscape information is obtained to guide the selection of mutation operators. Tian et al. [12] proposed a stochastic mixed mutation strategy by incorporating a cosine perturbation. In addition to the above mentioned, there are other similar documents, such as CoDE [13], LBLDE [14] and SRADE [15].

Some other researchers have developed new approaches to tune the control parameters. Brest et al. [16] encodes control parameters F and CR into the individuals and adjusted by introducing two new parameters $\tau 1$ and $\tau 2$. Qin et al. [17] proposed a self-adaptive DE (SaDE) which focuses on control parameters and trial vector generation strategy adaptation according to the previous information. Tanabe and Fukunaga [18] introduced a success-history based parameter adaptation scheme to revise the settings of both the scaling factor F and crossover rate CR in JADE [13]. Under the framework in Ref. [18], Tanabe and Fukunaga [19] further combined linear population size reduction. The NL-SHADE-RSP approach [20] is a further development of LSHADE [19], which combines several novel parameter control techniques.

3 The Proposed Approach

In this section, the proposed algorithm, namely ToHDE, is briefly described.

3.1 Linear Weighted Two-Operator Mutation Strategy

In mutation operator, the base individual can represent the center point of the searching area, the difference vector is applied to set the searching direction, and the scale factor is employed to control the step size. The strategy "DE/current-to-pbest/1" in [7] has characteristic of greed and benefit from the fast convergence by incorporating the best solution information in the evolutionary search. However, despite the archive, the best solution information may also cause problems such as premature convergence due to the resultant reduced population diversity. Inspired by [21], we design an improved weighted two-operator mutation strategy by using "DE/current-to-pbest-w/1" and "DE/pbad-to-pbest/1" and incorporating the cosine perturbation into the probability parameter σ, as following:

$$\begin{cases} V_i^G = X_i^G + F_{w,i}^G \cdot \left(X_{pbest}^G - X_i^G \right) + F_i^G \cdot \left(X_{r_1}^G - X_{r_2}^G \right), & \text{if } rand < \sigma \\ V_i^G = X_i^G + F_i^G \cdot \left(X_{pbest}^G - X_{pbad}^G \right), & \text{otherwise} \end{cases} \quad (4)$$

where F_w is calculated:

$$F^G_{w,i} = F^G_i * (0.8 + 0.4 * \left(\frac{\text{nfes}}{\text{max nfes}} \right)). \quad (5)$$

In the equation, the variable *nfes* denotes the current number of function evaluations, and *maxnfes* is the maximum number of function evaluations. The aim of the presented weighted mutation strategy is to apply a smaller factor F_w to multiply difference of vectors in which x^G_{pbest} appears at early stages of the evolutionary process, while in later stages Fw is equal to scaling factor F. In the first half of evolving process, the proposed algorithm focuses on improving the exploration ability and avoids the premature convergence. A small factor Fw might reduce the influence of the vector X^G_{pbest} on the direction of evolution. Obviously, the mutation strategy "DE/pbad-to-pbest/1" utilizes not only the good solution information but also the information of the bad solution to balance the exploration and exploitation.

Moreover, a cosine perturbation is applied to enhance the robustness and capability of jumping out of local optimum. The difference from [21] is that the current number of function evaluations (*nfes*) replaces the current generation g. The updated reason is that the maximal number of iterations is difficult to determine in advance at most of the time. The probability parameter σ is calculated as following:

$$\sigma = a \cdot \left(1 - \frac{nfes}{maxnfes} \right) + (1 - a) \cdot (1 + \cos(2 \cdot pi \cdot freq \cdot nfes/NP)/2). \quad (6)$$

where a is a weight coefficient and *freq* is the frequency of the cosine perturbation.

3.2 Hybrid with CMA-ES

As one of evolutionary algorithms, CMA-ES has revealed its capability to efficiently solve diverse types of optimization problems [5]. Recently, several techniques which hybridize DE with CMA-ES have been proposed [22]. In this paper, we applied the hybrid framework in Ref. [22]. Due to page restriction, more details are presented in [22].

3.3 A Two-Stage Population Size Reduction

To improve the performance of ToHDE, linear population size reduction (LPSR) [19] is used in the first half of the function evaluation. The linear equation is:

$$NP_{g+1} = Round[(\frac{NP_{min} - NP_{max}}{FES_{max}}) \cdot FEs + NP_{max}]. \quad (7)$$

In the second phase, non-linear population size reduction [20] is used to improve the convergence rate, as following:

$$NP_{g+1} = Round\left[(NP_{min} - NP_{max}) \cdot NFE_r^{1-NFE_r} + NP_{max} \right]. \quad (8)$$

where $NFE_r = \frac{FEs}{FES_{max}}$ is the ratio of current number of fitness evaluations, NP_{min} is set to 4, NP_{max} is the maximum initial size of the population, FEs is the current number of fitness evaluations, and FES_{max} is the maximum number of fitness evaluations.

3.4 Local Search

To increase the exploitation ability of ToHDE at later stages of evolution, the interior-point method [24] is used to the best individual found so far, with a probability of $prob_{ls} = 0.1$, and for up to a number of fitness evaluations. If the local search is not successful in finding a better solution, $prob_{ls}$ is set to a small value, i.e., 0.001, other it remains at 0.1. The pseudo-code of ToHDE algorithm is presented in Algorithm 1.

Algorithm. 1: Pseudo-code of ToHDE

---

01. Initialize D, NP_G = Ninit, Archive $A = \emptyset$, $nfes = 0$, set maxnfes = 10000*D.
02. Create a random initial population $\vec{x}_i^G$, $\forall i, i = 1, 2, \ldots, NP$,
03. Evaluate $f(\vec{x}_i^G)$, $\forall i, i = 1, 2, \ldots, NP$, $nfes = nfes + NP$.
04. Initialize CMA parameters;
05. While nfes < maxnfes
6.   for i = 1 to NP do
7. r_i = Select from [1, H] randomly;
8. $CR_{i,G}$ = randn($MCRr_i$, 0.1);
9. $FCP_{i,G} = MFCPr_i$;
10.   if nfes < max_nfes/2
11. $F_{i,G} = 0.45 + 0.1 * rand(0, 1)$;
12.   else
13. $F_{i,G} = randc(MFr_i, 0.1)$;
14.   end
15.   calculate F_w according to Eq. (5);
16.   end
17.   [PLSHADE$_G$, PCMA$_G$] = Split(PG, FCPG);
18.   Generate donor vectors using Eq. (4) and CMA, respectively;
19. V_G = Concatenate (donor vectors);
20. U_G = Generate trial vectors (V_G, CR_G) and Evaluate U_G;
21.   Update P_G, archive A, MCR, $MFCP$;
22.   If nfes > max_nfes/2, update MF;
23.   Calculate N_{G+1} according to Eq. (7) and Eq. (8);
24.   If nfes > max_nfes/2 apply the local to the best solution.;
24.   Update CMA parameters;
25. END

---

4 Experimental Study

ToHDE is tested in 30 benchmark test functions of IEEE CEC2014. These 30 test functions can be divided into four classes: unimodal functions $f1$–$f3$; multimodal functions $f4$–$f16$; hybrid functions $f17$–$f22$; composition functions $f23$–$f30$. Firstly, the adopted test functions are carried out on 30-D and 50-D. Moreover, the performance of ToHDE is compared with other state-of-the-art DE variants. We adopt the solution error measure $f(x) - f(x^*)$, where x^* is the well-known global optimum of each benchmark function. Error values and standard deviations smaller than 10^{-8} are taken as zero. Each algorithm runs for 10,000 × D functions evaluations for each of 51 independent runs. The parameter values of ToHDE algorithm are set as follow: The initial values of all μF, μCR are both set to 0.5, *pbest* individual rate (*Pbest*), Memory size (*H*), and archive rate (*Arc_rate*) were set to (0.11, 1.4, 5). For the hybridization, Probability Variable (*FCP*) was set to 0.5, and learning rate (*c*) was set 0.8. NP_{max} and NP_{min} are set to 18*D and 4, respectively, and set $a = 0.6$ and *freq* $= 0.01$ as it is described in [22].

4.1 Statistical Result

The statistical results of ToHDE on the benchmarks with 30-D and 50-D are reported in Tables 1 and 2, respectively. Each table contains the best, worst, median, mean values and the standard deviations over the 51runs of the error value between the best fitness values found in each run and the optimal value. In the case of 30-D, the statistical results in Table 1 show the ToHDE is successful able to give the optimal solution on f1–f4, and f6–f8. Similarly, ToHDE obtains the optimal solution on 5 (f1–f3 and f7–f8) and 3 (f2–f3, f7) test functions in 50-D, respectively. The phenomenon reveals that it is difficult to obtain the optimal solution when the dimension increases.

Table 1. Results of the 30*D* benchmark function, averaged over 51 independent runs.

	Best	Worst	Median	Mean	Std
F1	0.00E+00	0.00E+00	0.00E+00	0.00E+00	0.00E+00
F2	0.00E+00	0.00E+00	0.00E+00	0.00E+00	0.00E+00
F3	0.00E+00	0.00E+00	0.00E+00	0.00E+00	0.00E+00
F4	0.00E+00	0.00E+00	0.00E+00	0.00E+00	0.00E+00
F5	2.00E+01	2.00E+01	2.00E+01	2.00E+01	6.62E−05
F6	0.00E+00	0.00E+00	0.00E+00	0.00E+00	0.00E+00
F7	0.00E+00	0.00E+00	0.00E+00	0.00E+00	0.00E+00
F8	0.00E+00	0.00E+00	0.00E+00	0.00E+00	0.00E+00
F9	9.95E−01	5.97E+00	2.98E+00	3.39E+00	1.44E+00
F10	1.67E−01	6.45E+00	1.41E+00	1.53E+00	1.32E+00
F11	8.01E+02	1.83E+03	1.25E+03	1.25E+03	2.49E+02
F12	1.47E−02	2.47E−01	1.14E−01	1.17E−01	5.67E−02
F13	4.78E−02	1.43E−01	9.45E−02	9.32E−02	2.36E−02
F14	9.70E−02	2.21E−01	1.73E−01	1.68E−01	2.63E−02
F15	9.76E−01	3.36E+00	2.24E+00	2.21E+00	4.67E−01
F16	7.12E+00	9.64E+00	8.52E+00	8.60E+00	5.78E−01

(continued)

Table 1. (*continued*)

	Best	Worst	Median	Mean	Std
F17	5.00E+01	6.83E+02	1.98E+02	2.25E+02	1.43E+02
F18	3.62E+00	1.61E+01	7.46E+00	8.37E+00	3.38E+00
F19	1.04E+00	3.80E+00	2.54E+00	2.60E+00	6.77E−01
F20	9.75E−01	5.41E+00	2.32E+00	2.45E+00	1.12E+00
F21	2.00E+02	2.65E+02	4.92E+01	8.37E+01	7.78E+01
F22	2.09E+02	1.54E+02	2.74E+01	5.74E+01	5.29E+01
F23	3.15E+02	3.15E+02	3.15E+02	3.15E+02	2.35E−13
F24	2.00E+02	2.22E+02	2.00E+02	2.03E+02	7.59E+00
F25	2.00E+02	2.03E+02	2.03E+02	2.03E+02	3.63E−01
F26	1.00E+02	1.00E+02	1.00E+02	1.00E+02	1.86E−02
F27	4.90E+02	5.14E+02	5.05E+02	5.04E+02	4.94E+00
F28	8.17E+02	8.89E+02	8.42E+02	8.41E+02	1.60E+01
F29	7.14E+02	7.35E+02	7.18E+02	7.19E+02	5.08E+00
F30	4.39E+02	2.59E+03	1.09E+03	1.30E+03	4.88E+02

Table 2. Results of the $50D$ benchmark function, averaged over 51 independent runs.

	Best	Worst	Median	Mean	Std
F1	0.00E+00	0.00E+00	0.00E+00	0.00E+00	0.00E+00
F2	0.00E+00	0.00E+00	0.00E+00	0.00E+00	0.00E+00
F3	0.00E+00	0.00E+00	0.00E+00	0.00E+00	0.00E+00
F4	0.00E+00	9.84E+01	9.81E+01	7.31E+01	4.32E+01
F5	2.00E+01	2.00E+01	2.00E+01	2.00E+01	4.80E−05
F6	0.00E+00	5.19E−01	0.00E+00	2.04E−02	1.02E−01
F7	0.00E+00	5.92E+00	0.00E+00	0.00E+00	0.00E+00
F8	0.00E+00	0.00E+00	0.00E+00	0.00E+00	0.00E+00
F9	0.00E+00	7.96E+00	4.97E+00	4.21E+00	1.66E+00
F10	2.29E+00	1.56E+01	7.28E+00	7.82E+00	2.92E+00
F11	2.04E+03	4.51E+03	3.30E+03	3.27E+03	5.70E+02
F12	3.19E−02	3.23E−01	1.15E−01	1.30E−01	6.85E−02
F13	1.20E−01	2.44E−01	1.80E−01	1.79E−01	2.94E−02

(*continued*)

Table 2. (*continued*)

	Best	Worst	Median	Mean	Std
F14	1.21E−01	2.41E−01	1.68E−01	1.70E−01	2.38E−02
F15	3.24E+00	8.42E+00	5.39E+00	5.30E+00	1.09E+00
F16	1.56E+01	1.92E+01	1.78E+01	1.78E+01	8.53E+00
F17	1.16E+02	1.58E+03	6.21E+02	6.51E+02	3.14E+02
F18	8.29+00	4.80E+01	2.14E+01	2.23E+01	9.53E+00
F19	3.85E+00	7.40E+00	5.71E+00	5.76E+00	7.47E−01
F20	2.14E+00	8.61E+00	6.11E+00	6.01E+00	1.50E+00
F21	1.23E+02	1.02E+03	5.59E+02	5.58E+02	2.09E+02
F22	2.21E+01	3.67E+02	1.44E+02	1.28E+02	7.17E+01
F23	3.44E+02	3.44E+02	3.44E+02	3.44E+02	3.19E−13
F24	2.65E+02	2.71E+02	2.68E+02	2.68E+02	1.21E+00
F25	2.05E+02	2.05E+02	2.05E+02	2.05E+02	1.45E−01
F26	1.00E+02	1.00E+02	1.00E+02	1.00E+02	2.92E−02
F27	3.00E+02	3.45E+02	3.00E+02	3.07E+02	1.46E+01
F28	1.09E+03	1.24E+03	1.14E+03	1.15E+03	3.25E+01
F29	7.76E+02	9.08E+02	7.96E+02	8.08E+02	3.63E+01
F30	7.84E+03	9.67E+03	8.56E+03	8.54E+03	3.61E+02

4.2 Comparison with Other DE Variants on CEC2014

Further, ToHDE is compared with six other DE variants: jDE [16], JADE [7], SaDE [17], CoDE [13], SHADE [18], and LSHADE [19]. The same parameter values that were suggested in the original papers were used to run the tests when they were used in the comparison against the proposed work. For space limitation, the average mean values (in bracket) over 51 runs for 50D is only reported as shown in Table 3. The best error value is marked in boldface. In order to have statistically sound conclusions, a nonparametric statistical test, called the Wilcoxon's rank-sum test for independent sample, is conducted to judge the significance of the results at a 0.05 significance level. Signs "−", "+" and "≈" indicate that the corresponding comparative DE variant is significantly worse than, better than, and similar to ToHDE, respectively. The last row of Table 3 summarizes the results of this test at 0.05 significance level comparing each algorithm versus ToHDE on the 30 functions as (w/t/l) denoted w(win)/t(tie)/l(lose).

As shown in Table 3, ToHDE clearly has the best performance among all the algorithms. It demonstrates the efficiency of the two-operator weighted mutation, hybrid with CMA-ES, two-stage population size reduction and local search. These techniques provide a good balance between the exploration of new regions and the exploitation of current best solutions.

Table 3. Mean and standard deviation of the error values for function F1-F30 averaged over 51runs@50*D*. Best entries are marked in boldface.

	jDE	SaDE	JADE	CoDE	SHADE	LSHADE	ToHDE
F1	4.57E+05	9.24E+05	1.25E+04	2.51E+05	2.14E+04	5.23E+02	**0.00E+00**
	(1.55E+05)−	(3.58E+05)−	(6.13E+03)−	(9.50E+04)−	(1.61E+04)−	(5.23E+02)−	**(0.00E+00)**
F2	1.55E−08	1.93E+03	**0.00E+00**	3.91E+01	**0.00E+00**	**0.00E+00**	**0.00E+00**
	(3.15E−08)−	(2.40E+03)−	**(0.00E+00)≈**	(7.79E+01)−	**(0.00E+00)≈**	**(0.00E+00)≈**	**(0.00E+00)**
F3	4.64E−07	3.67E+03	4.24E+03	2.75E+01	**0.00E+00**	**0.00E+00**	**0.00E+00**
	(2.45E−06)−	(2.30E+03)−	(2.51E+03)−	(7.35E+01)−	**(0.00E+00)≈**	**(0.00E+00)≈**	**(0.00E+00)**
F4	7.78E+01	7.10E+01	**1.64E+01**	3.64E+01	2.95E+01	4.39E+01	7.31E+01
	(2.41E+01)≈	(3.79E+01)+	**(3.72E+01)+**	(3.90E+01)≈	(4.57E+01)≈	(4.66E+01)−	(4.32E+01)
F5	2.04E+01	2.07E+01	2.03E+01	2.00E+01	2.01E+01	2.02E+01	**2.00E+01**
	(2.48E−02)−	(4.67E−02)−	(3.70E−02)−	(6.00E−02)−	(1.64E−02)−	(3.40E−02)−	**(4.80E−05)**
F6	1.05E+01	1.85E+01	1.53E+01	8.24E+00	3.78E+00	2.60E−01	**2.40E−02**
	(1.10E+00)−	(2.42E+00)−	(6.74E+00)−	(2.80E+00)−	(1.90E+00)−	(5.51E−01)−	**(1.02E−01)**
F7	2.61E−13	1.01E−02	2.71E−03	1.15E−03	1.31E−03	**0.00E+00**	**0.00E+00**
	(1.31E−13)−	(1.22E−02)−	(5.42E−03)−	(3.02E−03)+	(3.47E−03)−	**(0.00E+00)≈**	**(0.00E+00)**
F8	**0.00E+00**	1.03E+00	**0.00E+00**	5.97E−01	**0.00E+00**	**0.00E+00**	**0.00E+00**
	(0.00E+00)≈	(9.23E−01)−	**(0.00E+00)≈**	(8.10E−01)−	**(0.00E+00)≈**	**(0.00E+00)≈**	**(0.00E+00)**
F9	9.36E+01	8.91E+01	5.31E+01	6.84E+01	3.07E+01	1.16E+01	**4.21E+00**
	(8.68E+00)−	(1.68E+01)−	(6.86E+00)−	(1.64E+01)−	(5.12E+00)−	(2.18E+00)−	**(1.66E+00)**
F10	**2.08E−03**	1.74E+00	7.91E−03	6.96E+00	4.16E−03	4.61E−02	7.82E+00
	(4.73E−03)+	(1.23E+00)+	(9.55E−03)+	(3.79E+00)≈	(5.99E−03)+	(2.30E−02)+	(2.92E+00)
F11	5.37E+03	6.77E+03	3.86E+03	4.33E+03	3.45E+03	3.35E+03	3.27E+03
	(4.53E+02)−	(1.51E+03)−	(3.08E+02)−	(7.25E+02)−	(5.49E+02)≈	(2.64E+02)≈	(5.70+02)
F12	4.76E−01	1.06E+00	2.49E−01	**1.02E−01**	1.56E−01	2.10E−01	1.30E−01
	(3.82E−02)−	(1.08E−01)−	(3.52E−02)−	**(4.96E−02)≈**	(2.02E−02)−	(3.15E−02)−	(6.85E−02)
F13	3.83E−01	3.96E−01	3.17E−01	3.31E−01	3.03E−01	**1.63E−01**	1.79E−01
	(3.58E−02)−	(5.28E−02)−	(4.02E−02)−	(5.13E−02)−	(5.49E−02)−	**(1.95E−02)+**	(2.94E−02)
F14	3.16E−01	3.12E−01	3.08E−01	2.71E−01	3.01E−01	3.05E−01	**1.70E−01**
	(2.72E−02)−	(2.77E−02)−	(8.78E−01)−	(3.63E−02)−	(8.25E−02)−	(2.33E−02)−	**(2.39E−02)**
F15	1.18E+01	1.46E+01	7.71E+00	6.81E+00	**5.46E+00**	5.05E+00	5.30E+00
	(1.19E+00)−	(3.81E+00)−	(8.78E−01)+	(1.48E+00)−	**(5.31E−01)≈**	(4.31E−01)≈	(1.09E+00)
F16	1.84E+01	2.01E+01	**1.77E+01**	1.82E+01	1.75E+01	1.69E+01	1.78E+01
	(4.43E−01)−	(3.41E−01)−	**(4.29E−01)≈**	(1.11E+00)≈	(4.47E−01)≈	(4.33E−01)−	(8.35E−01)
F17	2.60E+04	6.43E+04	2.49E+03	1.74E+04	2.50E+03	1.53E+03	**6.51E+02**
	(1.63E+04)−	(3.32E+04)−	(6.99E+02)−	(1.11E+04)−	(7.26E+02)−	(4.36E+02)−	**(3.14E+02)**
F18	4.32E+02	6.70E+02	1.77E+02	4.99E+02	1.55E+02	9.90E+01	**2.23E+01**
	(4.53E+02)−	(5.59E+02)−	(4.50E+01)−	(6.38E+02)−	(3.84E+01)−	(1.23E+00)−	**(9.53E+00)**
F19	1.20E+01	1.70E+01	1.23E+01	6.46E+00	8.86E+00	9.00E+00	**5.76E+00**
	(2.64E+00)−	(8.08E+00)−	(5.12E+00)	(1.21E+00)−	(2.86E+00)−	(1.93E+00)−	**(7.47E−01)**
F20	4.98E+01	8.90E+02	5.90E+03	2.09E+02	2.12E+02	**1.39E+01**	6.01E+00
	(1.92E+01)−	(7.58E+02)−	(5.99E+03)−	(2.19E+01)−	(6.86E+01)−	**(4.59E+00)−**	(1.50E+00)
F21	1.14E+04	7.41E+04	1.28E+03	9.73E+03	1.34E+03	4.94E+02	**5.58E+02**
	(1.20E+04)−	(5.96E+04)−	(3.78E+02)−	(8.80E+02)−	(4.23E+02)−	(1.55E+02)≈	**(2.09E+02)**
F22	5.10E+02	5.33E+02	5.73E+02	5.91E+02	3.90E+02	**1.12E+02**	1.28E+02
	(1.32E+02)−	(1.69E+02)−	(1.59E+02)−	(1.90E+02)−	(1.89E+02)−	**(7.04E+01)≈**	(7.17E+01)

(continued)

Table 3. (*continued*)

	jDE	SaDE	JADE	CoDE	SHADE	LSHADE	ToHDE
F23	3.44E+02	3.44E+02	3.44E+02	**3.44E+02**	**3.44E+02**	3.44E+02	3.44E+02
	(3.24E−13)−	(2.89E−13)+	(2.02E−13)+	**(1.73E−13)+**	**(1.73E−13)+**	(3.86E−13)≈	(3.19E−13)
F24	2.68E+02	2.76E+02	2.75E+02	2.71E+02	2.74E+02	**2.75E+02**	2.68E+02
	(2.83E+00)≈	(2.90E+00)−	(2.04E+00)−	(2.39E+00)−	(1.80E+00)−	**(6.27E−01)+**	(1.21E+00)
F25	2.08E+02	2.19E+02	2.17E+02	2.07E+02	2.07E+02	2.05E+02	2.05E+02
	(3.43E+00)−	(8.77E+00)−	(6.60E+00)−	(1.77E+00)−	(1.77E+00)−	(3.19E−01)−	(1.45E−01)
F26	1.00E+02	1.80E+02	1.00E+02	1.04E+02	1.00E+02	**1.00E+02**	1.00E+02
	(5.52E−02)−	(4.06E+01)−	(9.07E−02)−	(1.82E+01)−	(1.17E−01)−	**(1.55E−02)+**	(2.92E−02)
F27	4.09E+02	7.59E+02	4.60E+02	5.29E+02	4.27E+02	3.29E+02	**3.07E+02**
	(5.98E+01)−	(7.28E+01)−	(8.08E+01)−	(9.27E+01)−	(5.07E+01)−	(3.30E+01)−	**(1.46E+01)**
F28	1.12E+03	1.42E+03	1.14E+03	1.17E+03	1.15E+03	**1.11E+03**	1.15E+03
	(3.83E+01)−	(8.09E+01)−	(3.68E+01)≈	(6.10E+01)−	(8.33E+01)≈	**(2.55E+01)+**	(3.25E+01)
F29	1.01E+03	1.38E+03	9.19E+02	8.97E+02	8.65E+02	**8.00E+02**	8.08E+02
	(1.66E+02)−	(2.68E+02)−	(1.72E+02)−	(1.72E+02)−	(5.50E+01)−	**(3.32E+01)+**	(3.63E+01)
F30	**8.54E+03**	1.17E+04	9.98E+03	8.92E+03	9.52E+03	8.80E+03	**8.54E+03**
	(3.66E+02)≈	(1.65E+03)−	(5.92E+02)−	(4.77E+02)−	(7.58E+02)−	(4.26E+02)−	**(3.61E+02)**
w/t/l	1/3/26	2/0/28	4/4/22	2/4/24	2/8/20	6/9/15	

5 Conclusion

Differential evolution is an efficient and robust evolutionary algorithm for global numerical optimization and has become a hotspot. However, it is not completely free from the problems of premature convergence and stagnation. In order to alleviate these disadvantages and improvethe performance of DE, we present a DE variant, called ToHDE, which use a two-operator weighted mutation strategy and hybrid with CMA-ES. In ToHDE, a cosine perturbation is incorporated into the probability parameter to enhance the capability of jumping out of the local optimal. Additionally, a two-stage population size reduction and a local search is used to enhance the exploitation. The proposed method is tested on 30 benchmark functions of IEEE CEC2014 and compared with six state-of-the-art DE variants. The experimental results show that ToHDE has potential ability to obtain high quality solutions.

References

1. Storn, R., Price, K.: Differential Evolution-A simple and efficient adaptive scheme for global optimization over continuous spaces. J. Global Optim. **11**, 341–359 (1997)
2. Liang, J., Qin, A., Suganthan, P.: Comprehensive learning particle swarm optimizer for global optimization of multimodal functions. IEEE Trans. Evol. Comput. **10**(3), 281–295 (2006)
3. Das, S., Mullick, S., Suganthan, P.: Recent advances in differential evolution–an updated survey. Swarm Evol. Comput. **27**, 1–30 (2016)
4. Hansen, N.: The CMA evolution strategy: a tutorial (2011). https://www.lri.fr/~hansen/
5. Tian, M., Gao, X.: An improved differential evolution with information intercrossing and sharing mechanism for numerical optimization. Swarm Evol. Comput. **50**, 1–21 (2019)

6. Zhang, J., Sanderson, A.: JADE: adaptive differential evolution with optional external archive. IEEE Trans. Evol. Comput. **13**(5), 945–958 (2009)
7. Wang, Y., Lin, J., Sun, G., Pang, T.: Surrogate-assisted differential evolution with region division for expensive optimization problems with discontinuous responses. IEEE Trans. Evol. Comput. **26**(4), 780–792 (2022)
8. Li, Y., Liang, J.: Keenness for characterizing continuous optimization problems and predicting differential evolution algorithm performance. Complex Intell. Syst. **9**, 1–16 (2023)
9. Wang, P., Xue, B., Liang, J.: Feature clustering-assisted feature selection with differential evolution. Pattern Recogn. **140** (2023)
10. Sun, G., Lan, Y., Zhao, R.: Differential evolution with Gaussian mutation and dynamic parameter adjustment. Soft. Comput. **23**, 1615–1642 (2017)
11. Liang, J., Li, K., Yu, K.: A novel differential evolution algorithm based on local fitness landscape information for optimization problems. IEICE Trans. Inf. Syst. **E106-D**(5), 601–616 (2023)
12. Wang, Y., Cai, Z., Zhang, Q.: Differential evolution with composite trial vector generation strategies and control parameters. IEEE Trans. Evol. Comput. **15**(1), 55–66 (2011)
13. Qiao, K., Liang, J., Qu, B.: Differential evolution with level-based learning mechanism. Complex Syst. Model. Simul. **2**(1), 25–58 (2022)
14. Qiao, K., Liang, J., Yu, K.: Self-adaptative resources allocation-based differential evolution for constrained evolutionary optimization. Knowl.-Based Syst. **235** (2022)
15. Brest, J., Greiner, S., Boskovic, B.: Self-adapting control parameters in differential evolution: a comparative study on numerical benchmark problems. IEEE Trans. Evol. Comput. **10**(6), 646–657 (2006)
16. Qin, A., Huang, V., Suganthan, P.: Differential evolution algorithm with strategy adaption for global numerical optimization. IEEE Trans. Evol. Comput. **13**(2), 398–417 (2009)
17. Tanabe, R., Fukunage, A.: Success-history based parameter adaptation for differential evolution. In: 2013 IEEE Congress on Evolutionary Computation (CEC), pp. 71–78. IEEE (2013)
18. Tanabe, R., Fukunage, A.: Improving the search performance of SHADE using linear population size reduction. In: 2014 IEEE Congress on Evolutionary Computation (CEC), pp. 1658–1665. IEEE (2014)
19. Stanovov, V., Akhmedova, S., Semenkin, E.: NL-SHADE-RSP algorithm with adaptive archive and selective pressure for CEC2021 numerical optimization. In: 2021 IEEE Congress on Evolutionary Computation (CEC), pp. 809–816. IEEE (2021)
20. Brest, J., Mauces, M., Boskovic, B.: Single objective real-parameter optimization algorithm jSO. In: 2017 IEEE Congress on Evolutionary Computation (CEC), pp. 1311–1318. IEEE (2017)
21. Mohamed, A., Hadi, A., Fattouh, A.: LSHADE with semi-parameter adaptation hybrid with CMA-ES for solving CEC2017 benchmark problems. In: 2017 IEEE Congress on Evolutionary Computation (CEC), pp. 145–152. IEEE (2017)
22. Nesterov, Y., Nemirovskii, A., Ye, Y.: Interior-Point Polynomial Algorithms in Convex Programming, vol. 13. SLAM (1994)

Reinforcement Learning-Based Differential Evolution Algorithm with Levy Flight

Xiaoyu Liu, Qingke Zhang$^{(\boxtimes)}$, Hongtong Xi, Huixia Zhang, Shuang Gao, and Huaxiang Zhang

School of Information Science and Engineering, Shandong Normal University, Jinan 250358, China
tsingke@sdnu.edu.cn

Abstract. In this paper, a reinforcement learning-based differential evolution algorithm with levy flight strategy (RLLDE) for solving optimization problems is proposed. It introduces a novel mutation mode considering search directions is proposed firstly. Secondly, a levy flight strategy is employed to enhance the exploration capability of Differential Evolution (DE). Lastly, the Q-learning method from reinforcement learning is introduced to establish a switching mechanism between two different updating modes during the mutation stage. These strategies effectively improve the algorithm's convergence speed and accuracy. RLLDE is analyzed on CEC 2017 benchmark functions to validate its optimization performance. Compared to five basic DE and eight efficient optimizers, the experimental results demonstrate that the algorithm exhibits efficient and effective performance in solving optimization problems.

Keywords: Evolutionary computation · Differential evolution · Reinforcement learning · Levy flight

1 Introduction

In real-world scenarios, complex problems with numerous solution spaces are encountered in recent years [19]. These problems involve nonlinear constraints and non-convex optimization processes, demanding substantial computational resources and incurring significant costs, which complicates the problem-solving process due to the large number of constraints and variables [23]. Furthermore, classical approaches, while capable of providing approximate optimal solutions, cannot guarantee the best solutions for real-world optimization problems [8]. Consequently, researchers have developed and proposed numerous metaheuristic optimization algorithms that have had a significant impact on solving complex problems [13].

Among numerous evolutionary algorithms, Differential Evolution (DE) [6] is a robust and straightforward optimization algorithm capable of addressing non-linear, non-differentiable, continuous space optimization problems [20]. DE is characterized by its simplicity, a small number of control parameters, and the

© The Author(s), under exclusive license to Springer Nature Singapore Pte Ltd. 2024
L. Pan et al. (Eds.): BIC-TA 2023, CCIS 2061, pp. 142–156, 2024.
https://doi.org/10.1007/978-981-97-2272-3_11

ability for parallel computation, resulting in high convergence speed. DE has proven successful in solving engineering optimization problems across various engineering domains [9].

In recent decades, significant efforts have been made to extend the applicability of Differential Evolution (DE) to various problem domains, driven by its versatile nature [1]. Some of the research directions pursued in these efforts include the development of parameter adaptation techniques based on learning from past experiences, introduction of novel crossover and mutation schemes, exploration of ensembles comprising various mutation and crossover schemes, and population resizing during the search process [5].

DE-based algorithms have demonstrated successful application in constrained, multimodal, and high-dimensional optimization problems [2]. In recent years, DE algorithms have found extensive use in solving complex real-world optimization challenges in various domains, such as general engineering design [14], wireless sensor networks [4], and electrical networks [3], among others. In these research areas, DE-based algorithms are often combined with other techniques, leveraging DE's simplicity and robust performance within complex search spaces. The cumulative progress in DE and its variants has established a class of popular and robust algorithms for diverse optimization fields. These observations provide motivation for the current work.

The previously proposed variants of DE have improved searchability and accelerated the convergence process, but they still grapple with issues like premature convergence and insufficient convergence accuracy. In this paper, we introduce an enhanced DE algorithm to address optimization problems. Leveraging the characteristics of classical DE, Reinforcement Learning (RL) from machine learning is integrated into the mutation stage, enabling more suitable update strategies to be selected by the algorithm, thus training the search agent to perform more beneficial actions. Additionally, a Levy flight strategy is added after the entire update stage to expedite the exploration process and avoid local optima. The improved DE, incorporating both reinforcement learning and Levy flight strategy, is referred to as RLLDE.

The proposed approach's exploration and exploitation capabilities are assessed through benchmark testing using the CEC 2017 benchmark functions. Comparisons are made with several existing algorithms, including five classical DE algorithms and eight advanced methods. Experimental results demonstrate that the RLLDE method outperforms the comparative algorithms in terms of exploration and exploitation capabilities.

The reminder of this paper is organized as follows. Section 2 offers a concise overview of DE. Section 3 provides a detailed description of the proposed RLLDE algorithm. Section 4 includes experiments and result analysis. Finally, Sect. 5 presents the paper's conclusion.

2   Differential Evolution Algorithm

The Differential Evolution algorithm (DE) has parameters such as population size (N), crossover rate (CR), and scaling factor (F). The algorithm consists of four stages: generation, mutation, crossover, and selection. The process begins by generating a population of NP vectors in the search space, and the population generation process can be represented using Eq. (1).

$$x_{i,j}(t) = x_{\min} + rand \times (x_{\max} - x_{\min}) \tag{1}$$

where $x_{\min}$ and $x_{\max}$ represent the lower and upper bounds of decision variables. $x_{i,j}$ denotes the values of the ith individual at dimension j, where i ranges from 1 to N, and j ranges from 1 to D. N is the size of population and D is the number of variables.

The mutation step, creates a new vector, known as mutant vectors. Mutant vectors are generated for each member of the population using Eq. (2) to Eq. (6).

(1) DE/rand/1

$$v_{i,j}(t) = x_{r1,j}(t) + F \times (x_{r2,j}(t) - x_{r3,j}(t)) \tag{2}$$

(2) DE/rand/2

$$v_{i,j}(t) = x_{r1,j}(t) + F \times (x_{r2,j}(t) - x_{r3,j}(t)) + F \times (x_{r4,j}(t) - x_{r5,j}(t)) \tag{3}$$

(3) DE/best/1

$$v_{i,j}(t) = x_{best,j}(t) + F \times (x_{r1,j}(t) - x_{r2,j}(t)) \tag{4}$$

(4) DE/best/2

$$v_{i,j}(t) = x_{best,j}(t) + F \times (x_{r1,j}(t) - x_{r2,j}(t)) + F \times (x_{r3,j}(t) - x_{r4,j}(t)) \tag{5}$$

(5) DE/current_to_best/1

$$v_{i,j}(t) = x_{best,j}(t) + F \times (x_{best,j}(t) - x_{i,j}(t)) + F \times (x_{r1,j}(t) - x_{r2,j}(t)) \tag{6}$$

where $r1$, $r2$, $r3$, $r4$, and $r5$ are randomly selected from the range $[1, N]$, and they are different from the value of i. $x_{best,j}(t)$ represents the jth dimension of best individual in the current population.

After mutation, DE applies a crossover operator, resulting in a trial vector $u_{i,j}(t+1)$ as shown in Eq. (7).

$$u_{i,j}(t+1) = \begin{cases} v_{i,j}(t+1), if \ (rand < CR \ or \ j = jrand) \\ x_{i,j}(t+1), otherwise \end{cases} \tag{7}$$

where $rand$ is a uniformly distributed random number between 0 and 1, generated independently for each j, and $jrand$ is a random integers between 1 and D. The use of $jrand$ results in the trial vector $u_{i,j}(t+1)$ being distinct from its target vector $x_{i,j}(t+1)$. Then the selection operator determines which one, either the target vector $x_i(t+1)$ or the trial vector $x_i(t+1)$, is chosen to proceed to the next generation based on their relative quality.

$$\boldsymbol{x}_i(t+1) = \begin{cases} \boldsymbol{u}_i(t), if \ (f(\boldsymbol{u}_i(t)) < f(\boldsymbol{x}_i(t)))) \\ \boldsymbol{x}_i(t), otherwise \end{cases} \tag{8}$$

Algorithm 1: RLLDE

Input: $MaxFEs$, the size of the particle's population N
Output: x_{gbest} and $gbestfitness$
1 Initialization the particle's population $i \leftarrow 1, 2, ..., popsize$ and calculate all $fitness$;
2 Initialize the reward table and the $Qtable$;
3 $FEs = FEs + N$;
4 **while** $FEs < MaxFEs$ **do**
5   **for** $i \leftarrow 1$ to N **do**
6   **if** $rand < 0.5$ **then**
7   **if** $Qtable(1,1) > Qtable(1,2)$ **then**
8   Update the mutant vector by using Eq. (13);
9   **else**
10   Update the mutant vector by using Eq. (9);
11   **else**
12   **if** $Qtable(2,1) > Qtable(2,2)$ **then**
13   Update the mutant vector by using Eq. (13);
14   **else**
15   Update the mutant vector by using Eq. (9);
16   Boundary control;
17   Execute the mutation stage using Eq. (7);
18   Calculate the $f(\boldsymbol{u}_i)$;
19   $FEs = FEs + 1$;
20   Perform a greedy selection using Eq. (8);
21 Update the $Qtable$;

---

3 The Proposed RLLDE Algorithm

3.1 Levy Flight

Levy flight strategies have found widespread application in optimization problems and optimal search, offering substantial prospects for further development. Introducing the Levy flight strategy into the DE update formula disrupts the current best solution, enhancing its local escape capability. The improved formula significantly reduces the risk of individuals getting trapped in local optima while still fully utilizing local search capabilities. The mutant vector is calculated as Eq. (9).

$$\boldsymbol{v}_i(t) = \boldsymbol{x}_i(t) + F \times L(\lambda) \times (\boldsymbol{x}_{\text{gbest}} - \boldsymbol{x}_{\text{r1}}(t)) \tag{9}$$

where $L(\lambda)$ represents the random search path of Levy flight and its distribution equation is as follows.

$$L(s, \lambda) = \mu \sim s^{-\lambda}, \lambda \in (1, 3] \tag{10}$$

where s represents the random step length of Levy flight, which can be expressed as Eq. (11).

$$s = \frac{\mu}{|v|^{1/\beta}} \tag{11}$$

where u and v follow a random normal distribution, which can be expressed as Eq. (12).

$$\begin{cases} \mu \sim N\left(0, \sigma_\mu^2\right), \sigma_\mu = \left(\frac{\Gamma(1+\beta) \times \sin\left(\frac{\beta}{2} \times \pi\right)}{\Gamma((1+\beta)/2) \times \beta \times 2^{(\beta-1)/2}} \right) \\ v \sim N\left(0, \sigma_v^2\right), \sigma_v = 1 \end{cases} \tag{12}$$

where $\beta = 3/2$.

Furthermore, RLLDE employs an enhanced mutant vector generate formula based on the DE/rand/1 algorithm, which can be computed as Eq. (13).

$$\boldsymbol{v}_i(t) = \boldsymbol{x}_{r1}(t) + F \times sign\left(f\left(\boldsymbol{x}_{r2}(t)\right) - f\left(\boldsymbol{x}_{r3}(t)\right)\right) \times \left(\boldsymbol{x}_{r2}(t) - \boldsymbol{x}_{r3}(t)\right) \tag{13}$$

where $f\left(\boldsymbol{x}_{r2}(t)\right)$ and $f\left(\boldsymbol{x}_{r3}(t)\right)$ represent the fitness values of individuals r2 and r3, respectively. This approach allows for a more accurate search direction than DE/rand/1, thereby further enhancing the algorithm's convergence performance.

3.2   Reinforcement Learning

Machine learning is widely used for optimization [24]. It encompasses four categories: supervised, unsupervised, semi-supervised, and reinforcement learning (RL). In RL, agents learn optimal actions in complex environments, using training experience for future actions. RL has model-free and model-based approaches. Model-free divides into value-based and policy-based methods. Value-based RL is suitable for coordination with meta-heuristic algorithms, offering flexibility. Here, agents learn from their actions and environment experience, using rewards and penalties to measure success and make decisions.

To help individuals adjust their update modes more effectively, Q-Learning is introduced in RL for switching between update modes [26]. Q-Learning is a representative value-based RL algorithm. Individuals utilize Q-Table values to make decisions regarding update modes [15]. The Q-table is updated through a reward-penalty mechanism. By evaluating the benefit of various states, individuals select the update mode with the highest Q-values for their next step. After each step, individuals receive rewards or penalties based on their actions. This helps the agent decide between exploration and exploitation.

In this work, the reward table assigns either a positive $(+1)$ or negative (-1) value to each state-action pair. The Q-Table serves as the agent's experience, initially set to zero for all entries. Consequently, individuals update the Q-Table using the Eq. (14) to prepare it for the next iteration [22].

$$Q_{s_t,a_t}(t+1) = Q_{s_t,a_t}(t) + \lambda \times \left(r(t+1) + \gamma \times \max Q_{s_{t+1},a_t}(t) - Q_{s_t,a_t}(t)\right) \tag{14}$$

where s_t and s_{t+1} represent the current and the next state, $Q(t)$ and $Q(t+1)$ are the current Q-value and pre-estimated Q-value for the next state s_{t+1}, and at denotes the current action. λ and γ are the learning rate and discount factor, both in the range of 0 to 1. The learning rate determines the learning speed and convergence of the algorithm, while the discount factor influences how the algorithm values future rewards. $r(t + 1)$ signifies the immediate reward or penalty the agent receives for the current action.

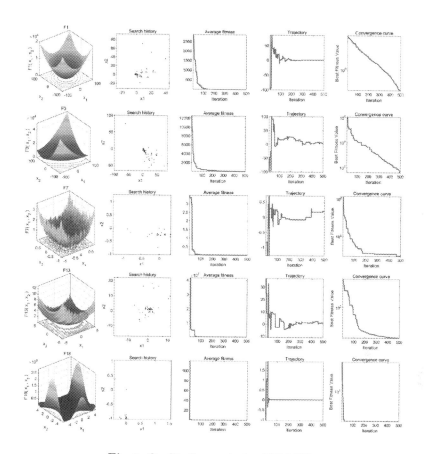

Fig. 1. Qualitative analysis of RLLDE.

In each iteration, the agent employs Eq. (9) to evaluate state-action pairs before selecting the best action with the highest potential for approaching the optimal solution. This RL strategy enables a switching mechanism between updating modes to identify the most suitable decision scheme.

The pseudocode of the proposed RLLDE algorithm is presented in Algorithm 1.

4   Experiments and Results

In this section, the advantages and distinctive features of the RLLDE algorithm are illustrated through a series of experiments. For the sake of fairness and reproducibility, all experiments conducted in this paper were executed within a consistent environment. In this experiment, the performance of the proposed RLLDE is evaluated using a set of 30 functions from CEC 2017 [25]. These functions are categorized into four groups: unimodal functions (F_1–F_3), multimodal functions (F_4–F_{10}), hybrid functions (F_{11}–F_{20}), and composition functions (F_{21}–F_{30}).

4.1 Qualitative Analysis of RLLDE

In this subsection, we employ the well-established set of 23 benchmark functions [27] to structure four experiments. These experiments are designed to provide a qualitative analysis of the RLLDE algorithm with a focus on agents, dimensional particles, fitness values, and iteration curves. In these experiments, the population size of RLLDE is fixed at 40, the number of iterations is set to 500, and 30 iterations are run independently in parallel.

Figure 1 demonstrates the outcomes of the RLLDE algorithm for the four qualitative experiments. In the first column of Fig. 1, the plots depict the location distribution of all solutions for each benchmark function, which serves as the solution space, where RLLDE seeks the optimal solution.

The second column of Fig. 1 represents the two-dimensional location distribution of RLLDE's search history. It's evident that a limited number of historically optimal solutions are scattered throughout the solution space, with the majority concentrated around the global optimal solution. This suggests that the RLLDE algorithm has the capability to quickly approximate the optimal solution within the solution space and transition to the exploitation stage sooner, thereby enhancing solution accuracy.

The third column in Fig. 1 documents the optimal fitness values achieved by the RLLDE algorithm at each iteration. Notably, the fitness values exhibit regular fluctuations in each iteration, which can be attributed to the dynamic adjustments in search positions made by the algorithm's search agents during their updates. This pattern suggests that the RLLDE algorithm is proficient at actively seeking the optimal solution through agent updates, aligning with the algorithm's intended design.

The fourth column of Fig. 1 tracks the evolution of the first dimension of RLLDE's agents over the iterations. Notably, in the early stages of the RLLDE algorithm, the agents take very long search steps, which prove advantageous for escaping local optima and discovering the global optimal solution. The reinforcement learning and levy flight strategies are instrumental in later iterations of the algorithm, leading to shorter search steps, thereby enhancing the accuracy of the optimal solution. Additionally, the dimensions of the agents exhibit less oscillation with an increasing number of iterations, facilitating rapid convergence during the exploitation phase of the algorithm.

The final column of Fig. 1 illustrates the overall convergence curve of RLLDE over iterations. Among most of the tested functions, the algorithm consistently enhances the solution quality as the number of iterations increases and effectively avoids getting stuck in local optima. It is noteworthy that the RLLDE algorithm demonstrates its efficacy not only on simple, single-peaked test functions like F_1

and F_3 but also on complex, multi-peaked, and composite functions such as F_7 and F_{13}. This is achieved through the synergistic action of the reinforcement learning and Levy flight strategies, enabling incremental search and development during iterations and minimizing the risk of falling into local optima.

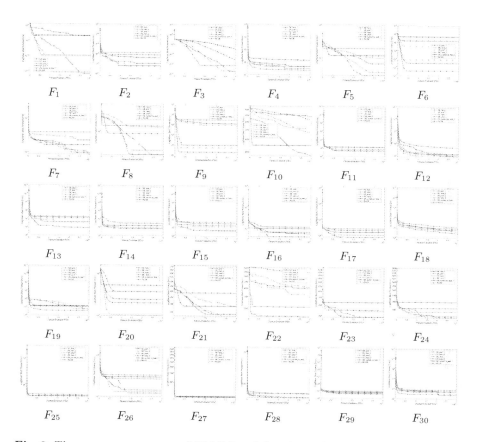

Fig. 2. The convergence curves of RLLDE and five classical DE obtained on the CEC 2017 benchmarks.

4.2   Performance Comparison Experiment of RLLDE

In this section, we conduct a comprehensive comparison of the RLLDE algorithm with five classical DE algorithms and eight advanced algorithms. The aim is to showcase the superiority of our proposed algorithm compared to its peer algorithms. It's worth noting that these algorithms are rigorously tested against a complete test set in the experiments to thoroughly highlight their performance advantages and distinctive characteristics.

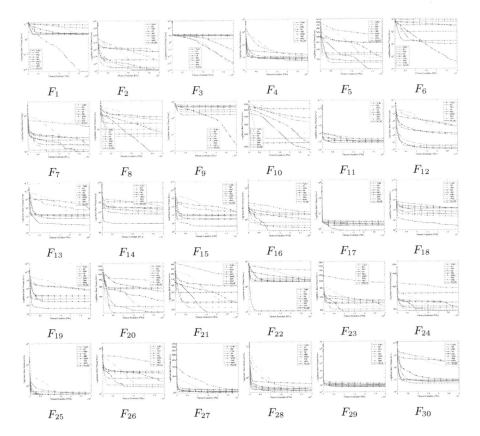

Fig. 3. The convergence curves of RLLDE and eight efficient optimizers obtained on the CEC 2017 benchmarks.

Performance Comparison with Five Classic de Algorithms. In this subsection, RLLDE is compared with five classic DE algorithms, which include DE/rand/1, DE/rand/2, DE/best/1, DE/best/2, and DE/current_to_best/1. In all versions of the Differential Evolution (DE) algorithm, CR (Crossover Rate) is consistently set to 0.9, and F (Scaling Factor) is set to 0.5.

Table 1 reveals that RLLDE outperforms other methods by achieving the lowest average values and standard deviations on 13 out of the 30 test functions. This demonstrates both high accuracy and stability. The performance on the CEC 2017 benchmark functions suggests that RLLDE excels in exploration and effectively avoids falling into local optima, due to the reinforcement learning (RL) strategy it incorporates.

Table 1. Results of RLLDE and five classical DE algorithms on CEC 2017 benchmark functions.

No.	Item	DE/best/1	DE/best/2	DE/rand/1	DE/rand/2	DE/current/1	RLLDE	No.	Item	DE/best/1	DE/best/2	DE/rand/1	DE/rand/2	DE/current/1	RLLDE
F_1	mean	2.62E-10	2.15E-21	4.31E+03	3.16E+03	1.65E+03	**3.86E-22**	F_{16}	mean	1.04E+03	6.76E+02	7.14E+02	1.22E+03	6.01E+02	**4.32E+02**
	std	5.85E-10	9.59E-22	4.15E+03	2.85E+03	2.61E+03	4.96E-22		std	2.10E+02	5.68E+02	4.67E+02	1.77E+02	3.74E+02	1.65E+01
	h			≈	+	+			h			+	+	+	
F_2	mean	1.56E-04	3.25E-06	1.07E+23	2.57E+18	4.42E+12	1.17E+01	F_{17}	mean	7.21E+02	4.23E+02	1.20E+02	4.43E+02	2.56E+02	6.26E+01
	std	2.03E-04	7.02E-06	2.39E+23	5.72E+18	8.31E+12	2.61E+01		std	2.63E+02	2.19E+02	8.12E+01	1.73E+02	5.48E+01	6.54E+01
	h			+	+	+			h			+	≈	+	
F_3	mean	1.08E-21	1.91E-27	1.35E-04	1.51E-05	1.81E-28	4.89E-16	F_{18}	mean	6.81E+03	1.51E+03	2.93E+03	1.18E+04	4.04E+03	6.85E+02
	std	1.44E-21	8.87E-28	2.97E-04	3.14E-05	2.00E-28	6.10E-16		std	1.44E+04	9.70E+02	2.38E+03	7.42E+03	5.09E+03	6.96E+02
	h				+				h			≈	+		
F_4	mean	3.15E+01	4.77E+01	4.24E+01	2.40E+01	1.49E+02	1.88E+01	F_{19}	mean	2.34E+02	1.93E+02	3.61E+01	2.99E+01	1.57E+02	1.56E+01
	std	3.53E+01	2.47E+01	3.48E+01	2.68E+01	5.05E+01	2.51E+01		std	1.22E+02	9.97E+01	4.85E+01	1.86E+01	6.96E+01	7.69E+00
	h				+				h			+	≈	+	
F_5	mean	1.89E+02	9.70E+01	3.73E+01	2.97E+01	1.18E+02	6.69E+01	F_{20}	mean	5.37E+02	3.74E+02	1.03E+02	3.74E+02	2.45E+02	2.03E+02
	std	5.12E+01	3.84E+01	1.79E+01	8.64E+00	3.07E+01	5.30E+01		std	2.27E+02	2.07E+02	1.20E+02	2.45E+02	1.07E+02	2.71E+02
	h					+			h			≈	≈	+	
F_6	mean	4.69E+01	3.38E+00	8.14E-02	3.78E-03	1.46E+01	3.19E-02	F_{21}	mean	3.87E+02	2.88E+02	2.27E+02	2.36E+02	2.86E+02	2.37E+02
	std	8.77E+00	8.53E-01	1.50E-01	8.16E-03	6.81E+00	5.50E-02		std	2.80E+01	3.10E+01	7.92E+00	1.63E+01	2.71E+01	3.43E+00
	h			≈		+			h			≈	≈	+	
F_7	mean	3.90E+02	2.26E+02	1.00E+02	8.68E+01	1.66E+02	9.60E+01	F_{22}	mean	2.12E+03	3.22E+03	4.07E+03	6.40E+03	4.08E+03	1.50E+03
	std	9.03E+01	9.84E+01	5.89E+01	5.75E+01	5.05E+01	5.81E+01		std	1.90E+03	2.31E+03	2.99E+03	1.42E+03	3.66E+03	3.13E+03
	h					+			h			≈	+	+	
F_8	mean	1.36E+02	1.28E+02	3.68E+01	2.91E+01	8.97E+01	2.89E+01	F_{23}	mean	6.62E+02	4.83E+02	3.80E+02	3.85E+02	4.99E+02	3.83E+02
	std	2.04E+01	7.74E+01	8.35E+00	3.94E+00	2.50E+01	8.02E+00		std	1.04E+02	6.40E+01	9.08E+00	1.22E+01	4.52E+01	3.48E+00
	h			+	≈	+			h			≈	+	+	
F_9	mean	3.03E+03	7.90E+02	5.18E+00	2.02E+00	1.23E+03	9.60E+00	F_{24}	mean	7.84E+02	5.51E+02	4.49E+02	4.55E+02	6.32E+02	4.72E+02
	std	5.25E+02	1.03E+03	4.11E+00	2.63E+00	8.42E+02	1.92E+01		std	8.11E+01	3.27E+01	1.15E+01	8.20E+00	5.79E+01	1.86E+01
	h			≈		+			h			≈	≈	+	
F_{10}	mean	3.51E+03	5.17E+03	4.38E+03	6.14E+03	6.79E+03	2.27E+03	F_{25}	mean	3.79E+02	3.92E+02	3.88E+02	3.77E+02	4.30E+02	3.88E+02
	std	3.53E+02	1.90E+03	2.31E+03	1.73E+03	5.40E+03	7.02E+02		std	6.20E+01	8.00E+00	8.29E-01	1.99E+00	3.60E+01	6.20E+01
	h			+	+	+			h			≈	≈	≈	
F_{11}	mean	1.85E+01	1.81E+02	2.81E+01	1.80E+01	1.69E+02	5.42E+01	F_{26}	mean	3.33E+03	2.69E+03	1.31E+03	1.21E+03	3.00E+03	1.44E+03
	std	6.21E+01	6.45E+01	2.75E+01	5.44E+00	1.09E+02	5.44E+01		std	1.90E+03	4.67E+02	9.91E+01	2.00E+02	7.25E+02	6.18E+01
	h				+	+			h			≈	≈	+	
F_{12}	mean	6.31E+03	7.96E+03	2.09E+04	1.84E+05	1.03E+04	1.59E+04	F_{27}	mean	5.09E+02	5.11E+02	5.07E+02	5.00E+02	5.00E+02	5.16E+02
	std	2.63E+03	5.46E+03	1.24E+04	2.99E+05	7.41E+03	1.46E+04		std	2.59E+02	1.42E+01	8.46E+00	1.48E-04	1.24E-04	8.97E+00
	h			≈		≈			h			≈	≈	≈	
F_{13}	mean	1.70E+03	1.44E+03	1.18E+04	3.27E+04	2.11E+04	5.82E+01	F_{28}	mean	3.66E+02	3.63E+02	3.70E+02	4.98E+02	4.78E+02	3.64E+02
	std	9.84E+02	9.10E+02	7.45E+03	4.04E+04	2.89E+04	6.37E+01		std	5.95E+01	7.20E+01	6.95E+01	5.30E+00	4.51E+01	5.87E+01
	h			+	+	+			h			≈	≈	+	
F_{14}	mean	9.60E+02	2.15E+02	4.61E+01	6.62E+01	2.51E+02	4.75E+01	F_{29}	mean	9.82E+02	7.83E+02	5.20E+02	4.34E+02	7.23E+02	5.36E+02
	std	1.61E+03	6.99E+01	2.53E+01	5.13E+01	1.65E+02	1.12E+01		std	1.83E+02	1.30E+02	8.73E+01	1.77E+02	1.02E+02	8.30E+01
	h			≈	+	+			h			≈	≈	+	
F_{15}	mean	9.13E+02	1.45E+02	2.09E+01	1.98E+02	3.01E+02	1.71E+01	F_{30}	mean	6.34E+03	2.69E+03	3.36E+03	8.22E+02	5.55E+02	2.18E+03
	std	1.42E+03	8.80E+01	1.84E+01	3.29E+03	1.77E+02	1.37E+01		std	8.37E+03	5.06E+02	4.66E+02	1.22E+03	2.89E+02	2.66E+02
	h			+	+	+			h			≈	≈	≈	

Figure 2 presents the convergence curves for RLLDE and the comparative algorithms. The convergence rate, as indicated by these curves, offers a more intuitive insight into the enhancement of exploration and exploitation. The Friedman rank test is computed based on the average fitness values of six algorithms, and the corresponding statistical results are presented in Table 2. From Table 2, it can be seen that RLLDE ranks first among the six algorithms overall. This demonstrates the superiority of RLLDE compared to classical DE algorithms.

Performance Comparison with Other Advanced Algorithms. In this subsection, RLLDE is compared against eight original advanced algorithms, which include Teaching-learning-based optimization (TLBO) [21], Equilibrium Optimizer (EO) [7], Particle Swarm Optimization (PSO) [12], Genetic Algorithm (GA) [18], Artificial Bee Colony (ABC) [11], Grey Wolf Optimizer (GWO) [17], Salp Swarm Algorithm (SSA) [16], and Harris Hawks Optimization (HHO) [10]. This comparison aims to highlight the superiority and rationality of the proposed algorithm. All swarm intelligence algorithms are initialized with a starting population size of 40, a maximum number of evaluations is set at 300,000, and 30 separate comparison experiments are conducted under these conditions to ensure the fairness of the experiments. The key parameters of the algorithms used for comparison were kept at their default values, and you can find the details of these parameters in Table 3.

Table 2. The wilcoxon signed-rank test results of CEC 2017.

Algorithms	w/t/l	Friedman rank
RLLDE	–	2.41
DE/best/1	18/10/2	4.37
DE/btst/2	16/13/1	3.8
DE/rand/1	5/24/1	3.02
DE/rand/2	12/15/3	3.17
DE/current_to_best/1	18/9/3	4.23

Table 3. Parameter settings for compared algorithms.

Algorithm	Time	Parameters settings
TLBO	2011	$T_F = 1 \ or \ 2$
EO	2020	$V = 1, GP = 0.5$
PSO	1995	$c_1 = 1.5, c_2 = 2.0, w = 1, w_p = 0.99$
GA	1975	$pc = 0.8, pm = 0.2$
ABC	2010	$L = round(0.6 \times N \times D)$
GWO	2014	$\alpha = 2 - (2 \times FEs/MaxFEs)$
SSA	2017	$c_1 = 2 \times \exp\left((4 \times FEs/MaxFEs)^2\right)$
HHO	2019	$E_0 = 2 \times rand - 1, E_1 = 2 - 2 \times (FEs/MaxFEs)$
RLLDE	2023	$\beta = 1.5$

Table 4 display the mean values for each function. Furthermore, the performance of each algorithm is intuitively compared by assigning rankings based on their results. The test results show that the proposed algorithm consistently secures the top rank across the CEC 2017 functions with 30 dimensions. This suggests that the proposed algorithm exhibits exceptional performance.

Furthermore, this experiment assesses the convergence performance of the RLLDE algorithm across all 30 functions. Figure 3 illustrates the convergence curves for the proposed algorithm along with those of the 8 comparison algorithms. These convergence curves serve as evidence of the efficacy of the strategies introduced in RLLDE and reflect its strong convergence performance.

Table 4. Results of RLLDE and eight efficient optimizers on CEC 2017 benchmark functions.

No.	Item	TLBO	EO	PSO	GA	ABC	GWO	SSA	HHO	RLLDE
F_1	mean	2.04E+03	9.10E+02	3.62E+03	4.33E+03	1.35E+03	1.59E+09	3.53E+03	1.37E+07	**3.86E−22**
	std	2.02E+03	4.62E+02	4.65E+03	2.78E+03	1.65E+03	1.04E+09	4.02E+03	3.94E+06	**4.96E−22**
	h	+	+	+	+	+	+	+	+	≈
F_2	mean	1.38E+15	2.45E+04	1.50E+02	**5.88E+00**	1.45E+27	7.48E+26	2.82E+01	9.96E+12	1.17E+01
	std	3.08E+15	2.85E+04	2.49E+02	**8.20E+00**	1.49E+27	1.67E+27	6.30E+01	8.93E+12	2.61E+01
	h	+	+	+	≈	+	+	+	+	≈
F_3	mean	5.67E−07	1.59E+02	**5.37E−17**	1.39E+05	4.63E+04	3.10E+04	4.14E−08	1.06E+04	4.89E−16
	std	1.12E−06	2.22E+02	**9.38E−17**	6.27E+04	2.35E+03	1.07E+04	1.06E−08	2.79E+03	6.10E−16
	h	+	+	−	+	+	+	+	+	≈
F_4	mean	7.65E+01	5.10E+01	8.97E+01	6.55E+01	9.42E+01	1.53E+02	9.75E+01	1.24E+02	**4.88E+01**
	std	1.21E+01	2.57E+01	1.88E+01	2.27E+01	1.25E+01	5.12E+01	1.48E+01	3.87E+01	**2.51E+01**
	h	+	≈	+	≈	+	+	+	+	≈
F_5	mean	9.23E+01	8.73E+01	1.19E+02	1.29E+02	1.97E+02	9.66E+01	1.05E+02	2.48E+02	6.69E+01
	std	2.70E+01	1.98E+01	3.66E+01	3.27E+01	6.03E+00	2.03E+01	3.00E+01	2.15E+01	5.30E+01
	h	≈	≈	≈	+	+	≈	+	+	≈
F_6	mean	1.03E+01	8.63E−01	3.35E+01	9.08E−03	1.31E−05	4.13E+00	3.75E+01	6.51E+01	**3.19E−02**
	std	3.28E+00	1.28E+00	1.14E+01	4.83E−03	2.96E−06	4.16E+00	1.18E+01	6.30E+00	**5.50E−02**
	h	+	+	≈	+	−	+	+	+	≈
F_7	mean	1.53E+02	1.01E+02	1.10E+02	2.05E+02	2.41E+02	1.41E+02	1.44E+02	4.87E+02	**9.60E+01**
	std	2.54E+01	3.06E+01	2.07E+01	5.31E+01	8.38E+00	2.26E+01	1.36E+01	6.06E+01	**5.81E+01**
	h	≈	≈	≈	+	+	≈	≈	+	≈
F_8	mean	7.52E+01	6.56E+01	9.79E+01	1.36E+02	1.94E+02	7.81E+01	1.11E+02	1.68E+02	**2.89E+01**
	std	1.36E+01	1.11E+01	2.91E+01	3.38E+01	4.31E+00	1.64E+01	2.19E+01	2.06E+01	**8.02E+00**
	h	+	+	+	+	+	+	+	+	≈
F_9	mean	2.70E+02	6.49E+01	2.89E+03	2.84E+03	**3.65E−09**	1.06E+03	3.98E+03	5.75E+03	9.60E+00
	std	1.43E+02	1.25E+02	2.27E+03	1.14E+03	**5.66E−09**	3.78E+02	1.91E+03	9.11E+02	1.82E+01
	h	+	≈	+	+	−	+	+	+	≈
F_{10}	mean	5.71E+03	3.61E+03	3.27E+03	3.25E+03	7.13E+03	3.05E+03	4.09E+03	5.16E+03	**2.27E+03**
	std	1.51E+03	6.30E+02	4.03E+02	6.32E+02	3.04E+02	8.54E+02	5.35E+02	1.14E+03	**7.02E+02**
	h	+	+	+	+	+	≈	+	+	≈
F_{11}	mean	1.44E+02	5.47E+01	1.06E+02	7.87E+01	1.16E+02	4.06E+02	1.39E+02	1.71E+02	**5.42E+01**
	std	4.41E+01	7.09E+01	3.94E+01	5.24E+01	1.23E+01	2.78E+02	1.38E+01	3.61E+01	**4.44E+01**
	h	+	≈	≈	≈	+	+	+	+	≈
F_{12}	mean	**1.18E+04**	3.53E+04	2.52E+04	1.04E+06	1.53E+06	1.66E+07	1.20E+06	1.55E+07	1.59E+04
	std	**4.01E+03**	2.13E+04	1.41E+04	5.01E+05	7.79E+05	1.15E+07	5.57E+05	1.09E+07	1.46E+04
	h	≈	+	+	+	+	+	+	+	≈
F_{13}	mean	2.22E+04	2.39E+04	1.29E+04	2.42E+03	8.49E+03	3.07E+05	1.10E+05	2.47E+05	**5.82E+01**
	std	1.35E+04	2.42E+04	1.45E+04	2.07E+03	8.25E+03	5.21E+05	4.22E+04	6.13E+04	**6.37E+01**
	h	+	+	+	+	+	+	+	+	≈
F_{14}	mean	3.14E+03	1.54E+04	6.84E+03	7.75E+05	3.42E+04	5.99E+04	4.24E+04	8.84E+04	**4.75E+01**
	std	4.47E+03	1.15E+04	3.47E+03	6.08E+05	3.52E+04	9.47E+04	6.69E+03	9.88E+04	**1.12E+01**
	h	+	+	+	+	+	+	+	+	≈
F_{15}	mean	4.11E+03	7.00E+03	1.09E+03	1.12E+04	8.36E+03	3.17E+04	5.01E+04	6.24E+04	**1.71E+01**
	std	1.80E+03	1.31E+04	7.05E+02	8.39E+03	5.56E+02	2.13E+04	2.13E+04	4.01E+04	**1.37E+01**
	h	+	+	+	+	+	+	+	+	≈
F_{16}	mean	5.42E+02	6.37E+02	8.01E+02	1.36E+03	1.42E+03	9.95E+02	9.67E+02	1.65E+03	**4.32E+02**
	std	3.36E+02	1.95E+02	1.66E+02	2.82E+02	1.68E+02	1.97E+02	2.81E+02	2.66E+02	**1.65E+02**
	h	+	≈	+	+	+	+	+	+	≈
F_{17}	mean	2.73E+02	2.63E+02	5.16E+02	8.09E+02	3.78E+02	3.10E+02	3.60E+02	8.62E+02	**6.26E+01**
	std	1.38E+02	1.65E+02	2.21E+02	2.28E+02	1.23E+02	1.00E+02	1.67E+02	2.76E+02	**6.54E+01**
	h	+	+	+	+	+	+	+	+	≈
F_{18}	mean	1.29E+05	1.29E+05	4.57E+04	1.85E+06	2.34E+06	3.84E+05	1.78E+05	1.56E+06	**6.85E+02**
	std	9.47E+04	6.57E+04	2.53E+04	2.01E+06	5.32E+05	3.47E+05	1.52E+05	9.38E+05	**6.96E+02**
	h	+	+	+	+	+	+	+	+	≈
F_{19}	mean	2.75E+03	1.27E+04	5.30E+03	1.94E+04	6.28E+02	4.21E+05	3.06E+05	2.58E+05	**1.56E+01**
	std	1.80E+03	2.22E+04	7.11E+03	1.99E+04	1.01E+03	5.16E+05	7.60E+04	1.35E+05	**7.69E+00**
	h	+	+	+	+	+	+	+	+	≈
F_{20}	mean	2.13E+02	2.64E+02	5.47E+02	5.66E+02	4.83E+02	3.26E+02	2.44E+02	7.55E+02	2.03E+02
	std	7.50E+01	9.31E+01	1.25E+02	1.72E+02	1.21E+02	9.20E+01	9.71E+01	1.81E+02	2.71E+02
	h	≈	≈	≈	≈	≈	≈	≈	+	≈
F_{21}	mean	2.83E+02	**2.57E+02**	2.89E+02	3.47E+02	3.92E+02	2.75E+02	2.94E+02	4.68E+02	2.37E+02
	std	1.39E+01	**1.51E+01**	3.07E+01	3.09E+01	6.87E+00	9.33E+00	1.24E+01	6.40E+01	3.43E+00
	h	+	+	+	+	+	+	+	+	≈
F_{22}	mean	**1.08E+02**	2.10E+03	2.04E+03	1.77E+03	1.54E+03	1.68E+03	2.39E+03	4.28E+03	1.50E+03
	std	**1.64E+01**	1.92E+03	2.67E+03	2.33E+03	3.21E+03	1.82E+03	2.17E+03	2.44E+03	3.13E+03
	h	≈	≈	≈	≈	≈	≈	≈	≈	≈
F_{23}	mean	4.94E+02	4.11E+02	5.08E+02	4.92E+02	5.37E+02	4.46E+02	4.41E+02	9.34E+02	**3.83E+02**
	std	5.68E+01	1.01E+01	3.61E+01	3.19E+01	6.44E+00	2.69E+01	4.72E+01	1.70E+02	**3.48E+02**
	h	+	+	+	+	+	+	+	+	≈
F_{24}	mean	5.22E+02	4.76E+02	5.46E+02	8.12E+02	6.08E+02	5.22E+02	5.05E+02	9.52E+02	**4.72E+02**
	std	2.03E+01	1.30E+01	2.94E+01	1.33E+02	6.01E+00	6.30E+01	1.52E+01	1.03E+02	**1.86E+01**
	h	+	+	+	+	+	≈	+	+	≈
F_{25}	mean	3.90E+02	3.98E+02	3.91E+02	3.88E+02	**3.87E+02**	4.65E+02	3.98E+02	4.04E+02	3.88E+02
	std	4.07E+00	2.33E+01	3.89E+00	1.87E+01	**1.06E−01**	3.27E+01	2.76E+01	2.29E+01	6.20E−01
	h	≈	≈	≈	≈	≈	+	≈	≈	

(continued)

Table 4. (*continued*)

No.	Item	TLBO	EO	PSO	GA	ABC	GWO	SSA	HHO	RLLDE
F_{26}	mean	2.14E+03	1.56E+03	2.84E+03	3.10E+03	2.80E+03	1.79E+03	1.34E+03	4.37E+03	1.44E+03
	std	1.08E+03	1.83E+02	5.64E+02	7.48E+02	4.57E+01	2.54E+02	9.72E+02	3.37E+02	6.18E+01
	h	≈	≈	+	+	+	+	≈	+	≈
F_{27}	mean	5.56E+02	5.23E+02	5.68E+02	4.85E+02	5.07E+02	5.49E+02	5.31E+02	5.54E+02	5.16E+02
	std	2.71E+01	9.08E+00	6.00E+01	6.98E+00	2.94E+00	2.65E+01	6.89E+00	7.28E+01	8.97E+00
	h	+	≈	+	−	≈	+	+	≈	≈
F_{28}	mean	3.66E+02	3.30E+02	3.86E+02	4.19E+02	4.17E+02	6.23E+02	3.88E+02	4.61E+02	3.64E+02
	std	6.87E+01	6.67E+01	5.75E+01	6.16E+01	1.18E+01	8.42E+01	5.45E+01	2.32E+01	5.87E+01
	h	≈	≈	≈	≈	≈	+	≈	+	≈
F_{29}	mean	8.03E+02	6.09E+02	7.35E+02	1.07E+03	1.14E+03	8.29E+02	9.96E+02	1.54E+03	5.36E+02
	std	1.12E+02	8.98E+01	8.13E+01	3.80E+02	1.57E+02	2.47E+02	1.81E+02	4.53E+02	8.30E+01
	h	+	≈	+	+	+	+	+	+	≈
F_{30}	mean	4.50E+03	6.48E+03	9.24E+03	5.01E+03	7.89E+03	6.87E+06	1.05E+06	1.98E+06	2.18E+03
	std	2.08E+03	4.50E+03	5.33E+03	3.99E+03	2.77E+03	3.52E+06	5.81E+05	7.55E+05	2.66E+02
	h	+	+	+	≈	+	+	+	+	
	w/t/l	20/10/0	14/16/0	21/8/1	17/12/1	22/6/2	24/6/0	22/8/0	27/3/0	0/30/0
	median Friedman	4.3	3.52	4.79	5.53	5.75	6.08	5.25	7.89	1.88

5 Conclusion

This study introduces an enhanced differential evolution algorithm (RLLDE) that integrates reinforcement learning (RL) and levy flight strategies. To address the deficiency in the exploitation capability of the Differential Evolution (DE) algorithm, an update formula based on Levy flight is proposed at the mutation stage. Furthermore, improvements were made to the existing DE/rand/1 update formula, making it more inclined to guide particles toward effective searching, thus significantly enhancing the algorithm's convergence performance. Additionally, these two update formulas are switched through the Q-learning mechanism in Reinforcement Learning (RL). This mechanism facilitates updates in a direction that is more conducive to convergence, thereby accelerating the convergence speed of RLLDE. In the experiments, this paper initially conducts qualitative analyses of the algorithm, including experiments involving agent historical positions, particle changes, fitness value changes, and iteration curves. These experiments serve to illustrate the algorithm's behavior in seeking the optimal solution. Subsequently, the RLLDE algorithm is compared with five classical DE algorithms and eight advanced algorithms to showcase its performance advantages on the CEC 2017 test suite. The experimental results demonstrate that the RLLDE algorithm excels in achieving a better balance between exploration and exploitation and exhibits several advantages over comparable algorithms.

Nevertheless, it's worth noting that this algorithm still encounters issues related to premature convergence when applied to various benchmark functions, which could be subject to further investigation in future research. Additionally, it's important to mention that RLLDE is limited to solving single-objective problems. In future research, exploring binary and multi-objective versions of RLLDE could be a promising avenue. Moreover, expanding the application of this algorithm to diverse fields presents valuable opportunities, including text clustering, scheduling problems, appliances management, parameter estimation, feature selection, test classification, image segmentation problems, network applications, sentiment analysis, and more.

Acknowledgements. This work is supported by the National Natural Science Foundation of China (No. 62006144), the Major Fundamental Research Project of Shandong, China (No. ZR2019ZD03), and the Taishan Scholar Project of Shandong, China (No. ts20190924).

Data Availibility Statement. Data will be made available on request.

CRediT Authorship Contribution Statement. Xiaoyu Liu: Conceptualization, Investigation, Methodology, Software, Data curation, Visualization, Writing - original draft. **Qingke Zhang:** Conceptualization, Methodology, Investigation, Writing - original draft, Writing - review & editing, Software, Supervision, Project administration. **Hongtong Xi:** Investigation, Methodology, Writing - review & editing, Resources. **Huixia Zhang:** Investigation, Writing - review & editing, Resources. **Shuang Gao:** Writing - review & editing. **Huaxiang Zhang:** Writing - review & editing, Resources.

Declaration of Competing Interest. The authors declare that they have no known competing financial interests or personal relationships that could have appeared to influence the work reported in this paper.

References

1. Ahmad, M.F., Isa, N.A.M., Lim, W.H., Ang, K.M.: Differential evolution: a recent review based on state-of-the-art works. Alex. Eng. J. **61**(5), 3831–3872 (2022)
2. Al-Dabbagh, R.D., Neri, F., Idris, N., Baba, M.S.: Algorithmic design issues in adaptive differential evolution schemes: Review and taxonomy. Swarm Evol. Comput. **43**, 284–311 (2018)
3. Biswas, P.P., Arora, P., Mallipeddi, R., Suganthan, P.N., Panigrahi, B.K.: Optimal placement and sizing of facts devices for optimal power flow in a wind power integrated electrical network. Neural Comput. Appl. **33**, 6753–6774 (2021)
4. Boussaïd, I., Chatterjee, A., Siarry, P., Ahmed-Nacer, M.: Hybridizing biogeography-based optimization with differential evolution for optimal power allocation in wireless sensor networks. IEEE Trans. Veh. Technol. **60**(5), 2347–2353 (2011)
5. Civicioglu, P., Besdok, E.: Bernstein-levy differential evolution algorithm for numerical function optimization. Neural Comput. Appl. **35**(9), 6603–6621 (2023)
6. Das, S., Suganthan, P.N.: Differential evolution: a survey of the state-of-the-art. IEEE Trans. Evol. Comput. **15**(1), 4–31 (2010)
7. Faramarzi, A., Heidarinejad, M., Stephens, B., Mirjalili, S.: Equilibrium optimizer: a novel optimization algorithm. Knowl.-Based Syst. **191**, 105190 (2020)
8. Ge, Y.F., Orlowska, M., Cao, J., Wang, H., Zhang, Y.: MDDE: multitasking distributed differential evolution for privacy-preserving database fragmentation. VLDB J. **31**(5), 957–975 (2022)
9. Georgioudakis, M., Plevris, V.: A comparative study of differential evolution variants in constrained structural optimization. Front. Built Environ. **6**, 102 (2020)
10. Heidari, A.A., Mirjalili, S., Faris, H., Aljarah, I., Mafarja, M., Chen, H.: Harris hawks optimization: algorithm and applications. Futur. Gener. Comput. Syst. **97**, 849–872 (2019)
11. Karaboga, D.: Artificial bee colony algorithm. Scholarpedia **5**(3), 6915 (2010)

12. Kennedy, J., Eberhart, R.: Particle swarm optimization. In: Proceedings of ICNN 1995-International Conference on Neural Networks, vol. 4, pp. 1942–1948. IEEE (1995)
13. Li, Y., Wang, S., Yang, H., Chen, H., Yang, B.: Enhancing differential evolution algorithm using leader-adjoint populations. Inf. Sci. **622**, 235–268 (2023)
14. Liao, T.W.: Two hybrid differential evolution algorithms for engineering design optimization. Appl. Soft Comput. **10**(4), 1188–1199 (2010)
15. Lingam, G., Rout, R.R., Somayajulu, D.V.: Adaptive deep q-learning model for detecting social bots and influential users in online social networks. Appl. Intell. **49**, 3947–3964 (2019)
16. Mirjalili, S., Gandomi, A.H., Mirjalili, S.Z., Saremi, S., Faris, H., Mirjalili, S.M.: Salp swarm algorithm: a bio-inspired optimizer for engineering design problems. Adv. Eng. Softw. **114**, 163–191 (2017)
17. Mirjalili, S., Mirjalili, S.M., Lewis, A.: Grey wolf optimizer. Adv. Eng. Softw. **69**, 46–61 (2014)
18. Mirjalili, S., Mirjalili, S.: Genetic algorithm. In: Evolutionary Algorithms and Neural Networks: Theory and Applications, pp. 43–55 (2019)
19. Mishra, S., Kumar, A., Singh, D., Kumar Misra, R.: Butterfly optimizer for placement and sizing of distributed generation for feeder phase balancing. In: Verma, N., Ghosh, A. (eds.) Computational Intelligence: Theories, Applications and Future Directions - Volume II. AISC, vol. 799, pp. 519–530. Singapore, Springer (2019). https://doi.org/10.1007/978-981-13-1135-2_39
20. Nadimi-Shahraki, M.H., Zamani, H.: DMDE: diversity-maintained multi-trial vector differential evolution algorithm for non-decomposition large-scale global optimization. Expert Syst. Appl. **198**, 116895 (2022)
21. Rao, R.V., Savsani, V.J., Vakharia, D.: Teaching-learning-based optimization: a novel method for constrained mechanical design optimization problems. Comput. Aided Des. **43**(3), 303–315 (2011)
22. Seyyedabbasi, A., Aliyev, R., Kiani, F., Gulle, M.U., Basyildiz, H., Shah, M.A.: Hybrid algorithms based on combining reinforcement learning and metaheuristic methods to solve global optimization problems. Knowl.-Based Syst. **223**, 107044 (2021)
23. Song, Y., et al.: Dynamic hybrid mechanism-based differential evolution algorithm and its application. Expert Syst. Appl. **213**, 118834 (2023)
24. Talbi, E.: Machine learning into metaheuristics: a survey and taxonomy of data-driven metaheuristics (2021)
25. Wu, G., Mallipeddi, R., Suganthan, P.N.: Problem definitions and evaluation criteria for the CEC 2017 competition on constrained real-parameter optimization. National University of Defense Technology, Changsha, Hunan, PR China and Kyungpook National University, Daegu, South Korea and Nanyang Technological University, Singapore, Technical Report (2017)
26. Xu, Y., Pi, D.: A reinforcement learning-based communication topology in particle swarm optimization. Neural Comput. Appl. **32**, 10007–10032 (2020)
27. Yao, X., Liu, Y., Lin, G.: Evolutionary programming made faster. IEEE Trans. Evol. Comput. **3**(2), 82–102 (1999)

Hierarchical Competitive Differential Evolution for Global Optimization

Hongtong Xi, Qingke Zhang[✉], Xiaoyu Liu, Huixia Zhang, Shuang Gao, and Huaxiang Zhang

School of Information Science and Engineering, Shandong Normal University, Jinan 250358, China
tsingke@sdnu.edu.cn

Abstract. Global search is a fundamental task in optimization, aiming to find the optimal solution across the entire search space. To address the challenges in global search, a hierarchical competitive differential evolution algorithm is proposed. It uniquely incorporates a hierarchical competition mechanism and an adaptive differential mutation strategy based on competition outcomes, substantially enhancing global search. The proposed algorithm benchmarks on a total of 30 international test functions of CEC 2017 benchmark functions. The convergence accuracy, coupled with the outcomes of two nonparametric statistical tests, the Friedman test and the Wilcoxon signed-rank test, clearly demonstrates that HCDE exhibits competitive performance when compared to the other 14 efficient optimizers.

Keywords: Swarm Intelligence · Differential Evolution · Competition

1 Introduction

Optimization, a realm of intricate problem-solving, involves navigating through a terrain brimming with potential solutions in the quest for the optimal one [5,18,25]. Nevertheless, as society and the economy advance, the complexity of challenges escalates, giving birth to fresh optimization dilemmas that necessitate the deployment of suitable optimization methodologies. To solve these problems, a class of algorithms has emerged over the past three decades, commonly referred to as metaheuristics [3].

The search process of each algorithm initiates with a random population and move between two phases: exploration and exploitation. The whole search space is explored and focuses on discovering possible search regions in the exploration phase. This is the reason why this phase is alternatively referred to as the global search phase. During the exploitation phase, a local search of the potential region within the search space is conducted. The objective is to identify more promising solutions in the already attained possible region. This phase is well known as a local search phase. To improve the quality of solution, a proper balance between these two phases is necessary.

© The Author(s), under exclusive license to Springer Nature Singapore Pte Ltd. 2024
L. Pan et al. (Eds.): BIC-TA 2023, CCIS 2061, pp. 157–171, 2024.
https://doi.org/10.1007/978-981-97-2272-3_12

In recent decades, the field of nature-inspired metaheuristic algorithms has made substantial progress. These algorithms can be broadly classified into four categories: evolution-based, population-based, nature-inspired, and human-inspired. In the forthcoming discussion, we will delve into some prominent and state-of-the-art algorithms from each of these categories (see Fig. 1).

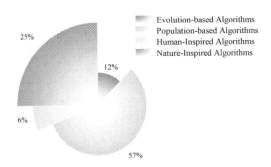

Fig. 1. Percentage Distribution of Metaheuristic Algorithm Categories in Recent Years

1. Evolution-based Algorithms: In the domain of optimization, evolution-based algorithms estimate the fundamental principles of biological evolution. These algorithms involve a population that use a series of iterative processes, including selection, and mutation. Notable examples in this category include Genetic Algorithms (GA) [17], Differential Evolution (DE) [24] and HJADE-GT [4].
2. Population-based Algorithms: Notable examples in this category include Particle Swarm Optimization (PSO) [22], Ant Colony Optimization (ACO) [6], Artificial Bee Colony (ABC) [10] and GA-GOA [2].
3. Human-Inspired Algorithms: Advancements in population research have brought human populations into focus. Human-inspired algorithms, like Teaching-Learning Based Optimization (TLBO) [20], Growth Optimizer (GO) [28] and Fans Optimizer (FO) [26].
4. Nature-Inspired Algorithms: These algorithms, inspired by nature, primarily rely on the physical and chemical principles found in nature. A prominent example in this category is Simulated Annealing (SA) [12]. Other algorithms in this group include Biogeography-Based Optimization (BBO) [23] and Sand Cat swarm optimization [21].

The No Free Lunch theorem emphasizes the absence of a universal algorithm, highlighting the necessity of adaptability in intricate, constantly evolving real-world scenarios. In nature, competition serves as a driving force for both biological evolution and social adaptation. Competitive algorithms thrive by balancing cooperation and competition.

Expanding on this premise, our research introduces an enhanced version of DE, named the Hierarchical Competitive Differential Evolution algorithm (HCDE). HCDE combines inspiration from population-based optimization and

DE. Its primary goal is to enhance global optimization of DE while addressing local optima. It achieves this by introducing a hierarchical competition mechanism, promoting collaboration among individuals, which divided the population into winners and losers. HCDE utilizes strategies like population stratification, differential mutation policies, adaptive control, and random perturbations. Population hierarchy creates competition among individuals in different layers, except for the elite, ensuring diversity and preventing local optima. The innovation of the proposed HCDE is summarized as follows:

1. Hierarchical Competition Mechanism: HCDE introduces a novel hierarchical competition mechanism by dividing the overall population into multiple layers, systematically guiding competition among individuals. This hierarchical structure aids in selectively retaining and fostering the possible solutions, thereby enhancing the algorithm's convergence and global search capabilities.
2. Adaptive Differential Mutation Strategy: HCDE uses an adaptive differential mutation strategy that adjusts the parameters of the differential mutation operation based on the results of hierarchical competition. This flexibility allows the algorithm to get a balance of exploration and exploitation.

The reminder of the paper is structured as follows: Sect. 2 reviews the DE process, Sect. 3 explains HCDE, Sect. 4 covers experiments, and Sect. 5 concludes and outlines future research.

2   The Process of DE

Differential Evolution. (DE) is a well-crafted evolutionary algorithm (EA) designed to optimize problems through a series of steps involving mutation, crossover, and selection [24]. Over its evolution, DE has demonstrated effectiveness in addressing a wide range of complex optimization challenges, including multi-objective, constrained, and multi-modal optimization.

Initialization: During initialization, population X is created, where $X = \{X_{i,j} : i = 1, 2, \ldots, \text{popsize}, j = 1, 2, \ldots, \text{dimension}\}$ with popsize as the population size. Each individual, denoted as X_{ij}, is a dimension-dimensional vector with components $X_{1,j}, X_{2,j}, \ldots, X_{D,j}$. These components are generated from uniform random numbers in the interval $[X_{\text{low}}, X_{\text{upp}}]$ as defined in Eq. (1). Here, X_{low} and X_{upp} set the lower and upper bounds of the search space S, $rand$ is a random number in the range $[0, 1]$. After the initial population is created, the algorithm proceeds to the next phase.

$$X_{i,j} = X_{low} + (X_{upp} - X_{low}) \cdot rand \qquad (1)$$

Mutation: In the mutation step, a mutant vector V_{ij} is generated for each target vector X_{ij}. This mutation is defined by Eq. (2).

$$V_{ij} = X_{r1} + F \cdot (X_{r2} - X_{r3}) \qquad (2)$$

Here, F represents a scaling factor, which varies between 0 and 1, and r_1, r_2, and r_3 are distinct, randomly selected vectors.

Crossover: Following the mutation phase, DE proceeds to the crossover operation, which creates a new vector known as the trial vector, represented as U_{ij}. This crossover operation blends the target vector X_{ij} and the mutant vector V_{ij} using a crossover probability C_R, which falls within the range [0, 1] as Eq. (3).

$$U_{i,j} = \begin{cases} V_{i,j} & \text{if } \text{rand}_j \leq C_R \\ X_{i,j} & \text{otherwise} \end{cases} \tag{3}$$

Here, C_R denotes the crossover probability, and rand_j represents a random number.

Selection: After the crossover phase, the selection step plays a critical role in deciding which vectors advance to the next generation. It involves comparing the fitness of the target vector and the trial vector, and selecting the one with superior fitness, as shown in Eq. (4).

$$X_i = \begin{cases} U_i & \text{if } f(U_i) \leq f(X_i) \\ X_i & \text{otherwise} \end{cases} \tag{4}$$

The mutation, crossover, and selection processes in the evolution phase are end until a predefined termination criterion is satisfied [24]. DE's simplicity, few control parameters, and adaptability to diverse optimization landscapes make it a widely used algorithm. Its capability to balance exploration and exploitation consistently delivers competition across various applications.

3   Hierarchical Competitive Differential Evolution Algorithm

The algorithm comprises two distinct phases. Initially, the population is ranked based on fitness and divided into four tiers: Elite, High-Level, Middle-Level, and Lower layers. Except for the Elite Layer, particles within each layer engage in random competitions to determine winners and losers. Population hierarchy creates competition among individuals in different layers, except for elite ones, ensuring diversity and preventing local optima.

During the second phase, adaptive differential mutation strategy is employed to provide customized updates to individuals based on their performance in competitions, enabling adaptive selection of strategies and improving global optima. This two-phase approach enhances the optimization process in a more effective and targeted manner. The competitive phase sustains diversity, identifies promising candidates, and update formulas for particles at different levels, fostering algorithm exploration. This approach enhances the algorithm's overall performance and efficiency across the entire population. Adaptive control mechanisms and random perturbations further enhance HCDE's resilience and global search abilities.

3.1   The Process of HCDE

HCDE initiates population categorization based on fitness. It then involves randomized competitions within each layer, carefully documenting outcomes into winners and losers from different layers except the Elite layer. In the subsequent phase, winners and losers from different layers employ distinct adaptive differential evolution operations, generating innovative solutions based on the competition results. These solutions undergo a process involving crossover and perturbation. HCDE uses a selective strategy to systematically update the succeeding population generation, with this iterative process continuing until termination conditions are met as Fig. 2 shown.

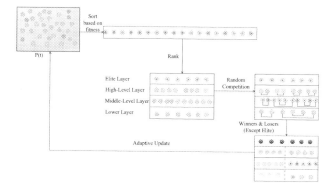

Fig. 2. The process of HCDE

Population Initialization: HCDE determines the population size as a number of NP individuals, based on the dimension of the problem. Following this, HCDE initializes a population consisting of 'popsize' individuals by Eq. (1). The formula for calculating NP is as shown in Eq. (5), where M is a predefined parameter, in general, it is set to 40 [27].

$$NP = M + \lfloor \frac{dimension}{10} \rfloor \qquad (5)$$

Fitness-Based Ranking: HCDE conducts a population-wide ranking based on the fitness evaluations of all individuals. The population is categorized into four layers: the Elite Layer with $rank1$ individuals, the High-Level Layer with $rank2$ individuals, the Middle-Level Layer with $rank3$ individuals, and the Lower Layer with $rank4$ individuals. This systematic ranking process categorizes individuals according to their performance in the solution space, distinguishing elite individuals from those with relatively lower performance.

Hierarchical Competition Mechanism: In this phase, HCDE randomly pairs selected individuals for competition. This competitive process culminates in winners and losers within the randomly selected pairs of particles, which contributes to the identification of the most promising candidates.

Elite Layer Update: Additionally, there's a probability for each dimension of elite-layer individuals to exchange information with other elite-layer particles sharing the same dimension. This enhancement significantly strengthens HCDE's global search capabilities, as represented by Eq. (6).

$$X_{i,j} = \begin{cases} X_{i,j} + r_1 \cdot (R_j - X_{i,j}) & \text{if } r_2 < Key \\ X_{i,j} & \text{otherwise} \end{cases} \tag{6}$$

R_j is an Elite Layer random particle, j corresponds to a specific dimension, Key is usually set to 0.3 [28]. And r_1, r_2 are random within the [0, 1].

Adaptive Differential Mutation Strategy: The algorithm tailors mutation strategies to individuals' success levels. Losers receive specialized strategies to promote diversity and enhance their search. In contrast, winners utilize a broader range of information, including data from individuals in different layers, for a more extensive exploration of the solution space. It's worth noting the algorithm's dynamic adjustment of the crucial scaling factor F in differential evolution as shown in Eq. (7).

$$F = \frac{|\text{fitness}(X_{r1}) - \text{fitness}(X_{r2})|}{\text{fitness}(i) - \text{gbestfitness}} \tag{7}$$

$$V_i = X_i + F \cdot (gbestx - X_i) + F \cdot (X_{r1} - X_{r2}) \tag{8}$$

In Eq. (8), X_i represents the position vector of the current individual. V_i represents the velocity vector of the individual. $gbestx$ represents the position vector of the global best individual. X_{r1} and X_{r2} represent the position vectors of random particles from popsize. For winners, r_1 and r_2 are randomly selected from all higher layers, current, and lower layers. For losers, they are chosen from the current and all higher layers. In the case of Lower Layer winners, r_1 and r_2 are randomly selected from Elite layer, while losers can be chosen from all population. The term $gbestx$ is associated with the position vector of the global best individual, and it is important for updating the velocity based on fitness comparison.

Crossover, Perturbation, and Selection: After the mutation phase, the algorithm adopts the crossover operations to individuals to generate new solutions, enhancing population diversity. The crossover operation use Eq. (3), with $C_R = 0.9$. Additionally, Gaussian perturbations are specifically introduced to losers, expanding and intensifying the search process as indicated Eq. (9), where r_3 are randomly selected from [1,$dimension$]. And Selection use Eq. (4) to update.

$$U_i = \begin{cases} U_i = U_i + 0.1 \cdot r_3 & \text{if } i \in losers \\ U_i & \text{otherwise} \end{cases} \tag{9}$$

Termination Criterion: When the current number of evaluations (FEs) reaches the maximum number of evaluations (MaxFEs), the program computes

$gbestx$, and subsequently halts its execution. In the absence of meeting this condition, the algorithm reverts to phase called Fitness-Based Ranking and Grading, ensuring the continuity of the optimization process, see Algorithm 1.

Algorithm 1: HCDE

Input: $MaxFEs$, M, X_{low}, X_{upp}.
Output: x_{gbest} and $gbestx$
1 use Eq. (5) to compute $popsize$ to initialization the particle's population $i \leftarrow 1, 2, ..., NP$ and calculate all $fitness$;
2 $FEs = FEs + popsize$;
3 **while** $FEs \leq MaxFEs$ **do**
4     $[fitness, ind] = sort(fitness)$
5     $X = X(ind, :)$
6     $V = V(ind, :)$
7   Based on $fitness$, the individuals are divided into four groups, namely: $Elite Layer, High - Level Layer, Middle - Level Layer, and Lower Layer$ by $rank1, rank2, rank3, rank4$.
8   Except for the first layer, each layer engages in random particle competition, categorizing the particles into $winners and losers$.
9   **for** $i \leftarrow 1$ **to** $popsize$ **do**
10   **if** $i \in rank1$ **then**
11   Update the mutant vector by using Eq. (6);
12   **else**
13   Select two distinct particles, r_1, r_2 according to requirements (Adaptive Mutation Strategies)
14   Update by using Eq. (7) Eq. (8) Eq. (3);
15   Boundary control;
16   Update by using Eq. (9) Eq. (4)
17        $FEs = FEs + 1$;
Output: x_{gbest} and $gbestx$

---

4 Experimental Results and Analysis

The complexity of the benchmark suites for CEC 2017, characterized by an abundance of local optima, dynamic uncertainties, and variable independence, demands empirical validation for metaheuristic algorithms. Consequently, we undertake numerical optimization experiments using HCDE and subject it to a comprehensive comparison with 14 metaheuristic algorithms. Subsequently, we perform statistical analysis on the experimental outcomes to ensure a precise assessment of HCDE's effectiveness. Table 1 offers a summary of the benchmark suites for CEC 2017, providing information about test suite names, function ranges, the number of functions, function types, search domains, and fitness errors. All functions have a standardized search range of $[-100, 100]$.

Table 1. The test benchmark functions of CEC 2017

Test suite	Number of function	Function types	Search range	Fitness error
CEC 2017	30	Unimodal functions (F1–F3)	$[-100, 100]$	0
		Simple Multimodal functions (F4-F10)	$[-100, 100]$	0
		Hybrid functions (F11–F20)	$[-100, 100]$	0
		Composition functions (F21–F30)	$[-100, 100]$	0

Table 2. The algorithm information and parameter settings

No	Algorithms	Proposed time	Parameters Settings	Citations	References
1	Genetic Algorithms (GA)	1975	pc = 0.8, pm = 0.2	17613	[17]
2	Particle Swarm Optimization (PSO)	1995	$w = 1$, $w_p = 0.99$, $c_1 = 1.5$, $c_2 = 2.0$	72995	[11]
3	Differential Evolution (DE)	1997	F = 0.5, CR = 0.9	29477	[24]
4	Equilibrium Optimizer (EO)	2020	$V = 1$, $a_1 = 2$, $a_2 = 1$, GP = 0.5	672	[7]
5	Teaching-Learning-Based Optimization (TLBO)	2011	$T_F = 1$ or 2	3423	[20]
6	Salp Swarm Algorithm (SSA)	2017	$c_1 = 2 \times exp((4 \times FEs/MaxFEs)^2)$	2413	[14]
7	Grey Wolf Optimizer (GWO)	2014	$\alpha = 2 - 2 \times (FEs/MaxFEs)$	8819	[16]
8	Artificial Bee Colony (ABC)	2007	$a = 1$, $L = round(0.6 \times N \times D)$	7212	[9]
9	Harris Hawks Optimization (HHO)	2019	$E_0 = 2 \times rand - 1$, $E_1 = 2 - 2 \times (FEs/MaxFEs)$	1897	[8]
10	Whale Optimization Algorithm (WOA)	2016	$b = 1$, $a_1 = 2 - (2 \times FEs/MaxFEs)$, $a_2 = -1 - (FEs/MaxFEs)$	5784	[15]
11	Competitive Swarm Optimizer (CSO)	2015	phi = 0.01	515	[19]
12	Reptile Search Algorithm (RSA)	2016	Alpha = 0.1, Beta = 0.005	5784	[1]
13	Elite Archives-driven Particle swarm optimization(EAPSO)	2023	no parameters	10	[29]
14	Comprehensive Learning Particle Swarm Optimizer(CLPSO)	2006	$c = 1.49445$	4043	[13]
15	Hierarchical competitive differential evolution algorithm(HCDE)		$M = 40$, $Key = 0.3$, $rank1 = rank4 = 0.1$, $rank3 = rank4 = 0.4$, $CR = 0.9$		

4.1   Comparison of HCDE with Other Efficient Opitimizers

To evaluate and validate the algorithm's strong search performance, a comprehensive series of experiments were conducted on HCDE. These experiments were consistently carried out with the same test parameter settings and environmental conditions.

Experimental Setup and Compared Algorithms: The experiments used MaxFEs as the termination criterion, collecting data from 30 benchmark problems on 30 and 50 dimensions with 30 runs each. Table 2 offers a comprehensive overview of the 14 metaheuristic algorithms, presenting their names, abbreviations, proposal years, parameter settings, citation counts, and references to relevant literature. In cases where decision variables violated constraints, HCDE constrained their values within the bounds, while other algorithms adhered to the methodologies outlined in relevant literature.

Convergence Results and Analysis: In this subsection, we present the convergence curves for all algorithms across various problem types, as illustrated in Fig. 3. The considered functions include the unimodal function F1, multimodal function F4, hybrid functions F14 and F16, as well as composition functions F25 and F28. These assessments are conducted in two dimensions, namely D = 30 and D = 50.

In both 30 and 50 dimensions, it can be seen that HCDE is most prominent in handling unimodal functions especially in F1. Its convergence curve is no tendency to fall into a local optimum. This underscores the algorithm's exceptional convergence and stability in addressing straightforward unimodal functions.

Table 3. The Experimental Results for the 30 dimensions of CEC 2017 benchmark suite

No.	Item	EO	EAPSO	CLPSO	CSO	PSO	DE/best 1	HHO	RSA	TLBO	SSA	GA	ABC	GWO	WOA	HCDE
F1	mean	7.70E+03	3.50E+03	1.50E+01	8.42E+03	2.01E+03	4.88E−09	1.27E+07	3.62E+10	4.72E+03	1.54E+03	4.10E+03	4.78E+03	2.71E+09	2.15E+06	3.25E−14
	std	6.02E+03	5.78E+03	2.30E+01	8.68E+03	3.85E+03	1.09E−08	1.45E+06	9.74E+09	4.81E+03	1.55E+03	5.51E+03	6.79E+03	1.70E+09	2.01E+06	4.14E−14
	h	+	+	+	+	+	≈	+	+	+	+	+	+	+	+	≈
F2	mean	1.38E+06	1.44E−04	5.70E+15	1.38E+09	1.50E+02	3.37E−04	6.46E+14	6.81E+44	6.36E+04	3.36E−04	4.94E+04	2.86E+28	2.23E+31	2.85E+20	2.26E−06
	std	3.03E+06	1.73E−04	1.13E+16	4.71E+09	2.49E+02	6.84E−04	1.23E+15	1.52E+45	1.03E+05	2.08E−04	1.11E+05	3.46E+28	4.98E+31	6.06E+20	3.23E−06
	h	+	+	+	+	+	≈	+	+	+	+	+	+	+	+	≈
F3	mean	1.00E+01	1.85E−26	1.47E+04	1.14E+04	5.37E−17	4.25E−21	1.06E+04	6.22E+04	4.22E−07	3.91E−08	1.29E+05	4.93E+04	3.54E+04	1.42E+05	3.46E−11
	std	1.27E+01	1.84E−26	7.70E+03	9.14E+03	9.38E−17	7.72E−21	3.62E+03	1.52E+04	8.22E−07	1.83E−08	3.51E+04	9.18E+03	7.13E+03	7.37E+04	4.11E−11
	h	+	−	+	+	−	≈	+	+	+	+	+	+	+	+	≈
F4	mean	4.18E+01	1.36E+01	7.70E+01	1.02E+02	8.97E+01	1.04E+01	1.29E+02	1.07E+04	6.61E+01	7.42E+01	5.05E+01	8.99E+01	1.99E+02	1.32E+02	2.40E+00
	std	3.48E+01	2.83E+01	5.74E+00	1.34E+01	1.88E+01	4.26E+00	2.81E+01	3.42E+03	3.05E+01	3.31E+01	3.78E+01	7.09E+00	7.61E+01	2.83E+01	2.19E+00
	h	+	≈	+	+	+	≈	+	+	+	+	+	+	+	+	≈
F5	mean	6.01E+01	8.76E+01	4.64E+01	3.86E+01	1.19E+02	1.75E+02	2.39E+02	3.69E+02	1.01E+02	1.26E+02	1.34E+02	1.97E+02	8.52E+01	2.69E+02	8.82E+01
	std	2.35E+01	2.18E+01	8.96E+00	5.60E+00	3.66E+01	4.74E+01	1.56E+01	1.06E+02	1.24E+01	4.75E+01	4.61E+01	8.85E+00	1.97E+01	4.42E+01	3.14E+01
	h	≈	+	≈	−	+	+	+	+	+	+	+	+	+	+	≈
F6	mean	1.55E−01	2.19E+00	1.73E−11	1.44E−01	3.35E−01	3.57E−01	5.92E+01	8.98E+01	6.15E+00	3.33E−01	5.81E−03	1.08E−05	8.80E+00	7.55E+01	1.58E−01
	std	2.12E−01	1.70E+00	1.17E−11	2.65E−01	1.14E−01	8.12E+00	3.51E+00	1.36E+01	4.72E+00	7.10E+00	2.24E−03	4.21E−06	4.95E+00	7.50E+00	9.71E−02
	h	≈	+	−	≈	+	+	+	+	+	+	−	−	+	+	≈
F7	mean	8.78E+01	1.21E+02	9.26E+01	5.60E+01	1.10E+02	5.77E+02	4.95E+02	6.35E+02	1.71E+02	1.65E+02	2.36E+02	2.36E+02	1.50E+02	5.37E+02	1.17E+02
	std	1.07E+01	2.45E+01	5.11E+00	4.70E+00	2.07E+01	2.31E+02	5.96E+01	2.00E+02	1.99E+01	8.37E+01	7.26E+01	9.32E+00	4.78E+01	9.49E+01	1.23E+01
	h	+	+	+	≈	+	+	+	+	+	+	+	+	+	+	≈
F8	mean	8.38E+01	9.19E+01	5.46E+01	2.85E+01	9.79E+01	1.47E+02	1.73E+02	2.15E+02	7.82E+01	1.14E+02	1.44E+02	2.01E+02	8.44E+01	1.90E+02	4.32E+01
	std	4.02E+01	3.45E+01	4.34E+00	6.73E+00	2.91E+01	1.27E+01	3.72E+01	7.93E+01	1.32E+01	1.18E+01	4.59E+01	4.52E+00	1.70E+01	1.90E+01	2.79E+01
	h	+	+	+	−	+	+	+	+	≈	+	+	+	+	+	≈
F9	mean	2.65E+01	8.63E+01	1.82E+01	4.18E+00	2.89E+03	3.52E+03	5.22E+03	8.09E+03	3.10E+02	9.03E+02	3.02E+03	3.40E−09	7.89E+02	8.43E+03	3.42E+02
	std	2.87E+01	4.56E+02	1.19E+01	4.85E+00	2.27E+03	4.91E+02	7.51E+02	4.48E+03	2.09E+02	6.73E+02	1.53E+03	5.72E−09	3.86E+02	2.81E+03	4.80E+02
	h	+	≈	+	−	+	+	+	+	≈	+	+	−	+	+	≈
F10	mean	3.36E+03	3.33E+03	2.31E+03	1.67E+03	3.27E+03	3.41E+03	4.74E+03	6.69E+03	5.60E+03	3.84E+03	3.28E+03	6.95E+03	2.71E+03	5.00E+03	3.74E+03
	std	7.14E+02	7.68E+02	1.39E+02	3.82E+02	4.03E+02	8.25E+02	3.66E+02	6.96E+02	1.05E+02	8.31E+02	6.46E+02	1.43E+02	9.44E+02	1.31E+03	1.16E+03
	h	+	+	−	−	≈	≈	+	+	+	+	≈	+	−	+	≈
F11	mean	5.98E+01	9.93E+01	5.27E+01	9.52E+01	1.06E+02	2.37E+02	1.72E+02	5.86E+03	1.26E+02	1.53E+02	7.96E+01	1.28E+02	7.11E+02	4.44E+02	7.74E+01
	std	2.43E+01	4.83E+01	1.14E+01	4.87E+01	3.94E+01	1.19E+02	6.98E+01	2.64E+03	3.06E+01	5.66E+01	2.24E+01	1.36E+01	9.29E+02	1.30E+02	2.24E+01
	h	≈	+	≈	+	+	+	+	+	+	+	≈	+	+	+	≈
F12	mean	6.32E+04	1.62E+04	3.79E+05	2.53E+05	2.52E+04	2.92E+03	1.04E+07	9.16E+09	4.24E+04	1.07E+06	1.80E+06	1.61E+06	3.59E+07	3.63E+07	1.94E+04
	std	4.52E+04	1.69E+04	2.97E+05	2.19E+05	1.41E+04	2.06E+03	4.50E+06	3.78E+09	3.72E+04	5.33E+05	1.95E+06	5.31E+05	4.01E+07	1.85E+07	3.49E+04
	h	+	≈	+	+	+	≈	+	+	≈	+	+	+	+	+	≈
F13	mean	1.47E+04	1.85E+04	3.19E+02	6.52E+03	1.29E+04	8.72E+02	2.60E+05	4.14E+09	1.77E+04	1.18E+05	1.91E+04	5.04E+03	1.36E+05	1.72E+05	1.13E+02
	std	1.19E+04	2.17E+04	1.11E+02	5.63E+03	1.45E+04	5.67E+02	9.81E+04	2.96E+09	1.54E+04	9.15E+04	1.46E+04	6.54E+03	1.19E+05	7.80E+04	4.58E+01
	h	+	+	+	+	+	+	+	+	+	+	+	+	+	+	≈
F14	mean	5.46E+03	4.13E+02	1.17E+04	7.80E+04	6.84E+03	1.36E+03	6.39E+04	5.37E+06	2.71E+03	4.69E+03	1.29E+04	2.71E+04	2.89E+05	7.51E+05	7.93E+01
	std	3.76E+03	3.54E+02	4.52E+03	6.56E+04	3.47E+03	2.47E+03	7.30E+04	6.44E+06	2.55E+03	2.10E+03	1.50E+06	1.47E+04	4.59E+05	8.26E+05	4.29E+01
	h	+	+	+	+	+	+	+	+	+	+	+	+	+	+	≈
F15	mean	2.66E+03	1.40E+04	1.55E+02	1.32E+04	1.09E+03	4.55E+02	4.84E+04	1.79E+08	2.45E+03	3.48E+03	1.46E+04	1.03E+03	3.80E+05	7.84E+04	5.75E+01
	std	2.67E+03	1.02E+04	8.77E+01	1.46E+04	7.05E+02	4.04E+02	2.62E+04	2.45E+08	2.32E+04	2.32E+04	1.09E+04	9.89E+02	7.98E+05	5.59E+04	3.68E+01
	h	+	+	+	+	+	+	+	+	+	+	+	≈	+	+	≈
F16	mean	7.40E+02	1.06E+03	5.21E+02	4.08E+02	8.01E+02	9.12E+02	1.73E+03	3.06E+03	5.87E+02	8.81E+02	9.79E+02	1.45E+03	8.99E+02	1.99E+03	4.02E+02
	std	3.07E+02	1.61E+02	6.80E+01	2.35E+02	1.66E+02	3.43E+02	3.91E+02	1.09E+03	2.27E+02	3.00E+02	3.17E+02	1.96E+02	3.16E+02	5.61E+02	1.77E+02
	h	+	+	+	≈	+	+	+	+	+	+	+	+	+	+	≈
F17	mean	1.79E+02	3.98E+02	1.66E+02	1.89E+02	5.16E+02	7.32E+02	7.80E+02	1.75E+03	2.25E+02	2.82E+02	6.60E+02	4.20E+02	1.60E+02	7.89E+02	1.96E+02
	std	8.78E+01	2.10E+02	7.02E+01	1.18E+02	2.21E+02	3.26E+02	2.97E+02	3.44E+02	1.06E+02	8.11E+01	1.21E+02	1.30E+02	7.97E+01	1.95E+02	9.49E+01
	h	≈	+	≈	≈	+	+	+	+	+	+	+	+	≈	+	≈
F18	mean	1.53E+05	1.89E+04	1.56E+05	4.50E+05	4.57E+04	1.45E+03	6.38E+05	2.67E+07	8.82E+04	1.15E+05	2.65E+06	1.90E+06	4.08E+05	4.00E+06	3.29E+03
	std	1.61E+05	2.48E+04	1.06E+05	5.24E+05	2.53E+04	2.40E+03	3.34E+05	4.72E+07	6.09E+04	8.72E+04	2.52E+06	6.12E+05	2.35E+05	5.46E+06	2.82E+03
	h	+	+	+	+	+	≈	+	+	+	+	+	+	+	+	≈
F19	mean	4.20E+03	9.84E+03	7.13E+01	8.34E+03	5.30E+03	3.14E+02	3.19E+05	8.77E+08	4.72E+03	1.68E+05	6.96E+05	1.19E+03	2.06E+05	2.01E+06	3.87E+01
	std	5.24E+03	9.69E+03	1.62E+01	1.25E+04	7.11E+03	2.52E+02	2.07E+05	8.19E+08	5.36E+03	1.56E+05	4.13E+03	9.63E+02	1.56E+05	2.17E+06	1.61E+01
	h	+	+	≈	+	+	+	+	+	+	+	+	≈	+	+	≈
F20	mean	4.46E+02	3.13E+02	2.21E+02	1.26E+02	1.26E+02	5.47E+02	6.36E+02	1.00E+03	2.93E+02	4.36E+02	6.44E+02	4.45E+02	3.90E+02	7.76E+02	1.77E+02
	std	1.84E+02	7.40E+01	6.28E+01	6.68E+01	1.25E+02	1.34E+02	3.29E+02	1.87E+02	3.64E+01	1.79E+02	2.91E+02	2.48E+02	1.74E+02	2.11E+02	1.22E+02
	h	+	+	+	≈	≈	+	+	+	+	+	+	+	+	+	≈
F21	mean	2.49E+02	2.88E+02	2.52E+02	2.35E+02	2.89E+02	3.74E+02	4.38E+02	6.30E+02	2.61E+02	3.08E+02	3.85E+02	3.89E+02	2.77E+02	4.89E+02	2.75E+02
	std	2.14E+01	4.10E+01	9.16E+00	3.60E+00	3.07E+01	6.74E+01	7.32E+01	8.54E+01	1.43E+01	2.91E+01	4.82E+01	1.00E+01	2.08E+01	6.65E+01	1.94E+01
	h	≈	+	≈	−	+	+	+	+	≈	+	+	+	≈	+	≈
F22	mean	1.48E+03	4.18E+03	5.76E+02	1.00E+02	2.04E+03	2.39E+03	5.33E+03	5.66E+03	1.00E+02	1.00E+02	3.37E+03	1.14E+03	9.77E+02	5.98E+03	1.01E+02
	std	2.03E+03	3.41E+03	9.97E+02	0.00E+00	2.67E+03	2.12E+03	5.22E+02	2.53E+03	1.10E+00	1.10E+00	4.18E+02	1.90E+03	1.34E+03	4.89E+02	1.94E+00
	h	+	+	+	≈	+	+	+	+	≈	≈	+	+	+	+	≈
F23	mean	4.00E+02	4.43E+02	4.06E+02	3.94E+02	5.08E+02	6.38E+02	7.01E+02	9.56E+02	4.84E+02	4.58E+02	5.10E+02	5.37E+02	4.74E+02	8.21E+02	3.70E+02
	std	1.43E+01	1.84E+01	3.74E+00	2.10E+01	3.61E+01	6.72E+01	1.32E+02	1.03E+02	3.42E+01	2.25E+01	6.13E+01	1.38E+01	7.26E+01	7.91E+01	2.10E+01
	h	+	+	+	+	+	+	+	+	+	+	+	+	+	+	≈
F24	mean	4.95E+02	4.89E+02	5.23E+02	4.68E+02	5.46E+02	7.29E+02	1.04E+03	1.14E+03	5.01E+02	5.29E+02	6.42E+02	6.02E+02	5.02E+02	7.23E+02	5.35E+02
	std	1.65E+01	2.09E+01	1.45E+01	1.22E+01	2.94E+01	1.36E+02	8.79E+01	2.59E+02	1.71E+01	1.92E+01	2.72E+01	1.75E+01	3.22E+01	8.58E+01	5.12E+01
	h	≈	≈	+	−	+	+	+	+	≈	+	+	+	≈	+	≈
F25	mean	3.87E+02	3.87E+02	3.87E+02	3.89E+02	3.91E+02	3.80E+02	4.26E+02	1.93E+03	4.04E+02	3.87E+02	3.85E+02	3.87E+02	4.92E+02	4.24E+02	3.79E+02
	std	2.43E+00	1.90E+00	2.74E−01	1.74E+00	3.89E+00	1.08E+00	2.55E+01	5.55E+02	1.24E+01	1.76E+00	1.78E+01	2.01E−01	4.10E+01	1.54E+01	1.46E+00
	h	≈	≈	+	+	+	≈	+	+	+	≈	≈	≈	+	+	≈
F26	mean	1.64E+03	1.48E+03	6.76E+02	1.46E+03	2.84E+03	4.21E+03	4.44E+03	7.73E+03	1.72E+03	1.25E+03	2.83E+03	2.86E+03	2.08E+03	5.52E+03	2.40E+03
	std	1.49E+02	6.86E+02	1.65E+02	1.29E+03	5.64E+02	1.03E+03	9.76E+02	1.28E+03	1.36E+03	9.25E+02	2.05E+02	1.66E+02	2.43E+02	8.47E+02	2.27E+02
	h		≈	+	≈	+	+	+	+	≈	+	+	+	≈	+	≈

(continued)

Table 3. (*continued*)

No.	Item	EO	EAPSO	CLPSO	CSO	PSO	DE/ best/ 1	HHO	RSA	TLBO	SSA	GA	ABC	GWO	WOA	HCDE
F27	mean	5.34E+02	5.39E+02	5.11E+02	5.34E+02	5.68E+02	5.00E+02	5.91E+02	1.28E+03	5.39E+02	5.27E+02	4.83E+02	5.05E+02	5.38E+02	7.23E+02	5.00E+02
	std	1.47E+01	1.72E+01	5.70E+00	1.14E+01	6.00E+01	2.87E−04	9.18E+01	5.37E+02	1.69E+01	1.90E+01	6.92E+00	5.64E+00	1.72E+01	1.69E+02	1.13E−04
	h	+	+	+	+	+	≈	+	+	+	−	−	≈	+	+	
F28	mean	3.44E+02	3.31E+02	4.22E+02	4.30E+02	3.86E+02	4.54E+02	4.53E+02	3.29E+03	3.94E+02	3.97E+02	3.48E+02	4.19E+02	5.74E+02	5.14E+02	5.00E+02
	std	6.03E+01	6.88E+01	3.98E+00	2.48E+01	5.75E+01	4.50E+01	2.13E+01	9.56E+02	1.17E+01	3.26E+01	7.31E+01	8.44E+00	4.19E+01	3.86E+01	1.28E−04
	h	+	+	+	+	+	≈	+	+	+	+	≈	+	+	+	
F29	mean	7.02E+02	8.86E+02	5.66E+02	6.52E+02	7.35E+02	1.28E+03	1.65E+03	3.55E+03	8.66E+02	9.75E+02	1.06E+03	1.03E+03	8.23E+02	2.17E+03	4.65E+02
	std	1.06E+02	1.94E+02	4.27E+01	8.66E+01	8.13E+01	4.59E+02	1.80E+02	1.07E+03	1.56E+02	1.85E+02	8.01E+01	1.17E+02	2.16E+02	4.58E+02	5.27E+01
	h	+	+	+	+	+	+	+	+	+	+	+	+	+	+	≈
F30	mean	5.75E+03	7.52E+03	6.84E+03	7.29E+03	9.24E+03	4.50E+03	1.47E+06	9.24E+08	4.00E+03	1.41E+06	2.81E+03	9.91E+03	4.34E+06	1.26E+07	2.94E+03
	std	2.47E+03	3.98E+03	2.86E+03	2.87E+03	5.33E+03	6.35E+01	1.24E+06	7.40E+08	2.72E+03	9.94E+05	2.59E+03	6.21E+03	3.82E+06	6.42E+06	1.68E+02
	w/t/l	13/14/3	12/14/4	15/9/6	14/7/9	17/10/3	23/5/2	28/1/1	30/0/0	16/13/1	20/8/2	22/4/4	23/4/3	20/9/1	28/2/0	0/30/0
	rank	5.55	5.9	4.67	5.33	7.23	7.09	12.01	14.49	6.47	7.68	8.49	8.81	9.34	13.16	3.81

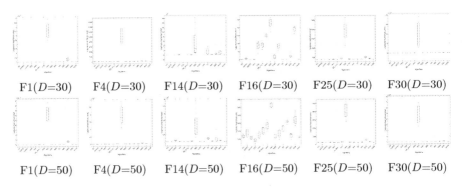

F1(D=30)	F4(D=30)	F14(D=30)	F16(D=30)	F25(D=30)	F30(D=30)

F1(D=50)	F4(D=50)	F14(D=50)	F16(D=50)	F25(D=50)	F30(D=50)

Fig. 3. The convergence curves of HCDE and 14 efficient optimizers obtained on the CEC 2017 benchmarks.

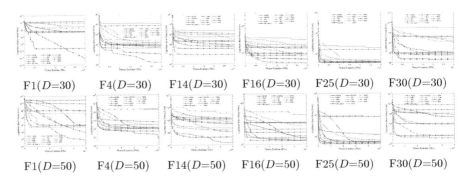

F1(D=30)	F4(D=30)	F14(D=30)	F16(D=30)	F25(D=30)	F30(D=30)

F1(D=50)	F4(D=50)	F14(D=50)	F16(D=50)	F25(D=50)	F30(D=50)

Fig. 4. The Box plot of HCDE and 14 efficient optimizers obtained on the CEC 2017 benchmarks.

When dealing with multimodal functions such as F4, HCDE exhibits a slight decline in convergence speed, but its performance remains outstanding. This success can be attributed to the algorithm's hierarchical competition mechanism and adaptive differential mutation strategy. The hierarchical competition mechanism enables HCDE to dynamically balance exploration and exploitation, allowing it to effectively navigate the complex landscape of multimodal functions. Additionally, The adaptive differential mutation strategy plays a crucial role in enhancing the algorithm's flexibility. This adaptability allows the algorithm to tailor its approach to the specific demands of the optimization process, contributing to its robust performance. This adaptability allows HCDE to fine-tune its search strategy, contributing to its robust performance in addressing the challenges posed by modal and multimodal functions.

With increasing problem complexity(like F14, F16, F25 and F30), HCDE outperforms the other 14 optimizers in convergence speed. However, its performance slightly declines as the problem dimensionality increases from 30 to 50. Despite this, HCDE maintains high convergence accuracy in both unimodal and multimodal functions, showcasing its overall superiority.

(a) 30 dimensions   (b) 50 dimensions

Fig. 5. Comparison of Friedman test for different algorithms in 30 and 50 dimensions

Statistical Results and Analysis: In Fig. 4, the statistical box plots present an analysis of algorithm stability across different function types. The selected functions include the unimodal function F1, multimodal function F4, hybrid function F14, and F16 and composition functions F25, and F30, each evaluated in two distinct dimensions. The box plot of HCDE exhibits a flat shape and a lower position than other optimizers. This indicates improved robustness and stability, highlighting HCDE's effectiveness.

The Friedman test involves the computation of average fitness values for 14 metaheuristic algorithms, and the outcomes are visually presented in Fig. 5. It can be seen from the table that HCDE ranks first, with an average ranking of 3.4, ahead of CLPSO, which ranks second. The smaller the rank result is, the higher the performance of the algorithm. From Fig. 5, we can notice that HCDE provides a precise search ability. Besides, from Table 3 and Table 4, we also find that HCDE get better results than other algorithms in $F1$, $F4$, $F12$, $F14$, $F16$, $F25$, $F30$. The Friedman test serves to identify broad difference among algorithms, while the Wilcoxon signed-rank test allows for a more detailed examination of specific distinctions. The Wilcoxon signed-rank test offers further insights into HCDE's performance relative to specific algorithms. In the majority of cases, HCDE performs similarly to the other without significant differences. However, in certain cases, HCDE may lag behind CSO and CLPSO. The indicator '$+/=$ $/-$' in Table 3 and Table 4 is haven based on the Wilcoxon signed-rank test, and it

Table 4. The Experimental Results for the 50 dimensions of CEC 2017 benchmark suite

No.	Item	EO	EAPSO	CLPSO	CSO	PSO	DE/best/1	HHO	RSA	TLBO	SSA	GA	ABC	GWO	WOA	HCDE
F1	mean	2.15E+03	2.25E+03	6.75E+01	2.08E+03	1.53E+04	8.66E+07	4.53E+07	6.54E+10	6.27E+03	8.18E+03	4.74E+03	2.34E+03	5.90E+09	1.38E+07	3.58E−01
	std	2.07E+03	1.21E+03	6.32E+01	2.64E+03	1.05E+04	1.94E+08	6.47E+06	1.64E+10	6.67E+03	5.59E+03	5.89E+03	1.76E+03	1.36E+09	1.87E+07	8.00E−01
	h		+	+	+	+	+	+	+	+	+	+	+	+	+	
F2	mean	2.18E+18	5.43E−05	3.69E+29	3.06E+32	1.88E+16	1.61E+01	3.06E+29	1.54E+69	1.97E+17	1.06E+03	6.64E+03	7.37E+59	2.31E+49	8.48E+50	6.38E−06
	std	4.65E+18	3.05E−05	6.84E+29	6.78E+32	3.22E+16	3.59E+01	6.80E+29	2.96E+69	4.19E+17	1.71E+03	1.22E+04	1.41E+60	2.63E+49	1.88E+51	5.60E−06
	h	+	≈	+	+	+	+	+	+	+	+	+	+	+	+	
F3	mean	4.56E+03	1.32E+00	5.35E+04	5.03E+04	9.82E−02	1.69E−14	1.90E+04	1.08E+05	5.96E+02	1.51E−07	2.75E+05	1.45E+05	7.89E+04	7.08E+04	4.38E−03
	std	4.18E+03	2.94E+00	7.75E+03	2.15E+04	1.23E−01	3.78E−14	5.84E+03	4.36E+03	4.23E+02	1.46E−08	7.19E+04	1.30E+04	1.12E+04	1.12E+04	7.17E−03
	h	+	≈	+	+	+	−	+	+	+	+	+	+	+	+	
F4	mean	9.33E+01	4.59E+01	9.34E+01	1.65E+02	9.37E+01	4.96E+01	2.62E+02	1.95E+04	1.13E+02	1.29E+02	9.57E+01	1.43E+02	6.00E+02	2.91E+02	6.65E+00
	std	4.23E+01	6.13E+01	1.71E+01	4.29E+01	3.51E+01	2.74E+01	3.97E+01	7.10E+03	3.10E+01	7.11E+01	5.76E+01	3.26E+01	2.80E+02	5.72E+01	4.41E+00
	h	+	≈	+	+	+	+	+	+	+	+	+	+	+	+	
F5	mean	1.87E+02	1.76E+02	1.23E+02	7.08E+01	2.30E+02	3.83E+02	3.88E+02	5.49E+02	1.84E+02	2.48E+02	2.73E+02	4.14E+02	2.08E+02	4.89E+02	1.48E+02
	std	2.00E+01	3.04E+01	1.32E+01	1.24E+01	3.57E+01	1.01E+02	2.91E+01	1.47E+02	2.52E+01	6.18E+01	4.24E+01	1.87E+01	6.32E+00	1.24E+02	3.17E+01
	h	+	≈	−	−	+	+	+	+	+	+	+	+	+	+	
F6	mean	9.91E−01	1.83E+01	4.55E−12	5.70E−01	3.97E+01	5.34E+01	7.23E+01	8.71E+01	1.98E+01	3.59E+01	6.47E−03	8.90E−02	1.67E+01	7.20E+01	2.80E−01
	std	7.56E−01	8.67E+00	1.93E−12	5.46E−01	7.09E+00	2.12E+00	4.86E+00	1.61E+01	4.17E+00	4.04E+00	2.03E−03	1.84E−02	5.99E+00	5.02E+00	1.53E−01
	h	+	+	−	+	+	+	+	+	+	+	−	+	+	+	
F7	mean	1.94E+02	3.67E+02	1.74E+02	1.24E+02	2.89E+02	1.17E+03	1.12E+03	1.35E+03	3.83E+02	4.25E+02	4.69E+02	4.81E+02	3.72E+02	9.43E+02	2.95E+02
	std	1.60E+01	8.81E+01	9.34E+00	1.69E+01	7.41E+01	1.81E+02	4.13E+01	1.44E+02	1.22E+02	1.82E+02	1.48E+02	1.21E+01	8.63E+01	1.10E+02	3.35E+01
	h	≈	+	≈	−	+	+	+	+	≈	+	+	+	≈	+	
F8	mean	1.56E+02	2.18E+02	1.27E+02	7.00E+01	2.28E+02	2.83E+02	3.84E+02	6.29E+02	1.94E+02	2.49E+02	2.32E+02	4.21E+02	2.18E+02	3.44E+02	1.55E+02
	std	1.35E+01	5.94E+01	7.09E+00	1.58E+01	2.78E+01	2.61E+01	2.58E+01	1.39E+02	4.33E+01	6.52E+01	5.75E+01	1.90E+01	3.25E+01	4.14E+01	4.09E+01
	h	≈	+	−	−	+	+	+	+	+	+	+	+	+	+	
F9	mean	2.60E+02	2.39E+03	3.40E+02	6.89E+01	8.91E+03	1.45E+04	1.84E+04	3.37E+04	2.45E+03	9.20E+03	8.41E+03	1.48E+02	4.84E+03	1.80E+04	4.34E+03
	std	2.52E+02	5.28E+02	1.76E+02	4.75E+01	2.98E+03	5.14E+03	2.89E+03	1.28E+04	1.07E+03	2.45E+03	2.67E+03	3.96E+01	3.04E+03	8.53E+03	2.47E+03
	h	−	+	≈	−	+	+	+	+	≈	+	+	−	+	+	
F10	mean	5.85E+03	6.54E+03	4.55E+03	3.36E+03	6.00E+03	7.46E+03	7.37E+03	1.10E+04	9.58E+03	6.31E+03	5.96E+03	1.33E+04	5.53E+03	9.13E+03	9.85E+03
	std	1.35E+03	6.83E+02	3.87E+02	1.00E+03	7.04E+02	7.13E+02	1.26E+03	1.31E+03	3.92E+03	6.83E+02	7.28E+02	1.47E+02	7.34E+02	1.68E+03	3.43E+02
	h	≈	≈	−	−	≈	+	+	+	+	+	≈	+	≈	+	
F11	mean	1.74E+02	1.77E+02	1.02E+02	2.21E+02	2.11E+02	3.29E+02	3.68E+02	1.18E+04	1.93E+02	2.57E+02	2.30E+03	2.34E+02	4.97E+03	6.19E+02	1.46E+02
	std	5.28E+01	6.53E+01	1.42E+01	9.10E+01	2.88E+01	1.51E+02	5.96E+01	4.20E+03	3.85E+01	2.64E+01	4.06E+03	1.07E+01	2.58E+03	1.38E+02	5.16E+01
	h	+	+	−	+	+	+	+	+	≈	+	+	+	+	+	
F12	mean	5.59E+05	6.77E+04	4.48E+06	1.26E+06	2.17E+05	2.01E+05	6.66E+07	5.43E+10	2.55E+05	8.73E+06	1.95E+06	5.67E+07	5.84E+08	9.29E+07	1.49E+05
	std	2.51E+05	4.49E+04	1.44E+06	7.94E+05	1.56E+05	3.31E+05	4.69E+07	8.05E+09	1.87E+05	3.95E+06	1.37E+06	3.86E+07	5.02E+08	3.91E+07	9.36E+04
	h	+	≈	+	+	+	+	+	+	+	+	+	+	+	+	
F13	mean	1.29E+04	1.73E+04	6.33E+04	5.53E+03	3.16E+03	1.31E+04	2.51E+06	3.35E+10	4.71E+03	8.70E+04	8.88E+03	3.60E+03	1.70E+08	1.36E+05	2.24E+03
	std	1.62E+04	1.15E+04	1.02E+02	5.15E+03	1.76E+03	8.78E+03	2.12E+06	8.45E+09	2.16E+03	4.64E+04	7.03E+03	6.45E+03	9.60E+07	5.50E+04	2.36E+03
	h	+	+	+	+	≈	+	+	+	≈	+	+	≈	+	+	
F14	mean	2.25E+04	2.12E+03	8.04E+05	1.46E+05	2.42E+04	1.78E+03	3.67E+05	1.18E+07	5.03E+04	1.50E+04	1.97E+06	2.24E+05	1.89E+06	6.45E+05	2.40E+02
	std	1.41E+04	8.89E+02	3.48E+05	7.01E+04	1.01E+04	2.64E+03	2.43E+05	1.01E+07	3.22E+04	9.04E+03	5.61E+05	9.45E+04	3.58E+06	4.56E+05	1.36E+02
	h	+	+	+	+	+	+	+	+	+	+	+	+	+	+	
F15	mean	1.42E+04	7.52E+03	2.18E+02	3.52E+03	5.12E+03	2.13E+04	3.63E+05	5.76E+09	5.75E+03	3.69E+04	1.29E+04	6.32E+02	8.99E+06	6.37E+04	1.94E+03
	std	5.47E+03	8.15E+03	8.29E+01	4.42E+03	4.87E+03	2.28E+04	6.76E+04	2.49E+09	7.41E+03	2.16E+04	8.78E+03	2.89E+02	1.09E+07	3.13E+04	1.52E+03
	h	+	≈	−	≈	+	+	+	+	≈	+	+	−	+	+	
F16	mean	1.38E+03	1.68E+03	1.14E+03	8.39E+02	1.47E+03	1.96E+03	2.69E+03	3.28E+03	1.60E+03	2.33E+03	3.16E+03	1.39E+03	3.36E+03	1.22E+03
	std	2.13E+02	4.61E+02	1.58E+02	2.59E+02	3.32E+02	4.14E+02	5.55E+02	2.81E+02	2.54E+02	5.41E+02	3.62E+02	2.33E+02	3.04E+02	7.24E+02	8.86E+01
	h	+	+	≈	−	+	+	+	+	+	+	+	≈	+	≈	
F17	mean	1.14E+03	1.24E+03	7.06E+02	8.83E+02	1.21E+03	1.83E+03	1.99E+03	4.03E+03	1.17E+03	1.30E+03	1.96E+03	1.98E+03	8.51E+02	2.22E+03	5.94E+02
	std	2.74E+02	2.83E+02	1.36E+02	1.39E+02	3.74E+02	4.42E+02	4.05E+02	1.97E+03	1.13E+02	2.53E+02	4.43E+02	1.03E+02	2.79E+02	5.75E+02	1.16E+02
	h	+	+	≈	≈	+	+	+	+	+	+	+	+	≈	+	
F18	mean	2.17E+05	1.49E+04	9.34E+05	1.60E+06	6.40E+04	4.72E+04	2.66E+06	3.42E+07	3.66E+05	2.05E+05	3.27E+06	6.30E+06	3.19E+06	7.14E+06	3.75E+04
	std	1.54E+05	6.15E+03	6.18E+05	8.34E+05	4.87E+04	4.16E+03	2.04E+06	2.23E+07	3.19E+05	7.61E+04	2.49E+06	3.43E+06	2.40E+06	2.25E+06	3.95E+04
	h	+	≈	+	+	+	≈	+	+	+	+	+	+	+	+	
F19	mean	3.06E+04	1.70E+04	8.33E+04	2.18E+04	1.22E+04	3.63E+03	7.22E+05	2.05E+09	9.32E+03	4.15E+05	2.50E+04	2.39E+04	8.72E+05	2.81E+06	1.49E+04
	std	1.26E+04	1.37E+04	2.28E+01	1.40E+04	9.54E+03	4.07E+03	3.04E+05	1.47E+09	5.99E+03	2.68E+05	1.56E+04	6.37E+03	5.33E+05	1.96E+06	1.34E+04
	h	≈	≈	+	≈	−	−	+	+	−	+	+	≈	+	+	
F20	mean	7.94E+02	1.23E+03	4.58E+02	7.38E+02	1.09E+03	1.18E+03	1.37E+03	1.60E+03	5.26E+02	1.21E+03	1.86E+03	1.76E+03	7.87E+02	1.68E+03	3.94E+02
	std	2.39E+02	3.27E+02	7.51E+01	2.66E+02	3.35E+02	2.70E+02	3.02E+02	1.90E+02	2.22E+02	1.96E+02	4.92E+02	1.79E+02	1.73E+02	3.15E+02	1.06E+02
	h	+	+	≈	+	+	+	+	+	≈	+	+	+	+	+	
F21	mean	3.54E+02	3.79E+02	3.48E+02	2.76E+02	4.52E+02	5.57E+02	7.50E+02	1.08E+03	3.64E+02	4.72E+02	4.78E+02	6.08E+02	4.03E+02	7.45E+02	3.88E+02
	std	2.46E+01	2.69E+01	1.88E+01	1.80E+01	3.49E+01	7.98E+01	6.15E+01	7.03E+01	4.20E+01	3.91E+01	6.30E+01	8.92E+00	3.37E+01	1.21E+02	5.15E+01
	h	≈	≈	≈	−	+	+	+	+	≈	+	+	+	+	+	
F22	mean	5.87E+03	7.16E+03	5.56E+03	4.43E+03	5.29E+03	7.04E+03	8.53E+03	1.21E+04	8.05E+03	7.72E+03	6.36E+03	1.35E+04	6.68E+03	8.83E+03	9.07E+03
	std	3.81E+02	7.79E+02	3.80E+02	1.22E+03	3.10E+03	8.03E+02	1.55E+03	1.36E+03	5.57E+03	9.26E+02	8.32E+02	2.62E+02	8.32E+02	1.34E+03	1.15E+03
	h	−	+	−	−	−	+	+	+	≈	+	≈	+	−	≈	
F23	mean	5.52E+02	6.25E+02	5.65E+02	5.21E+02	8.62E+02	1.05E+03	1.39E+03	1.83E+03	7.58E+02	7.35E+02	7.27E+02	8.27E+02	6.87E+02	1.33E+03	6.58E+02
	std	1.96E+01	3.99E+01	2.23E+01	2.19E+01	1.06E+02	1.28E+02	1.86E+02	3.21E+02	8.98E+01	8.40E+01	2.95E+01	1.52E+01	1.12E+02	1.41E+02	4.84E+01
	h	−	≈	−	−	+	+	+	+	+	+	+	+	≈	+	
F24	mean	6.21E+02	6.93E+02	7.40E+02	5.99E+02	8.98E+02	1.31E+03	1.84E+03	2.12E+03	7.45E+02	6.91E+02	1.18E+03	8.78E+02	7.88E+02	1.22E+03	8.15E+02
	std	4.19E+01	2.30E+01	3.47E+01	2.29E+01	1.10E+02	1.36E+02	2.99E+02	2.78E+02	6.91E+01	5.70E+01	1.57E+02	1.98E+01	1.43E+02	1.38E+02	5.89E+01
	h	−	−	−	−	+	+	+	+	≈	−	+	+	≈	+	
F25	mean	5.81E+02	5.50E+02	5.41E+02	5.57E+02	5.42E+02	4.92E+02	6.28E+02	8.10E+03	5.60E+02	5.27E+02	4.56E+02	5.30E+02	1.16E+03	6.53E+02	4.44E+02
	std	5.09E+01	2.61E+01	9.12E+00	3.96E+01	3.06E+01	3.29E+01	3.21E+01	1.94E+03	3.38E+01	4.18E+01	2.28E+01	1.98E+01	1.81E+02	5.18E+01	1.74E+01
	h	+	+	+	+	+	+	+	+	+	+	≈	+	+	+	
F26	mean	2.59E+03	3.29E+03	2.08E+03	2.24E+03	1.88E+03	6.87E+03	4.78E+03	1.20E+04	6.16E+03	1.82E+03	4.56E+03	1.83E+03	3.68E+03	1.05E+04	4.60E+03
	std	3.99E+02	5.99E+02	8.11E+02	1.51E+02	2.16E+03	1.40E+03	3.77E+03	2.76E+03	1.51E+03	2.11E+03	9.70E+02	1.19E+02	9.38E+02	1.87E+03	9.84E+02
	h	−	≈	−	−	−	+	+	+	≈	−	+	−	+	+	

(continued)

Table 4. (*continued*)

No.	Item	EO	EAPSO	CLPSO	CSO	PSO	DE/best/1	HHO	RSA	TLBO	SSA	GA	ABC	GWO	WOA	HCDE
F27	mean	7.20E+02	7.16E+02	6.39E+02	7.75E+02	8.11E+02	5.00E+02	6.57E+02	3.66E+03	1.01E+03	6.66E+02	7.69E+02	5.58E+02	8.45E+02	1.34E+03	5.00E+02
	std	1.33E+02	7.36E+01	4.62E+01	6.55E+01	8.54E+01	4.41E-04	3.50E+02	4.95E+02	2.31E+02	2.30E+01	1.36E+02	2.51E+01	1.08E+02	1.63E+02	1.20E-04
	h	+	+	+	+	+	≈	+	+	+	+	+	+	+	+	
F28	mean	5.10E+02	5.17E+02	5.26E+02	5.50E+02	4.94E+02	4.94E+02	5.50E+02	6.08E+03	5.11E+02	4.92E+02	4.47E+02	4.79E+02	1.19E+03	6.39E+02	5.00E+02
	std	3.20E+01	4.01E+01	6.08E+00	4.41E+01	2.39E+01	1.28E+01	3.25E+01	1.52E+03	5.37E+01	2.62E+01	6.39E+00	2.16E+01	2.51E+01	5.27E+01	2.32E-01
	h	≈	≈	+	≈	≈	≈	+	+	+	-	≈	+	+	+	
F29	mean	1.19E+03	1.41E+03	8.29E+02	8.40E+02	1.66E+03	2.55E+03	2.27E+03	2.18E+04	1.81E+03	1.91E+03	1.67E+03	2.03E+03	1.45E+03	3.65E+03	9.22E+02
	std	4.37E+02	2.06E+02	1.23E+02	1.76E+02	3.95E+02	5.08E+02	3.80E+02	9.65E+03	5.65E+02	4.27E+02	3.28E+02	2.20E+02	3.02E+02	8.48E+02	3.78E+02
	h	≈	≈	≈	≈	+	+	+	+	+	+	+	≈	+	+	
F30	mean	1.24E+06	7.93E+05	7.51E+05	1.00E+06	8.74E+05	3.64E+03	1.26E+07	4.98E+09	9.42E+05	2.16E+07	4.11E+06	9.01E+05	5.99E+07	1.10E+08	1.17E+03
	std	2.60E+05	1.37E+05	1.09E+05	1.18E+05	5.66E+04	3.02E+03	1.05E+07	1.75E+09	9.72E+04	3.17E+06	3.05E+05	1.79E+05	2.74E+07	2.36E+07	9.89E+02
	h	+	+	+	+	≈	+	+	+	≈	+	+	≈	+	+	
	w/t/l	14/9/7	10/17/3	11/11/8	11/8/11	14/14/2	20/6/4	27/2/1	29/1/0	12/18/0	20/7/3	18/8/4	22/6/2	19/9/2	28/2/0	0/30/0
	rank	5.56	5.79	4.44	4.99	6.82	7.51	11.52	14.65	6.95	7.78	8.13	9.32	9.59	12.71	4.25

shows the number of test functions in which the compared algorithms performed better, similar, or worse compared to HCDE in the condition of $\alpha = 0.05$. We can notice that HCDE and TLBO have the lowest number of significant differences.

In summary, the findings from the box plot, Friedman test, and Wilcoxon signed-rank test underscore the robustness, stability, exceptional performance, and statistical significance of HCDE across diverse optimization problems within the domain of intelligent optimization algorithms.

5   Conclusion

This research introduces a new algorithm, HCDE, characterized by its innovative approach. It combines a Hierarchical Competition Mechanism and an Adaptive Differential Mutation Strategy, based on competition outcomes, significantly bolstering the algorithm's global search capabilities.

Through comprehensive comparisons involving CEC 2017 benchmark, the results emphasize HCDE's exceptional performance in contrast to conventional algorithms such as GA and PSO. HCDE also demonstrates competitiveness when compared to contemporary algorithms like CSO and WOA, validating the substantial enhancements made to HCDE through combined efforts in algorithm exploitation and exploration. In addition to demonstrating statistical competitiveness through the Friedman and Wilcoxon signed-rank tests, HCDE's robustness and stability across various functions and dimensions are further confirmed by box plots. HCDE's strength is particularly apparent in its effective handling of unimodal and multimodal problems, solving the challenges associated with local optima since Hierarchical Competition Mechanism and Adaptive Differential Mutation Strategy lead algorithm to enhance diversity to promote algorithm convergence.

Certainly, HCDE exhibits certain limitations, including an increase in time overhead as the complexity of the problem and dimensionality grows. Despite conducting experiments across multiple dimensions, HCDE may not deliver superior results for better problems. Future research endeavours should prioritize enhancing the algorithm's robustness and expanding its practical applications to further elevate HCDE's performance, particularly in the realm of complex system optimization.

Acknowledgements. This work is supported by the National Natural Science Foundation of China (No. 62006144), the Major Fundamental Research Project of Shandong, China (No. ZR2019ZD03), and the Taishan Scholar Project of Shandong, China (No. ts20190924).

Data Availibility Statement. Data will be made available on request.

CRediT Authorship Contribution Statement. Hongtong Xi: Investigation, Methodology, Writing - original draft. Qingke Zhang: Conceptualization, Methodology, Investigation, Writing - original draft, Writing - review & editing, Supervision, Funding, Project administration. **Xiaoyu Liu:** Investigation, Methodology, Software, Visualization. **Huixia Zhang:** Investigation, Writing - review & editing, Resources. **Shuang Gao:** Writing - review & editing. **Huaxiang Zhang:** Writing - review & editing, Resources.

Declaration of Competing Interest. The authors declare that they have no known competing financial interests or personal relationships that could have appeared to influence the work reported in this paper.

References

1. Abualigah, L., Abd Elaziz, M., Sumari, P., Geem, Z.W., Gandomi, A.H.: Reptile search algorithm (RSA): a nature-inspired meta-heuristic optimizer. Expert Syst. Appl. **191**, 116158 (2022)
2. Arrif, T., Hassani, S., Guermoui, M., Sánchez-González, A., Taylor, R.A., Belaid, A.: GA-GOA hybrid algorithm and comparative study of different metaheuristic population-based algorithms for solar tower heliostat field design. Renew. Energy **192**, 745–758 (2022)
3. Azizi, M., Talatahari, S., Gandomi, A.H.: Fire hawk optimizer: a novel metaheuristic algorithm. Artif. Intell. Rev. **56**(1), 287–363 (2023)
4. Chen, H., Li, S., Li, X., Zhao, Y., Dong, J.: A hybrid adaptive differential evolution based on gaussian tail mutation. Eng. Appl. Artif. Intell. **119**, 105739 (2023)
5. Dehghani, M., Montazeri, Z., Trojovská, E., Trojovský, P.: Coati optimization algorithm: a new bio-inspired metaheuristic algorithm for solving optimization problems. Knowl.-Based Syst. **259**, 110011 (2023)
6. Dorigo, M., Di Caro, G.: Ant colony optimization: a new meta-heuristic. In: Proceedings of the 1999 Congress on Evolutionary Computation-CEC99 (Cat. No. 99TH8406), vol. 2, pp. 1470–1477. IEEE (1999)
7. Faramarzi, A., Heidarinejad, M., Stephens, B., Mirjalili, S.: Equilibrium optimizer: a novel optimization algorithm. Knowl.-Based Syst. **191**, 105190 (2020)
8. Heidari, A.A., Mirjalili, S., Faris, H., Aljarah, I., Mafarja, M., Chen, H.: Harris hawks optimization: algorithm and applications. Futur. Gener. Comput. Syst. **97**, 849–872 (2019)
9. Karaboga, D., Basturk, B.: A powerful and efficient algorithm for numerical function optimization: artificial bee colony (ABC) algorithm. J. Global Optim. **39**(3), 459–471 (2007)
10. Karaboga, D., et al.: An idea based on honey bee swarm for numerical optimization. Technical report, Technical report-tr06, Erciyes University, Engineering Faculty, Computer (2005)

11. Kennedy, J., Eberhart, R.: Particle swarm optimization. In: Proceedings of ICNN 1995 - International Conference on Neural Networks, vol. 4, pp. 1942–1948. IEEE (1995)

12. Kirkpatrick, S., Gelatt, C.D., Jr., Vecchi, M.P.: Optimization by simulated annealing. Science **220**(4598), 671–680 (1983)

13. Liang, J.J., Qin, A.K., Suganthan, P.N., Baskar, S.: Comprehensive learning particle swarm optimizer for global optimization of multimodal functions. IEEE Trans. Evol. Comput. **10**(3), 281–295 (2006)

14. Mirjalili, S., Gandomi, A.H., Mirjalili, S.Z., Saremi, S., Faris, H., Mirjalili, S.M.: Salp swarm algorithm: a bio-inspired optimizer for engineering design problems. Adv. Eng. Softw. **114**, 163–191 (2017)

15. Mirjalili, S., Lewis, A.: The whale optimization algorithm. Adv. Eng. Softw. **95**, 51–67 (2016)

16. Mirjalili, S., Mirjalili, S.M., Lewis, A.: Grey wolf optimizer. Adv. Eng. Softw. **69**, 46–61 (2014)

17. Mitchell, M.: An Introduction to Genetic Algorithms. MIT Press, Cambridge (1998)

18. Pierezan, J., Coelho, L.D.S.: Coyote optimization algorithm: a new metaheuristic for global optimization problems. In: 2018 IEEE Congress on Evolutionary Computation, pp. 1–8. IEEE (2018)

19. Cheng, R., Jin, Y.: A competitive swarm optimizer for large scale optimization. IEEE Trans. Cybern. **45**(2), 191–204 (2015)

20. Rao, R.V., Savsani, V.J., Vakharia, D.: Teaching-learning-based optimization: a novel method for constrained mechanical design optimization problems. Comput. Aided Des. **43**(3), 303–315 (2011)

21. Seyyedabbasi, A., Kiani, F.: Sand cat swarm optimization: a nature-inspired algorithm to solve global optimization problems. Eng. Comput. **39**(4), 2627–2651 (2023)

22. Shi, Y.: Particle swarm optimization. IEEE Connections **2**(1), 8–13 (2004)

23. Simon, D.: Biogeography-based optimization. IEEE Trans. Evol. Comput. **12**(6), 702–713 (2008)

24. Storn, R., Price, K.: Differential evolution–a simple and efficient heuristic for global optimization over continuous spaces. J. Global Optim. **11**(4), 341–359 (1997)

25. Wang, F., Wang, X., Sun, S.: A reinforcement learning level-based particle swarm optimization algorithm for large-scale optimization. Inf. Sci. **602**, 298–312 (2022)

26. Wang, X., Xu, J., Huang, C.: Fans optimizer: a human-inspired optimizer for mechanical design problems optimization. Expert Syst. Appl. **228**, 120242 (2023)

27. Wang, Y., Li, B., Weise, T., Wang, J., Yuan, B., Tian, Q.: Self-adaptive learning based particle swarm optimization. Inf. Sci. **181**(20), 4515–4538 (2011)

28. Zhang, Q., Gao, H., Zhan, Z.H., Li, J., Zhang, H.: Growth optimizer: a powerful metaheuristic algorithm for solving continuous and discrete global optimization problems. Knowl.-Based Syst. **261**, 110206 (2023)

29. Zhang, Y.: Elite archives-driven particle swarm optimization for large scale numerical optimization and its engineering applications. Swarm Evol. Comput. **76**, 101212 (2023)

A Hybrid Response Strategy for Dynamic Constrained Multi-objective Optimization

Jinhua Zheng⬿, Wang Che[✉], Yaru Hu⬿, and Juan Zou

Key Laboratory of Intelligent Computing and Information Processing,
Ministry of Education, School of Computer Science and School
of Cyberspace Science of Xiangtan University, Xiangtan 411100, China
chewang2023@163.com

Abstract. Multi-objective optimization problems are widely present in the real world. When the objective functions of these problems change over time and are subject to constrained limitations, they become dynamic constrained multi-objective optimization problems (DCMOPs). As the problems become more complex, multi-objective optimization algorithms face greater challenges. In this paper, a hybrid response strategy for dynamic constrained multi-objective optimization algorithm (HRDC) is proposed to address these challenges. Specifically, a classical constraint coevolutionary framework (CCMO) is used to handle constraints. In addition, two independent response strategies are included. The first strategy is a region-based sampling strategy, which analyzes the previous environment's optimal population, divides the decision space into uniform grids, and performs random sampling within each grid to improve the distribution of the initial population in the new environment. The second strategy is the classification prediction strategy. Firstly, the CCMO is employed to obtain a feasible priority population and an unconstrained population. Then, prediction is performed separately on each population, with the unconstrained population assisting evolution from the perspective of infeasible solutions. The effectiveness of the algorithm is validated through two sets of test instances. The experimental results demonstrate that compared to several state-of-the-art algorithms, HRDC exhibits strong competitiveness in handling DCMOPs.

Keywords: Dynamic constrained multi-objective optimization ·
Evolutionary algorithms · Change response strategies

1 Introduction

Multi-objective optimization has significant application value in various fields, such as controller optimization [1], mechanical design [2], scheduling [3], etc. However, in practical scenarios, there is a class of problems where the objective function, relevant parameters, and constraints may changes over time. These problems are called dynamic constrained multi-objective optimization problems

© The Author(s), under exclusive license to Springer Nature Singapore Pte Ltd. 2024
L. Pan et al. (Eds.): BIC-TA 2023, CCIS 2061, pp. 172–184, 2024.
https://doi.org/10.1007/978-981-97-2272-3_13

(DCMOPs), which consist of a series of evolving environments where the population has a limited amount of time to search for optimal solutions to the problem. Such problems introduce greater complexity and difficulty in the solution using evolutionary algorithms. Because it requires the population to quickly track (i.e., converge) to the optimal positions in dynamic environments, also known as the pareto-optimal front (PF) and pareto-optimal set (PS) [4], while also ensuring the feasibility of individuals. Constraints have a significant impact on algorithms, especially in dynamic environments. Therefore, researching dynamic constrained multi-objective optimization algorithms (DCMOEAs) holds essential theoretical and practical significance.

To solve DCMOPs, a few DCMOEAs have recently been proposed. Azzouz et al. [5,6] and Chen et al. [7,8] utilized an adaptive penalty function approach to handle infeasible solutions, but this method suffers from inaccurate evaluation. Azzouz et al. [6] employed a feasibility-driven strategy, while Chen et al. [7] introduced random migration and memory mechanisms. However, we note that the dynamic response of these algorithms primarily focuses on feasible solutions, with less emphasis on infeasible solutions. Additionally, the population diversity cannot be guaranteed when responding to environmental changes, making it prone to getting trapped in local optima.

This paper aims to design an optimization algorithm to solve constrained multi-objective optimization problems in complex dynamic environments. In summary, the contributions of this paper are as follows:

1) A region-based sampling strategy is proposed to increase the diversity of the population.
2) A classification prediction strategy is proposed to respond to environmental changes from the perspective of infeasible solutions.

This paper will be organized as follows: In Sect. 2, the dynamic constrained multi-objective optimization problem will be modeled, and relevant definitions will be provided. Section 3 will provide a detailed description of the designed dynamic constrained multi-objective optimization algorithm, including the algorithm framework and the design principles of critical strategies. Section 4 will present the algorithm experiments and analysis, evaluating its performance through experiments comparing and discussing it with other algorithms. Finally, in Sect. 5, the research work will be summarized, and future directions for further research will be proposed.

2   Preliminary

2.1   Definition of DCMOP

Without loss of generality, a standard DCMOP can be formulated as follows.

$$\min F(x,t) = (f_1(x,t), ..., f_m(x,t))^T$$
$$s.t. \begin{cases} x \in \Omega \\ h_k(x,t) = 0, \ k = 1, ..., a \\ g_k(x,t) \leq 0, \ k = 1, ..., b \end{cases} \quad (1)$$

where m is the number of objectives, $x = (x_1, ..., x_n)^T$ denotes n-dimensional decision variables in decision space Ω, and $h_k(x, t)$ and $g_k(x, t)$ are kth equality or inequality constraint at the time t, respectively. The discrete time parameter t is defined [9] as the following mathematical form:

$$t = \frac{1}{n_t} \left\lfloor \frac{\tau}{\tau_t} \right\rfloor \tag{2}$$

where τ is the iteration counter, n_t is the number of distinct time steps in t, and τ_t is the number of iterations where t remains the same.

2.2 Definition of PF and PS

For two solutions $x_1(t)$ and $x_2(t)$ at the time t, if and only if $\forall i \in \{1, ..., m\}$, $f_i(x_1(t)) \leq f_i(x_2(t))$ and $\exists j \in \{1, ..., m\}$, $f_j(x_1(t)) < f_j(x_2(t))$, $x_1(t)$ is considered to pareto-dominate $x_2(t)$, denoted as $x_1(t) \prec x_2(t)$. If none of the solutions dominate $x_1(t)$, then $x_1(t)$ is called a pareto-optimal solution [10]. The set composed of all pareto-optimal solutions at the time t is called the current pareto-optimal set (PS). The objective space corresponding to all pareto solutions is called the pareto-optimal front (PF).

2.3 Definition of Dynamic Feasible Region

The feasible region at time t is defined [11] as follows:

$$O(t) = \{x \in \Omega | CV(x, t) = 0\} \tag{3}$$

$$CV(x, t) = \sum_{i=1}^{a+b} CV_i(x, t) \tag{4}$$

$$CV_i(x, t) = \begin{cases} \max(0, g_p(x, t)), \ p = 1, ..., a \\ \max(0, |h_q(x, t)| - \eta), \ q = 1, ..., b \end{cases} \tag{5}$$

where CV is the degree of constraint violation, and η is a very small positive value. e.g., $\eta = 1e - 4$.

2.4 Classification of DCMOP

DCMOP can generally be classified into the following three categories: [1]:
   Type I: Objective functions change over time and constraints are fixed.
   Type II: Objective functions are fixed while constraints change over time.
   Type III: Both objective functions and constraints change over time.

3   Proposed Algorithm

In this section, we provide a detailed description of the proposed HRDC algorithm. Firstly, the overall framework of the algorithm is presented. Then, the details of the change response strategies are provided, including a region-based sampling strategy and a classification prediction strategy.

3.1   The Framework of HRDC

To improve the likelihood of the population approaching the PS in the new environment, as well as to utilize helpful infeasible solutions to quickly track the PS, a DCMOEA called HRDC is proposed, which employs a hybrid response strategy. Algorithm 1 presents the basic framework of HRDC. Initially, when there is no change in the environment, a classical cooperative coevolution framework called CCMO [12] is used as the static algorithm (line 3 of Algorithm 1). In this framework, two populations evolve cooperatively, with one population prioritizing feasibility and another disregarding constraints and prioritizing better objective values. It is important to note that the cooperation between these two populations is weak cooperation, meaning that they are independent of each other during their respective mating processes and only exchange information about the offspring after their generation. If a change in the environment is detected, a hybrid response strategy is employed, which includes a region-based sampling strategy and a classification prediction strategy (line 6-7 of Algorithm 1). Once a change in the environment occurs, the populations obtained from these two strategies will be used as the initial populations in the new environment (line 8 of Algorithm 1). The following sections will provide detailed explanations of these two strategies.

Algorithm 1. The framework of HRDC

Input: N (population size), the stopping criterion
Output: A series of approximated populations P
1: Random initialize a population P_0 with size N;
2: **while** the stopping criterion is not met **do**
3:     Pop1, Pop2 $\leftarrow$ Perform CCMO;
4:     **if** change is detected **then**
5:         $t = t + 1$;
6:         $P_1 \leftarrow$ Grid-based initialization strategy(Pop1);
7:         $P_2 \leftarrow$ Classification prediction strategy(Pop1, Pop2);
8:         $P_t = P_1 \cup P_2$;
9:     **end if**
10: **end while**

3.2   Region-Based Sampling Strategy

The main idea of this strategy is to increase the diversity of the population by analyzing the information on the previous environment's optimal solutions and conducting region-based sampling in the decision space. Firstly, based on the distribution of non-dominated feasible solutions from the previous environment in the decision space, the decision variables are divided into two categories: significant dimensions and insignificant dimensions. The significant dimensions refer to the dimensions with the most considerable variance in the population distribution. For ease of subsequent region partitioning, all dimensions except the significant dimensions are considered insignificant dimensions.

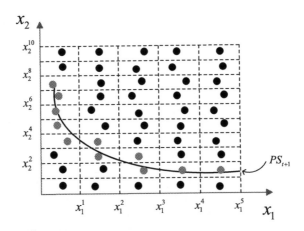

Fig. 1. Schematic diagram of region-based sampling in 2D space.

As shown in Fig. 1, which illustrates an example of region-based sampling in a two-dimensional decision space, x_1 represents the significant dimension while x_2 represents the insignificant dimension. Next, the significant dimension is uniformly divided into $h1$ regions, and the insignificant dimension is uniformly divided into h_2 regions, resulting in a total of $h_1 \geq h_2$ grid regions. In the example, $h_1 = 5$ and $h_2 = 10$. The next step is to perform random sampling in each grid region. The following formula represents the sampling for the significant dimension:

$$x_{IS}^i = L_{IS} + (i + rand) \times \frac{U_{IS} - L_{IS}}{h_1}, \, i = 1, 2, ..., h_1 \tag{6}$$

where x_{IS}^i represents the i-th sampled value for the significant dimension, and rand is a random number uniformly distributed between 0 and 1. L_{IS} and U_{IS} represent the minimum and maximum boundaries of the significant dimension, respectively. The sampling formula for the insignificant dimension is:

$$x_S^j = L_S + (j + rand) \times \frac{U_S - L_S}{h_2}, \, j = 1, 2, ..., h_2 \tag{7}$$

where x_S^j represents the j-th sampled value for the insignificant dimension, and L_S and U_S represent the minimum and maximum boundaries of the insignificant dimension, respectively. The sampled points are ultimately composed of $x = (x_{IS}^i, x_S^j)$, where $i = 1, 2, ..., h_1, j = 1, 2, ..., h_2$.

From Fig. 1, it can be observed that in the insignificant dimension, PS_{t+1} has a more minor variance. To increase the chances of sampling individuals exploring the vicinity of PS_{t+1}, compared to the significant dimension, we divide the insignificant dimension into more regions. Therefore, while ensuring $h_1 * h_2 = N$, we set $h_2 \geq 2 * h_1$. Algorithm 2 provides the specific steps for region-based sampling.

Algorithm 2. Region-based sampling strategy

Input: Pop1
Output: P_1

1: Based on the optimal feasible solutions from the previous environment, the decision variables are divided into significant dimensions and insignificant dimensions;
2: Uniformly divide the two dimensions into segments to form a grid region with a uniform distribution;
3: To sample N individuals' positions in the decision space according to Eq. (6) and Eq. (7).
4: Evaluate these individuals as the initialized population P_1.

3.3 Classification Prediction Strategy

In many cases, environmental changes exhibit linearity and regularity, where the current environment may have the same or similar changes as the previous environment [13]. In these cases, we assume that the movement trend of the central point represents the movement trend of the entire population. In other words, other individuals in the population should have a similar movement trend to the central point. Therefore, predicting the movement trend of the central point is crucial. At the same time, we should consider that compared to a feasibility-driven population, an unconstrained population has a more excellent distribution. Using individuals from this part of the population can be beneficial in predicting potential Pareto solutions. Thus, when detecting environmental changes, we utilize the movement trends of the central points in the previous two environments to predict the two populations separately. The main idea of the classification prediction strategy is represented in Algorithm 3.

In Algorithm 3, first, the center point and movement trend Dir_t of the non-dominated feasible solutions population at time t are calculated using Eqs. 9 and 8 (lines 1-2 of Algorithm 3) [14]. Then, Eqs. 10 and 11 are used separately to obtain the populations P_{t+1}^1 and P_{t+1}^2 predicted for the next time step for these two classes of populations (lines 3-4 of Algorithm 3). Finally, the individuals predicted from these two classes are merged to form the population P_2 (line 5 of Algorithm 3).

Algorithm 3. Classification prediction strategy

Input: Pop1, Pop2
Output: P_2
1: Calculate the population centers according to Eq.9;
2: Predict the moving direction according to Eq.8;
3: $P^1_{t+1} \leftarrow$ generate new solutions at time period t+1; //refer to Eq.10
4: $P^2_{t+1} \leftarrow$ generate new solutions at time period t+1; //refer to Eq.11
5: $P_2 = P^1_{t+1} \cup P^2_{t+1}$;

Assume $PS_t = \{x^t_1, x^t_2, ..., x^t_D\}$ represents all feasible non-dominated individuals obtained at time t, and let C^t_{Pop1} represent the center point of the population composed of these feasible non-dominated individuals in the decision space. The movement trend of the center point at time $t+1$ can be represented as follows:

$$Dir_t = \left\{ Dir^t_1, Dir^t_2, ..., Dir^t_i, ..., Dir^t_D \right\}, i = 1, 2, 3, ..., D \tag{8}$$

where

$$Dir^t_i = C^t_{Pop1} - C^{t-1}_{Pop1}$$
$$C^t_{Pop1} = \frac{\sum_{x^t \in PS_t} x^t_i}{|PS_t|}, i = 1, 2, ..., D \tag{9}$$

where Dir^t_i represents the movement trend of the i-th dimension of the decision variable population at time t, and x^t_i represents the decision variable values of all individuals in the population at time t in the i-th dimension.

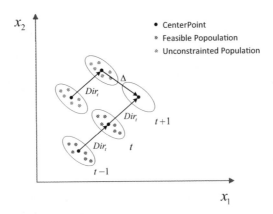

Fig. 2. Schematic diagrams of predictions using two populations.

The prediction formulas for the two populations are as follows:

$$PS^1_{t+1} = PS^1_t + Dir_t \tag{10}$$

$$PS^2_{t+1} = PS^2_t + Dir_t + \alpha\Delta \tag{11}$$

where PS_{t+1}^1 and PS_{t+1}^2 represent the decision space values of the predicted populations at the next time step by Pop1 and Pop2, respectively, α is a random number obtained from a uniform distribution $\cup(0.5, 1.5)$. Here, $\Delta = C_{Pop1}^t - C_{Pop2}^t$, and C_{Pop2}^t represents the center point of Pop2 in the decision space. The vector Δ represents the distance vector between the center points of Pop1 and Pop2 in the decision space. In other words, the prediction of Pop2 is adjusted based on the prediction of the center point, aiming to improve the assistance of the unconstrained population in predicting the next PF by guiding it with the distance vector Δ between the center points of the two populations. Figure 2 illustrates the schematic diagram of predicting the population at time $t+1$ using the two populations when the environment changes at time t.

4   Experimental Setup

4.1   Test Instances and Comparative Algorithms

Ten test instances are used in our proposed algorithm to compare with other algorithms. It contains five DCTP instances (DCTP4-DCTP8) [5], and five Ins instances (Ins4-Ins8) [7]. These test problems belong to Type III. Four popular algorithms are used for comparison in this paper. They are DC-NSGA-II-A [5], DC-NSGA-II-B [5], DC-MOEA [6], and dCMOEA [7] respectively, with different dynamic change response mechanisms.

4.2   Parameter Settings

The settings of the relevant parameters are as follows:
  1) Population size: N = 100;
  2) The change frequency τ_t was set to values of 10, 20, and 30;
  3) The number of evaluations per generation is N*τ_t;
  4) Significant dimension has $h_1 = 5$ segments, while non-significant dimension has $h_2 = 20$ segments;
  5) Each algorithm performs 20 independent runs on each test instance, and DCTP and Ins runs consist of 5 and 20 environmental changes, respectively.

4.3   Performance Indicators

To evaluate the performance of the algorithm, mean inverted generational distance (MIGD) [15] and mean Hypervolume (MHV) [16] are used as performance metrics. The MIGD can comprehensively reflect the distribution and convergence of Pareto optimal solutions obtained under the total environments [17]. The MHV is a representative metric that measures the comprehensive performance of algorithm [18]. The MIGD value is calculated as follows:

$$MIGD = \frac{1}{T} \sum_{t=1}^{T} IGD(t) \tag{12}$$

$$IGD(t) = \frac{1}{|PF_t|} \sum_{x \in PF_t^*} \min_{y \in PF_t} dis(x, y) \tag{13}$$

where T is the number of environments in a run, PF_t^* represent a uniformly distributed true PF at time t, PF_t represents the set of solutions obtained by the algorithm at time t, $dis(x, y)$ represents the Euclidean distance between a point x in PF_t^* and a point y in PF_t. The MHV value is calculated as follows:

$$MHV = \frac{1}{T} \sum_{1}^{T} HV(t) \tag{14}$$

$$HV = Vol\left(\bigcup_{i=1}^{|PF_t|} v_i\right) \tag{15}$$

where $Vol(\cdot)$ denotes the Lebesgue measure, $|PF_t|$ represents the number of non-dominated solutions in PF_t, v_i represents the hypervolume formed by the reference point and the i-th solution in PF_t.

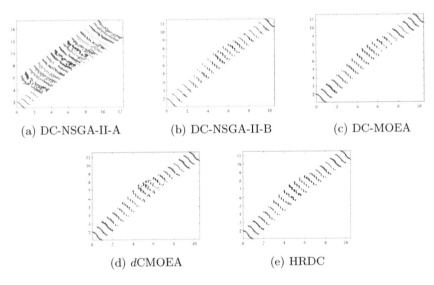

(a) DC-NSGA-II-A   (b) DC-NSGA-II-B   (c) DC-MOEA

(d) dCMOEA   (e) HRDC

Fig. 3. The scatter plot of the population obtained by HRDC and other peer algorithms(mean MIGD values) on Ins4 for time period t from 0 to 20. The black line represents the true PF under different environments. The points of other colors represent the solutions obtained by each algorithm (Color figure online).

4.4   Experimental Results and Analysis

Under the parameter configuration described in the previous section, HRDC is compared with other state-of-the-art dynamic multi-objective algorithms. The statistical data (the mean and standard deviation, where standard deviation displays in brackets) for each indicator (MIGD and MHVD) are calculated and listed in Tables 1–2. To make the comparison results in Tables 1–2 more evident, the best results for each test instance are bolded.

Tables 1 and 2 show that HRDC demonstrates strong competitiveness compared to DC-NSGA-II-A, DC-NSGA-II-B, DC-MOEA, and dCMOEA regarding the number of cases achieving the best results. Among the 30 cases shown in Table 1, HRDC achieves the best MIGD results in 21 cases. However, its average performance on the DCTP4 and DCTP5 test problems with $\tau_t = 10$. This is mainly because the truePF of these two test problems is controlled by constraint boundaries, resulting in a tiny truePF. It poses a significant challenge for the CCMO framework, and the effect of the prediction strategy needs to be evident due to the test problems having only five variations. Additionally, HRDC is slightly inferior to DC-MOEA on the Ins6 test problem with taut = 30. This may be due to the CCMO framework being relatively weak in late-stage convergence, indicating the need to develop a more advanced algorithm to handle constraints. In Table 2, HRDC achieves the best MHV results in 24 cases, which is an improvement compared to MIGD. This indicates that the algorithm proposed in this paper performs better regarding overall distribution and convergence.

Table 1. MIGD indicators of five algorithms on DCTP and Ins test suites.

Problem	τ_t	DC-NSGA-II-A	DC-NSGA-II-B	DC-MOEA	dCMOEA	HRDC
DCTP4	10	5.1358e−1 (1.03e−1) −	**4.9145e−1 (1.40e−1)** +	8.6756e−1 (2.88e−1) −	8.9875e−1 (1.97e−1) −	5.3462e−1 (7.40e−2)
	20	**3.8156e−1 (6.52e−2)** +	3.9391e−1 (1.32e−1) +	7.8003e−1 (3.15e−1) −	7.3388e−1 (1.62e−1) −	4.4800e−1 (4.37e−2)
	30	**3.3486e−1 (7.11e−2)** +	3.5088e−1 (1.41e−1) +	1.0685e+0 (2.84e−1) −	5.9423e−1 (1.43e−1) −	4.1362e−1 (7.69e−2)
DCTP5	10	1.8536e−1 (1.09e−1) −	**8.1421e−2 (4.65e−2)** +	4.7738e−1 (1.55e−1) −	3.7911e−1 (1.30e−1) −	1.0902e−1 (2.41e−2)
	20	2.3423e−1 (1.56e−1) −	**7.0315e−2 (6.08e−2)** +	5.4593e−1 (1.22e−1) −	3.7800e−1 (1.32e−1) −	8.6482e−2 (1.91e−2)
	30	2.4762e−1 (1.93e−1) −	**5.9585e−2 (4.25e−2)** +	5.6894e−1 (1.39e−1) −	3.3554e−1 (1.12e−1) −	8.1061e−2 (2.94e−2)
DCTP6	10	3.7285e+0 (2.11e+0) −	1.1809e+1 (5.20e+0) −	2.6562e+0 (1.17e+0) −	1.1521e+0 (4.56e−1) −	**6.2064e−1 (3.77e−1)**
	20	3.0899e+0 (1.72e+0) −	1.2149e+1 (4.76e+0) −	2.1865e+0 (1.22e+0) −	3.7107e−1 (1.29e−1) −	**2.8230e−1 (3.05e−1)**
	30	2.6584e+0 (1.79e+0) −	1.1668e+1 (5.12e+0) −	2.0270e+0 (1.16e+0) −	3.9225e−1 (2.11e−1) −	**2.2612e−1 (1.42e−1)**
DCTP7	10	2.4508e−1 (1.55e−1) −	1.9606e−1 (1.25e−1) −	2.1550e−1 (1.51e−1) −	2.2987e−1 (1.08e−1) −	**5.2366e−2 (4.79e−2)**
	20	1.4228e−1 (1.30e−1) −	8.5661e−2 (4.46e−2) −	1.3036e−1 (9.13e−2) −	1.3254e−1 (8.40e−2) −	**3.6809e−2 (4.47e−2)**
	30	9.5123e−2 (9.24e−2) −	7.0595e−2 (5.81e−2) −	1.0023e−1 (1.03e−1) −	1.2054e−1 (1.20e−1) −	**2.1377e−2 (1.44e−2)**
DCTP8	10	3.7133e+0 (1.61e+0) −	1.1561e+1 (4.53e+0) −	2.3174e+0 (7.91e−1) −	1.4211e+1 (5.15e−1) −	**1.0533e+0 (5.16e−1)**
	20	3.1940e+0 (1.78e+0) −	1.0459e+1 (5.60e+0) −	2.6728e+0 (9.93e−1) −	4.6513e−1 (2.37e−1) −	**2.3283e−1 (2.03e−1)**
	30	2.5980e+0 (1.41e+0) −	1.3398e+1 (1.41e+0) −	2.2679e+0 (9.52e−1) −	4.7410e−1 (2.26e−1) −	**3.4992e−1 (3.70e−1)**
Ins4	10	6.7945e−2 (5.43e−3) −	5.8151e−2 (3.10e−3) −	5.8862e−2 (6.28e−3) −	8.1992e−2 (5.38e−3) −	**2.5063e−2 (1.78e−3)**
	20	2.8806e−2 (7.59e−3) −	2.3448e−2 (2.41e−3) −	2.6770e−2 (5.22e−3) −	4.0098e−2 (5.98e−3) −	**1.9059e−2 (2.32e−3)**
	30	1.8152e−2 (3.94e−3) =	1.7299e−2 (3.12e−3) −	2.2337e−2 (6.78e−3) −	3.3892e−2 (6.09e−3) −	**1.6174e−2 (2.26e−3)**
Ins5	10	9.2194e−2 (1.32e−2) −	7.5791e−2 (3.43e−3) −	7.4257e−2 (7.06e−3) −	9.8128e−2 (7.98e−3) −	**3.2360e−2 (2.11e−3)**
	20	2.8806e−2 (7.59e−3) −	2.3448e−2 (2.41e−3) −	2.6770e−2 (5.22e−3) −	4.0098e−2 (5.98e−3) −	**1.9059e−2 (2.32e−3)**
	30	2.1349e−2 (8.10e−3) −	1.9564e−2 (2.39e−3) −	2.0983e−2 (4.74e−3) −	3.2328e−2 (2.67e−3) −	**1.7869e−2 (1.95e−3)**
Ins6	10	5.1009e−2 (3.50e−3) −	4.3805e−2 (2.96e−3) −	4.0600e−2 (2.32e−3) −	5.5949e−2 (2.97e−3) −	**2.7247e−2 (1.62e−3)**
	20	2.0394e−2 (1.07e−3) −	2.0182e−2 (1.39e−3) −	1.9667e−2 (1.19e−3) −	2.2780e−2 (1.80e−3) −	**1.8235e−2 (1.41e−3)**
	30	1.5445e−2 (1.31e−3) −	1.4954e−2 (1.23e−3) −	**1.4110e−2 (8.86e−4)** +	1.8002e−2 (1.50e−3) −	1.5349e−2 (1.65e−3)
Ins7	10	5.4089e−2 (2.76e−3) −	4.8304e−2 (2.95e−3) −	4.3995e−2 (2.15e−3) −	6.0480e−2 (3.54e−3) −	**3.0018e−2 (1.30e−3)**
	20	2.2230e−2 (1.28e−3) −	2.2086e−2 (1.65e−3) −	2.0863e−2 (1.23e−3) −	2.4441e−2 (1.72e−3) −	**1.9532e−2 (1.50e−3)**
	30	1.6260e−2 (1.30e−3) −	1.5915e−2 (1.41e−3) −	**1.4407e−2 (1.12e−3)** =	1.8558e−2 (1.16e−3) −	1.5128e−2 (1.72e−3)
Ins8	10	5.4089e−2 (2.76e−3) −	4.8304e−2 (2.95e−3) −	4.3995e−2 (2.15e−3) −	6.0480e−2 (3.54e−3) −	**3.0018e−2 (1.30e−3)**
	20	3.2961e−2 (1.30e−2) −	2.6008e−2 (2.41e−3) −	3.0781e−2 (4.61e−3) −	3.6911e−2 (2.72e−3) −	**2.2715e−2 (2.29e−3)**
	30	2.3390e−2 (1.14e−2) −	**1.9270e−2 (2.23e−3)** =	2.4205e−2 (9.07e−3) −	3.0272e−2 (2.15e−3) −	1.9751e−2 (2.31e−3)
+/−/=		2/25/3	6/21/3	1/28/1	0/30/0	

Table 2. MHV indicators of five algorithms on DCTP and Ins test suites.

Problem	τ_t	DC-NSGA-II-A	DC-NSGA-II-B	DC-MOEA	dCMOEA	HRDC
DCTP4	10	2.8549e-1 (1.80e−2) +	**3.1713e−1 (1.51e−2)** +	2.4354e−1 (3.04e−2) −	2.5025e−1 (2.39e−2) −	2.7000e−1 (2.12e−2)
	20	3.3673e−1 (1.18e−2) +	**3.5298e−1 (1.72e−2)** +	2.7040e−1 (3.63e−2) −	2.8762e−1 (2.07e−2) −	3.0694e−1 (1.41e−2)
	30	3.5484e−1 (1.30e−2) +	**3.7097e−1 (2.13e−2)** +	2.3766e−1 (3.93e−2) −	3.1877e−1 (2.00e−2) −	3.1911e−1 (1.38e−2)
DCTP5	10	2.9828e−1 (4.98e−2) −	**3.5802e−1 (1.87e−2)** +	1.8706e−1 (4.14e−2) −	1.9959e−1 (3.15e−2) −	3.3081e−1 (1.34e−2)
	20	2.9827e−1 (5.97e−2) −	**3.6583e−1 (2.93e−2)** +	1.7335e−1 (3.55e−2) −	2.0265e−1 (3.50e−2) −	3.5148e−1 (1.24e−2)
	30	2.9850e−1 (7.41e−2) −	**3.7207e−1 (1.85e−2)** +	1.6881e−1 (4.53e−2) −	2.1496e−1 (3.28e−2) −	3.5758e−1 (1.89e−2)
DCTP6	10	2.0485e−1 (8.46e−2) −	6.7353e−2 (1.53e−1) −	2.4090e−1 (6.42e−2) −	3.1891e−1 (4.01e−2) −	**3.4626e−1 (3.64e−2)**
	20	2.4677e−1 (8.28e−2) −	5.7077e−2 (1.48e−1) −	2.7521e−1 (8.12e−2) −	3.9935e−1 (1.12e−2) −	**4.1744e−1 (3.01e−2)**
	30	2.7393e−1 (8.43e−2) −	6.8009e−2 (1.56e−1) −	2.8488e−1 (7.77e−2) −	4.0061e−1 (1.98e−2) −	**4.2531e−1 (1.62e−2)**
DCTP7	10	6.1284e−1 (4.11e−2) −	6.2604e−1 (3.37e−2) −	6.2087e−1 (4.16e−2) −	6.0954e−1 (3.35e−2) −	**6.6682e−1 (2.24e−2)**
	20	6.4289e−1 (3.82e−2) −	6.5880e−1 (1.61e−2) −	6.4755e−1 (2.71e−2) −	6.4463e−1 (2.53e−2) −	**6.7645e−1 (2.11e−2)**
	30	6.6100e−1 (2.70e−2) −	6.6506e−1 (1.73e−2) −	6.5797e−1 (3.47e−2) −	6.4971e−1 (3.39e−2) −	**6.8454e−1 (7.70e−3)**
DCTP8	10	1.3143e−1 (5.32e−2) −	3.5435e−2 (8.27e−2) −	1.6697e−1 (5.14e−2) −	2.1930e−1 (2.68e−2) =	**2.2405e−1 (4.17e−2)**
	20	1.5635e−1 (6.04e−2) −	6.4283e−2 (1.15e−1) −	1.3657e−1 (4.33e−2) −	2.8732e−1 (1.97e−2) −	**3.1513e−1 (2.27e−2)**
	30	1.8575e−1 (5.59e−2) −	4.0497e−3 (2.22e−2) −	1.3406e−1 (5.03e−2) −	2.9172e−1 (2.02e−2) −	**3.2025e−1 (2.48e−2)**
Ins4	10	2.2401e+0 (1.69e−2) −	2.2579e+0 (4.36e−3) −	2.2578e+0 (4.92e−3) −	2.2365e+0 (1.32e−2) −	**2.2838e+0 (1.49e−3)**
	20	2.2816e+0 (3.24e−3) −	2.2846e+0 (7.78e−4) −	2.2827e+0 (1.95e−3) −	2.2793e+0 (2.19e−3) −	**2.2896e+0 (8.72e−4)**
	30	2.2885e+0 (1.18e−3) −	2.2893e+0 (7.96e−4) −	2.2870e+0 (2.77e−3) −	2.2837e+0 (2.83e−3) −	**2.2919e+0 (5.13e−4)**
Ins5	10	2.2243e+0 (9.89e−3) −	2.2371e+0 (2.05e−3) −	2.2380e+0 (2.92e−3) −	2.2257e+0 (6.83e−3) −	**2.2583e+0 (1.12e−3)**
	20	2.2604e+0 (1.21e−3) −	2.2613e+0 (1.01e−3) −	2.2593e+0 (3.38e−3) −	2.2577e+0 (1.93e−3) −	**2.2653e+0 (8.22e−4)**
	30	2.2658e+0 (2.12e−3) −	2.2666e+0 (6.69e−4) −	2.2655e+0 (1.65e−3) −	2.2610e+0 (2.86e−3) −	**2.2677e+0 (7.45e−4)**
Ins6	10	2.2661e+0 (4.09e−3) −	2.2716e+0 (2.36e−3) −	2.2740e+0 (1.75e−3) −	2.2610e+0 (8.12e−3) −	**2.2861e+0 (8.96e−4)**
	20	2.2883e+0 (7.49e−4) −	2.2886e+0 (6.80e−4) −	2.2890e+0 (5.09e−4) −	2.2898e+0 (5.41e−4) −	**2.2932e+0 (5.13e−4)**
	30	2.2925e+0 (5.48e−4) −	2.2928e+0 (4.02e−4) −	2.2930e+0 (4.98e−4) −	2.2942e+0 (4.09e−4) −	**2.2956e+0 (3.82e−4)**
Ins7	10	2.2428e+0 (1.57e−3) −	2.2463e+0 (1.63e−3) −	2.2486e+0 (1.21e−3) −	2.2413e+0 (1.76e−3) −	**2.2587e+0 (9.91e−4)**
	20	2.2620e+0 (7.91e−4) −	2.2623e+0 (4.92e−4) −	2.2625e+0 (7.65e−4) −	2.2636e+0 (6.04e−4) −	**2.2660e+0 (5.16e−4)**
	30	2.2660e+0 (6.84e−4) −	2.2662e+0 (4.42e−4) −	2.2665e+0 (4.50e−4) −	2.2677e+0 (5.12e−4) −	**2.2688e+0 (4.78e−4)**
Ins8	10	2.2387e+0 (1.46e−2) −	2.2611e+0 (5.30e−3) −	2.2633e+0 (2.99e−3) −	2.2431e+0 (1.45e−2) −	**2.2859e+0 (1.28e−3)**
	20	2.2856e+0 (3.61e−3) −	2.2879e+0 (1.08e−3) −	2.2859e+0 (2.04e−3) −	2.2852e+0 (1.28e−3) −	**2.2920e+0 (1.29e−3)**
	30	2.2916e+0 (3.18e−3) −	2.2932e+0 (7.00e−4) −	2.2912e+0 (2.63e−3) −	2.2907e+0 (1.14e−3) −	**2.2951e+0 (8.85e−4)**
+/−/=		3/27/0	6/24/0	0/30/0	0/28/2	

To better understand the tracking capability of these algorithms, we also plotted the approximate PFs obtained by each algorithm on Ins4. For this test problem, τ_t was set to 20, and the number of environmental changes was 20. Figure 3 show that, except for the proposed HRDC algorithm, the approximate PFs obtained by the other algorithms on this test case do not accurately fit the truePF. This indicates that HRDC exhibits strong responsiveness to environmental changes and can better track the truePF.

5 Conclusion

This paper proposes a new hybrid response mechanism called HRDC, which consists of a partitioned sampling strategy and a classification prediction strategy. The partitioned sampling strategy focuses on diversity and can effectively handle unpredictable changes. The classification prediction strategy responds quickly to predictable changes from the perspective of a subset of infeasible solutions. Experimental results on multiple test problems demonstrate that HRDC, combined with the CCMO framework, can generate a well-distributed set of solutions. However, using some infeasible solutions in this study has yet to be fully explored, leaving room for improvement. Future research should focus on exploring helpful infeasible solutions to achieve faster and more accurate responses to changes.

Acknowledgment. This work was supported in part by the National Natural Science Foundation of China (Grant No. 62176228, Grant No. 62276224 and Grant No. 62306262), in part by the Natural Science Foundation of Hunan Province, China (Grant No. 2022JJ40452, and 2023JJ40637), in part by the Project of Hunan Education Department (Grant No. 21C0077, 21A0444, and 22B0139), Xiaohe Sci-Tech Talents Special Funding under Hunan Provincial Sci-Tech Talents Sponsorship Program Grant No. 2023TJ-X77.

References

1. Huang, L., Suh, I.H., Abraham, A.: Dynamic multi-objective optimization based on membrane computing for control of time-varying unstable plants. Inf. Sci. **181**(11), 2370–2391 (2011)
2. Roy, R., Mehnen, J.: Dynamic multi-objective optimisation for machining gradient materials. CIRP Ann. **57**(1), 429–432 (2008)
3. Deb, K., Rao N., U.B., Karthik, S.: Dynamic multi-objective optimization and decision-making using modified NSGA-II: a case study on hydro-thermal power scheduling. In: Obayashi, S., Deb, K., Poloni, C., Hiroyasu, T., Murata, T. (eds.) EMO 2007. LNCS, vol. 4403, pp. 803–817. Springer, Heidelberg (2007). https://doi.org/10.1007/978-3-540-70928-2_60
4. Jiang, S., Zou, J., Yang, S., Yao, X.: Evolutionary dynamic multi-objective optimisation: a survey. ACM Comput. Surv. **55**(4), 1–47 (2022)
5. Azzouz, R., Bechikh, S., Ben Said, L.: Multi-objective optimization with dynamic constraints and objectives: new challenges for evolutionary algorithms. In: Proceedings of the 2015 Annual Conference on Genetic and Evolutionary Computation, pp. 615–622 (2015)
6. Azzouz, R., Bechikh, S., Said, L.B., Trabelsi, W.: Handling time-varying constraints and objectives in dynamic evolutionary multi-objective optimization. Swarm Evol. Comput. **39**, 222–248 (2018)
7. Chen, Q., Ding, J., Yang, S., Chai, T.: A novel evolutionary algorithm for dynamic constrained multiobjective optimization problems. IEEE Trans. Evol. Comput. **24**(4), 792–806 (2019)
8. Chen, Q., Ding, J., Yen, G.G., Yang, S., Chai, T.: Multi-population evolution based dynamic constrained multiobjective optimization under diverse changing environments. IEEE Trans. Evol. Comput. (2023)
9. Zheng, J., Zhou, Y., Zou, J., Yang, S., Ou, J., Hu, Y.: A prediction strategy based on decision variable analysis for dynamic multi-objective optimization. Swarm Evol. Comput. **60**, 100786 (2021)
10. Deb, K., Pratap, A., Agarwal, S., Meyarivan, T.: A fast and elitist multi-objective genetic algorithm: NSGA-II. IEEE Trans. Evol. Comput. **6**(2), 182–197 (2002)
11. Chen, G., Guo, Y., Wang, Y., Liang, J., Gong, D., Yang, S.: Evolutionary dynamic constrained multiobjective optimization: test suite and algorithm. IEEE Trans. Evol. Comput. (2023)
12. Tian, Y., Zhang, T., Xiao, J., Zhang, X., Jin, Y.: A coevolutionary framework for constrained multiobjective optimization problems. IEEE Trans. Evol. Comput. **25**(1), 102–116 (2020)
13. Chen, Y., Zou, J., Liu, Y., Yang, S., Zheng, J., Huang, W.: Combining a hybrid prediction strategy and a mutation strategy for dynamic multiobjective optimization. Swarm Evol. Comput. **70**, 101041 (2022)

14. Zhou, A., Jin, Y., Zhang, Q.: A population prediction strategy for evolutionary dynamic multiobjective optimization. IEEE Trans. Cybern. **44**(1), 40–53 (2013)
15. Sun, H., Cao, A., Hu, Z., Li, X., Zhao, Z.: A novel quantile-guided dual prediction strategies for dynamic multi-objective optimization. Inf. Sci. **579**, 751–775 (2021)
16. Zitzler, E., Thiele, L.: Multiobjective evolutionary algorithms: a comparative case study and the strength pareto approach. IEEE Trans. Evol. Comput. **3**(4), 257–271 (1999)
17. Zheng, J., Zhou, F., Zou, J., Yang, S., Hu, Y.: A dynamic multi-objective optimization based on a hybrid of pivot points prediction and diversity strategies. Swarm Evol. Comput. **78**, 101284 (2023)
18. Zheng, J., Wu, Q., Zou, J., Yang, S., Hu, Y.: A dynamic multi-objective evolutionary algorithm using adaptive reference vector and linear prediction. Swarm Evol. Comput. **78**, 101281 (2023)

An Adaptive Knowledge Transfer Strategy for Evolutionary Dynamic Multi-objective Optimization

Donghui Zhao[1], Xiaofen Lu[1(✉)], and Ke Tang[1,2]

[1] Guangdong Provincial Key Laboratory of Brain-Inspired Intelligent Computation, Department of Computer Science and Engineering, Southern University of Science and Technology, Shenzhen 518055, China
12132376@mail.sustech.edu.cn, {luxf,tangk3}@sustech.edu.cn
[2] The Research Institute of Trustworthy Autonomous Systems, Southern University of Science and Technology, Shenzhen 518055, China

Abstract. Dynamic multi-objective optimization problems (DMOPs) are optimization problems involve multiple conflicting objectives, and these objectives change over time. The challenge in solving DMOPs is how to quickly track the Pareto optimal solution set when the environment changes. Recently, dynamic multi-objective evolutionary algorithms (DMOEAs) combined with transfer learning (TL) have been proven to be promising in solving DMOPs. TL-based DMOEAs showed advantages in reusing historical information and predicting high-quality solutions in the new environment. Various TL techniques have been employed to DMOEAs, which learn and transfer knowledge either in decision space or in objective space to predict the Pareto optimal solutions. However, problems usually have different types of change in decision and objective spaces. A single knowledge learning and transfer strategy may be unsuitable for all types of DMOPs. In this paper, a DMOEA with an adaptive knowledge learning and transfer strategy is proposed to solve DMOPs. It first estimates the change type of the problem when the environment changes, i.e., whether there exists change in decision or objective spaces, and then based on the change type, it adaptively chooses to learn and transfer knowledge in the decision space or objective space or both to generate an initial population that guides the search in new environment. A comprehensive empirical study is conducted to evaluate the performance of the proposed method. The method is compared to six state-of-the-art prediction-based DMOEAs on widely used DMOP benchmarks. Experimental results demonstrate that the proposed method outperforms or achieves comparable results to the compared algorithms on most of the test problems.

Keywords: Dynamic multi-objective optimization · Evolutionary algorithm · Transfer learning · Change type

© The Author(s), under exclusive license to Springer Nature Singapore Pte Ltd. 2024
L. Pan et al. (Eds.): BIC-TA 2023, CCIS 2061, pp. 185–199, 2024.
https://doi.org/10.1007/978-981-97-2272-3_14

1 Introduction

Multi-objective optimization problems (MOPs) are problems with multiple conflicting objectives that cannot be optimized simultaneously. The optimal solutions for a MOP constitute a set representing different tradeoffs among the objectives. This solution set is called a Pareto-optimal set (POS), and its corresponding objective vectors form a Pareto-optimal front (POF) [5]. Dynamic multi-objective optimization problems (DMOPs) represent a specialized category of MOPs, which introduces time-varying objective functions or constraints. DMOPs widely exist in many real-world application scenarios, such as path planning, job scheduling, and resource allocation.

In recent years, many researchers employed evolutionary algorithms (EAs) to solve DMOPs, namely dynamic multi-objective EAs (DMOEAs). According to the used dynamic handling mechanism, DMOEAs can be categorized into diversity-based methods, memory-based methods, and prediction-based methods [1]. Diversity-based DMOEAs increase the population diversity when the environment changes or maintain the population diversity during the search process. Memory-based DMOEAs store the optimal solutions found in the historical environment and reuse them when a similar environment is detected. Prediction-based DMOEAs learn prediction models using historical information and then predict high-quality solutions in the new environment. There are also some DMOEAs that use multiple dynamic handling mechanisms together.

Recently, there is an increasing interest in conducting research on evolutionary transfer optimization, which integrates EAs with knowledge learning and transfer across related domains [21]. The transfer learning (TL) techniques do not require the independent and identically distributed assumption and can be used as prediction strategies in DMOEAs. Jiang et al. first proposed a TL-based DMOEA, Tr-DMOEA [11], which uses the transfer component analysis (TCA) to predict high-quality solutions in the new environment. Inspired by the promising performance of Tr-DMOEA, more TL techniques are introduced to solve DMOPs including manifold TL (MTL) [14], TrAdaBoosting [12], and autoencoder [7,25]. Most existing TL-based DMOEAs transfer knowledge either in decision space or objective space. However, problems may have different types of changes in decision space or objective space. A single knowledge transfer strategy may be unsuitable for all types of DMOPs.

To address this issue, a DMOEA with an adaptive knowledge transfer strategy is proposed to solve DMOPs based on the change type of problems, namely change type estimation based DMOEA (CTE-DMOEA). It adaptively chooses to transfer knowledge in decision space or objective space or both according to the estimated change type of an environmental change. To summarize, the major contributions of this work are as follows:

1) A change type estimation method is proposed to estimate the change patterns of POS and POF for every environmental change.
2) A change type based TL method using a kernelized autoencoding model is proposed, which can adaptively choose to employ prediction in decision or objective space according to the change type.

The remainder of this paper is organized as follows. Section 2 presents the definition of DMOP, the kernelized autoencoding model and a review of the existing prediction-based DMOEAs. In Sect. 3, the details of the proposed method are presented. Section 4 presents the experimental studies and analysis. Section 5 gives the conclusion and future work directions.

2   Related Work

2.1   Dynamic Multi-objective Optimization Problems

For a minimization problem, DMOP can be defined as follows:

$$\min F(\mathbf{x}, t) = \{f_1(\mathbf{x}, t), f_2(\mathbf{x}, t), ..., f_m(\mathbf{x}, t)\}$$
$$\text{s.t.} \quad \mathbf{x} \in \Omega \tag{1}$$

where $\mathbf{x} = (x_1, x_2, ..., x_d)$ is the d-dimensional decision vector in decision space $\Omega \subseteq \mathbb{R}^d$. t is the time or environment variable. $f_i(\mathbf{x}, t) : \Omega \to \mathbb{R}(i = 1, ..., m)$ defines the ith objective function at time t and m is the number of objectives. The aim of solving DMOPs is to get a set of diverse solutions that are close to the true POF under each environment as soon as possible. There are several definitions for DMOPs according to [1].

Definition 1 (Dynamic Pareto Dominance): Given two decision vectors $\mathbf{x}_1$ and $\mathbf{x}_2$, $\mathbf{x}_1$ is said to dominate $\mathbf{x}_2$ at time step t, which is denoted by $\mathbf{x}_1 \succ_t \mathbf{x}_2$, if and only if

$$\begin{cases} \forall i \in \{1, ..., m\}, \quad f_i(\mathbf{x}_1, t) \leq f_i(\mathbf{x}_2, t) \\ \exists i \in \{1, ..., m\}, \quad f_i(\mathbf{x}_1, t) < f_i(\mathbf{x}_2, t) \end{cases} \tag{2}$$

Definition 2 (Dynamic Pareto-Optimal Set, DPOS). If there is no other decision vector dominating a decision vector $\mathbf{x}^*$ at time t, $\mathbf{x}^*$ is called a nondominated solution or Pareto-optimal solution. All Pareto-optimal solutions constitute the DPOS:

$$\text{DPOS} = \{\mathbf{x}^* | \nexists \mathbf{x}, \mathbf{x} \succ_t \mathbf{x}^*\} \tag{3}$$

Definition 3 (Dynamic Pareto-Optimal Front, DPOF). The set of objective vectors of solutions in DPOS at time t is called a DPOF:

$$\text{DPOF} = \{F(\mathbf{x}^*, t) | \mathbf{x}^* \in \text{DPOS}\} \tag{4}$$

The dynamic of objective functions in DMOPs lead to the changing POS or POF. Farina et al. [6] categorized DMOPs into the following four types:

- Type I: POS changes over time while POF remains unchanged;
- Type II: Both POS and POF change over time;
- Type III: POS is fixed while POF changes over time;
- Type IV: Both POS and POF are fixed, although the objective functions may change over time.

2.2   Kernelized Autoencoding Evolutionary Search

Autoencoder (AE) [2] is a type of neural network used for unsupervised learning, which aims to compress and then reconstruct the input. AE have also been used as a transfer learning technique, which learns representations from source problem and transfer to help in solving target problem [17]. Feng et al. [7] introduced AE to solve optimization problems and proposed the autoencoding evolutionary search. Zhou et al. [24] further proposed a kernelized autoencoding method to handle the nonlinear relationship.

In particular, given two different optimization problems $\mathbf{OP}_1$ and $\mathbf{OP}_2$, and the optimal solution sets of two problems $\mathbf{P} \in \mathbb{R}^{d \times N}$ and $\mathbf{Q} \in \mathbb{R}^{d \times N}$, where N is the number of solutions and d is the dimension of decision vectors. Then a mapping from $\mathbf{OP}_1$ to $\mathbf{OP}_2$, denoted by $\mathbf{M} \in \mathbb{R}^{d \times d}$, can be learnt by a kernelized autoencoder model that minimizes the loss function

$$L(\mathbf{M}) = \frac{1}{2N} \sum_{i=1}^{N} ||\mathbf{q}_i - \mathbf{M}\Phi(\mathbf{p}_i)||^2 \tag{5}$$

where $\Phi(\cdot)$ maps $\mathbf{P}$ into a Reproducing Kernel Hilbert Space (RKHS) $\mathcal{H}$. The matrix form can be expressed as

$$L(\mathbf{M}) = \frac{1}{2N} tr[(\mathbf{Q} - \mathbf{M}\Phi(\mathbf{P}))^T (\mathbf{Q} - \mathbf{M}\Phi(\mathbf{P}))] \tag{6}$$

where $tr(\cdot)$ denotes the trace of a matrix. The mapping can be expressed as $\mathbf{M} = \mathbf{M}_k \Phi(\mathbf{P})^T$, which represents a linear combination of the mapped solutions in $\mathcal{H}$. The loss function thus becomes

$$L(\mathbf{M}_k) = \frac{1}{2N} tr[(\mathbf{Q} - \mathbf{M}_k\mathbf{K}(\mathbf{P},\mathbf{P}))^T (\mathbf{Q} - \mathbf{M}_k\mathbf{K}(\mathbf{P},\mathbf{P}))] \tag{7}$$

where $\mathbf{K}(\mathbf{P},\mathbf{P}) = \Phi(\mathbf{P})^T\Phi(\mathbf{P})$ is the kernel matrix built by a kernel function $k(\cdot,\cdot)$. Equation(7) holds a closed-form solution

$$\mathbf{M}_k = \mathbf{Q}\mathbf{K}(\mathbf{P},\mathbf{P})^T (\mathbf{K}(\mathbf{P},\mathbf{P})\mathbf{K}(\mathbf{P},\mathbf{P})^T)^{-1} \tag{8}$$

A solution set $\mathbf{S}$ can be transferred by

$$\mathbf{S}_{tr} = \mathbf{M}\Phi(\mathbf{S}) = \mathbf{M}_k\mathbf{K}(\mathbf{P},\mathbf{S}) \tag{9}$$

2.3   Prediction-Based Dynamic Multi-objective Evolutionary Algorithms

Prediction-based DMOEAs learn the experience from the past environments and then predict an initial population to guide the search when the environment changes. This approach enables the learning of dynamic tendency in DMOPs and facilitates the adaption to changing environments. Various prediction-based DMOEAs have been presented by researchers. Zhou et al. [23] introduced a population prediction strategy(PPS) using an autoregression model, which predicts a manifold of POS in the new environment. Muruganantham et al. [20] proposed a prediction-based DMOEA that integrates MOEA/D with a Kalman filter. In this method, each dimension of the decision vector is predicted independently. Cao et al. [4] proposed to predict new optimal solutions with a support vector regression (SVR) model by training the model on a series of solutions from consecutive time steps. Although these prediction-based DMOEAs have shown superiority in obtaining high-quality solutions, most of them assume the POS distributions in different environments obey an independent and identical distribution (IID) condition. As a result, these methods meet challenges when solving DMOPs with non-identical distribution of optimal solutions [11].

To address this issue, transfer learning techniques are introduced as prediction methods to solve DMOPs. TL-based methods can release the IID condition, assuming the distributions of optimal solutions from changing environments are different but related. Jiang et al. [11] first introduced TL techniques to DMOEAs and proposed the Tr-DMOEA. Tr-DMOEA constructs a latent space using the TCA technique. Then the POF, optimal solutions in objective space, from previous environment is transferred to new environment through the latent space. Several researches further extended Tr-DMOEA and also transferred solutions in objective space. Jiang et al. [13] proposed the KT-DMOEA, which only transferred some representative individuals called knee points to reduce the computational cost. Liu and Wang [19] used the solutions predicted by PPS to construct the transfer model, which improved the performance.

In addition, there are TL-based DMOEAs that transfer solutions in decision space. In [14], Jiang et al. proposed a prediction strategy using the manifold transfer learning (MTL) technique, called MMTL-DMOEA. A manifold transfer strategy is combined with historical solutions to predict an initial population in the new environment. In [12], an individual-based transfer method (IT-DMOEA) was proposed to predict the initial solutions using the TrAdaBoosting technique. A prediction model is trained using the selected high-quality solutions in new environment and POS in previous environment, which can effectively predict whether a solution performs well in the new environment. In AE-DMOEA [8], an autoencoder was used to learn the mapping between POS in previous two environments. The mapping matrix is then used to predict the moving directions of POS in the new environment.

Most existing TL-based DMOEAs transfer solutions either in decision space or in objective space. However, there are different types of DMOPs that have dynamic POS, dynamic POF or both POS and POF change over time. Zhou et al. [25] proposed a multi-view prediction method (MV-DMOEA), which transferred solutions in both views of decision and objective space using a kernelized autoencoding model. The predicted solutions in both spaces are combined to construct an initial population in the new environment. However, DMOPs with different change patterns require different transfer strategies. For instance, the transfer in objective space may bring negative effects to DMOPs with stationary POF or tiny changes in POF.

3   The Proposed Method

In this section, the details of the proposed change type estimation based method for solving DMOPs, CTE-DMOEA, are introduced. The algorithm is constructed through a change type estimation method and a change type based prediction strategy.

3.1   Change Type Estimation

As mentioned in Sect. 2.1, DMOPs can be divided into four types according to the change of POS and POF. Since there is no prior knowledge about optimal solutions in the new environment, the type of current environmental change is estimated using non-dominated solutions found in previous two environments, denoted by $\mathbf{NDS}_{t-1}$ and $\mathbf{NDS}_t$. According to [10], most of DMOPs show similar change type during two adjacent environmental changes.

A set of reference points, denoted by $\mathbf{RP}$, is first selected from the solution set according to [26]. $\mathbf{RP}$ should be a set of solutions that represent the location and shape of the POS or POF. The details of the reference point selection is described in Algorithm 1. Some boundary solutions are selected and added to $\mathbf{RP}$. The boundary solution for the lth objective is defined as follow

$$x_l = \arg\min_{x \in NDS} \max_{1 \le k \le m} \{\lambda_{lk} | f_k(x) - z_k^* |\} \tag{10}$$

where $f_k(x)$ represents the kth objective value of solution x. $z_k^* = \min\{f_k(x) | x \in POS\}$ is the minimum value of the kth objective. λ_l represents the lth weight vector for lth boundary solution. Each axis in objective space can be defined as a weight vector and the elements of lth weight vector are defined by

$$\lambda_{lk} = \begin{cases} 1, & l = k \\ 1e-2, & l \ne k \end{cases} \quad 1 \le k \le m \tag{11}$$

One element in the weight vector is set to 1, while the others are set to a small value (i.e. $1e-2$) rather than zero. This setting avoids selecting solutions that far away from other solutions in $\mathbf{NDS}$. m boundary solutions are selected and constitutes the initial set of reference points $\mathbf{RP} = \{x_l, 1 \le l \le m\}$.

Then, the Cosine similarity metric is employed to select the other solutions for $\mathbf{RP}$. The Cosine similarity between a solution x_r and the solution set $\mathbf{RP}$ is given by:

$$d(x_r, RP) = \min_{x_i \in RP} \{\arccos(\cos(F(x_i), F(x_r)))\}$$

$$\cos(F(x_i), F(x_r)) = \frac{F(x_i) \cdot F(x_r)}{|F(x_i)||F(x_r)|} \tag{12}$$

A solution x_r that has the maximal Cosine similarity is selected and appended to $\mathbf{RP}$ repeatedly until the number of reference points reaches a predefined number.

$$x_r = \arg\max_{x_r \in NDS-RP} \{d(x_r, RP)\} \tag{13}$$

Algorithm 1. Reference Point Selection

Input: NDS: the set of nondominated solutions; N_{RP}: the number of reference points.
Output: RP: the set of reference points.
1: Select the boundary points of **NDS** according to Eq.(10);
2: Remove boundary points from **NDS** and add them to **RP**;
3: **while** the number of solutions in **RP** is smaller than N_{RP} **do**
4:     Select a reference point x_r from NDS according to Eq.(13);
5:     Remove x_r from **NDS** and add it to **RP**;
6: **end while**
7: Sort **RP** according to the objective values;
8: **return RP**

Algorithm 2. Change Type Estimation

Input: NDS$_{t-1}$ and **NDS**$_t$: nondominated solutions found in time $t-1$ and t; ε_{POS} and ε_{POF}: thresholds that determine the change patterns of POS and POF.
Output: POS_change, POF_change: logical values that indicate whether the POS and POF change respectively.
1: **RP**$_{t-1}$, **RP**$_t$ ← Select reference points from **NDS**$_{t-1}$ and **NDS**$_t$;
2: d^t_{POS}, d^t_{POF} ← Calculate the distance between **RP**$_{t-1}$ and **RP**$_t$ according to Eq.(14);
3: $POS\_change = boolean(d_{POS} > \varepsilon_{POS})$;
4: $POF\_change = boolean(d_{POF} > \varepsilon_{POF})$;
5: **return** POS_change, POF_change

After two reference points sets **RP**$_{t-1}$ and **RP**$_t$ are extracted from **NDS**$_{t-1}$ and **NDS**$_t$, solutions in each set are sorted according to one objective. The moving distance of POS and POF can be calculated by

$$d^t_{POS} = \frac{1}{N_{RP}} \sum_{i=1}^{N_{RP}} ||x^t_i - x^{t-1}_i||_2$$

$$d^t_{POF} = \frac{1}{N_{RP}} \sum_{i=1}^{N_{RP}} ||F_t(x^t_i) - F_{t-1}(x^{t-1}_i)||_2 \tag{14}$$

where N_{RP} is the number of reference points. x^t_i and x^{t-1}_i represent the ith solution in **RP**^{t} and **RP**$^{t-1}$, respectively. $F_t(\cdot)$ and $F_{t-1}(\cdot)$ are the objective functions at time t and $t-1$. The movement of POS and POF are estimated by defining two threshold values $\varepsilon_{POS}, \varepsilon_{POF} > 0$. If $d^t_{POS} < \varepsilon_{POS}$, the POS is considered to be stationary from time $t-1$ to t. If $d^t_{POF} < \varepsilon_{POF}$, it is believed that POF does not change. The detailed change type estimation method is presented in Algorithm 2.

3.2 Change Type Based Prediction Method

In order to track the dynamic POS and POF of DMOPs, CTE-DMOEA makes the predictions in both decision space and objective space, using a kernelized

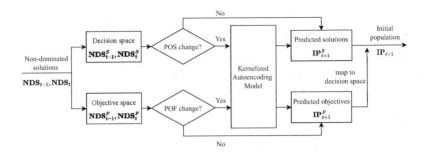

Fig. 1. Outline of the change type based prediction method.

autoencoding model. The outline of the change type based prediction method is shown in Fig. 1.

In particular, when a dynamic change occurs at time $t + 1$, $\mathbf{NDS}_{t-1}$ and $\mathbf{NDS}_t$ will be used to predict an initial population. The proposed change type based prediction method is detailed in Algorithm 3. Before the prediction, the nondominated solutions are preprocessed to fit the autoencoding model. Solutions in $\mathbf{NDS}_{t-1}$ and $\mathbf{NDS}_t$ are first sorted with respect to the crowd distance and the best $N_l = \min(N_{t-1}, N_t)$ solutions are preserved. Then solutions are sorted again according to one objective. After the preprocessing, two solution sets have equal sizes and solutions in $\mathbf{NDS}_{t-1}$ and $\mathbf{NDS}_t$ are matched to each other according to their objective values.

Algorithm 3. Change Type based Prediction

Input: $\mathbf{NDS}_{t-1}$ and $\mathbf{NDS}_t$: the set of nondominated solutions found in time $t - 1$ and t.

Output: $\mathbf{IP}_{t+1}$: predicted initial population for the search in new environment.

1: $N_l = \min(N_{t-1}, N_t)$, the minimum solution number in $\mathbf{NDS}_{t-1}$ and $\mathbf{NDS}_t$;

2: Sort solutions in $\mathbf{NDS}_{t-1}$ and $\mathbf{NDS}_t$ using crowd distance and preserve the best N_l solutions;

3: $\mathbf{NDS}^S_{t-1}$, $\mathbf{NDS}^S_t$ ← Randomly select a half of the solutions in $\mathbf{NDS}_{t-1}$ and $\mathbf{NDS}_t$;

4: $\mathbf{NDS}^F_{t-1}$, $\mathbf{NDS}^F_t$ ← The objective values of the remainder half of solutions in each solution set;

5: Estimate the **change type** by conducting **Algorithm 2** on $\mathbf{NDS}_{t-1}$ and $\mathbf{NDS}_t$;

6: $\mathbf{IP}^S_{t+1}$ ← The predicted initial population in decision space according to Eq.(15);

7: $\mathbf{IP}^F_{t+1}$ ← The predicted objective values according to Eq.(16);

8: Map the $\mathbf{IP}^F_{t+1}$ to decision space using Eq.(17);

9: $\mathbf{IP}_{t+1} = \mathbf{IP}^S_{t+1} \cup \mathbf{IP}^F_{t+1}$.

10: **return** $\mathbf{IP}_{t+1}$

Then a half of solutions are randomly selected from $\mathbf{NDS}_{t-1}$ and $\mathbf{NDS}_t$ respectively and archived in $\mathbf{NDS}^S_{t-1}$ and $\mathbf{NDS}^S_t$. The objective values of the remainder half of solutions are archived in $\mathbf{NDS}^F_{t-1}$ and $\mathbf{NDS}^F_t$, respectively.

According to the change type obtained by Algorithm 2, the predicted population in decision space $\mathbf{IP}_{t+1}^{S}$ is given by:

$$\mathbf{IP}_{t+1}^{S} = \begin{cases} \mathbf{M}_{k}^{S}\mathbf{K}(\mathbf{NDS}_{t-1}^{S}, \mathbf{NDS}_{t}^{S}), & \text{if POS changes;} \\ \mathbf{NDS}_{t}^{S}, & \text{otherwise.} \end{cases} \tag{15}$$

where $\mathbf{M}_{k}^{S}$ is the mapping matrix learned using Eq.(8), configuring $\mathbf{NDS}_{t-1}^{S}$ as $\mathbf{P}$, and $\mathbf{NDS}_{t}^{S}$ as $\mathbf{Q}$. Similarly, the predicted solutions in objective space, denoted as $\mathbf{IP}_{t+1}^{F}$, is given by:

$$\mathbf{IP}_{t+1}^{F} = \begin{cases} \mathbf{M}_{k}^{F}\mathbf{K}(\mathbf{NDS}_{t-1}^{F}, \mathbf{NDS}_{t}^{F}), & \text{if POF changes;} \\ \mathbf{NDS}_{t}^{F}, & \text{otherwise.} \end{cases} \tag{16}$$

It should be noticed that the solutions obtained by Eq.(16) are predicted objective values. The objective vectors in $\mathbf{IP}_{t+1}^{F}$ are mapped to decision space using a single objective optimization algorithm

$$x^{*} \leftarrow \min ||F(x^{*}, t+1) - q^{*}||, \quad q \in \mathbf{IP}_{t+1}^{F} \tag{17}$$

The problem can be solved by Interior Point Algorithm [3], which was also used in [11]. Combining the predicted solutions both in decision space and objective space, an initial population is obtained to guide the search in new environment. The overall framework of CTE-DMOEA is given in Algorithm 4.

Algorithm 4. CTE-DMOEA

Input: $F_t(x)$: the objective vector of target DMOP; SMOEA: a static MOEA
Output: POS: the approximate POS at different time steps.
1: **POS** $= \emptyset$; $t = 0$;
2: $\mathbf{IP}_t \leftarrow$ Initialize a population;
3: **while** termination criterion not met **do**
4:     $\mathbf{NDS}_t = \mathbf{SMOEA}(F_t(x), \mathbf{IP}_t)$;
5:     **POS** = **POS** $\cup \mathbf{NDS}_t$;
6:     **if** environment changes **then**
7:         $\mathbf{IP}_{t+1} =$ Change Type based Prediction($\mathbf{NDS}_{t-1}, \mathbf{NDS}_t$);
8:         $t = t + 1$;
9:     **end if**
10: **end while**
11: **return POS**

---

4 Experimental Study

4.1 Benchmark Problems

The DF benchmark [16] was used in the experiments, which contains 14 Type II benchmark problems with different dynamic characteristics. To test on problems of different change types, additional benchmark problems were used, including FDA1-FDA5 [6], dMOP1 [9], DIMP1-DIMP2 [18] and JY1-JY10 [15]. In total, there are 32 DMOPs in the test benchmark suite, including DMOPs of Type I, II, III and hybrid change types (i.e. JY9 and JY10). The dynamics of them was controlled by a time variable t, which was set as:

$$t = \frac{1}{n_t} \lfloor \frac{\tau}{\tau_t} \rfloor \tag{18}$$

where τ is the maximum generation, n_t denotes the change severity and τ_t denotes the change frequency. A small value of n_t indicates a severe change, while a small value of τ_t means a rapid change. In this study, all the benchmark problems were tested under four commonly used dynamic configurations [13, 14], where $(n_t, \tau_t) = \{(5, 10), (10, 10), (5, 5), (10, 5)\}$ respectively.

4.2   Performance Metric

In this paper, Mean Inverted Generational Distance (MIGD) was employed to evaluate the performance of algorithms.

$$\begin{aligned} \text{MIGD} &= \frac{1}{|T|} \sum_{t \in T} \text{IGD}(P^{t*}, P^t) \\ \text{IGD}(P^{t*}, P^t) &= \frac{\sum_{v^* \in P^{t*}} \min_{v \in P^t} ||v^* - v||}{|P^{t*}|} \end{aligned} \tag{19}$$

where P^{t*} is the true pareto optimal set, P^t is the obtained optimal set at time t. $|T|$ refers to the number of environmental changes in a run. A smaller MIGD represents the obtained solution set has better convergence and higher diversity.

4.3   Compared Algorithms and Parameter Settings

The proposed CTE-DMOEA was compared to six state-of-the-art prediction-based DMOEAs, including PPS [23], MOEA/D-KF [20], Tr-DMOEA [11], MMTL-DMOEA [14], AE-DMOEA [8] and MV-MOEAD [25]. For a fair comparison, the basic multi-objective optimization algorithm in all comparison algorithms was MOEA/D [22]. Moreover, the parameters of them were set according to their original references. Detailed parameter setting are given as follows.

1) The dimension of decision vectors was set to 10. The population size of bi-objective and tri-objective problems was set to 100 and 150, respectively.
2) There were 30 environmental changes in a run and 50 generations are executed before the first environmental change. Each algorithm was run 30 times independently on each test instance.
3) For parameters in CTE-DMOEA, the number of reference points was set to $2m + 1$, where m is the number of objectives. Thresholds parameters ε_{POS} and ε_{POF} were set to 0.2 and 0.05 respectively. The kernel function used in kernelized autoencoding model was a Gaussian RBF kernel $k(x_i, x_j) = exp(-\frac{||x_i - x_j||^2}{2\sigma^2})$, where $\sigma = 0.75$.

4.4 Experimental Results and Discussion

The average and standard deviations of MIGD results are presented in Tables 1 and 2. The minimum mean value on each test instance is highlighted in bold. The significance of the result difference was examined using Wilcoxon rank-sum test at a confidence level of 0.05. The symbols $(+)$, $(-)$, and $(\approx)$ indicate that the proposed method outperforms, underperforms, or performs similarly to the compared algorithms, respectively.

Table 1. Mean and standard deviation values of migd metric obtained by compared algorithms for DF problems

Problem	n_t,τ_t	CTE-MOEA/D	PPS-MOEA/D	MOEA/D-KF	Tr-MOEA/D	MMTL-MOEA/D	AE-MOEA/D	MV-MOEA/D
DF1	5,10	0.0353±1.28e−02	0.0465±5.14e−03(+)	0.0422±4.07e−03(+)	**0.0242±1.42e−03**(−)	0.0575±1.24e-02(+)	0.1593±2.28e−02(+)	0.0455±1.61e-02(+)
	10,10	0.0179±2.46e−03	0.0303±8.73e−03(+)	**0.0160±9.35e−04**(−)	0.0235±1.76e−03(+)	0.0476±4.46e−03(+)	0.0483±1.20e−02(+)	0.0254±5.22e−03(+)
	5,5	**0.0522±8.94e−03**	0.1925±2.65e−02(+)	0.1521±1.88e−02(+)	0.0563±9.45e−03(≈)	0.1386±2.39e−02(+)	0.3640±2.97e−02(+)	0.0859±1.32e−02(+)
	10,5	**0.0382±3.35e−03**	0.1776±3.18e−02(+)	0.0480±4.46e−03(+)	0.0556±1.03e−02(+)	0.1414±1.21e−02(+)	0.1861±2.57e−02(+)	0.0644±1.43e−02(+)
DF2	5,10	0.0275±4.99e−03	0.1061±1.02e−02(+)	0.0658±5.60e−03(+)	**0.0209±1.14e−03**(−)	0.0647±1.21e−02(+)	0.1449±1.26e−02(+)	0.0373±5.27e−03(+)
	10,10	**0.0181±3.59e−03**	0.0893±9.21e−03(+)	0.0393±4.57e−03(+)	0.0199±1.14e−02(+)	0.1092±1.48e−02(+)	0.1005±8.63e−03(+)	0.0402±9.99e−03(+)
	5,5	0.0453±4.59e−03	0.2171±1.36e−02(+)	0.1760±2.11e−02(+)	**0.0275±6.68e−03**(−)	0.1610±2.76e−02(+)	0.2739±2.15e−02(+)	0.0675±7.74e−03(+)
	10,5	0.0416±5.61e−03	0.1757±1.54e−02(+)	0.0954±9.51e−03(+)	**0.0220±2.57e−03**(−)	0.2047±2.66e−02(+)	0.1961±1.75e−02(+)	0.0580±7.90e−03(+)
DF3	5,10	0.2630±4.36e−02	0.3771±4.78e−02(+)	**0.1198±4.34e−02**(−)	0.2782±1.83e−02(≈)	0.3407±4.70e−02(+)	0.3688±4.42e−02(+)	0.2637±4.69e−02(≈)
	10,10	0.2125±4.44e−02	0.3023±9.76e−02(+)	**0.0825±3.68e−02**(−)	0.2737±1.50e−02(+)	0.3024±4.84e−02(+)	0.2715±7.80e−02(+)	0.2477±4.72e−02(+)
	5,5	0.2960±5.04e−02	0.4942±3.87e−02(+)	0.3026±5.50e−02(+)	**0.2437±1.99e−02**(−)	0.4140±5.68e−02(+)	0.4878±3.43e−02(+)	0.3021±6.82e−02(≈)
	10,5	0.2669±5.76e−02	0.3965±6.69e−02(+)	**0.2304±9.57e−02**(−)	0.2510±1.90e−02(≈)	0.4003±4.42e−02(+)	0.4291±5.76e−02(+)	0.3039±7.33e−02(+)
DF4	5,10	**0.0821±4.74e−03**	0.1161±6.43e−03(+)	0.0955±2.06e−03(+)	0.1433±1.03e−02(+)	0.1303±7.57e−03(+)	0.1140±5.18e−03(+)	0.0833±5.23e−03(≈)
	10,10	**0.0771±4.62e−03**	0.1090±7.71e−03(+)	0.0882±2.40e−03(+)	0.1451±7.57e−03(+)	0.1315±7.70e−03(+)	0.1100±4.08e−03(+)	0.0786±1.25e−03(≈)
	5,5	**0.1233±7.20e−03**	0.2329±3.20e−02(+)	0.2506±2.89e−02(+)	0.3642±1.78e−02(+)	0.2388±1.93e−02(+)	0.1844±1.77e−02(+)	0.1355±9.97e−03(+)
	10,5	0.1208±7.32e−03	0.2047±2.85e−02(+)	0.1736±1.11e−02(+)	0.3766±2.93e−02(+)	0.2268±1.49e−02(+)	0.1542±9.97e−03(+)	**0.1187±1.11e−02**(≈)
DF5	5,10	0.0212±8.77e−03	**0.0135±4.02e−03**(−)	0.0216±1.85e−03(+)	0.0573±5.87e−03(+)	0.0508±7.73e−03(+)	0.0366±4.35e−03(+)	0.0249±4.86e−03(+)
	10,10	0.0136±1.61e−03	0.0138±5.54e−03(≈)	**0.0129±7.01e−04**(≈)	0.0598±6.42e−03(+)	0.0343±4.64e−03(+)	0.0211±9.71e−04(+)	0.0206±7.47e−03(+)
	5,5	**0.0374±4.84e−03**	0.0766±2.91e−02(+)	0.0768±1.18e−02(+)	0.0625±7.86e−03(+)	0.1276±1.29e−02(+)	0.1688±1.81e−02(+)	0.0544±5.38e−03(+)
	10,5	**0.0296±3.39e−03**	0.0632±2.44e−02(+)	0.0392±2.36e−02(+)	0.0626±4.52e−03(+)	0.1014±2.00e−02(+)	0.0837±2.24e−02(+)	0.0401±3.94e−03(+)
DF6	5,10	**0.4619±1.17e−01**	0.7430±1.64e−01(+)	1.3345±4.92e−01(+)	0.6487±5.22e−02(+)	0.7660±3.57e−01(+)	0.7306±2.24e−01(+)	0.5147±1.23e−01(+)
	10,10	**0.7100±1.52e−01**	1.1206±2.77e−01(+)	3.0754±9.31e−01(+)	0.7652±1.05e−01(≈)	0.8905±2.81e−01(+)	1.3265±4.73e−01(+)	0.8849±2.31e−01(+)
	5,5	**0.7663±1.58e−01**	1.3897±1.97e−01(+)	3.4461±7.08e−01(+)	1.2185±1.59e−01(+)	0.9696±1.58e−01(+)	1.5579±5.68e−01(+)	0.8291±1.77e−01(≈)
	10,5	1.1171±2.73e−01	1.6875±3.28e−01(+)	3.2658±9.43e−01(+)	1.1542±1.32e−01(≈)	1.2837±4.05e−01(≈)	1.6558±5.78e−01(+)	**1.1110±2.04e−01**(≈)
DF7	5,10	0.3080±5.60e−02	0.4427±3.77e−02(+)	**0.2776±5.64e−02**(−)	0.3670±2.83e−02(+)	0.4983±3.33e−02(+)	0.4623±5.55e−02(+)	0.3196±4.59e−02(≈)
	10,10	0.1629±7.38e−02	0.2962±2.57e−02(+)	**0.1457±3.46e−02**(−)	0.2334±2.36e−02(+)	0.3342±6.72e−02(+)	0.2221±3.50e−02(+)	0.2492±5.00e−02(+)
	5,5	**0.3189±6.49e−02**	0.4867±4.02e−02(+)	0.3693±4.50e−02(+)	0.4612±2.11e−02(+)	0.5164±4.06e−02(+)	0.4857±6.01e−02(+)	0.3536±4.83e−02(+)
	10,5	**0.2180±6.13e−02**	0.3292±2.31e−02(+)	0.2353±6.72e−02(≈)	0.3452±2.63e−02(+)	0.3441±6.41e−02(+)	0.2490±5.12e−02(+)	0.2872±3.05e−02(+)
DF8	5,10	**0.0210±4.16e−03**	0.0490±9.61e−03(+)	0.0274±1.09e−03(+)	0.0461±5.43e−03(+)	0.1427±1.48e−02(+)	0.0690±1.04e−02(+)	0.0246±6.06e−03(+)
	10,10	**0.0223±7.58e−03**	0.0500±8.50e−03(+)	0.0247±1.27e−03(+)	0.0542±6.87e−03(+)	0.1385±1.67e−02(+)	0.0601±8.16e−03(+)	0.0351±7.14e−03(+)
	5,5	**0.0333±7.60e−03**	0.0868±1.29e−02(+)	0.0739±7.66e−03(+)	0.1141±1.62e−02(+)	0.1969±1.58e−02(+)	0.0881±1.34e−02(+)	0.0410±6.08e−03(+)
	10,5	**0.0352±6.65e−03**	0.0884±1.60e−02(+)	0.0597±6.01e−03(+)	0.1149±7.96e−03(+)	0.1983±2.08e−02(+)	0.0892±1.14e−02(+)	0.0418±7.66e−03(+)
DF9	5,10	**0.0948±1.21e−02**	0.1981±1.29e−02(+)	0.4828±8.23e−02(+)	0.2103±1.98e−02(+)	0.1199±1.49e−02(+)	0.4767±5.45e−02(+)	0.0968±1.01e−02(≈)
	10,10	**0.0790±1.17e−02**	0.1832±2.15e−02(+)	0.1092±2.44e−02(+)	0.2111±1.70e−02(+)	0.1400±3.63e−02(+)	0.3270±8.96e−02(+)	0.0927±9.78e−03(≈)
	5,5	**0.1772±1.89e−02**	0.3597±2.15e−02(+)	1.1415±1.40e−01(+)	0.2498±2.23e−02(+)	0.2466±3.53e−02(+)	0.6198±7.39e−02(+)	0.2013±2.40e−02(+)
	10,5	**0.1432±2.35e−02**	0.3282±3.78e−02(+)	0.3165±8.29e−02(+)	0.2423±2.90e−02(+)	0.2368±2.89e−02(+)	0.4627±6.71e−02(+)	0.1814±1.93e−02(+)
DF10	5,10	**0.1550±1.21e−02**	0.2776±2.45e−02(+)	0.2115±1.65e−02(+)	0.2013±1.97e−02(+)	0.2017±1.80e−02(+)	0.2383±3.76e−02(+)	0.1578±1.36e−02(≈)
	10,10	**0.2052±2.54e−02**	0.3294±3.67e−02(+)	0.2104±2.10e−02(+)	0.2394±2.43e−02(+)	0.2257±1.47e−02(+)	0.3111±2.42e−02(+)	0.2079±1.27e−02(≈)
	5,5	0.1815±1.43e−02	0.3025±1.95e−02(+)	0.2836±2.66e−02(+)	0.3091±1.58e−02(+)	0.2307±1.50e−02(+)	0.2657±4.15e−02(+)	**0.1767±1.14e−02**(≈)
	10,5	**0.2325±2.60e−02**	0.3496±4.26e−02(+)	0.3125±4.76e−02(+)	0.3464±2.13e−02(+)	0.2540±1.81e−02(+)	0.3376±4.37e−02(+)	0.2341±2.71e−02(≈)
DF11	5,10	0.0707±2.05e−03	0.0997±2.04e−02(+)	0.0878±1.87e−03(+)	0.0937±3.32e−03(+)	0.0946±4.28e−03(+)	**0.0702±6.39e−04**(≈)	0.0735±5.61e−03(+)
	10,10	0.0655±3.69e−03	0.1268±4.31e−02(+)	0.0872±1.93e−03(+)	0.0822±2.88e−03(+)	0.1154±4.28e−03(+)	0.0672±6.55e−04(≈)	**0.0652±4.16e−03**(≈)
	5,5	0.1319±2.17e−02	0.2580±7.32e−02(+)	0.1286±4.21e−03(≈)	0.1401±4.94e−03(+)	0.1477±5.66e−03(+)	**0.1215±2.54e−03**(−)	0.1267±1.88e−02(≈)
	10,5	0.1205±2.63e−02	0.2451±4.24e−02(+)	0.1259±4.42e−03(≈)	0.1250±6.32e−03(≈)	0.1797±1.02e−02(+)	**0.1128±1.75e−03**(−)	0.1260±2.83e−02(≈)
DF12	5,10	0.1560±2.80e−02	0.3191±5.86e−02(+)	0.1851±1.81e−02(+)	0.2957±1.95e−02(+)	0.5599±3.55e−02(+)	0.2456±1.44e−02(+)	**0.1455±3.07e−02**(≈)
	10,10	**0.1436±3.67e−02**	0.2729±1.16e−01(+)	0.1610±6.67e−03(+)	0.2930±1.35e−02(+)	0.5550±2.92e−02(+)	0.2043±1.21e−02(+)	0.1730±5.93e−02(≈)
	5,5	0.2139±3.01e−02	0.4029±5.13e−02(+)	0.2421±2.32e−02(+)	0.4177±2.97e−02(+)	0.6020±2.03e−02(+)	0.5032±1.80e−01(+)	0.2173±3.81e−02(≈)
	10,5	0.2738±5.31e−02	0.4676±1.26e−01(+)	**0.2206±8.82e−03**(−)	0.4294±1.75e−02(+)	0.6338±1.43e−02(+)	0.2539±5.80e−02(−)	0.2615±3.99e−02(≈)
DF13	5,10	0.1199±6.71e−03	0.2319±9.40e−03(+)	0.2297±3.59e−03(+)	0.1733±9.12e−03(+)	0.2287±8.38e−03(+)	0.2309±1.02e−02(+)	**0.1200±7.23e−03**(≈)
	10,10	**0.1132±6.22e−03**	0.2103±8.06e−03(+)	0.2021±2.61e−03(+)	0.1691±9.20e−03(+)	0.2129±4.92e−03(+)	0.2102±6.04e−03(+)	0.1162±6.54e−03(≈)
	5,5	**0.1693±1.02e−02**	0.2644±4.85e−02(+)	0.2727±8.56e−03(+)	0.4355±3.22e−02(+)	0.2771±1.31e−02(+)	0.2857±2.61e−02(+)	0.1709±1.11e−02(≈)
	10,5	**0.1656±1.12e−02**	0.2870±6.02e−02(+)	0.2477±5.58e−03(+)	0.4553±5.47e−02(+)	0.2519±1.08e−02(+)	0.2592±3.33e−02(+)	0.1691±1.09e−02(≈)
DF14	5,10	0.0550±3.90e−03	0.0615±3.93e−03(≈)	0.0545±1.01e−02(≈)	0.0607±2.30e−03(+)	0.1101±3.06e−02(+)	0.0769±9.26e−03(+)	**0.0543±3.61e−03**(≈)
	10,10	**0.0478±1.45e−03**	0.0599±1.24e−03(+)	0.0524±5.30e−04(+)	0.0648±8.95e−03(+)	0.0687±3.11e−03(+)	0.0649±3.13e−03(+)	0.0514±5.30e−04(+)
	5,5	**0.0797±9.53e−03**	0.0817±1.19e−02(≈)	0.0842±3.25e−03(+)	0.0878±3.67e−03(+)	0.1475±5.43e−02(+)	0.1060±9.85e−03(+)	0.0803±6.67e−03(≈)
	10,5	0.0792±5.92e−03	0.0833±1.45e−02(≈)	**0.0725±1.79e−03**(−)	0.0887±3.35e−03(+)	0.0943±8.98e−03(+)	0.0867±9.28e−03(+)	0.0782±7.70e−03(≈)
$+/\approx/-$		-/-/-	52/2/2	42/7/7	45/6/5	55/1/0	52/2/2	30/26/0

Table 2. Mean and standard deviation values of MIGD metric obtained by compared algorithms for additional problems

Problem	n_t,τ_t	CTE-MOEA/D	PPS-MOEA/D	MOEA/D-KF	Tr-MOEA/D	MMTL-MOEA/D	AE-MOEA/D	MV-MOEA/D
FDA1	5,10	**0.0229±4.28e−03**	0.0309±1.36e−02(+)	0.0240±2.46e−03(+)	0.0704±6.31e−03(+)	0.0547±6.95e−03(+)	0.0915±1.35e−02(+)	0.0667±1.68e−02(+)
	10,10	**0.0180±4.77e−03**	0.0284±1.17e−02(+)	0.0185±1.00e−03(+)	0.0746±6.40e−03(+)	0.0460±7.37e−03(+)	0.0320±8.94e−03(+)	0.0494±1.17e−02(+)
	5,5	**0.0627±1.27e−02**	0.1579±4.32e−02(+)	0.0880±8.05e−03(+)	0.1876±1.64e−02(+)	0.1526±1.52e−02(+)	0.2148±2.90e−02(+)	0.0749±1.34e−02(+)
	10,5	**0.0389±6.87e−03**	0.1832±5.05e−02(+)	0.0622±6.03e−03(+)	0.2116±1.68e−02(+)	0.1122±1.23e−02(+)	0.1153±2.46e−02(+)	0.0466±7.60e−03(+)
FDA2	5,10	0.0128±4.62e−03	0.0338±2.96e−02(+)	0.0227±5.22e−03(+)	0.0262±3.11e−03(+)	0.0214±4.13e−03(+)	**0.0115±6.59e−04**(+)	0.0266±1.12e−02(+)
	10,10	0.0086±1.09e−03	0.0145±9.98e−03(+)	0.0280±7.29e−03(+)	0.0288±4.34e−03(+)	0.0191±3.61e−03(+)	**0.0086±5.86e−04**(+)	0.0160±4.98e−03(+)
	5,5	**0.0265±7.09e−03**	0.0844±3.01e−02(+)	0.1097±2.48e−02(+)	0.0706±1.19e−02(+)	0.0419±5.00e−03(+)	0.0294±3.41e−03(+)	0.0284±1.07e−02(+)
	10,5	0.0157±2.10e−03	0.0667±3.00e−02(+)	0.1137±1.23e−02(+)	0.0756±1.29e−02(+)	0.0355±1.18e−02(+)	**0.0155±1.22e−03**(≈)	0.0484±2.09e−02(+)
FDA3	5,10	0.0332±5.02e−03	0.0499±8.62e−03(+)	0.0407±8.11e−03(+)	0.0209±2.48e−03(−)	**0.0113±1.44e−03**(−)	0.0612±7.34e−03(+)	0.0333±5.92e−03(+)
	10,10	0.0218±6.98e−03	0.0317±6.42e−03(+)	0.0219±5.24e−03(≈)	**0.0177±1.91e−03**(−)	0.0192±2.41e−03(≈)	0.0364±5.83e−03(+)	0.0269±6.74e−03(+)
	5,5	0.0541±7.92e−03	0.1021±1.75e−02(+)	0.1181±2.64e−02(+)	0.0523±5.96e−03(≈)	**0.0254±2.92e−03**(−)	0.1178±1.83e−02(+)	0.0560±1.18e−02(+)
	10,5	0.0409±7.91e−03	0.0662±1.29e−02(+)	0.0559±1.01e−02(+)	0.0446±6.59e−03(≈)	**0.0373±4.05e−03**(−)	0.0659±7.49e−03(+)	0.0391±8.17e−03(≈)
FDA4	5,10	0.0670±2.88e−03	0.0849±3.19e−03(+)	0.0874±3.23e−03(+)	0.1108±7.31e−03(+)	0.0682±2.39e−03(+)	0.0925±4.14e−03(+)	**0.0661±1.20e−03**(≈)
	10,10	**0.0626±1.26e−03**	0.0673±1.65e−03(+)	0.0727±1.67e−03(+)	0.1174±7.76e−03(+)	0.0731±1.44e−03(+)	0.0742±1.55e−03(+)	0.0634±1.38e−03(+)
	5,5	**0.0934±4.93e−03**	0.1765±1.90e−02(+)	0.1705±9.51e−03(+)	0.2317±1.52e−02(+)	0.1089±7.24e−03(+)	0.1768±1.12e−02(+)	0.1034±7.02e−03(+)
	10,5	**0.0848±4.58e−03**	0.1199±1.66e−02(+)	0.1189±5.73e−03(+)	0.2406±1.86e−02(+)	0.1194±4.69e−03(+)	0.1170±3.95e−03(+)	0.0917±5.29e−03(+)
FDA5	5,10	0.0704±1.22e−02	0.1071±2.86e−02(+)	0.0662±4.19e−03(≈)	0.0847±7.09e−03(+)	**0.0545±4.73e−03**(−)	0.0786±6.23e−03(+)	0.0682±1.19e−02(≈)
	10,10	**0.0516±5.77e−03**	0.0704±1.53e−02(+)	0.0517±2.96e−03(≈)	0.0840±5.85e−03(+)	0.0592±4.88e−03(+)	0.0691±4.91e−03(+)	0.0550±1.15e−02(+)
	5,5	0.0855±1.06e−02	0.1412±1.81e−02(+)	0.1147±7.16e−03(+)	0.1087±7.02e−03(+)	**0.0625±5.34e−03**(−)	0.1275±7.61e−03(+)	0.0825±9.39e−03(+)
	10,5	0.0630±1.08e−02	0.1133±1.50e−02(+)	0.0732±5.09e−03(+)	0.1007±7.37e−03(+)	0.0744±4.76e−03(+)	0.0986±7.30e−03(+)	**0.0623±8.46e−03**(≈)
dMOP1	5,10	0.0092±1.11e−03	0.0206±1.43e−02(+)	0.0345±3.38e−03(+)	0.1929±1.86e−02(+)	0.0265±1.66e−02(+)	**0.0054±4.48e−04**(−)	0.0262±1.43e−02(+)
	10,10	0.0092±1.55e−03	0.0220±3.45e−02(+)	0.0338±2.28e−03(+)	0.1969±2.31e−02(+)	0.0200±1.01e−02(+)	**0.0055±5.88e−04**(−)	0.0183±8.02e−03(+)
	5,5	0.0172±7.39e−03	0.1992±1.08e−01(+)	0.1219±2.54e−02(+)	0.4594±5.67e−02(+)	0.0298±2.44e−02(+)	**0.0062±9.91e−04**(−)	0.0501±2.53e−02(+)
	10,5	0.0224±1.76e−02	0.1876±1.17e−01(+)	0.1092±2.19e−02(+)	0.4791±5.25e−02(+)	0.0270±1.83e−02(+)	**0.0059±1.08e−03**(−)	0.0534±2.11e−02(+)
DIMP1	5,10	**0.0279±1.69e−02**	0.0587±2.69e−02(+)	0.0374±2.45e−03(+)	0.0689±9.46e−03(+)	0.0458±4.95e−03(+)	0.1924±3.37e−02(+)	0.0413±1.89e−02(+)
	10,10	**0.0194±4.48e−03**	0.0339±1.32e−02(+)	0.0225±9.94e−04(+)	0.0682±5.50e−03(+)	0.0579±4.20e−03(+)	0.0378±9.05e−03(+)	0.0318±1.69e−02(+)
	5,5	**0.0648±1.57e−02**	0.3415±7.01e−02(+)	0.1515±3.26e−02(+)	0.2267±2.78e−02(+)	0.1319±1.27e−02(+)	0.3834±3.27e−02(+)	0.0702±1.43e−02(+)
	10,5	**0.0388±3.53e−03**	0.1986±5.26e−02(+)	0.0804±6.79e−03(+)	0.2254±1.71e−02(+)	0.1468±8.68e−03(+)	0.1878±4.34e−02(+)	0.0426±5.35e−03(+)
DIMP2	5,10	1.6767±1.31e−01	2.3254±2.02e−01(+)	3.0701±6.22e−01(+)	2.1342±1.88e−01(+)	**1.1766±1.75e−01**(−)	2.4086±1.72e−01(+)	1.6246±1.23e−01(≈)
	10,10	1.4327±1.49e−01	2.0810±1.90e−01(+)	1.2483±5.93e−01(−)	2.1615±1.78e−01(+)	**1.0432±2.01e−01**(−)	1.6347±2.47e−01(+)	1.3941±1.50e−01(≈)
	5,5	2.6093±2.05e−01	3.0028±2.09e−01(+)	6.3550±6.46e−01(+)	2.7320±1.60e−01(+)	**1.3024±1.66e−01**(−)	2.9407±2.52e−01(+)	2.6449±1.72e−01(≈)
	10,5	2.3376±1.95e−01	2.8605±2.23e−01(+)	3.5723±8.44e−01(+)	2.9002±2.41e−01(+)	**1.5999±1.77e−01**(−)	2.6299±2.13e−01(+)	2.3600±1.74e−01(≈)
JY1	5,10	0.0227±4.43e−03	**0.0185±6.93e−03**(−)	0.0294±2.94e−03(+)	0.0655±8.72e−03(+)	0.1718±2.98e−02(+)	0.0315±1.31e−03(+)	0.0304±1.01e−02(+)
	10,10	**0.0165±3.00e−03**	0.0204±2.79e−02(−)	0.0237±1.93e−03(+)	0.0684±7.27e−03(+)	0.2079±3.06e−02(+)	0.0210±1.34e−03(+)	0.0216±5.37e−03(+)
	5,5	**0.0414±5.60e−03**	0.1071±6.16e−02(+)	0.1737±2.20e−02(+)	0.2440±3.11e−02(+)	0.4475±7.48e−02(+)	0.1224±3.44e−02(+)	0.0581±6.49e−03(+)
	10,5	**0.0329±3.69e−03**	0.1382±8.15e−02(+)	0.1551±1.96e−02(+)	0.2420±3.19e−02(+)	0.4954±8.67e−02(+)	0.0655±9.13e−03(+)	0.0432±5.62e−03(+)
JY2	5,10	0.0239±6.50e−03	**0.0153±7.37e−03**(−)	0.0170±7.11e−04(−)	0.0608±8.13e−03(+)	0.0381±8.23e−03(+)	0.0284±2.07e−03(+)	0.0276±7.31e−03(+)
	10,10	0.0186±4.54e−03	**0.0125±6.36e−03**(−)	0.0130±3.97e−04(−)	0.0635±8.12e−03(+)	0.0686±1.18e−02(+)	0.0186±1.12e−03(≈)	0.0230±6.71e−03(+)
	5,5	**0.0426±4.52e−03**	0.0971±3.06e−02(+)	0.0539±4.61e−03(+)	0.2489±3.28e−02(+)	0.1151±1.41e−02(+)	0.1033±1.13e−02(+)	0.0537±7.15e−03(+)
	10,5	**0.0336±5.30e−03**	0.0860±3.38e−02(+)	0.0391±4.21e−03(+)	0.2580±4.05e−02(+)	0.1932±2.68e−02(+)	0.0519±4.00e−03(+)	0.0389±6.76e−03(+)
JY3	5,10	0.0976±1.08e−03	0.1011±1.69e−02(+)	**0.0698±2.29e−02**(−)	0.1053±1.63e−03(+)	0.0961±9.47e−03(+)	0.0985±6.54e−04(+)	0.0979±1.87e−03(≈)
	10,10	0.0952±5.28e−03	0.1023±1.37e−02(+)	**0.0707±2.14e−02**(−)	0.1041±1.20e−03(+)	0.0945±1.45e−02(+)	0.0973±5.57e−04(+)	0.0963±2.74e−03(≈)
	5,5	0.0992±1.79e−03	0.1163±1.13e−02(+)	**0.0885±1.66e−02**(−)	0.1407±8.31e−03(+)	0.0977±1.16e−02(≈)	0.0970±1.61e−02(≈)	0.1005±2.00e−03(+)
	10,5	0.0980±3.09e−03	0.1221±1.64e−02(+)	**0.0795±2.01e−02**(−)	0.1373±6.05e−03(+)	0.0980±2.46e−03(≈)	0.0960±9.72e−03(−)	0.0989±1.66e−03(+)
JY4	5,10	**0.0294±1.86e−03**	0.0601±1.44e−02(+)	0.0342±4.05e−03(+)	0.1040±9.69e−03(+)	0.2003±3.62e−02(+)	0.0472±4.40e−03(+)	0.0331±2.92e−03(+)
	10,10	**0.0282±2.90e−03**	0.0612±1.69e−02(+)	0.0316±3.16e−03(+)	0.1110±9.63e−03(+)	0.0941±1.08e−02(+)	0.0348±2.24e−03(+)	0.0320±3.01e−03(+)
	5,5	**0.0649±6.36e−03**	0.2234±5.17e−02(+)	0.0958±9.22e−03(+)	0.3121±3.64e−02(+)	0.4022±5.94e−02(+)	0.1476±1.22e−02(+)	0.0732±5.34e−03(+)
	10,5	**0.0578±5.03e−03**	0.1899±5.96e−02(+)	0.0903±1.42e−02(+)	0.3357±3.04e−02(+)	0.2593±4.71e−02(+)	0.0978±5.79e−03(+)	0.0620±5.44e−03(+)
JY5	5,10	0.0076±1.23e−03	0.0092±9.77e−03(−)	0.0151±8.61e−04(+)	0.0184±1.22e−03(+)	0.0122±2.48e−03(+)	**0.0053±1.70e−04**(−)	0.0164±7.06e−03(+)
	10,10	0.0063±4.68e−04	0.0086±5.01e−03(+)	0.0143±6.49e−04(+)	0.0188±1.84e−03(+)	0.0102±1.06e−03(+)	**0.0053±1.90e−04**(−)	0.0098±4.50e−03(+)
	5,5	0.0112±1.21e−03	0.0179±1.12e−02(≈)	0.1314±3.35e−02(+)	0.0678±1.22e−02(+)	0.0201±5.06e−03(+)	**0.0057±4.17e−04**(−)	0.0378±1.79e−02(+)
	10,5	0.0096±8.18e−04	0.0211±1.00e−02(+)	0.0598±9.45e−03(+)	0.0755±1.52e−02(+)	0.0175±2.06e−03(+)	**0.0057±2.38e−04**(−)	0.0525±3.61e−02(+)
JY6	5,10	0.9106±9.74e−02	0.9900±1.30e−01(+)	**0.7703±7.78e−02**(−)	1.6787±1.37e−01(+)	1.2392±1.38e−01(+)	1.0101±9.30e−02(+)	0.9142±1.02e−01(≈)
	10,10	0.8719±1.16e−01	0.7583±1.14e−01(−)	**0.4847±8.76e−02**(−)	1.6515±1.36e−01(+)	0.8519±8.45e−02(≈)	0.7847±8.54e−02(−)	0.9376±1.10e−01(+)
	5,5	**1.8293±1.86e−01**	2.6474±3.26e−01(+)	1.9356±2.53e−01(≈)	3.2532±3.51e−01(+)	2.5392±2.88e−01(+)	2.2181±1.89e−01(+)	1.9041±2.33e−01(≈)
	10,5	**1.5574±1.98e−01**	2.2638±2.89e−01(+)	1.6048±2.69e−01(≈)	3.2796±3.02e−01(+)	1.9398±2.77e−01(+)	1.8829±1.95e−01(+)	1.5717±2.49e−01(≈)
JY7	5,10	**0.7498±1.50e−01**	1.2306±1.38e−01(+)	1.1403±6.23e−01(+)	1.2874±1.36e−01(+)	0.9069±1.46e−01(+)	1.6157±2.08e−01(+)	0.7523±1.48e−01(≈)
	10,10	**0.9123±2.07e−01**	1.4504±1.99e−01(+)	1.2195±1.02e+00(+)	1.2452±1.53e−01(+)	1.1188±2.28e−01(+)	1.4446±3.35e−01(+)	1.1046±2.33e−01(+)
	5,5	1.3189±1.44e−01	2.1346±1.95e−01(+)	1.3299±9.14e−01(≈)	2.0637±2.29e−01(+)	**1.3004±2.24e−01**(≈)	2.3230±2.97e−01(+)	1.3968±2.55e−01(+)
	10,5	1.4801±3.16e−01	2.3763±2.60e−01(+)	2.2291±1.08e+00(+)	2.0185±2.68e−01(+)	1.5024±2.99e−01(≈)	2.1547±3.61e−01(+)	**1.4532±2.25e−01**(≈)
JY8	5,10	0.0230±4.87e−03	0.0248±9.76e−03(≈)	0.0134±9.05e−04(−)	0.0246±1.80e−03(+)	0.0324±1.08e−02(+)	**0.0073±1.39e−03**(−)	0.0310±4.37e−03(+)
	10,10	0.0191±4.90e−03	0.0328±9.63e−03(+)	0.0125±8.54e−04(−)	0.0244±2.10e−03(+)	0.0368±1.26e−02(+)	**0.0069±1.35e−03**(−)	0.0401±6.20e−03(+)
	5,5	0.0203±3.66e−03	0.0441±1.15e−02(+)	0.0430±6.34e−03(+)	0.0636±8.29e−03(+)	0.0605±1.22e−02(+)	**0.0081±1.48e−03**(−)	0.0300±5.26e−03(+)
	10,5	0.0191±4.01e−03	0.0313±9.88e−03(+)	0.0415±5.50e−03(+)	0.0626±7.27e−03(+)	0.0564±1.87e−02(+)	**0.0073±1.52e−03**(−)	0.0413±8.06e−03(+)
JY9	5,10	**0.0425±1.16e−02**	0.1995±1.24e−01(+)	0.3665±5.33e−02(+)	0.3800±5.24e−02(+)	0.1484±7.56e−02(+)	0.3410±6.02e−02(+)	0.0601±1.50e−02(+)
	10,10	**0.0341±1.01e−02**	0.1060±1.86e−02(+)	0.2136±3.68e−02(+)	0.2844±2.47e−02(+)	0.0720±2.22e−02(+)	0.2024±3.34e−02(+)	0.0389±1.21e−02(+)
	5,5	**0.0646±1.02e−02**	0.5104±1.19e−01(+)	0.9014±1.21e−01(+)	1.0072±1.10e−01(+)	0.3136±1.52e−01(+)	0.6834±8.09e−02(+)	0.0869±1.48e−02(+)
	10,5	**0.0605±9.08e−03**	0.3432±5.62e−02(+)	0.6417±6.59e−02(+)	0.7875±1.03e−01(+)	0.1502±5.45e−02(+)	0.3444±1.32e−01(+)	0.0779±1.20e−02(+)
JY10	5,10	**0.0508±1.66e−02**	0.1498±1.35e−01(+)	0.0960±6.35e−02(+)	0.2013±6.63e−02(+)	0.1554±1.44e−01(+)	0.2468±1.61e−01(+)	0.0669±1.72e−02(+)
	10,10	**0.0457±1.46e−02**	0.1184±8.97e−02(+)	0.1686±1.55e−01(+)	0.1938±7.33e−02(+)	0.1997±3.14e−01(+)	0.1796±3.49e−01(+)	0.0612±1.84e−02(+)
	5,5	**0.0849±1.95e−02**	0.3888±1.90e−01(+)	0.3965±2.94e−01(+)	0.5381±1.81e−01(+)	0.3712±1.72e−01(+)	0.3774±3.29e−01(+)	0.0943±2.46e−02(≈)
	10,5	**0.0670±1.56e−02**	0.3013±2.08e−01(+)	0.4544±3.28e−01(+)	0.4737±1.42e−01(+)	0.4954±3.08e−01(+)	0.3308±2.16e−01(+)	0.0828±2.08e−02(+)
+/≈/−		-/-/-	64/2/6	55/5/12	68/2/2	57/6/9	53/5/14	50/22/0

Comparison Results. As shown in Table 1 and Table 2, CTE-DMOEA performed the best on more than a half of test instances. In detail, CTE-DMOEA won on problems that have various change severity on both POS and POF in different environment changes (i.e. DF7-9, JY9-10). However, when there is a complex change in decision space or objective space, the prediction may negatively influence the results (i.e. DF2 has a switch of the position-related variables in POS). For JY1-3 and JY6, there is a small number of non-dominated solutions and methods that predict using the whole population, such as PPS and MOEA/D-KF, got better performance sometimes. The prediction method of CTE-DMOEA is similar to MV-DMOEA when both POS and POF are estimated to be changed. According to the results, CTE-DMOEA performs better or approximately compared to MV-DMOEA on almost all of the test instance. This demonstrates the superiority of the adaptive transfer strategy in CTE-DMOEA.

Parameter Sensitivity Analysis. Additional Experiments were conducted to find the proper parameter settings. The number of reference points was set as $N_{RP} = C * m + 1$, where C is a multiplier of the objective number m. Figure 2 shows the average accuracy of the change type estimation on some representative benchmarks, which indicates the accuracy is relatively higher when $C = 2$ or $C = 3$. The performance effect of thresholds ε_{POS} and ε_{POF} on CTE-DMOEA on some representative problems are shown in Fig. 3. It can be seen from Fig. 3 that setting ε_{POS} to 0.2 and ε_{POF} to 0.05 can bring better results in general.

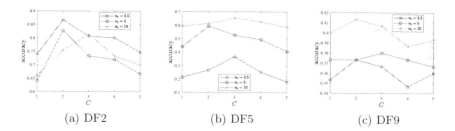

(a) DF2   (b) DF5   (c) DF9

Fig. 2. The average estimation accuracy over 10 runs with different C.

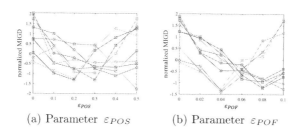

(a) Parameter ε_{POS} (b) Parameter ε_{POF}

Fig. 3. The normalized MIGD with different ε_{POS} and ε_{POF}.

5 Conclusion

In this paper, a DMOEA based on an adaptive knowledge transfer strategy, CTE-DMOEA, was presented to deal with DMOPs of different types of change. It first applies a change type estimation method to determine the change type of a DMOP and then based on the change type adaptively chooses to transfer knowledge in decision space or objective space or both. To evaluate the performance of the CTE-DMOEA, experiments have been conducted to test the algorithm on various benchmark problems and the results were compared to those obtained by six state-of-the-art prediction-based DMOEAs. Statistical test results showed the superiority of CTE-DMOEA. There are also several issues need to be further improved, including how to more accurately estimate the current environmental change and how to extract more information when the obtained optimal solutions are limited.

Acknowledgements. This work was supported by the National Natural Science Foundation of China (Grant No. 61906082), Guangdong Provincial Key Laboratory (Grant No. 2020B121201001), the Research Institute of Trustworthy Autonomous Systems (RITAS).

References

1. Azzouz, R., Bechikh, S., Ben Said, L.: Dynamic multi-objective optimization using evolutionary algorithms: a survey. In: Bechikh, S., Datta, R., Gupta, A. (eds.) Recent Advances in Evolutionary Multi-objective Optimization. ALO, vol. 20, pp. 31–70. Springer, Cham (2017). https://doi.org/10.1007/978-3-319-42978-6_2
2. Bengio, Y., Lamblin, P., Popovici, D., Larochelle, H.: Greedy layer-wise training of deep networks. In: Advances in Neural Information Processing Systems, vol. 19, pp. 153–160 (2006)
3. Byrd, R.H., Hribar, M.E., Nocedal, J.: An interior point algorithm for large-scale nonlinear programming. SIAM J. Optim. **9**(4), 877–900 (1999)
4. Cao, L., Xu, L., Goodman, E.D., Bao, C., Zhu, S.: Evolutionary dynamic multiobjective optimization assisted by a support vector regression predictor. IEEE Trans. Evol. Comput. **24**(2), 305–319 (2019)
5. Coello, C.C.: Evolutionary multi-objective optimization: a historical view of the field. IEEE Comput. Intell. Mag. **1**(1), 28–36 (2006)
6. Farina, M., Deb, K., Amato, P.: Dynamic multiobjective optimization problems: test cases, approximations, and applications. IEEE Trans. Evol. Comput. **8**(5), 425–442 (2004)
7. Feng, L., Ong, Y.S., Jiang, S., Gupta, A.: Autoencoding evolutionary search with learning across heterogeneous problems. IEEE Trans. Evol. Comput. **21**(5), 760–772 (2017)
8. Feng, L., Zhou, W., Liu, W., Ong, Y.S., Tan, K.C.: Solving dynamic multiobjective problem via autoencoding evolutionary search. IEEE Transactions on Cybernetics **52**(5), 2649–2662 (2020)
9. Goh, C.K., Tan, K.C.: A competitive-cooperative coevolutionary paradigm for dynamic multiobjective optimization. IEEE Trans. Evol. Comput. **13**(1), 103–127 (2008)

10. Helbig, M., Engelbrecht, A.P.: Benchmarks for dynamic multi-objective optimisation algorithms. ACM Comput. Surv. (CSUR) **46**(3), 1–39 (2014)
11. Jiang, M., Huang, Z., Qiu, L., Huang, W., Yen, G.G.: Transfer learning-based dynamic multiobjective optimization algorithms. IEEE Trans. Evol. Comput. **22**(4), 501–514 (2017)
12. Jiang, M., Wang, Z., Guo, S., Gao, X., Tan, K.C.: Individual-based transfer learning for dynamic multiobjective optimization. IEEE Trans. Cybern. **51**(10), 4968–4981 (2020)
13. Jiang, M., Wang, Z., Hong, H., Yen, G.G.: Knee point-based imbalanced transfer learning for dynamic multiobjective optimization. IEEE Trans. Evol. Comput. **25**(1), 117–129 (2020)
14. Jiang, M., Wang, Z., Qiu, L., Guo, S., Gao, X., Tan, K.C.: A fast dynamic evolutionary multiobjective algorithm via manifold transfer learning. IEEE Trans. Cybern. **51**(7), 3417–3428 (2020)
15. Jiang, S., Yang, S.: Evolutionary dynamic multiobjective optimization: benchmarks and algorithm comparisons. IEEE Trans. Cybern. **47**(1), 198–211 (2016)
16. Jiang, S., Yang, S., Yao, X., Tan, K.C., Kaiser, M., Krasnogor, N.: Benchmark functions for the CEC'2018 competition on dynamic multiobjective optimization. Newcastle University, Technical report (2018)
17. Kandaswamy, C., Silva, L.M., Alexandre, L.A., Sousa, R., Santos, J.M., de Sá, J.M.: Improving transfer learning accuracy by reusing stacked denoising autoencoders. In: 2014 IEEE International Conference on Systems, Man, and Cybernetics (SMC), pp. 1380–1387. IEEE (2014)
18. Koo, W.T., Goh, C.K., Tan, K.C.: A predictive gradient strategy for multiobjective evolutionary algorithms in a fast changing environment. Memetic Comput. **2**, 87–110 (2010)
19. Liu, Z., Wang, H.: Improved population prediction strategy for dynamic multiobjective optimization algorithms using transfer learning. In: 2021 IEEE Congress on Evolutionary Computation (CEC), pp. 103–110. IEEE (2021)
20. Muruganantham, A., Tan, K.C., Vadakkepat, P.: Evolutionary dynamic multiobjective optimization via Kalman filter prediction. IEEE Trans. Cybern. **46**(12), 2862–2873 (2015)
21. Tan, K.C., Feng, L., Jiang, M.: Evolutionary transfer optimization-a new frontier in evolutionary computation research. IEEE Comput. Intell. Mag. **16**(1), 22–33 (2021)
22. Zhang, Q., Li, H.: MOEA/D: a multiobjective evolutionary algorithm based on decomposition. IEEE Trans. Evol. Comput. **11**(6), 712–731 (2007)
23. Zhou, A., Jin, Y., Zhang, Q.: A population prediction strategy for evolutionary dynamic multiobjective optimization. IEEE Trans. Cybern. **44**(1), 40–53 (2013)
24. Zhou, L., Feng, L., Gupta, A., Ong, Y.S.: Learnable evolutionary search across heterogeneous problems via kernelized autoencoding. IEEE Trans. Evol. Comput. **25**(3), 567–581 (2021)
25. Zhou, W., Feng, L., Tan, K.C., Jiang, M., Liu, Y.: Evolutionary search with multi-view prediction for dynamic multiobjective optimization. IEEE Trans. Evol. Comput. **26**(5), 911–925 (2021)
26. Guo, Y., Chen, G., Jiang, M., Gong, D., Liang, J.: A knowledge guided transfer strategy for evolutionary dynamic multiobjective optimization. IEEE Trans. Evol. Comput. **27**(6), 1750–1764 (2023)

Multi-objective Biological Survival Optimizer with Application in Engineering Problems

Xueliang Fu and Qingyang Zhang$^{(\boxtimes)}$ (iD)

Jiangsu Normal University, Xuzhou 221116, China
sweqyian@126.com

Abstract. Biological survival optimizer is a recent proposed swarm-based optimization algorithm, which is inspired by the nature behavior of prey and has two significant components, escape phase and adjustment phase. This paper proposes multi-objective Biological survival optimizer (MOBSO) based on non-dominated framework, in which an external set is utilized to store the obtained non-dominated solutions for guiding search. Besides that, elite selection mechanism is also employed to choose promising solutions from parent and offspring agents based on the non-dominated levels. The performance of MOBSO is also evaluated on a suite of benchmark problems with various features and three classical engineering design problems. Simulation comparison results considering different indicators show that MOBSO can generate competitive results compared with other state-of-art optimization techniques.

Keywords: Biological survival optimizer · Multi-objective optimization · Engineering problems

1 Introduction

Multi-objective optimization problems (MOPs) usually involve multiple conflicting objectives, and they need to be optimized simultaneously [1]. The following provides the multi-objective optimization model considering minimization problem.

$$
\begin{aligned}
&Min.F(x) = \{f_1(x), f_2(x), ..., f_o(x)\} \\
s.t. \quad &g_i(x) \leqslant 0, i = 1, 2, ..., m \\
&h_j(x) = 0, j = 1, 2, ..., p \\
&Lb_k \leqslant x_k \leqslant Ub_k, k = 1, 2, ..., D
\end{aligned}
\tag{1}
$$

where o is the number of objective functions, m and p denote the number of inequality and equality constraints, respectively, D is the dimension of objective functions, Lb_k and Ub_k are the boundaries of k-th variable.

Obviously, compared with single objective problem, MOPs employ the concept of Pareto dominance to distinguish the quality of individuals, and the optimal solution of the problem will no longer be a single optimal value, but an

© The Author(s), under exclusive license to Springer Nature Singapore Pte Ltd. 2024
L. Pan et al. (Eds.): BIC-TA 2023, CCIS 2061, pp. 200–212, 2024.
https://doi.org/10.1007/978-981-97-2272-3_15

optimal solution set containing tradeoff solutions. The related definitions are provided as follows.

Definition 1 (Pareto Dominance). For two given decision vectors (x, y), vector x is said to dominate vector y (denote as $x \prec y$) if and only if:

$$\forall i \in \{1, 2, ..., o\} : f_i(x) \leq f_i(y) \land \exists p \in \{1, 2, ..., o\} : f_p(x) < f_p(y) \qquad (2)$$

Definition 2 (Pareto Optimality). A feasible solution x is called Pareto-optimal if and only if:

$$\neg \exists y : y \prec x \qquad (3)$$

Definition 3 (Pareto optimal set). The set all Pareto-optimal solutions is called Pareto set as follows:

$$POS : \forall x, y, \neg \exists y \prec x \ \ or \ \ x \prec y \qquad (4)$$

Definition 4 (Pareto optimal front). The set includes the objective value of POS:

$$POF : F(x), x \in POS \qquad (5)$$

2 Literature Review

Over the course of past decade, as one of the most attractive and popular areas in intelligent computing field, the existing Multi-objective optimization algorithms can be classified into three categories as follows. The first class is Pareto ranking-based algorithms, which are designed based on the dominated relationships among population individuals. Some representative algorithms include the non-dominated sorting genetic algorithm II (NSGA-II) [2], and strength Pareto evolutionary algorithm (SPEA2) [3]. Besides that, some classic and recent proposed efficient swarm intelligence algorithms inspired by different nature behaviors have also used to solve MOPs, such as multi-obejctve particle swarm optimization (MOPSO) [4], Multi-Objective Grasshopper Optimisation Algorithm (MOGOA) [5], Multi-Objective Multi-Verse Optimizer (MOMVO) [6], Multi-Objective Ant Lion Optimizer [7], and Multi-objective Salp Swarm Algorithm (MOSSA) [8], and so on. Although the non-dominant ranking strategy can well screen out excellent individuals, it also produces marginal individuals, which generate negative effect on the whole optimization process. These algorithms can obtain good local optimal solutions, but it is difficult to achieve ideal global optimal solutions.

The second class is indicator-Based algorithms, which are designed based on the performance indicators. The hypervolume [9], the epsilon indicator and the R2 one are the most utilized for proposed various algorithms, such as, indicator-based EA (IBEA) [10], S-metric selection EMO algorithm [11], R2 EMO algorithm (R2EMOA) [12], and approximation-guided EMO (AGE) [13].

The last class is decomposition-Based algorithms, which aim to decompose the MOP into some optimization sub-problems and solve them simultaneously. The most used algorithms are NSGA-III [14], and MOEA based on decomposition (MOEA/D) [15]. Although this kind of algorithm is efficient, the division of sub-problems depends on the weight deeply.

As mentioned before, two main tasks of multi-objective optimization are the approximations of the true Pareto optimal sets and the well-distributed degree of the obtained solutions. In general, there exists more than one feasible individual in MOPs, so two solutions cannot be compared based on the method employed in single objective problems. Here, the concept of Pareto dominance is employed for comparing the feasible solutions, and an stored archive is also set to save the obtained Pareto optimal solutions. Actually, the most task in designing MOBSO is to find the best or worst individual in evolution population, which benefit for leading the search agent towards promising areas of the search space. In the single objective optimization, it is easily to find them by comparing the corresponding fitness of problems obtained so far. In multi-objective optimization, the Pareto optimal solutions are usually selected from a set of Pareto optimal solutions. Meanwhile, the chosen agent has better distribution in the archive, so the crowding distance strategy is utilized for computing the relevant distance among them. These individual with better crowded neighbourhood will selected to avoid premature convergence and to enhance the distribution of the feasible regions to some extent.

However, there is a limitation that how to define the capacity or the maximum of the archive or upper number of stored solutions. As the iteration progresses, the number of non-dominated solutions increases greatly, which improves the computational complexity of the algorithm. Therefore, it is necessary to control the size of archive for reducing the complexity to some extent. Here, these solutions with crowded neighbourhood tend to be removed from archive, which guarantees the quality of archive commendably.

3 Multi-objective Biological Survival Optimizer (MOBSO)

Biological survival optimizer is a recent proposed swarm-based optimization algorithm, which is inspired by the nature behavior of prey and has two significant components, escape phase and adjustment phase [16]. This section proposed the multi-objective version of BSO based on the non-dominated framework widely involved in various multi-objective algorithms, the basic flowchart of MOBSO is presented as follows (Fig. 1).

3.1 Group Generation

It is assumed that the search population (Pop) is comprised of N feasible solutions ($X_i, i = 1, 2, 3..., N$), which are generated randomly using the following equations.

$$X_i(t) = (x_i^1(t), x_i^2(t), ..., x_i^D(t)), i = 1, 2, ..., N \quad (6)$$

$$x_i^k(t) = LB^k + r \times (UB^k - LB^k), k = 1, 2, ..., D \quad (7)$$

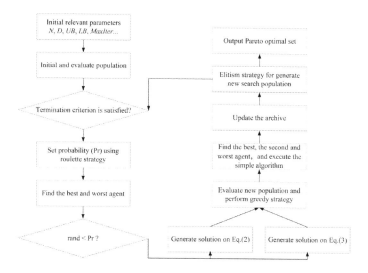

Fig. 1. The flowchart of the proposed MOBSO

$$Pop(t) = (X_1(t), X_2(t), ..., X_N(t))$$

where N and D refers to the member of search members and dimension of issue, respectively. t denotes the current number of generation, r denotes a random number generated from $(0, 1)$. LB and UB are the lower and upper boundaries of variables, respectively.

3.2   Escape Phase

This section mainly simulates the behavior that the leaders help other species members away from the hunting. BSO uses the roulette strategy to set probability for determining whether the current agent will be caught. Obviously, if the current individual is caught, it will be replaced by a new one. Besides that, each agent modifies its position according to the best individual and its neighbors. To define the neighborhood structure of an individual, a ring topology structure (see Fig. 2) is utilized. If the index of individual is i, the index of the selected neighbor is $i + 1$. If the index of agent is N, the index of the selected neighbor is 1. Here, we assume that the best individual has the best fitness, the worst member is opposite. The following provides the calculation equations.

$$newX_i = X_i + \delta_i(pop^{best} - pop^{worst}) +$$
$$\varphi_i(X_{neigh,i} - X_i) \tag{8}$$

where each number of δ_i and φ_i are generated from $(0,1)$, pop^{best} and pop^{worst} are the best and worst individual of population, respectively, $X_{neigh,i}$ is the neighbor of X_i.

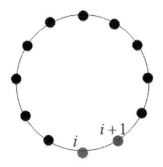

Fig. 2. The ring topology structure

3.3   Adjustment Phase

The phase main simulates the phenomenon that the leaders guide the other members in the weak side of the population toward safer place. This action is commanded by the two best individuals of the population. From the view of optimization, this phase aims to make the worst individual better with the assistance of two best solutions of the population. Therefore, this course is implemented using the simplex algorithm (Fig. 3), which is a popular technique launched by the some existing literatures.

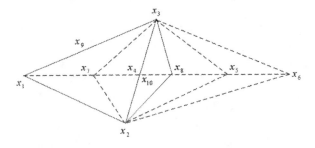

Fig. 3. Schematic view of the simplex method

4   Results on UF Problems

This section aims to evaluate the effectiveness of MOBSO by solving a set of UF test problems with various characteristics.

4.1 Benchmark Problems

The UF benchmark problems is very suitable to evaluate the performance of an optimization algorithm comprehensively, since it has diverse features including unconstrained, multimodal, constrained and different variables subcomponents. More details about the basic characteristics of these problems can be found in corresponding literature [17].

Table 1. Mean and standard deviation results of three indicators obtained by all algorithms

Function	Index	MOALO	MOSSA	MOMVO	MODA	MOBSO
f_1	IGD	2.19e−01(5.72e−02)	2.08e−01(6.3571e−02)	1.29e−01(3.36e−02)	3.53e−01(9.38e−02)	1.15e−01(3.02e−02)
	HV	1.42e+00(1.11e−01)	1.43e+00(1.4355e−01)	1.60e+00(8.34e−02)	1.17e+00(1.76e−01)	1.66e+00(8.62e−02)
	SP	6.58e−03(8.69e−03)	2.80e−02(2.0050e−02)	5.18e−02(7.80e−02)	4.70e−02(5.25e−02)	2.12e−04(5.80e−04)
f_2	IGD	2.42e−01(3.82e−02)	6.73e−02(8.8771e−03)	5.35e−02(6.28e−03)	1.38e−01(4.58e−02)	5.57e−02(9.10e−03)
	HV	1.39e+00(8.34e−02)	1.74e+00(3.1887e−02)	1.81e+00(2.03e−02)	1.61e+00(1.07e−01)	1.80e+00(3.50e−02)
	SP	5.55e−03(1.69e−03)	1.05e−02(6.2538e−03)	1.71e−02(9.20e−03)	2.80e−02(1.14e−02)	3.32e−04(1.14e−03)
f_3	IGD	4.28e−01(8.63e−02)	5.62e−01(1.0108e−01)	6.00e−01(1.54e−01)	5.76e−01(1.46e−01)	4.97e−01(1.23e−01)
	HV	1.12e+00(1.48e−01)	7.28e−01(1.4899e−01)	6.89e−01(1.96e−01)	7.85e−01(2.94e−01)	8.87e−01(2.25e−01)
	SP	5.05e−02(5.37e−02)	4.09e−02(3.6146e−02)	5.53e−02(2.84e−02)	1.06e−01(8.41e−02)	2.20e−05(8.69e−05)
f_4	IGD	8.77e−02(7.23e−03)	9.28e−02(7.0021e−03)	8.96e−02(8.42e−03)	9.54e−02(6.96e−03)	9.46e−02(1.28e−02)
	HV	1.39e+00(1.44e−02)	1.39e+00(1.3504e−01)	1.38e+00(3.33e−02)	1.38e+00(1.12e−02)	1.36e+00(3.53e−02)
	SP	4.97e−03(2.23e−03)	9.24e−03(4.1611e−03)	1.49e−02(4.37e−03)	1.44e−02(4.60e−03)	3.61e−04(1.01e−03)
f_5	IGD	1.47e+00(2.75e−01)	1.23e+00(2.0762e−01)	4.85e−01(1.22e−01)	1.12e+00(2.93e−01)	5.37e−01(1.64e−01)
	HV	1.95e−02(4.31e−02)	4.25e−02(8.3911e−02)	6.66e−01(2.62e−01)	8.19e−02(1.18e−01)	8.27e−01(2.91e−01)
	SP	1.29e−02(7.69e−03)	4.28e−02(3.1335e−02)	6.24e−02(5.60e−02)	8.49e−02(5.83e−02)	1.43e−04(3.99e−04)
f_6	IGD	1.55e+00(3.80e−01)	1.69e+00(4.1259e−01)	1.04e+00(4.45e−01)	1.57e+00(4.44e−01)	8.59e−01(3.13e−01)
	HV	1.90e−02(5.60e−02)	1.11e−02(3.3652e−02)	2.46e−01(2.64e−01)	1.48e−02(4.40e−02)	2.97e−01(2.75e−01)
	SP	1.25e−02(1.13e−02)	1.03e−01(1.0732e−01)	2.49e−01(4.46e−01)	1.90e−01(1.45e−01)	3.71e−04(1.35e−03)
f_7	IGD	3.15e−01(8.50e−02)	2.13e−01(9.7398e−02)	2.46e−01(1.19e−01)	3.67e−01(1.28e−01)	1.77e−01(9.38e−02)
	HV	1.06e+00(1.89e−01)	1.29e+00(1.9812e−01)	1.20e+00(2.64e−01)	9.84e−01(2.49e−01)	1.38e+00(2.16e−01)
	SP	8.63e−03(6.18e−03)	2.82e−02(2.1904e−02)	3.06e−02(3.10e−02)	5.13e−02(4.28e−02)	1.79e−04(4.75e−04)
f_8	IGD	9.15e−01(3.17e−01)	4.58e−01(1.5172e−01)	4.47e−01(2.44e−01)	8.26e−01(4.05e−01)	3.50e−01(6.04e−02)
	HV	3.00e−01(2.22e−01)	1.14e+00(4.3581e−01)	1.40e+00(6.42e−01)	5.81e−01(4.35e−01)	1.72e+00(2.89e−01)
	SP	1.37e−01(5.53e−02)	3.64e−01(1.7873e−01)	3.12e−01(2.51e−01)	3.60e−01(3.61e−01)	3.75e−03(7.52e−03)
f_9	IGD	1.05e+00(2.18e−01)	5.41e−01(1.4978e−01)	4.85e−01(2.52e−01)	9.27e−01(4.05e−01)	3.29e−01(5.47e−02)
	HV	3.35e−01(2.63e−01)	1.27e+00(3.8078e−01)	1.54e+00(4.84e−01)	6.54e−01(4.69e−01)	1.91e+00(2.11e−01)
	SP	1.49e−01(8.51e−02)	3.84e−01(2.0672e−01)	3.22e−01(1.81e−01)	2.74e−01(1.20e−01)	1.38e−03(3.18e−03)
f_{10}	IGD	5.13e+00(1.33e+00)	2.09e+00(4.5830e−01)	2.56e+00(1.18e+00)	3.76e+00(1.35e+00)	1.70e+00(7.11e−01)
	HV	1.35e−03(8.61e−03)	1.97e−03(1.0773e−02)	1.11e−02(4.59e−02)	1.55e−03(2.75e−03)	6.59e−02(1.51e−01)
	SP	4.91e−01(1.83e−01)	3.86e−03(3.1070e−01)	1.32e+00(1.03e+00)	1.57e+00(9.11e−01)	1.92e−03(7.50e−03)

4.2 Evaluation Indicator

Three widely employed evaluated indicators are utilized for assess the performances of different algorithms [18] [19].

Inverted Generational Distance (IGD). The first performance indicator is IGD, which is utilized to evaluate the convergence and diversity of solutions obtained by an algorithm, and the mathematical equation is provided as follows.

$$IGD(POF_t^*, POF_t^{ob}) = \frac{\sum_{g \in POF_t^*} d(g, POF_t^{ob})}{|POF_t^*|} \tag{9}$$

where POF_t^* is the true POF solutions, POF_t^{ob} represents a POF approximation, $d(g, POF_t^{ob})$ is the minimum Euclidian distance between g and the points in POF_t^{ob}, and $|POF_t^*|$ is the number of solution in POF_t^*.

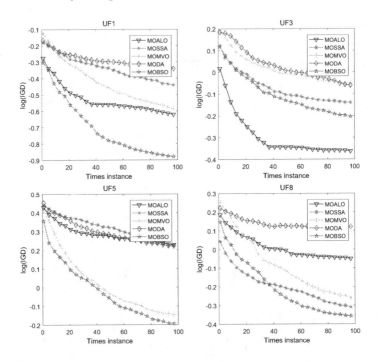

Fig. 4. Four convergence curves of all algorithms on UFs

Hypervolume Metric. The second performance indicator is Hypervolume (HV), which is a important metric for evaluating solutions. Different from the other indicators mentioned above, HV needs to set a reference vector dominated by any points in the POF_t^*.

$$HV_t = HV(POF_t^{ob}), \tag{10}$$

where $HV(POF_t^{ob})$ refers to the hypervolume of set POF_t^{ob}. The reference point for the computation of hypervolume is $(z_j + 0.5, j = 1, ..., m)$, where z_j is the maximum value of the jth objective of true POF.

Schott's Spacing Metric (SP). The last performance indicator is the Schott's spacing metric, which is used to measure the distribution of the obtained solutions POF_t^{ob} using the following formula:

$$SP(POF_t^{ob}) = \sqrt{\frac{1}{|POF_t^{ob}| - 1}(\sum_{i=1}^{|POF_t^{ob}|} (D_i - \overline{D}))} \tag{11}$$

where D_i represents the Euclidean distance between the ith point in POF_t^{ob} and its nearest point in POF_t^{ob}. $\overline{D}$ is the average value of D_i.

4.3   Parameter Settings

The stopping condition is set to the maximum number of iterations ($MaxIter$), which is defined as 100 for all algorithms. 30 runs are independently carried out to reduce the computation error. In addition, eight existing optimization algorithms are utilized to compare the performance of MOBSO, four existing multi-objective optimization algorithms are cited and the relevant descriptions and involved parameters are consistent with the original literature, MOMVO [6], MOALO [7], MOSSA [8] and MODA [20].

4.4   Experiment Results

Table 1 records all results of the five optimization algorithms considering IGD, HV and SP indicators. It can be seen that MOBSO has the best values compared with its peers for most of the benchmark functions in terms of IGD values. However, for two functions UF3 and UF4, MOALO obtains the best values, but the difference is not significant greatly. As shown in Table 1, which summarizes the MHV values of all the algorithms, although MOBSO has great better values than the other techniques on a majority of the problems, it is not slight effective enough for solving UF2, UF3 and UF4. In addition, it is also observed from SP results obtained by all the algorithms that although MOBSO can obtain the best solution on most of bi-objective and three objective problems, e.g., UF1, UF5 and UF7. Actually, well-distributed solutions does not mean that they approximate the true POF, it is necessary to evaluate overall performance of algorithm considering on all indicators Comprehensively. Besides that, Fig. 4 presents some convergence graphs of the IGD values for four benchmark functions. It is obvious that MOBSO converges faster than other competitors except UF3 problems. For UF3, MOBSO performs weaker than MOALO, but it has significant advantage over its peers. Totally, All results demonstrate that MOBSO is able to generate promising results on majority of the functions and track the true POF closely.

5   Engineering Optimization Problems

To further explore the effectiveness of the proposed MOBSO in practice, three widely used multi-objective engineering designed problems are adopted [7,8]. Note that the stopping condition, the number of runs and the other relevant settings keep the same as that in previous Subsect. 4.3. The same constraint handling technique used in CEC2009 is also utilized to deal with the constrains involved in problems. In addition, different from the test function, the optimal solution set cannot be known in advance in the actual problem, so IGD and SP cannot be used as the index. In this case, HV can only be used as the evaluation index, and the extreme value of each objective function is set as the reference point to calculate the corresponding results. Tables 2, 3 and 4 summarize the experiment results considering the mean HV (HV) and standard deviation (std) on the corresponding statistical results.

5.1   Case 1

The four-bar truss design problem is a well-known problem in the structural optimization field [6], in which structural volume (f_1) and displacement (f_2) of a 4-bar truss should be minimized. As can be seen in the following equations, there are four design variables ($x_1 - x_4$) related to cross sectional area of members 1, 2, 3, and 4.

$$Min \ \ f_1(x) = 200(2x_1 + \sqrt{2x_2} + \sqrt{x_3} + x_4)$$

$$Min \ \ f_2(x) = 0.01(\frac{2}{x_2} + \frac{2\sqrt{2}}{x_2} - (\frac{2\sqrt{2}}{x_3} + \frac{2}{x_1}))$$

$$s.t. \ \ 1 \leqslant x_1 \leqslant 3$$

$$\sqrt{2} \leqslant x_2 \leqslant 3$$

$$\sqrt{2} \leqslant x_3 \leqslant 3$$

$$1 \leqslant x_4 \leqslant 3$$

$$(12)$$

Table 2. Experimental results of all methods for four-bar truss design problem

Index	MOALO	MOSSA	MOMVO	MODA	MOBSO
HV	38.2016538	38.1364344	38.3261434	38.4490444	38.4492683
std	4.6970e−01	2.5774e−01	5.4017e−01	2.3815e−01	3.4325e−01

In this problem, the computation results listed in Table 2 clearly show that the performance of MOBSO is better than other optimizers in terms of HV metric. Although the standard deviation values of BSO is a little weaker than that of MODA and MOSSA, it generates the best 'HV' values, which means that the proposed MOBSO is able to obtain a set of optimal design with minimum weight compared to other competitors.

5.2   Case 2

As one of the most popular problems in the field of mechanical engineering, the speed reducer design problem aims to optimize the weight (f_1) and stress (f_2), which contains of eleven constraints and involves seven decision variables as follows, gear face width (x_1), teeth module (x_2), number of teeth of pinion (x_3), distance between bearings 1 (x_4), distance between bearings 2 (x_5), diameter of shaft 1 (x_6), and diameter of shaft 2 (x_7).

Table 3. Experimental results of all methods for speed reducer design problem

Index	MOALO	MOSSA	MOMVO	MODA	MOBSO
HV	2.5919e+06	2.5943e+06	2.6019e+06	2.5963e+06	5.5965e+06
std	7.1701e+03	5.7104e+03	6.2886e+03	5.7681e+03	6.2233e+03

$$Min \quad f_1(x) = 0.7854x_1x_2^2(10/3x_3^2 + 14.9334x_3 - 49.0934)$$
$$- 1.508x_1(x_6^2 + x_7^2) + 7.4777(x_6^3 + x_7^3)$$
$$+ 0.7854(x_4x_6^2 + x_5 + x_7^2)$$

$$Min \quad f_2(x) = \frac{\sqrt{(745x_4/x_2x_3)^2 + 1.69 \times 10^7}}{0.1x_6^3}$$

$$s.t. \quad g_1(x) = \frac{1}{x_1x_2^2x_3} - \frac{1}{27} \leqslant 0$$

$$g_2(x) = \frac{1}{x_1x_2^2x_3^2} - \frac{1}{397.5} \leqslant 0$$

$$g_3(x) = \frac{x_4^3}{x_2x_3x_6^4} - \frac{1}{1.93} \leqslant 0$$

$$g_4(x) = \frac{x_5^3}{x_2x_3x_7^4} - \frac{1}{1.93} \leqslant 0 \tag{13}$$

$$g_5(x) = x_2x_3 - 40 \leqslant 0$$

$$g_6(x) = x_1/x_2 - 12 \leqslant 0$$

$$g_7(x) = 5 - x_1/x_2 \leqslant 0$$

$$g_8(x) = 1.9 - x_4 + 1.5x_6 \leqslant 0$$

$$g_9(x) = 1.9 - x_5 + 1.1x_7 \leqslant 0$$

$$g_{10}(x) = \frac{\sqrt{(745x_4/x_2x_3)^2 + 1.69 \times 10^7}}{0.1x_6^3} \leqslant 1300$$

$$g_{11}(x) = \frac{\sqrt{(745x_5/x_2x_3)^2 + 1.515 \times 10^8}}{0.1x_7^3} \leqslant 1100$$

$$2.6 \leqslant x_1 \leqslant 3.6, \, 0.7 \leqslant x_2 \leqslant 0.8, \, 17 \leqslant x_3 \leqslant 28, \, 7.3 \leqslant x_4 \leqslant 8.3, \, 7.3 \leqslant x_5 \leqslant 8.3,$$
$$2.9 \leqslant x_6 \leqslant 3.9, \, 5.0 \leqslant x_7 \leqslant 5.5$$

Table 3 clearly testifies that the results delivered by the MOBSO is slight weaker than that of MOMVO under the same run circumstance and termination criterion, but the superiority is statistically significant with respective to other competitors considering the obtained *HV* and standard deviation results. That means MOBSO is able to generate a set of parameter combinations such that the optimization objective is optimal.

Table 4. Experimental results of all methods for disk brake design problem

Index	MOALO	MOSSA	MOMVO	MODA	MOBSO
HV	62.0121665	62.0236935	62.0426595	N/A	62.0764383
std	5.2557e−02	2.2214e−02	1.5084e−02	N/A	4.6685e−02

5.3   Case 3

The disk brake design problem has mixed constraints and was proposed by Ray and Liew. The objectives to be minimized are: stopping time (f_1) and mass of a brake (f_2) of a disk brake. As can be seen in following equations, there are four design variables: the inner radius of the disk (x_1), the outer radius of the disk (x_2), the engaging force (x_3), and the number of friction surfaces (x_4) as well as five constraints.

$$Min \ f_1(x) = 4.9 \times 10^{-5}(x_2^2 - x_1^2)(x_4 - 1)$$

$$Min \ f_2(x) = 9.85 \times 10^6 \frac{x_2^2 - x_1^2}{x_2^3 - x_1^3 x_4 x_3}$$

$$s.t. \quad g_1(x) = 20 + x_1 - x_2 \leqslant 0$$

$$g_2(x) = 2.5(x_4 + 1) - 30 \leqslant 0$$

$$g_3(x) = \frac{x_3}{3.14(x_2^2 - x_1^2)^2} - 0.4 \leqslant 0 \qquad (14)$$

$$g_4(x) = \frac{2.22 \times 10^{-3} x_3 (x_2^3 - x_1^3)}{(x_2^2 - x_1^2)^2} - 1 \leqslant 0$$

$$g_5(x) = 900 - \frac{2.66 \times 10^{-2} x_3 x_4 (x_2^3 - x_1^3)}{x_2^2 - x_1^2} \leqslant 0$$

$$55 \leqslant x_1 \leqslant 80, \ 75 \leqslant x_2 \leqslant 110, \ 1000 \leqslant x_3 \leqslant 3000, \ 2 \leqslant x_4 \leqslant 20$$

According to the experimental results listed in Table 4, although the standard deviation value of MOBSO is weaker than that of MOSSA and MOMVO, it sufficiently outperform other optimization algorithms with the same termination criterion and run circumstance. Such evidence indicates that the proposed algorithm will be an attractive alternative optimizer for generating satisfactory results on challenging optimization problems in future.

6   Conclusion

In this paper, a multi-objective version of BSO (MOBSO)is proposed based on non-dominate framework. To evaluate the performance of MOBSO, a test suite of benchmark functions with different characteristics and three popular engineering design problems are utilized. The simulation results compared with some existing optimization algorithms show the effectiveness of MOBSO is competitive. In future work, MOBSO will be further improved or modified as tool for addressing diverse practical applications in real world.

Acknowledgements. This work is supported by the National Natural Science Foundation of China under Grants 62006103, in part by the Postgraduate research and practice innovation program of Jiangsu province under Grand KYCX22_2858, in part by Xuzhou Basic Research Program under Grand KC23025, in part by the Royal Society International Exchanges Scheme IEC\NSFC\211404, and in part by China Scholarship Council under Grand 202310090064.

References

1. Liang, J.F., Ban, X.S.: A survey on evolutionary constrained multiobjective optimization. IEEE Trans. Evol. Comput. **27**(2), 201–221 (2022)
2. Deb, K.F., Pratap, A.S., Agarwal, S.T., Meyarivan, T.F.: A fast and elitist multiobjective genetic algorithm: NSGA-II. IEEE Trans. Evol. Comput. **6**, 182–197 (2002)
3. Zitzler, E.F., Laumanns, M.S., Thiele, L.T.: SPEA2: improving the strength pareto evolutionary algorithm. Technical Report 103, Computer Engineering and Networks Laboratory (TIK), Switzerland (2001)
4. Coello Coello, C.A.F., Lechuga, M.S. .: MOPSO: a proposal for multiple objective particle swarm optimization. In: Proceedings of the 2002 Congress on Evolutionary Computation, pp. 1051–1056. CEC'02, USA (2002)
5. Mirjalili, S.F., Mirjalili, S.S., Saremi, S.T., Aljarah, I.F.: Grasshopper optimization algorithm for multi-objective optimization problems. Appl. Intell. **48**, 805–820 (2018)
6. Mirjalili, S.F., Jangir, P.S., Mirjalili, S.T., Saremi, S.F., Trivedi, I.F.: Optimization of problems with multiple objectives using the multi-verse optimization algorithm. Knowl.-Based Syst. **134**, 50–71 (2017)
7. Mirjalili, S.F., Jangir, P.S., Saremi, S.T.: Multi-objective ant lion optimizer: a multi-objective optimization algorithm for solving engineering problems. Appl. Intell. **46**, 79–95 (2017)
8. Mirjalili, S.F., Gandomi, A.S., Mirjalili, S.T., Saremi, S.F., Faris, H.F., Mirjalili, S.S.: Salp swarm algorithm: a bio-inspired optimizer for engineering design problems. Adv. Eng. Softw. **114**, 163–191 (2017)
9. Auger, A.F., Bader, J.S., Brockhoff, D.T., Zitzler, E.F.: Hypervolume-based multiobjective optimization: theoretical foundations and practical implications. Theoret. Comput. Sci. **425**(1), 75–103 (2012)
10. Phan, D. F., Suzuki, J. S.: R2-IBEA: R2 indicator based evolutionary algorithm for multiobjective optimization. In: 2013 IEEE Congress on Evolutionary Computation, pp. 1836–1845. CEC'13, Mexico (2013)
11. Zhang, Q.F., Li, H.S.: MOEA/D: a multiobjective evolutionary algorithm based on decomposition. IEEE Trans. Evol. Comput. **11**(6), 712–731 (2007)
12. Saxena, D.F., Duro, J.S., Tiwari, A.T., Deb, K.F., Zhang, Q.F.: Objective reduction in many-objective optimization: linear and nonlinear algorithms. IEEE Trans. Evol. Comput. **17**(1), 77–99 (2013)
13. Bringmann, K.F., Friedrich, T.S., Neumann, F.T., Wagner, M.F.: Approximation guided evolutionary multi-objective optimization. In: Proceedings of the Twenty-Second International Joint Conference on Artificial Intelligence, pp. 1198–1203. Spain (2011)
14. Deb, K.F., Jain, H.S.: An evolutionary many-objective optimization algorithm using reference-point based non-dominated sorting approach, part I: solving problems with box constraints. IEEE Trans. Evol. Comput. **18**(4), 577–601 (2014)

15. Ke, L.F., Zhang, Q.S., Battiti, R.T.: MOEA/D-ACO: a multiobjective evolutionary algorithm using decomposition and ant colony. IEEE Trans. Cybern. **43**(6), 1845–1859 (2013)
16. Wang, L.F., Zhang, Q.S.: Biological survival optimization algorithm with its engineering and neural network applications. Soft Comput. **27**, 6437–6463 (2023)
17. Zhang, Q.F., Zhou, A.S., Suganthan, P.N.T.: Multiobjective optimization test instances for the CEC2009 Special session and competition. Mechanical Engineering (2008)
18. Zhang, Q.F., Yang, S.S., Wang, R.T.: Novel prediction strategies for dynamic multiobjective optimization. IEEE Trans. Evol. Comput. **24**(2), 260–274 (2020)
19. Zhang, Q.F., He, X.S., Yang, S.T., Dong, Y.F., Song, H.F., Jiang, S.S.: Solving dynamic multi-objective problems using polynomial fitting-based prediction algorithm. Inform. Sci. **610**, 868–886 (2022)
20. Mirjalili, S.F.: Dragonfly algorithm: a new meta-heuristic optimization technique for solving single-objective, discrete, and multi-objective problems. Neural Comput. Appl. **27**(4), 1053–1073 (2016)

Dynamic Constrained Robust Optimization over Time for Operational Indices of Pre-oxidation Process

Yilin Fang[1]([✉])(iD), Ziheng Zhao[1](iD), and Liang Jin[2]

[1] School of Information Engineering, Wuhan University of Technology, Wuhan, China
`fangspirit@whut.edu.cn`
[2] School of Mechanical and Electronic Engineering, Wuhan University of Technology, Wuhan, China

Abstract. The pre-oxidation process of carbon fiber, as one of its important production processes, can affect its physical properties directly. In this paper, in order to provide more valuable theoretical guidance for the actual pre-oxidation process, we construct a dynamic constrained multi-objective optimization model of the pre-oxidation process. For the treatment of dynamic constraints, we use the strategy based on penalty function, extending the constraint violation value in the objective space from a single feasibility deviation value to a weighted sum of feasibility deviation value and non-dominated deviation value, and propose the dynamic constrained multi-objective evolutionary algorithm considering non-dominated deviation (DCMOEA-ND). We then incorporate the new robustness definition and propose the dynamic constrained robust optimization over time (DCROOT), which is designed to obtain high-quality filaments and reduce energy consumption while reducing the switching cost of solutions. Experimental results show that DCMOEA-ND can obtain Pareto optimal set (POS) with better convergence and distribution, and the robust solutions obtained by DCROOT have better performance than other algorithms.

Keywords: Pre-oxidation process · Dynamic constrained multi-objective optimization · Robust optimization over time

1 Introduction

Carbon fiber is a high-performance material with high strength, high-temperature resistance, friction resistance, etc. Its diameter is equivalent to about one-tenth of a hairline, but its strength is four times more than that of aluminum alloy. With these excellent characteristics, carbon fiber is widely used in automotive equipment, aerospace equipment, construction materials, bridge composite materials, and other fields. In general, the production of polyacrylonitrile (PAN) carbon fibers can be divided into four different stages, which are the polymerization process, the spinning process, the pre-oxidation process, and

© The Author(s), under exclusive license to Springer Nature Singapore Pte Ltd. 2024
L. Pan et al. (Eds.): BIC-TA 2023, CCIS 2061, pp. 213–227, 2024.
https://doi.org/10.1007/978-981-97-2272-3_16

the carbonization process [14]. Among them, the pre-oxidation process is also known as the thermal stabilization process in some research, which changes the internal structure and can directly affect the physical properties of the carbon fiber, so it is a very important process.

In [9], we constructed a dynamic multi-objective optimization model of the pre-oxidation process. In order to obtain high-quality filaments while reducing the energy consumption, we proposed the modified robust optimization over time (mROOT) to solve the model, aiming to find the solution with less switching cost while still meeting the above requirements. However, this model does not consider some constraints, resulting in the final solution meeting the theoretical requirements but not the field requirements. Therefore, in this paper, we add some constraints that are feedback to us from the production site and construct a dynamic constrained multi-objective optimization model with the PAN modulus and the energy consumption of the pre-oxidation process as objective, the pre-oxidation time of each temperature zone as the decision variable, the temperature of each temperature zone as the dynamic environment, and the equivalent time of each environment as the dynamic constraints. To solve the model, we propose the dynamic constrained multi-objective evolutionary algorithm considering non-dominated deviation (DCMOEA-ND) firstly to handle the dynamic constraints. Then we propose the dynamic constrained robust optimization over time (DCROOT) to find deployment solutions that are less expensive to switch, maximize PAN modulus and minimize energy consumption, and meet the field constraints, aiming to provide more valuable theoretical guidance for the actual production of carbon fiber.

The rest of this paper is organized as follows: In Sect. 2, dynamic constrained multi-objective optimization problems and robust optimization over time are introduced. Section 3 introduces the model and proposed related algorithm in the pre-oxidation process of carbon fiber in detail. The experimental results are given and analyzed in Sect. 4. Finally, the conclusions are summarized in Sect. 5.

2   Related Work

In this section, we focus on the theory and recent related work of dynamic constrained multi-objective optimization problems and robust optimization over time.

2.1   Dynamic Constrained Multi-objective Optimization Problems

In real life, there are many engineering optimization problems, such as logistics scheduling and path planning, which often need to optimize two or more conflicting objectives at the same time, and the corresponding problem parameters or constraints will change with the external environment [8]. This class of problems can be called dynamic constrained multi-objective optimization problems (DCMOPs) and can be expressed by the following mathematical expressions:

$$\min F(\vec{x}_i, \theta^{(t)}) = \{f_1(\vec{x}_i, \theta^{(t)}), f_2(\vec{x}_i, \theta^{(t)}), \ldots, f_m(\vec{x}_i, \theta^{(t)})\}$$

$$\text{s.t.} \begin{cases} g_j(\vec{x}_i, \theta^{(t)}) \geq 0, j = 1, 2, \ldots, n_g \\ h_k(\vec{x}_i, \theta^{(t)}) = 0, k = 1, 2, \ldots, n_h \\ x_{\min} \leq \vec{x}_i \leq x_{\max}, i = 1, 2, \ldots, n \end{cases} \quad (1)$$

where at time t, $F(\vec{x}_i, \theta^{(t)})$ is the m-dimension objective function of the problem, $\vec{x}_i$ is the decision variable whose corresponding dimension is n, and the upper and lower bounds are $x_{\min}$ and $x_{\max}$. $\theta^{(t)}$ is the environmental parameter, and $f_m(\vec{x}_i, \theta^{(t)})$ is the mth objective function. $g_j(\vec{x}_i, \theta^{(t)})$ and $h_k(\vec{x}_i, \theta^{(t)})$ denote the jth inequality constraint and the kth equality constraint, and the number of them are n_g and n_h, respectively.

DCMOPs are generally characterized by two features, one is that the feasible domain of the objective function changes with the external environment, and the other is that the Pareto optimal front (POF) and POS of the problem change with the external environment. A solution that is feasible in the current environment is likely to become infeasible in the next environment because of the change of objective functions or constraints. Therefore, it is a challenge to deal with the relationship between feasible and infeasible solutions and obtain feasible solutions with good distribution and convergence.

Azzouz [2,3] designed some test problems for the DCMOPs and used a dynamic non-dominated sequential genetic algorithm based on penalty functions to solve them, Chen [7] added some new challenging test problems to the previous test functions for DCMOPs and proposed a novel evolutionary algorithm to solve. Aliniya [1] proposed a hybrid RM-RPM response mechanism and used three methods to generate an initial population of new environments to help create new solutions with acceptable variability near the optimal solution, and Wang [15] proposed a new dynamic constrained multi-objective evolutionary algorithm based on a constraint handling strategy of penalty function, combining constraint bias values in the objective space and constraint bias values in the decision space.

2.2 Robust Optimization over Time

For dynamic optimization problems, the current mainstream solutions can be divided into two categories, one is tracking the moving optima (TMO) and the other is robust optimization over time (ROOT). The main idea of TMO is to restart the algorithm and find the optimal solution under the new environment whenever the external environment changes. When the external environment changes frequently or the objective function of the problem is complex to evaluate, it may be difficult to find the optimal solution within a limited period of time by using TMO, especially in real life, frequent changes in deployment solutions may result in larger costs or simply be impractical.

In order to solve the problem caused by frequent replacement of deployment solutions, Yu [20] firstly proposed the concept of ROOT, finding some robust solutions that are not only applicable in the current environment but also applicable in some successive environments in the future to some extent. Jin [13] proposed a generalized solution framework for ROOT, including an optimizer, an approximator, a predictor, and a database. Fu [10] proposed two robustness definitions, survival time and average fitness, respectively. The goal of the former is to maximize the number of current and future environments to which the solution can be applied while ensuring the solution meets a preset threshold, and the goal of the latter is to allow the solution to maintain a good fitness value within a preset time window. Huang [12] proposed the switching cost of a solution by calculating the Euclidean distance between the new solution that will be applied to and the old solution that will be replaced. Chen [6] extended ROOT from single-objective to multi-objective and proposed robust Pareto-optima over time (RPOOT), aiming to find a set of robust Pareto-optimal solution sets, and proposed two robustness definitions, time robustness, and performance robustness, respectively in [4]. Wei [16] introduced dynamic constraints to RPOOT and proposed a new dynamic constrained multi-objective evolutionary algorithm for solving the benchmark function. Nowadays, most of the research on ROOT [5,11,18,19] focuses on solving artificial benchmark problems with continuous search space, such as moving peak benchmarks or benchmark functions [17], and little research has been done on solving the application problems [9,21].

3   Proposed Method

In this section, we first construct a new dynamic constrained multi-objective optimization model, then propose the DCMOEA-ND for dealing with the dynamic constraints in it, and finally combine the new robustness definition and DCMOEA-ND to propose the DCROOT.

3.1   Optimization Modeling

We constructed a dynamic multi-objective optimization model with the energy consumption of the pre-oxidation process and PAN modulus as the objective, the pre-oxidation time of each temperature zone as the decision variable, the temperature of each temperature zone as the dynamic environment, and provided some guidance for the actual production of carbon fiber in [9]. In the actual pre-oxidation process, the workers need to focus on the temperature matching and reaction time of each temperature zone in the pre-oxidation furnace, and these two variables will directly affect the final physical properties of the pre-oxidized filaments, so they need to be considered comprehensively. After a long time of experimentation and production in the factory, the workers summarized an empirical formula that can comprehensively consider the temperature and time of each temperature zone in the pre-oxidation furnace, i.e., the equivalent time, which is expressed in Eq. (2), where x_i denotes the time of the ith temperature zone, $T_i(t)$ denotes the temperature of the ith temperature zone at moment t.

$$T_{240}(t) = \sum_i x_i \times (1.76)^{\frac{T_i(t)-240}{10}} \tag{2}$$

The change of temperature or time in each temperature zone can lead to the change of equivalent time, so in the actual production, the workers can adjust the equivalent time to design the temperature and time parameters, providing a certain reference for obtaining high-quality filaments. Through the actual production, it is found that controlling the equivalent time in the interval of $80 \sim 100$ is a better choice. Since the temperature of each temperature zone in the pre-oxidation process can not be kept constant as the set value but will fluctuate up and down in a small range, the equivalent time will also change in each environment. In this paper, we still select a set of temperature zone gradient data commonly used in industry, which contains data from ten temperature zones, as shown in Table 1.

Table 1. A group of classic ten-temperature gradient data

Temperature Zone	T_1	T_2	T_3	T_4	T_5	T_6	T_7	T_8	T_9	T_{10}
Temperature /°C	190	200	210	220	230	240	250	260	265	270

Therefore, after adding the equivalent time, we get a dynamic constrained multi-objective optimization model for the pre-oxidation process and the expression is as follows:

$$J \sim \{ \min EC(\vec{x}_i, T_i, t), \min -M(\vec{x}_i, T_i, t) \}$$

$$\begin{cases} EC(\vec{x}_i, T_i, t) = \sum_i \frac{[T_i(t) - T_0] c_a V_a \rho}{3595} x_i \times \eta + \sum_i \sqrt{3} UI |\cos\theta| x_i \\ M(\vec{x}_i, T_i, t) = w_1 \frac{\int_0^{x_1+x_2}(ES_\tau + F_0 v_1)\,dx}{S(L + \int_0^{x_1+x_2} v_1 dx)} + w_2 \frac{\sum_i Ae^{-\frac{E_a}{RT_i(t)}} x_i F_0}{S} \\ I = I_E + I_R + 2I_D \\ T_i(t) = T_i(t_0) + \varphi \times N(\mu, \sigma^2) \\ T_{240}(t) = \sum_i x_i \times (1.76)^{\frac{T_i(t)-240}{10}} \\ 80 \le T_{240}(t) \le 100 \\ 3 \le x_i \le 9, i = 1, 2, \ldots, 10 \end{cases} \tag{3}$$

where $EC(\vec{x}_i, T_i, t)$ is the energy consumption of the pre-oxidation process, $M(\vec{x}_i, T_i, t)$ is the PAN modulus, $T_i(t_0)$ denotes the temperature of the ith temperature zone in Table 1. φ denotes the magnitude of temperature change in each temperature zone, $N(\mu, \sigma^2)$ denotes the generation of random numbers obeying a standard normal distribution, i.e., mean $\mu = 0$ and standard deviation $\sigma^2 = 1$, so $\varphi \times N(\mu, \sigma^2)$ is used to simulate small fluctuations in the temperature of each temperature zone. Table 2 describes the symbols used in the above model, their corresponding meanings, and values.

Table 2. Symbols, meanings, and their values in the model

Symbol	Meaning	Value
T_0	Room temperature	$25\,^{\circ}\mathrm{C}$
c_a	Specific heat capacity of air	$1.01\,\mathrm{kJ/kg}\cdot^{\circ}\mathrm{C}$
ρ	Density of air	$1.29\,\mathrm{kg/m^3}$
V_a	Flow of air	$10\,\mathrm{m^3/min}$
η	Air heater efficiency	75%
U	Three-phase voltage	$380\,\mathrm{V}$
I_R	Recycling fan current	$5\,\mathrm{A}$
I_D	Drive motor current	$0.4\,\mathrm{A}$
I_E	Exhaust fan current	$1.5\,\mathrm{A}$
θ	Phase difference	120°
E_a	Activation energy	$124.7\,\mathrm{kJ/mol}$
R	Molar gas constant	$8.314\,\mathrm{J/mol}\cdot\mathrm{K}$
A	Pre-exponential factor	4.15×10^5
τ	Pre-oxidation draft ratio	5%
v_1	Speed of the first drive roll	$0.5\,\mathrm{m/s}$
E	Young's modulus	$230\,\mathrm{N/m^2}$
S	Filament cross-sectional area	$38.48\,\mathrm{\mu m^2}$
L	Distance between the drive rollers	$50\,\mathrm{m}$

3.2   Proposed Algorithm

Since the model in this paper involves dynamic constraints, the algorithm in
[9] is no longer applicable to solve. We need to do extra processing on dynamic
constraints to ensure that the solution is close to the real POF and in the feasible
regions in the end.

In the process of population evolution, many types of individual will appear.
As Fig. 1 shows, in the objective vector space, the gray regions is feasible and

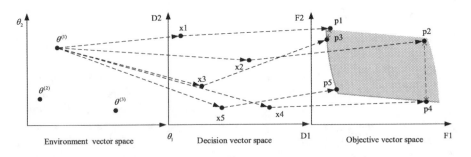

Fig. 1. The diagram of the various types of individuals that appear during the evolution
of a population in an environment.

the blue line is the POF with the environmental parameter $\theta^{(1)}$, p3, p4 and p5 are on the POF. p1 is not in the feasible regions but close to the POF and p2 is in the feasible regions but not close to the POF. These two types of individuals are particularly important. The former is already close to the edge of the POF and can guide the population to explore the POF although it is in the infeasible regions, while the latter is far away from the POF, but has entered into the feasible regions and can guide the population to find more feasible individuals. According to the constraint handling strategy based on the penalty function, since p1 is an infeasible solution, the corresponding penalty value will be larger than p2. In this way, p1 is more likely to be eliminated in the process of population evolution, while p2 relies on a smaller penalty value and obtains a larger probability of retention. This is something we don't want to see, because these two types of individuals should not be compared only on the basis of whether they are feasible or not, after all, the individual who is not in the feasible regions but close to the POF is in a good position to help the population find the POF.

Therefore, considering the above problems, we propose a new constraint processing method, which not only considers the viability deviation value of the individual itself $cv_f(\vec{x}_i, \theta^{(t)})$, as shown in Eq. (4), but also considers the deviation value of the individual with respect to the non-dominated population $cv_{non}(\vec{x}_i, \theta^{(t)})$, as shown in Eq. (5), and then by weighting the two, the individual deviation value $cv(\vec{x}_i, \theta^{(t)})$ is calculated together, as shown in Eq. (6).

$$cv_f(\vec{x}_i, \theta^{(t)}) = \sum_{j=1}^{n_g} G_j(\vec{x}_i, \theta^{(t)}) + \sum_{k=1}^{n_h} H_k(\vec{x}_i, \theta^{(t)}) \tag{4}$$

$$cv_{non}(\vec{x}_i, \theta^{(t)}) = \min \left\| \overline{f_m}(\vec{x}_i, \theta^{(t)}) - \overline{f_m}(\vec{x}_j^{non}, \theta^{(t)}) \right\|, j = 1, ..., |\mathrm{P}^{non}| \tag{5}$$

$$cv(\vec{x}_i, \theta^{(t)}) = \varphi_c \times cv_{non}(\vec{x}_i, \theta^{(t)}) + (1 - \varphi_c) \times \overline{cv_f}(\vec{x}_i, \theta^{(t)}) \tag{6}$$

For the decision variable $\vec{x}_i$ with the environmental parameter $\theta^{(t)}$, $G_j(\vec{x}_i, \theta^{(t)})$ and $H_k(\vec{x}_i, \theta^{(t)})$ are the jth inequality constraint and the kth equation constraint violation values, respectively. $\overline{f_m}(\vec{x}_i, \theta^{(t)})$ is the normalization of the mth objective function, $\overline{f_m}(\vec{x}_j^{non}, \theta^{(t)})$ is the normalization of the non-dominated individuals of the population at the mth objective function. φ_c is the product of the proportion of viable individuals in the population φ_f and the proportion of non-dominated individuals contained in the population φ_{non}. After calculating the degree of constraint violation of an individual, the distance term $d_m(\vec{x}_i, \theta^{(t)})$, the penalty term $p_m(\vec{x}_i, \theta^{(t)})$ and the modified fitness value $f'(\vec{x}_i, \theta^{(t)})$ of an individual under the mth objective are calculated using Eqs. (7–10), respectively. Combining the above ideas, we propose the dynamic constrained multi-objective evolutionary algorithm considering non-dominated deviation (DCMOEA-ND), and it is shown in Algorithm 1.

$$d_m(\vec{x}_i, \theta^{(t)}) = \sqrt{\overline{f_m}\left(\vec{x}_i, \theta^{(t)}\right)^2 + \overline{cv_f}\left(\vec{x}_i, \theta^{(t)}\right)^2} \tag{7}$$

$$p_m(\vec{x}_i, \theta^{(t)}) = (1 - \varphi_f) \times \overline{cv}(\vec{x}_i, \theta^{(t)}) + \varphi_f \times Y_m(\vec{x}_i, \theta^{(t)}) \quad (8)$$

$$Y_m(\vec{x}_i, \theta^{(t)}) = \begin{cases} 0, & \text{if } \overline{cv_f}(\vec{x}_i, \theta^{(t)}) = 0 \\ \overline{f_m}(\vec{x}_i, \theta^{(t)}), & \text{otherwise} \end{cases} \quad (9)$$

$$f'(\vec{x}_i, \theta^{(t)}) = d_m(\vec{x}_i, \theta^{(t)}) + p_m(\vec{x}_i, \theta^{(t)}) \quad (10)$$

Firstly, we generate N individuals randomly in P, calculate $f_m(\vec{x}_i, \theta^{(t)})$, $cv_f(\vec{x}_i, \theta^{(t)})$ of each individual and normalize them. Then non-dominated sorting of P is performed to obtain P_{non}, and calculate φ_f and φ_{non}. After getting φ_f and φ_{non}, we can calculate $cv_{non}(\vec{x}_i, \theta^{(t)})$, $cv(\vec{x}_i, \theta^{(t)})$, $f'(\vec{x}_i, \theta^{(t)})$, and $L(\vec{x}_i, \theta^{(t)}, \delta)$ for each individual in P. After that, we judge whether the iteration termination condition is reached, if not, enter the main loop, otherwise exit the loop. In the main loop, we do the crossover and mutation operation on P to obtain Q, and calculate $cv(\vec{x}_i, \theta^{(t)})$, $F(\vec{x}_i, \theta^{(t)})$ and $L(\vec{x}_i, \theta^{(t)}, \delta)$ for each individual in Q. Then we combine P and Q to obtain W. To get better individuals in W, we sort W in ascending order by non-dominated rank firstly, and then sort W in descending order by $L(\vec{x}_i, \theta^{(t)}, \eta)$. If there are individuals with same $L(\vec{x}_i, \theta^{(t)}, \eta)$, we sort them in descending order by crowding distance. After doing these steps, we select the first N individuals in W and assign them to P. Finally, we update φ_f, φ_{non} and P_{non}.

In ROOT, the survival time $L(\vec{x}_i, \theta^{(t)}, \delta)$ refers to the number of current and future continuous environments to which a solution can be applied and the time window tw is designed to allow the decision maker to define the desired survival time of the solution. But the actual survival time of solutions in the final POS is larger or smaller than the preset time window. Meanwhile, the dynamic constraints added in this paper make the model itself difficult to find solutions with a large survival time. Since it is a practical application and we end up deploying a specific solution rather than a POS, we take the robustness definition in mROOT [9] as a reference, combine the survival time and time window to give a new robustness definition, as shown in Eq. (11),

$$\begin{cases} L(\vec{x}_i, \theta^{(t)}, \delta) = \min\{ \max_{\vec{x}_i \in PS^{(t)}} \{L \mid \Delta\left(l\left(\vec{x}_i\right)\right) \leq \delta, \forall j, 0 \leq j \leq L\}, tw \} \\ \Delta(l\left(\vec{x}_i\right)) = \dfrac{\|F(\vec{x}_i, \theta^{(t)}) - \widehat{F}(\vec{x}_i, \theta^{(t+k)})\|}{\|F(\vec{x}_i, \theta^{(t)})\|} \\ F(\vec{x}_i, \theta^{(t)}) = (f_1^{avg}(\vec{x}_i, \theta^{(t)}), \ldots, f_m^{avg}(\vec{x}_i, \theta^{(t)})) \\ f_m^{avg}(\vec{x}_i, \theta^{(t)}) = \dfrac{1}{tw}(f_m(\vec{x}_i, \theta^{(t)}) + \sum_{j=1}^{tw} f_m'(\vec{x}_i, \theta^{(t+j)})) \end{cases} \quad (11)$$

where δ and tw are the preset robustness threshold and time window by decision maker, respectively. $F(\vec{x}_i, \theta^{(t)})$ is the individual's performance robustness, i.e., the fitness value. It should be noted that we need to convert the above mentioned modified fitness value $f'(\vec{x}_i, \theta^{(t)})$ to $F(\vec{x}_i, \theta^{(t)})$ by $f_m^{avg}(\vec{x}_i, \theta^{(t)})$. $\widehat{F}(\vec{x}_i, \theta^{(t+k)})$ is the individual's fitness value predicted by the predictor at the future moment

Algorithm 1. DCMOEA-ND

Input: population size N, number of iterations gen, time window tw, robustness threshold δ, parent population $P = \varnothing$, child population $Q = \varnothing$, merged population $W = \varnothing$, non-dominated population $P_{non} = \varnothing$, environment parameter $\theta^{(t)}$.

Output: P in the current time window.

1: Generate N individuals randomly and add to P;
2: **for** $i = 1 : N$ **do**
3:    Calculate $f_m(\vec{x}_i, \theta^{(t)})$, $cv_f(\vec{x}_i, \theta^{(t)})$ and normalize them;
4: **end for**
5: Undominated ordering of P to get P_{non};
6: Calculate φ_f and φ_{non} in P;
7: **for** $i = 1 : N$ **do**
8:    Calculate $cv_{non}(\vec{x}_i, \theta^{(t)})$, $cv(\vec{x}_i, \theta^{(t)})$, $f'(\vec{x}_i, \theta^{(t)})$, and $L(\vec{x}_i, \theta^{(t)}, \delta)$ in P;
9: **end for**
10: $g \leftarrow 0$;
11: **while** g < gen **do**
12:    $Q \leftarrow$ crossover and mutation P;
13:    Calculate $cv(\vec{x}_i, \theta^{(t)})$, $F(\vec{x}_i, \theta^{(t)})$ and $L(\vec{x}_i, \theta^{(t)}, \delta)$ in Q;
14:    $W \leftarrow P \cup Q$;
15:    Ascending order of W by non-dominated rank;
16:    Descending order of W by $L(\vec{x}_i, \theta^{(t)}, \eta)$;
17:    Descending order of individuals with same $L(\vec{x}_i, \theta^{(t)}, \delta)$ by crowding distance;
18:    $P \leftarrow$ select the first N individuals in W;
19:    Update φ_f, φ_{non} and P_{non};
20:    $g \leftarrow g + 1$;
21: **end while**

$t + k$. For the predictor, we use the same online prediction model based on lightGBM in [9].

Using the DCMOEA-ND will eventually get the POS, from which we need to select a suitable solution for deployment since it is a practical application. The original intention of ROOT is to reduce the problems arising from frequent switching of the deployment solution, such as the high switch cost. In order to integrate DCMOEA-ND and apply it in practice better, we propose the dynamic constrained robust optimization over time (DCROOT), and the framework of it is shown in Algorithm 2. When the external environment changes, if the usage time of the current deployment solution does not exceed the corresponding survival time, then we continue to use the current deployment solution, otherwise, we need to use DCMOEA-ND to obtain the POS in the new environment. To find the most suitable deployment solution from POS, we select the individual with the largest survival time firstly. If there are two and more individuals with the same largest survival time, we select the one with the smallest switching cost as the next deployment solution. The expression of the switching cost is as follows:

$$SC_i = \|\vec{x}_i - \vec{x}_i^*\| \tag{12}$$

where $\vec{x}_i$ is the new solution to be applied, and $\vec{x}_i^*$ is the old solution.

Algorithm 2. DCROOT

Input: time window tw, robustness threshold δ, number of environment changes t_{end}.
Output: robust solution sequence $< S_1, \ldots, S_K >$.

1: **while** $t < t_{end}$ **do**
2:     **if** the external environment changes **then**
3:         $t \leftarrow t + 1$;
4:         **if** the usage time of the current deployment solution does not exceed the corresponding survival time **then**
5:             Continue to use the current deployment solution;
6:         **else**
7:             $P \leftarrow$ DCMOEA-ND(tw, δ);
8:             Select the individual with the greatest survival time from P as the next deployment solution;
9:             **if** more than one individual with the same greatest survival time **then**
10:                 Select the individual with the smallest SC as the next deployment solution;
11:             **end if**
12:             Update the database;
13:             Retrain the predictor;
14:         **end if**
15:     **end if**
16: **end while**

---

4   Experimental Results and Analysis

In ROOT, we focus on how many external environments the deployment solution can be applied to and whether its switching cost is reduced. Therefore, in this paper, we first choose the average survival time [10] as the first performance indicator, as shown in Eq. (13), where K is the length of the robust solution sequence under all dynamic environments. Obviously, the larger average survival time is, the better performance of the robust solution sequence is.

$$P = \frac{1}{K} \sum_{i=1}^{K} L(\vec{x}_i, \theta^{(t)}, \delta) \tag{13}$$

In order to measure whether the proposed algorithm is able to reduce the switching cost, we choose the average switching cost [12] as another performance indicator, as shown in Eq. (14), where t_{end} denotes the number of changes in the external environment, and SC_i denotes the switching cost incurred at the ith switching from the old solution to the new solution.

$$C = \frac{1}{t_{end}} \sum_{i=1}^{t_{end}} SC_i \tag{14}$$

Meanwhile, we use hypervolume to evaluate the convergence and distribution of the algorithm, as shown in Eq. (15), where S represents the POS solved by the algorithm and R represents the reference point. Obviously, the algorithm with a larger HV has better convergence and distribution.

$$HV(S, R) = volume(\bigcup_{i=1}^{|S|} v_i) \tag{15}$$

4.1 Experimental Settings

In order to verify the effectiveness of the proposed algorithms and the reasonableness of the robustness definition, we design three sets of experiments in this section. One is to compare the performance of different constraint processing algorithms, one is to compare the performance of different algorithms under different robustness thresholds with the same time window, and the other is to compare the performance of different algorithms under different time windows with the same robustness threshold. Since the model in this paper involves dynamic constraints, the comparison algorithms are TMO, which adopts the same constraint processing strategy as this paper, and RPOOT, which adopts its own constraint processing strategy in [16], respectively. The statistical results provided in this section are based on 15 independent runs in 100 environments. For statistical analysis, we use comparison tests using the Wilcoxon rank-sum test with $\alpha = 0.05$, and the best results are bolded in each table. For the parameters of the algorithm, the population size N is set to 100, the number of iterations gen is set to 500, the crossover probability p_c is set to 1, the mutation probability p_m is set to 0.1, and the predictor training data length is set to 15.

4.2 Comparison with Other Algorithms

Firstly, in order to compare the performance of the constraint processing algorithms DC-NSGA-II [2], DCMOEA in RPOOT [16] and DCMOEA-ND in DCROOT, we take HV as the performance indicator, the comparison experiments of the three algorithms are carried out in five different environments, and the experimental results are shown in Table 3. In different environments, the HV obtained by the three algorithms are all very close, indicating that they can handle the constraints in this paper. Meanwhile, the HV obtained by DCMOEA-ND is slightly larger than DCMOEA and DC-NSGA-II, indicating that the POS obtained by DCMOEA-ND has better convergence and distribution.

We then set the time window tw to 2 uniformly and compare three algorithms under the robustness threshold δ of 0.2, 0.3, 0.4, and 0.5, respectively, and the experimental results are shown in Table 4. It can be seen from the result that the change of the robustness threshold does not affect the results of TMO, because whenever the environment changes, TMO restarts the algorithm solving and switches the deployment solution, so the survival time of the solution is 1. The over-frequent switching of the deployment solution also leads to the corresponding switching cost being larger than that of the other two algorithms. As δ increases, the average survival time of the robust solutions derived

Table 3. Result of HV for different algorithms

Environment	DC-NSGA-II	DCMOEA	DCMOEA-ND
Environment1	0.8709(0.8968)	0.8626(0.0104)	**0.8759(0.0130)**
Environment2	0.8568(0.0079)	0.8628(0.0106)	**0.8632(0.0090)**
Environment3	0.8783(0.0121)	0.8737(0.0131)	**0.8790(0.0130)**
Environment4	0.8801(0.0128)	0.8788(0.0151)	**0.8856(0.0071)**
Environment5	0.7836(0.0203)	0.7814(0.0197)	**0.7860(0.0210)**

by RPOOT and DCROOT rises, and the corresponding number of switching decreases, resulting in the decrease in the average switching cost. This is because the increase in δ represents the decision maker's tolerance for the solution rises, and some solutions that cannot satisfy the condition when δ is small may satisfy the condition when δ is large. Since DCROOT, after solving the POS in a certain environment, will find the solution with the largest survival time among them as the final deployment solution, and if there are more than one optional solution, we will pick the one with the smallest switching cost as the final deployment solution, whereas RPOOT will choose the solution with the smallest survival time as the survival time of the whole POS, resulting in the average survival time of the solutions obtained by DCROOT being larger than that of RPOOT, and the corresponding average switching cost is smaller.

Table 4. Result of indicators for algorithms under different robustness thresholds

δ	Method	P	C
0.2	TMO	1.0000(0.0000)	6.3881(0.3541)
	RPOOT	1.0594(0.0128)	4.4561(0.2942)
	DCROOT	**1.8846(0.0203)**	**1.1571(0.0850)**
0.3	TMO	1.0000(0.0000)	6.3881(0.3541)
	RPOOT	1.1890(0.0223)	3.7108(0.3092)
	DCROOT	**1.9207(0.0162)**	**1.1073(0.0904)**
0.4	TMO	1.0000(0.0000)	6.3881(0.3541)
	RPOOT	1.3715(0.0249)	3.2074(0.2265)
	DCROOT	**1.9356(0.0177)**	**1.0119(0.1767)**
0.5	TMO	1.0000(0.0000)	6.3881(0.3541)
	RPOOT	1.5139(0.0229)	2.9196(0.2482)
	DCROOT	**1.9973(0.0097)**	**1.1889(0.0415)**

Finally, we set the robustness threshold δ to 0.2 uniformly and compare three algorithms under the time windows tw of 2, 3, and 4, respectively, and the experimental results are shown in Table 5. We can see from the experimental results

Table 5. Result of indicators for algorithms under different time windows

tw	Method	P	C
2	TMO	1.0000(0.0000)	6.3881(0.3541)
	RPOOT	1.0594(0.0128)	4.4561(0.2942)
	DCROOT	**1.8846(0.0203)**	**1.1571(0.0850)**
3	TMO	1.0000(0.0000)	6.3881(0.3541)
	RPOOT	1.0461(0.0104)	4.4694(0.2378)
	DCROOT	**1.8682(0.0214)**	**1.0719(0.0912)**
4	TMO	1.0000(0.0000)	6.3881(0.3541)
	RPOOT	1.0477(0.0153)	4.3535(0.3359)
	DCROOT	**1.8566(0.0212)**	**1.2001(0.1107)**

that the change of time window tw still does not change the results of TMO, for the same reason as experiment two. And when the time window tw increases, the average survival time of RPOOT and DCROOT has a tendency to decrease, which may be due to the fact that as the time window tw increases, the feasible solutions fluctuate more in the future environment, resulting in fewer numbers that can adapt to the future environment. Meanwhile, at a certain robustness threshold, the average survival time of the solutions derived by RPOOT and DCROOT does not exceed 3 even if the tw is set to 3 or 4, indicating that the decision maker expects to get a solution with survival time of 3 or 4, respectively, but the model itself may not have a solution that satisfies the decision maker's expectation.

5   Conclusions

In this paper, we add some dynamic constraints to the previous model and construct a dynamic constrained multi-objective optimization model with the energy consumption of the pre-oxidation process and PAN modulus as the objective, the time of temperature zone as the decision variable, the temperature of temperature zone as the dynamic environment, and equivalent time as the dynamic constraint. To solve the dynamic constraints in the model, we propose the DCMOEA-ND that considers the value of non-dominated deviations. To reduce the switching cost of the deployment solution and make better use of DCMOEA-ND, we give the new robustness definition and propose the DCROOT, using DCMOEA-ND as the solver in it. Under the condition of same time window tw, the solution derived by DCROOT has longer average survival time and smaller average switching cost compared to RPOOT, and the average survival time of the solution derived by both algorithms rises with the increase of robustness threshold δ. Under the same robustness threshold, the solution derived by DCROOT also has better performance, and the average survival time of the solution derived by both algorithms tends to decrease as the time window increases. The experimental results prove that DCMOEA-ND can handle the constraints of the model in this paper, DCROOT can better accomplish the goal of obtaining high-quality filaments and reducing energy consumption while reducing the switching cost of the solution than TMO and RPOOT, and it also provides more valuable theoretical references for the actual production of carbon fibers.

Acknowledgements. This research has been supported by the National Major Technology Equipment Research Program of China (No. 2021-1635-01) and the National Natural Science Foundation of China (Grant No. 52075402).

References

1. Aliniya, Z., Khasteh, S.H.: Dynamic constrained multi-objective optimization with combination response mechanism. Available at SSRN 4123450 (2022)
2. Azzouz, R., Bechikh, S., Ben Said, L.: Multi-objective optimization with dynamic constraints and objectives: new challenges for evolutionary algorithms. In: Proceedings of the 2015 Annual Conference on Genetic and Evolutionary Computation, pp. 615–622 (2015)
3. Azzouz, R., Bechikh, S., Said, L.B., Trabelsi, W.: Handling time-varying constraints and objectives in dynamic evolutionary multi-objective optimization. Swarm Evol. Comput. **39**, 222–248 (2018)
4. Chen, M., Guo, Y., Gong, D., Yang, Z.: A novel dynamic multi-objective robust evolutionary optimization method. Acta Automatica Sinica **43**(11), 2014–2032 (2017)
5. Chen, M., Guo, Y., Jin, Y., Yang, S., Gong, D., Yu, Z.: An environment-driven hybrid evolutionary algorithm for dynamic multi-objective optimization problems. Complex Intell. Syst. **9**(1), 659–675 (2023)
6. Chen, M., Guo, Y., Liu, H., Wang, C., et al.: The evolutionary algorithm to find robust pareto-optimal solutions over time. Math. Probl. Eng. **2015** (2015)
7. Chen, Q., Ding, J., Yang, S., Chai, T.: A novel evolutionary algorithm for dynamic constrained multiobjective optimization problems. IEEE Trans. Evol. Comput. **24**(4), 792–806 (2019)
8. Deb, K., Udaya Bhaskara Rao, N., Karthik, S.: Dynamic multi-objective optimization and decision-making using modified NSGA-II: a case study on hydro-thermal power scheduling. In: Obayashi, S., Deb, K., Poloni, C., Hiroyasu, T., Murata, T. (eds.) EMO 2007. LNCS, vol. 4403, pp. 803–817. Springer, Heidelberg (2007). https://doi.org/10.1007/978-3-540-70928-2_60
9. Fang, Y., Zhao, Z., Jin, L., Li, K.: Modified robust optimization over time for process parameter optimization in pre-oxidation process of carbon fiber production. In: 2023 IEEE Congress on Evolutionary Computation (CEC), pp. 1–8. IEEE (2023)
10. Fu, H., Sendhoff, B., Tang, K., Yao, X.: Finding robust solutions to dynamic optimization problems. In: Esparcia-Alcázar, A.I. (ed.) EvoApplications 2013. LNCS, vol. 7835, pp. 616–625. Springer, Heidelberg (2013). https://doi.org/10.1007/978-3-642-37192-9_62
11. Guzmán-Gaspar, J.Y., Mezura-Montes, E., Domínguez-Isidro, S.: Differential evolution in robust optimization over time using a survival time approach. Math. Comput. Appl. **25**(4), 72 (2020)

12. Huang, Y., Jin, Y., Hao, K.: Decision-making and multi-objectivization for cost sensitive robust optimization over time. Knowl.-Based Syst. **199**, 105857 (2020)
13. Jin, Y., Tang, K., Yu, X., Sendhoff, B., Yao, X.: A framework for finding robust optimal solutions over time. Memetic Comput. **5**, 3–18 (2013)
14. Newcomb, B.A.: Processing, structure, and properties of carbon fibers. Compos. A Appl. Sci. Manuf. **91**, 262–282 (2016)
15. Wang, F., Huang, M., Yang, S., Wang, X.: Penalty and prediction methods for dynamic constrained multi-objective optimization. Swarm Evol. Comput. **80**, 101317 (2023)
16. Wei, S.: Dynamic constrained robust evolutionary optimization method. China University of Mining and Technology (2021)
17. Yazdani, D., Cheng, R., Yazdani, D., Branke, J., Jin, Y., Yao, X.: A survey of evolutionary continuous dynamic optimization over two decades-part B. IEEE Trans. Evol. Comput. **25**(4), 630–650 (2021)
18. Yazdani, D., et al.: Robust optimization over time: a critical review. IEEE Trans. Evol. Comput. (2023)
19. Yazdani, D., Yazdani, D., Branke, J., Omidvar, M.N., Gandomi, A.H., Yao, X.: Robust optimization over time by estimating robustness of promising regions. IEEE Trans. Evol. Comput. (2022)
20. Yu, X., Jin, Y., Tang, K., Yao, X.: Robust optimization over time-a new perspective on dynamic optimization problems. In: IEEE Congress on Evolutionary Computation, pp. 1–6. IEEE (2010)
21. Zhang, X., Fang, Y., Liu, Q., Yazdani, D.: Multi-objective robust optimization over time for dynamic disassembly sequence planning. Int. J. Precis. Eng. Manuf. 1–20 (2023)

Comparison of CLPSO, ECLPSO and ACLPSO on CEC2013 Multimodal Benchmark Functions

Yi Zhang[1] , Xiang Yu[1(✉)] , and Kaiwen Xu[2]

[1] School of Information Engineering, Nanchang Institute of Technology, Nanchang 330099, China
xyuac@connect.ust.hk
[2] School of Civil and Architectural Engineering, Nanchang Institute of Technology, Nanchang 330099, China

Abstract. Particle swarm optimization (PSO) is a class of modern generalized intelligent optimization algorithms. Comprehensive learning PSO (CLPSO) is a powerful PSO variant that is good at exploration and works well on many multimodal problems. Enhanced CLPSO (ECLPSO) and adaptive CLPSO (ACLPSO) are improved versions of CLPSO that we have previously proposed. ECLPSO enhances the exploitation performance of CLPSO, and ACLPSO strengthens the exploration performance of ECLPSO. We have compared CLPSO, ECLPSO and ACLPSO on 16 benchmark functions in our previous study. To further understand the generalization performance of the three algorithms, this paper compares them on the well-known CEC2013 test set of multimodal benchmark functions. Compared with the benchmark functions employed in the previous study, though the number of dimensions for the CEC2013 benchmark functions are quite smaller, there are also a number of local optima in the search space. In addition, the CEC2013 test set contains composition functions that mix different characteristics of various basic functions, causing the search space to have a huge quantity of local optima and is very complex. Experimental results demonstrate that the accuracy of the solution obtained by ECLPSO is slightly better than that by CLPSO on just a few functions and is similar on all the other functions; both ECLPSO and CLPSO fail to derive the global optimum or a near-optimum on most of the composition functions; and ACLPSO is well balanced between exploration and exploitation, as it outperforms ECLPSO and CLPSO by being able to find the global optimum or a near-optimum with high accuracy on almost all the functions.

Keywords: Particle Swarm Optimization · Comprehensive Learning · CEC2013 · Multimodal Benchmark Function

1 Introduction

Traditional optimization methods, e.g. linear programming, gradient descent method, and Newton's method are based on mathematical analysis. These methods require problems to possess characteristics such as linearity, continuity, differentiability, and convexity. In contrast, modern generalized intelligent optimization algorithms do not impose

© The Author(s), under exclusive license to Springer Nature Singapore Pte Ltd. 2024
L. Pan et al. (Eds.): BIC-TA 2023, CCIS 2061, pp. 228–240, 2024.
https://doi.org/10.1007/978-981-97-2272-3_17

such specific requirements on the problems, thus they are able to address various problems in different areas. However, these intelligent optimization algorithms lack a solid mathematical theoretical foundation. They solve the problem by iterative probabilistic search in the search space, and the solution obtained may not be the global optimum or a near-optimum. To further understand the generalization performance of these intelligent algorithms, it is necessary to evaluate them on a significant number of benchmark functions with different characteristics.

Particle swarm optimization (PSO) is a class of intelligence optimization algorithms that simulates foraging behaviour of a flock of birds [1]. In PSO, flock and birds are respectively termed as swarm and particles. All particles "fly" in the search space, where each particle is associated with a velocity and a position, and remembers its best historical position (i.e. the position that exhibits the best fitness value). PSO randomly initializes particles' velocities and positions, and then iteratively updates them, trying to converge to the global optimum. PSO needs to be balanced between exploration and exploitation. Exploration refers to probing the search space in order to identify position(s) that hopefully are close to the global optimum or a small region that hopefully contains the global optimum, while exploitation means searching around the hopeful position(s) or the small region in order to dig out the global optimum or a near-optimum.

Comprehensive Learning PSO (CLPSO), proposed by Liang et al. in 2006 [2], is a powerful PSO variant that is good at exploration and works well on many multimodal problems. CLPSO encourages particles to comprehensively learn from different exemplars on different dimensions of the search space when updating their velocities. Each particle possesses a learning probability that controls whether the position of the exemplar on different dimension is the historical best position of that particle or the historical best position of other particles. CLPSO can effectively preserves the swarm's diversity, hence it can find the global optimum or a near-optimum on certain multimodal problems. However, researchers have compared some PSO variants including CLPSO on many benchmark functions, and the results demonstrate that CLPSO achieves low solution accuracy on most functions [3, 4].

Yu and Zhang proposed enhanced CLPSO (ECLPSO) [5] to enhance the exploitation performance of CLPSO. There are two enhancements employed in ECLPSO. Firstly, ECLPSO add a perturbation term into each particle's velocity update procedure to achieve high performance exploitation. Normative knowledge about dimensional bounds of particle's best historical positions is used to appropriately activate the perturbation based exploitation. Additionally, the particles' learning probabilities are determined adaptively based on not only rankings of particle's best fitness values but also the dimensions' exploitation progress to facilitate convergence.

Though ECLPSO significantly improves the exploitation performance of CLPSO, both of them cannot derive the global optimum or a near-optimum solution on some benchmark functions according to the experimental results in [5]. Therefore, Yu and Qiao proposed adaptive CLPSO (ACLPSO). ACLPSO assigns independent inertia weight and independent acceleration coefficient to each dimension in the search space, where inertia weight and acceleration coefficient are parameters in the particle velocity update procedure; proposes adaptively updated dimensional maximum velocity based on the size of dimensional normative interval, where dimensional maximum velocity is used to constrain the particle's velocity to facilitate convergence; and assigns independent learning probabilities for each particle on each dimension. ACLPSO is well balanced between exploration and exploitation. Experimental results reported in [6] demonstrate that ACLPSO is capable of finding the global optimum or a near-optimum solution on all the benchmark functions.

In our previous study, we have evaluated and compared CLPSO, ECLPSO, and ACLPSO on just 16 benchmark functions, including unimodal, multimodal, shifted, rotated, and shifted rotated functions. To further understand the generalization performance of these three algorithms, this paper evaluates and compares their performance on CEC2013 multimodal benchmark functions. In contrast with 16 functions employed in [6], though the number of dimensions for the CEC2013 benchmark functions are quite smaller, there are also a number of local optima in the search space. In addition, the CEC2013 test set contains composition functions that mix different characteristics of various basic functions, causing the search space to have a huge quantity of local optima and is very complex.

2   Relevant Algorithms

2.1   CLPSO

Suppose N particles move in a D-dimensional search space. For each particle i ($1 \leq i \leq N$), i's velocity is $V_i = (V_{i,1}, V_{i,2},..., V_{i,D})$; i's position is $P_i = (P_{i,1}, P_{i,2},..., P_{i,D})$; i's historical best position is $B_i = (B_{i,1}, B_{i,2}, ..., B_{i,D})$; i's exemplar is $E_i = (E_{i,1}, E_{i,2},..., E_{i,D})$ and i's fitness value is $f(P_i)$ that indicating the optimization performance of P_i, where $f()$ is a function or numerical simulation procedure that outputs the fitness value of a given position. V_i and P_i are randomly initialized, and then are iteratively updated on each dimension d ($1 \leq d \leq D$) according to Eqs. (1) and (2).

$$V_{i,d} = wV_{i,d} + cr\left(E_{i,d} - P_{i,d}\right) \tag{1}$$

$$P_{i,d} = P_{i,d} + V_{i,d} \tag{2}$$

where $c = 1.5$ is the acceleration coefficient; w is the inertia weight; and r is a random number uniformly distributed in [0, 1]. $V_{i,d}$ is clamped by $[-V_d^{\max}, V_d^{\max}]$, with $V_d^{\max}$ being maximum velocity bound on dimension d, usually set to 20% of the search boundary. If $V_{i,d} > V_d^{\max}$, then set $V_{i,d} = V_d^{\max}$, or if $V_{i,d} < -V_d^{\max}$, then set $V_{i,d} = -V_d^{\max}$. $E_{i,d}$ is i's exemplar position on dimension d, and the determination of $E_{i,d}$ is controlled by i's learn probability L_i. Specifically, a random number uniformly distributed in [0, 1] is generated, if $L_{i,d}$ is no greater than the random number, then $E_{i,d} = B_{i,d}$, otherwise $E_{i,d} = B_{j,d}$, with $j \neq i$, and j is the particle with better personal best fitness (i.e. $f(B_j)$) out of two randomly selected particles that are different and neither of which is i. The inertia weight w is updated according to Eq. (3) and the learning probability L_i are set as described in Eqs. (4).

$$w = w_{\max} - \frac{k}{K}(w_{\max} - w_{\min}) \tag{3}$$

$$L_i = 0.05 + 0.45\frac{\exp(10(i-1)/(N-1))-1}{\exp(10)-1} \tag{4}$$

where $w_{max} = 0.9$ and $w_{min} = 0.4$ are the maximum and minimum inertia weight respectively; k is the iteration counter and K is the predefined maximum number of iterations.

CLPSO lets i to learn from the same exemplar E_i until i's personal best fitness stops improving for a refreshing gap of 7 consecutive iterations and then determines a new exemplar E_i.

CLPSO does not repair the position of particles that move out of the search boundary. It calculate the fitness of P_i only if P_i is within the search space on all the dimensions. Since the position of E_i on each dimension is within the search space, the particle will eventually return to the search space.

2.2 ECLPSO

ECLPSO defines the dimensional normative interval $[B_d^{min}, B_d^{max}]$, with B_d^{min} and B_d^{max} being the lowest and highest dimensional historical best positions of all the particles, respectively. In order to achieve more efficient exploitation, ECLPSO delineates an exploitation region on each dimension according to the size of the dimensional normative interval. Once the interval $[B_d^{min}, B_d^{max}]$ becomes small enough as determined by Eq. (5), ECLPSO conclude the exploration stage and commence the exploitation stage on dimension d, and $V_{i,d}$ will then be updated iteratively according to Eq. (6) instead of Eq. (1).

$$B_d^{max}-B_d^{min} \leq \alpha(P_d^{max}-P_d^{min}) \text{ and } B_d^{max}-B_d^{min} \leq \beta \tag{5}$$

$$V_{i,d} = w_{pe}V_{i,d} + c_{pe}r(E_{i,d} + \eta(\frac{B_d^{min}+B_d^{max}}{2}-E_{i,d}) - P_{i,d}) \tag{6}$$

where $\alpha = 0.01$ is the relative ratio; and $\beta = 2$ is the small absolute threshold; P_d^{max} and P_d^{min} are respectively upper and lower search bound on each dimension; $w_{pe} = 0.5$ and $c_{pe} = 1.5$ are the inertia weight and acceleration coefficient exclusively used for perturbation exploitation; $\eta((B_d^{min}+B_d^{max})/2-E_{i,d})$ is perturbation term and η is a perturbation coefficient randomly generated from a Gaussian distribution with a mean of 1 and a standard deviation of 0.65. What is more, η is clamped within the range defined by 10 times the standard deviation on both sides of the mean. The perturbation term ensures thorough exploitation within the relatively narrow normative interval of each dimension. Note that $V_{i,d}$ updated by Eq. (6) is not limited by the dimensional maximum velocity.

ECLPSO adaptively determines the learning probability of particles based on the ranking of personal best fitness values and the exploitation progress of dimensions. As shown in Eq. (7).

$$L_i = L_{min} + (L_{max} - L_{min})\frac{\exp(10(T_i-1)/(N-1))-1}{\exp(10)-1} \tag{7}$$

where T_i is i's rank, and if i gives the best personal best fitness value, then $T_i = 1$; L_{min} and L_{max} are the minimum and maximum learning probabilities, respectively, with L_{min} being set to 0.05, and L_{max} being adaptively updated based on the equation provided below.

$$L_{max} = L_{min} + 0.25 + 0.45\log_{(D+1)}(M_k + 1) \tag{8}$$

where M_k is the number of dimensions that have entered the exploitation stage. L_{max} increases logarithmically with the particles' exploitation progress, and the more dimensions undergoing exploitation, the larger L_{max} to facilitate convergence.

2.3 ACLPSO

ACLPSO adaptively adjusts the maximum velocity V_d^{max} on each dimension according to the size of the dimensional normative interval; and also assigns an independent inertia weight w_d and an independent acceleration coefficient c_d to each dimension of the search space, as well as an independent learning probability $L_{i,d}$ for each particle on each dimension.

ACLPSO adaptively adjust V_d^{max} according to the following equation:

$$V_d^{max} = s(B_d^{max} - B_d^{min}) \tag{9}$$

where s is the interval scaling coefficient. The large normative interval in the early stage corresponds to the large dimension maximum velocity, that is, the search granularity is larger, which is conducive to exploration. The later normative interval is small, which corresponds to a small dimension maximum velocity, and the search granularity is smaller, which is conducive to exploitation.

When $[B_d^{min}, B_d^{max}]$ does not meet the conditions of Eq. (5), the dimensional inertia weight is updated according to Eq. (10).

$$w_d = \frac{B_d^{max} - B_d^{min}}{P_d^{max} - P_d^{min}} + (1 - u)(w_{max} - \frac{k}{K}(w_{max} - w_{min})) \tag{10}$$

where $u = 0.3$ is the tradeoff coefficient that balances the influence between the dimensional normative interval and the iteration counter. w_d also is constrained within the interval $[w_{min}, w_{max}]$, if $w_d < w_{min}$, then set $w_d = w_{min}$, or if $w_d > w_{max}$, then set $w_d = w_{max}$. w_d and c_d must satisfy the following so-called stability condition [7–9]:

$$1 - w_d \geq 0 \text{ and } w_d + 1 - rc_d \geq 0 \tag{11}$$

Hence ACLPSO takes the following value for c_d.

$$c_d = w_d + 1 \tag{12}$$

ACLPSO assigns a learning probability $L_{i,d}$ for each particle on dimension d according to Eq. (13).

$$L_{i,d} = v\log_K k + \frac{B_d^{max} - B_d^{min}}{P_d^{max} - P_d^{min}} \frac{\exp{(D(T_i - 1)/(N-1))} - 1}{\exp(D) - 1} \tag{13}$$

where v is learning probability tradeoff coefficient, and similar to w_d, $L_{i,d}$ also limited by $[L_{min}, L_{max}]$ similar to w_d, with $L_{min} = 0.05$ and $L_{max} = 0.75$ being respectively the minimum and maximum learning probabilities.

Different from CLPSO and ECLPSO, ACLPSO repairs the position of the particle that move out of the search boundary on each dimension. This is achieved by using the Eq. (14).

$$P_{i,d} = \min(P_d^{max}, \max(P_d^{min}, P_{i,d})) \qquad (14)$$

The interval scaling coefficient s in Eq. (9) and the learning probability tradeoff coefficient v in Eq. (13) are key parameters dominating convergence. ACLPSO statically sets the values of s and v, where s has two empirical values of 0.1 and 1.1, respectively. There are also two empirical values for v, 0.05 and 0.3, respectively. ACLPSO combines different s and v in pairs to handle different problems.

3　CEC2013 Test Set

The CEC2013 test set contains 20 benchmark functions with different characteristics (including some identical functions with different dimension size, as shown in Table 1). All benchmark functions are formulated as maximization problems. The first 10 benchmark functions are well-known and widely used functions, the remaining benchmark functions are more complex and follow the paradigm of composition functions defined in the IEEE CEC 2005 special session on real-parameter optimization [10]. Table 1 lists the expressions of first 10 multimodal functions and the basic functions used to construct the remaining composition functions in the CEC2013 test set.

In order to facilitate a more intuitive comparison of the convergence accuracy of different algorithms on various functions, we employed f_bias to shift the peaks of certain functions with non-zero to zero. The f_bias values for each function in the experiments are presented in Table 1, with $f(x^*)$ represents the global optimum value of the function after shifting its peak by f_bias. The dimensions from f_1 to f_5 are non-scalable, and are all multimodal functions in one or two dimensions. f_1 to f_3 are all one-dimensional multimodal functions. They all have five peaks in the search space. The peak heights of f_1 and f_3 are uneven, while the peak height of f_2 is completely equal and evenly distributed. f_1 and f_3 have three and four local optima in the search space, respectively, while f_2 has no local optima. f_4 has no local optimum in the search space, it has four global optima with two closer to each other than the other two. f_5 has two local optima and two global optima. There is an obvious deep valley between the two highest peaks, and the valley bottom is located in the center of the function plot. F_6 to F_8 are multimodal functions with scalable dimensions, and the number of global optimal solutions for each function is dependent on the dimensions. f_6 and f_7 represent the 2D and 3D forms of F_6, with f_6 having 18 global optima and f_7 having 81 global optima. Additionally, both functions possess numerous local optima. f_8 and f_9 represent the 2D and 3D forms of F_7, and both have 6^D global optima and no local optima. f_{10} is 2D form of F_8, and it has 12 global optima and no local optima.

f_{11} to f_{20} are composition functions tested on different dimensions from G_1 to G_4. G_1 to G_4 are constructed from several other basic functions and mix the characteristics

Table 1. All benchmark functions.

Function	D	Expression	f_bias	$f(x^*)$	Search space
f_1	1	$F_1(x) = \begin{cases} 80(2.5-x) & \text{for } 0 \le x < 2.5, \\ 64(x-2.5) & \text{for } 2.5 \le x < 5.0, \\ 64(7.5-x) & \text{for } 5.0 \le x < 7.5, \\ 28(x-7.5) & \text{for } 7.5 \le x < 12.5, \\ 28(17.5-x) & \text{for } 12.5 \le x < 17.5, \\ 32(x-17.5) & \text{for } 17.5 \le x < 22.5, \\ 32(27.5-x) & \text{for } 22.5 \le x < 27.5, \\ 80(x-27.5) & \text{for } 27.5 \le x < 30. \end{cases}$	-200.0	0	$x \in [0, 30]$
f_2	1	$F_2(x) = \sin^6(5\pi x)$	-1.0	0	$x \in [0, 1]$
f_3	1	$F_3(x) = \exp(-2 \log (2.0) \left(\frac{x-0.08}{0.854}\right)^2)\sin^6(5\pi(x^{3/4}- 0.05))$	-1.0	0	$x \in [0, 1]$
f_4	2	$F_4(x, y) = 200 - (x^2+y-11)^2 - (x+y^2-7)^2$	-200.0	0	$x, y \in [-6, 6]$
f_5	2	$F_5(x, y) = -4\left[\left(4-2.1x^2+\frac{x^4}{3}\right)x^2+xy+(4y^2-4)y^2\right]$	-1.03163	0	$x \in [-1.9,1.9]$ $y \in [-1.1,1.1]$
f_6	2	$F_6(x) = -\prod_{d=1}^{D}\sum_{p=1}^{5} j \cos[(p+1)x_d+p]$	-186.713	0	$[0.25, 10]^D$
f_7	3		-2709.0935	0	
f_8	2	$F_7(x) = \frac{1}{D}\sum_{d=1}^{D} \sin(10\log(x_d))$	-1.0	0	$[-10, 10]^D$
f_9	3		-1.0	0	
f_{10}	2	$F_8(x)= -\sum_{d=1}^{D} (10+9 \cos(2\pi k_d x_d))$, where k_1=3, k_2=4	-2.0	0	$[0, 1]^D$
f_{11}	2	$G_1 \begin{cases} g_1\text{-}g_2 \text{: Grienwank's function} \\ g_3\text{-}g_4 \text{: Weierstrass function,} \\ g_5\text{-}g_6 \text{: Sphere function.} \end{cases}$	0	0	$[-5, 5]^D$
f_{12}	2	$G_2 \begin{cases} g_1\text{-}g_2 \text{: Rastrigin's function,} \\ g_3\text{-}g_4 \text{: Weierstrass function,} \\ g_5\text{-}g_6 \text{: Griewank's function,} \\ g_7\text{-}g_8 \text{: Sphere function.} \end{cases}$	0	0	$[-5, 5]^D$
f_{13}	2	$G_3 \begin{cases} g_1\text{-}g_2 \text{: EF8F2 function,} \\ g_3\text{-}g_4 \text{: Weierstrass function,} \\ g_5\text{-}g_6 \text{: Griewank's function.} \end{cases}$	0	0	$[-5, 5]^D$
f_{14}	3		0	0	
f_{15}	5		0	0	
f_{16}	10		0	0	
f_{17}	3	$G_4 \begin{cases} g_1\text{-}g_2 \text{: Rastrigin's function,} \\ g_3\text{-}g_4 \text{: EF8F2 function,} \\ g_5\text{-}g_6 \text{: Weierstrass function,} \\ g_7\text{-}g_8 \text{: Griewich's function.} \end{cases}$	0	0	$[-5, 5]^D$
f_{18}	5		0	0	
f_{19}	10		0	0	
f_{20}	20		0	0	

of these functions. The composition function G_m ($1 \le m \le 4$) is calculated according to the following equation:

$$G_m(x) = \sum_{h=1}^{H} \omega_h(\hat{g}_h((x - o_h)/\lambda_h \cdot M_h) + b_h) \qquad (15)$$

where n is the number of basic functions used to construct the composition function; $\hat{g}_h$ denotes a normalization of the h-th basic function, h belong to 1 to n; ω_h is the corresponding weight; o_h is the new shifted optimum of each $\hat{g}_h$; M_h is the linear transformation (rotation) matrix of each $\hat{g}_h$; and λ_h is a parameter which is used to stretch or compress each $\hat{g}_h$. CEC2013 set $b_h = 0$.

The basic functions that used to construct the composition functions are the Sphere function, Grienwank's function, Rastrigin's function, Weierstrass function, and Expanded Griewank's plus Rosenbrock's function. All their expressions can be found in [11].

The number of global optima in all the composition functions is independent of the dimension D and depends only on the number of basic functions that construct the composition function. G_1 and G_3 have 6 global optima in the search space, while G_2 and G_4 have 8 global optima in the search space. The search space of all composite functions is very complex because the characteristics of different basic functions are mixed, resulting in a huge amount of local optima in the search space.

4 Experiments

4.1 Experimental Setup

As mentioned in Sect. 2.3 of this paper, the interval scaling coefficient s and the learning probability tradeoff coefficient v are key parameters that significantly influence convergence. ACLPSO with different combinations of s and v are named ACLPSO-1 ($s = 0.1$, $v = 0.05$), ACLPSO-2 ($s = 1.1$, $v = 0.05$), ACLPSO-3 ($s = 0.1$, $v = 0.3$), and ACLPSO-4 ($s = 1.1$, $v = 0.3$). For CLPSO, ECLPSO, and ACLPSO, they are run 25 times on each function. We use different the number of particles N and the maximum function evaluations (MaxFEs) on different functions, as shown in Table 2.

Table 2. Parameters settings

Parameter	f_1	f_2	f_3	f_4	f_5	f_6	f_7	f_8	f_9	f_{10}
MaxFEs	600	600	600	1200	1200	1200	3000	1200	3000	1200
N	12	12	12	12	12	12	20	12	20	12
Parameter	f_{11}	f_{12}	f_{13}	f_{14}	f_{15}	f_{16}	f_{17}	f_{18}	f_{19}	f_{20}
MaxFEs	1500	1500	1500	3000	10000	30000	3000	10000	30000	90000
N	12	12	12	20	20	30	20	30	30	40

4.2 Experimental Results and Discussion

Table 3 lists mean and standard deviation (SD) results of the 25 runs of CLPSO, ECLPSO and ACLPSO under four combinations of s and v on CEC2013 multimodal functions, and the best fitness value on each function is marked in bold. Figure 1 shows the changes of the global best fitness value during the search process of CLPSO, ECLPSO and ACLPSO with the best combination of s and v in the best run.

Table 3. Mean and standard deviation (SD) global best fitness value results of CLPSO, ECLPSO and ACLPSO on all multimodal functions.

Function	Result	CLPSO	ECLPSO	ACLPSO-1	ACLPSO-2	ACLPSO-3	ACLPSO-4
f_1	Mean	2.61E+00	2.06E+00	4.80E+00	**0.00E+00**	1.60E+00	0.00E+00
	SD	8.34E+00	7.97E+00	1.33E+01	**0.00E+00**	8.00E+00	0.00E+00
f_2	Mean	1.18E−09	**2.15E−12**	6.21E−10	3.00E−10	1.28E−09	7.02E−09
	SD	5.72E−09	**1.06E−11**	1.04E−09	6.35E−10	1.75E−09	1.68E−08
f_3	Mean	2.14E−02	3.23E−02	4.17E−02	**2.98E−05**	2.16E−02	2.51E−04
	SD	4.80E−02	6.34E−02	4.59E−02	**1.00E−04**	2.53E−02	9.78E−04
f_4	Mean	3.08E−01	1.68E−01	7.18E−03	6.06E−03	2.72E−03	**2.65E−03**
	SD	6.27E−01	3.93E−01	1.78E−02	1.06E−02	3.79E−03	**4.24E−03**
f_5	Mean	1.05E−02	9.07E−03	2.37E−04	5.37E−04	**1.34E−05**	2.83E−05
	SD	2.21E−02	1.88E−02	4.20E−04	1.36E−03	**1.55E−05**	4.78E−05
f_6	Mean	1.32E+01	1.36E+00	6.10E−01	9.10E−01	**1.19E−01**	9.39E−01
	SD	2.80E+01	3.24E+00	1.63E+00	1.30E+00	**2.03E−01**	1.43E+00
f_7	Mean	3.02E+02	1.55E+02	5.16E+01	1.61E+02	**2.22E+01**	9.51E+01
	SD	3.88E+02	2.49E+02	5.94E+01	2.22E+02	**2.64E+01**	8.96E+01
f_8	Mean	2.42E−04	2.31E−04	**2.77E−06**	4.05E−05	5.67E−06	8.18E−06
	SD	5.45E−04	5.83E−04	**3.77E−06**	6.66E−05	8.53E−06	1.56E−05
f_9	Mean	5.32E−04	7.45E−04	6.95E−05	1.12E−04	**2.90E−05**	7.38E−05
	SD	9.70E−04	1.70E−03	1.66E−04	1.50E−04	**5.67E−05**	9.43E−05
f_{10}	Mean	2.30E−02	1.73E−03	7.47E−05	3.34E−04	**6.18E−05**	5.54E−04
	SD	8.26E−02	3.21E−03	9.67E−05	5.24E−04	**7.02E−05**	9.06E−04
f_{11}	Mean	3.76E−01	1.46E−01	7.64E−02	4.43E−02	3.81E−02	**7.02E−03**
	SD	7.48E−01	3.51E−01	2.04E−01	8.53E−02	1.56E−01	**1.52E−02**
f_{12}	Mean	2.52E+01	3.92E+01	1.49E+00	1.45E+01	**5.51E−01**	7.93E−01
	SD	5.30E+01	7.74E+01	4.14E+00	3.52E+01	**1.25E+00**	1.26E+00
f_{13}	Mean	2.57E+00	8.19E−01	1.43E−01	1.37E−01	**4.12E−03**	9.31E−02
	SD	7.87E+00	1.65E+00	3.18E−01	4.64E−01	**8.40E−03**	2.35E−01
f_{14}	Mean	1.68E+01	1.18E+01	1.57E+00	9.80E+00	**8.78E−01**	3.55E+00
	SD	2.57E+01	2.13E+01	1.65E+00	1.70E+01	**1.06E+00**	5.09E+00
f_{15}	Mean	1.02E+01	5.97E+00	6.92E+00	3.38E+01	**8.92E−01**	2.73E+01
	SD	1.82E+01	1.38E+01	1.04E+01	4.55E+01	**1.64E+00**	3.92E+01

(continued)

Table 3. (*continued*)

Function	Result	CLPSO	ECLPSO	ACLPSO-1	ACLPSO-2	ACLPSO-3	ACLPSO-4
f_{16}	Mean	1.55E+01	5.67E+00	1.28E+01	1.68E+02	**6.75E−05**	1.86E+02
	SD	1.60E+01	6.90E+00	1.80E+01	8.69E+01	**2.99E−04**	1.37E+02
f_{17}	Mean	2.19E+01	1.53E+01	5.54E+00	4.78E+00	1.91E+00	**7.61E−01**
	SD	2.89E+01	1.76E+01	6.46E+00	4.77E+00	2.26E+00	**1.27E+00**
f_{18}	Mean	1.29E+01	6.71E+00	1.28E+01	8.61E+00	**2.19E+00**	3.45E+00
	SD	1.49E+01	1.43E+01	1.50E+01	7.96E+00	**2.48E+00**	2.65E+00
f_{19}	Mean	1.31E+01	9.90E+00	1.38E+01	2.08E+01	**2.42E+00**	7.12E+00
	SD	4.92E+00	4.21E+00	4.86E+00	6.50E+00	**2.12E+00**	1.53E+01
f_{20}	Mean	4.84E−01	**2.88E−01**	1.25E+00	6.69E+00	5.19E−01	1.79E+00
	SD	6.20E−01	**3.13E−01**	9.62E−01	2.63E+00	9.66E−01	4.65E+00

As we can be seen from Table 3, CLPSO performs poorly in terms of convergence accuracy on $f_1, f_6, f_7, f_{12}, f_{13}, f_{14}, f_{15}, f_{16}, f_{17}, f_{18}$, and f_{19}. In comparison to CLPSO, ECLPSO demonstrates slightly improved accuracy on functions such as f_2, f_{10}, and f_{16}, but overall does not show significant improvement. Compared with CLPSO, ECLPSO only exhibits slightly improved convergence accuracy in f_2, f_{10}, and f_{16}, while showing similar convergence accuracy to CLPSO in other functions, demonstrating no significantly pronounced improvement in convergence accuracy overall. By observing the bold results in Table 3, it can be seen that ACLPSO with the best combination of s and v achieves better convergence accuracy than CLPSO and ECLPSO on 17 functions out of 20 functions.

CLPSO and ECLPSO can achieve a near-optimum just on $f_2, f_3, f_4, f_5, f_8, f_9, f_{10}$, f_{11} and f_{20}. Through analyzing the results of 25 independent runs, it is found that both ECLPSO and CLPSO are prone to fall into local optima at runtime, resulting in inferior convergence outcomes on $f_1, f_6, f_7, f_{12}, f_{14}, f_{15}, f_{17}, f_{18}$, and f_{19}. This may indicates that to achieve better convergence results on the CEC2013 multimodal functions, the algorithm needs to possess good exploration performance to avoid falling avoid falling into local optima. While ECLPSO enhances the exploration performance of CLPSO and approaches its exploration performance, the convergence outcomes do not exhibit significant differences. In contrast, ACLPSO with the best value pair of s and v can obtain the global optimum or a near-optimum solution on benchmark functions except for f_7, f_{18} and f_{19}, which outperforms CLPSO and ECLPSO in terms of quantity. The overall performance of the ACLPSO indicates that ACLPSO is well balanced between exploration and exploitation, enabling it to jump out the local optima and achieves higher convergence accuracy. The convergence results of CLPSO, ECLPSO, and ACLPSO on CEC2013 multimodal functions indicate that ACLPSO possesses generalization performance superior to CLPSO and ECLPSO.

As shown in Fig. 1, CLPSO is prone to falling into local optima on certain functions, leading to get stuck in premature stagnancy, such as f_4, f_{10}, f_{14} and f_{17}. ECLPSO can

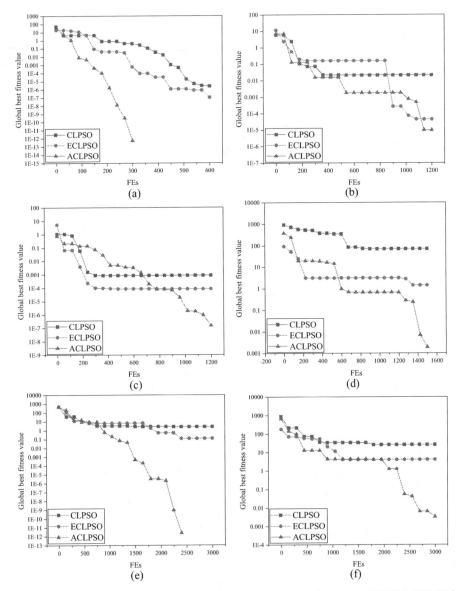

Fig. 1. The changes of the global best fitness value during the search process of CLPSO, ECLPSO and ACLPSO with the best (s, v) combination in the best run. (a) f_1, (b) f_4, (c) f_{10}, (d) f_{12}, (e) f_{14}, (f) f_{17}.

maintain good convergence speed on most functions, but there may still be stagnation in the later iteration stage on certain functions, such as f_4, f_{10} and f_{17}. Compared with CLPSO and ECLPSO, ACLPSO may fall into local optima in the early or middle iteration of functions f_4, f_{12}, and f_{17}, but it still jumps out of unsatisfactory local optima in the subsequent iterations. ACLPSO performs well on functions f_1, f_{12}, f_{14} and f_{17}, which not only has faster convergence speed in the early iteration, but also obtains higher convergence accuracy in the later iteration. It can be seen that ACLPSO is well balanced between exploration and exploitation as it can jump out in time when falling into a local optimum.

5 Conclusions

In this paper, we have compared and evaluated CLPSO, ECLPSO, and ACLPSO on the CEC2013 multimodal benchmark functions, with the aim of further understanding the generalization performance of these three algorithms. In contrast with the 16 benchmark functions employed in [6], though the number of dimension for the CEC2013 benchmark functions is quite smaller, there are also a large number of local optima in the search space, especially for the composite functions. The composite functions not only have a huge number of local optima, but also mix the characteristics of many basic functions, which makes the solution space very complex. The experimental results indicate that CLPSO and ECLPSO exhibit similar exploratory performance, as their convergence results are close on most benchmark functions, with ECLPSO slightly outperforming CLPSO on a few functions. In contrast, ACLPSO achieves well balanced between exploration and exploitation, exhibiting superior generalization performance by attaining the global optimum or a near-optimal solution with high accuracy on almost all functions. Furthermore, it outperforms CLPSO and ECLPSO in terms of convergence results for 17 out of 20 functions.

Acknowledgement. This work was supported by the Jiangxi Province Natural Science Foundation (20224BAB204071).

References

1. Kennedy, J., Eberhart, R.C.: Particle swarm optimization. In: IEEE International Conference on Neural Networks, vol. 4, pp. 1942–1948. Piscataway, NJ (1995)
2. Liang, J.J., Qin, A.K., Suganthan, P.N., Baskar, S.: Comprehensive learning particle swarm optimizer for global optimization of multimodal functions. IEEE Trans. Evol. Comput. **10**(3), 281–295 (2006)
3. Nasir, M., Das, S., Maity, D., Sengupta, S., Halder, U., Suganthan, P.N.: A dynamic neighborhood learning based particle swarm optimizer for global numerical optimization. Inf. Sci. **209**, 16–36 (2012)
4. Zhan, Z.H., Zhang, J., Li, Y., Shi, Y.H.: Orthogonal learning particle swarm optimization. IEEE Trans. Evol. Comput. **6**(15), 832–847 (2011)
5. Yu, X., Zhang, X.Q.: Enhanced comprehensive learning particle swarm optimization. Appl. Math. Comput. **242**, 265–276 (2014)
6. Yu, X., Qiao, Y.: Enhanced comprehensive learning particle swarm optimization with dimensional independent and adaptive parameters. Comput. Intell. Neurosci. **2021**, 1–16 (2021)
7. Taherkhani, M., Safabakhsh, R.: A novel stability-based adaptive inertia weight for particle swarm optimization. Appl. Soft Comput. **38**, 281–295 (2016)

8. García-Gonzalo, E., Fernández-Martínez, J.L.: Convergence and stochastic stability analysis of particle swarm optimization variants with generic parameter distributions. Appl. Math. Comput. **249**, 286–302 (2014)
9. Tian, D.P.: A review of convergence analysis of particle swarm optimization. Int. J. Grid Distrib. Comput. **6**(6), 117–128 (2013)
10. Suganthan, P.N., et al.: Problem definitions and evaluation criteria for the CEC 2005 special session on real-parameter optimization. KanGAL report **2005005**, 2005 (2005)
11. Li, X., Engelbrecht, A., Epitropakis, M.G.: Benchmark functions for CEC2013 special session and competition on niching methods for multimodal function optimization. RMIT University, Evolutionary Computation and Machine Learning Group, Australia, Technical report (2013)

A Non Dominant Sorting Algorithm with Dual Population Dynamic Collaboration

Cong Zhu[1,2], Yanxiang Yang[1,2], Li Jiang[1,2], and Yongkuan Yang[1,2(✉)]

[1] School of Electrical Engineering and Automation, Xiamen University of Technology, Xiamen 361024, China
yangyongkuanneu@sina.com
[2] Xiamen Key Laboratory of Frontier Electric Power Equipment and Intelligent Control, Xiamen 361024, China

Abstract. Solving multi-objective optimization problems with constraints is a complex task, involving the delicate balance of interrelated and conflicting objective functions and constraint values. While numerous evolutionary algorithms have been developed to tackle this issue, they often struggle to effectively balance population feasibility, convergence, and diversity. In this paper, we introduce the bi-population dynamic coordinated non-dominated sorting algorithm (DDCNDS), which maintains two dynamically coordinated populations and can adjust population size at different evolution stages to balance feasibility and convergence. During the early stages of evolution, the focus is primarily on convergence, with the first population having a larger size to facilitate convergence to the feasible domain across infeasible regions. In the later stages, the emphasis shifts to feasibility, with the second population having a larger size and utilizing local search to explore the feasible domain. Additionally, we propose a non-dominated criterion sorting method to select better individuals, adjusting the non-dominated level based on the proportion of feasible solutions in the input population. Experimental results on three well-known benchmark suites demonstrate the competitiveness of the proposed algorithm.

Keywords: Evolutionary algorithms · Constrained multi-objective optimization problems · Dual-population

1 Introduction

Essentially, many engineering problems can be classified as constrained multi-objective optimization problems (CMOP) [1–4] Due to the conflicting nature of multi-objectives and constraints, multi-objective optimization does not result in a single solution, but rather a set of Pareto-optimal solutions (PS). The projection of the PS on the objective space is called the Pareto front (PF). Thus, the ultimate goal of solving CMOPs is to obtain a set of feasible solutions lying on the Pareto front with good convergence and diversity.

Evolutionary algorithms (EAs) are algorithms that perform an iterative search by modeling the evolutionary process in nature and gradually approach

© The Author(s), under exclusive license to Springer Nature Singapore Pte Ltd. 2024
L. Pan et al. (Eds.): BIC-TA 2023, CCIS 2061, pp. 241–253, 2024.
https://doi.org/10.1007/978-981-97-2272-3_18

the optimal solution of a problem in the form of a population. As a heuristic algorithm, [5] EAs have gained the attention of researchers because their effectiveness in finding Pareto frontiers. However, when faced with complex CMOPs, traditional EAs often fail to obtain feasible solutions. Therefore, researchers have started to combine constraint handling techniques (CHTs) with algorithms to deal with CMOPs more efficiently. After more than two decades of development, existing evolutionary algorithms can usually be categorized into the following five types based on CHTs: (1) sorting-based CHTs, (2) penalization-based CHTs, (3) multi-objective-based CHTs, (4) multi-stage CHTs, and (5) multi-cluster CHTs.

(1) Sort-based CHT [6] involves sorting the population individuals based on the objective function value or the constraint violation value. A classical example of sort-based CHT is the Constraint Dominance Criterion (CDP) [7]. CDP compares two solutions and selects the feasible solution among them. When both individuals are feasible, the one with the smaller objective value is chosen. If both solutions are infeasible, the individual with the smaller constraint violation is selected.

(2) Penalty function-based CHT transforms a constrained optimization problem into an unconstrained one. This strategy was initially proposed by Raziyeh in 2003 and has been further developed [8]. For instance, Li [9] introduced an s-type dynamic penalty factor in differential search to balance exploration and development. However, penalty-based CHT requires specific parameter analysis for individual problems; otherwise, it may easily lead to local optimality.

(3) Multi-objective-based CHT considers the relationship between constraint-violating solutions and the objective function, handling constraints by identifying Pareto solutions. Peng [10] uses a guiding strategy that guides the search while skipping the infeasible region and staying on the feasible region. Jiao [11] uses a dynamic weight adjustment method to guide infeasible solutions with better convergence into the feasible region by adjusting their weights and keeping the weights of feasible solutions constant. By increasing the weight of infeasible solutions in the population, multi-objective-based CHT demonstrates competitiveness in solving CMOPs but is still prone to falling into local optimization.

(4) Multi-stage CHT divides the search into multiple processing stages, employing different CHTs in each stage. For instance, Fan [12] divides evolution into two stages: push and pull search. It utilizes unconstrained search in push search and employs a constraint processing mechanism in pull search. Li [13] divides constraints into multiple levels and uses a multi-stage step-by-step solution. The drawback of multi-stage CHT is that it may lead to longer computation times.

(5) Multiple population-based CHT divides the population into multiple independent groups, each handling a portion of the constraints. By merging the results of the populations, a solution meeting all the constraints can be obtained. For example, Li [14] balances the convergence and diversity of the evolutionary process by leveraging the strengths of dominance and decomposition-based approaches. M [15] by analyzing the role of variables,

influences the decision variable decomposition. Multiple cluster CHTs rely on parameter settings, and inappropriate parameters are detrimental to the algorithm's ability to find an optimal solution.

Based on the above information, we can summarize that different CHTs have their respective advantages and disadvantages. No single CHT can comprehensively address how to balance the convergence, diversity, and feasibility of solutions. Therefore, in this paper, we try to design a new multiple swarm CMOEA based called Dual Swarm Dynamic Coordinated Non-Dominated Sorting Algorithm (DDCNDS), which contributes as follows:

1. A dynamically coordinated population size mechanism is introduced, which can dynamically adjust the size of the dual populations according to different stages of evolution. In the early stages of evolution, the algorithm focuses on convergence, allowing the first population using unconstrained search to have a larger population size. In the later stages of evolution, the algorithm emphasizes feasibility, enabling the second population with constrained search to obtain a larger population size.
2. A new non-dominance criterion ranking is presented that can change the non-dominance rank of population individuals based on the proportion of feasible solutions in the input population. When the proportion of feasible solutions is high, individuals with higher crowding and fitness will receive a higher non-dominance rank. Conversely, when the proportion of feasible solutions is low, the non-dominance rank of high fitness individuals will decrease.

The remainder of the paper is organized as follows: a detailed description of the algorithm is provided in Sect. 2, the algorithm's performance is tested using several benchmark problems in Sect. 3, and a final summary of the algorithm is given in Sect. 4.

2   Related Backgrounds

2.1   Definition of Multi-objective Optimization Algorithms

Many engineering problems are constraints multi-objective optimization problems(CMOP) essentially. Due to the conflict of multiple objectives and constraints, there is no single solution in multi-objective optimization, but a series of optimal solutions. The mathematical expression of a minimization CMOP with n objectives is:

$$\begin{cases} \min y = F(x) = (f_1(x), f_2(x), \cdots, f_n(x)) \\ \quad \text{s.t. } g_i(x) = 0, i = 1, 2, \cdots, u \\ \quad \quad h_j(x) \leq 0, j = 1, 2, \cdots, v \end{cases} \tag{1}$$

In these functions, $x = (x_1, x_2, x_m)^T \in R\ m$ is the m-dimensional decision vector of the CMOP, where R^m is the decision space. The decision vector is mapped by the function F to an m-dimensional objective vector y, where $y \in Y^m$ and Y^m is the objective space. The $g_i x$ denotes the equality constraint conditions, $h_j x$ represents the inequality constraint conditions, the quantities of u and v constraints, respectively. If at least one of them are no zero, the problem is classified as a CMOP; otherwise, it is an multi-objective optimization problems(MOP).

The difference between the CMOP and the constraints single-objective optimization problem is that CMOP's objectives are conflict with each other generally, and the solution cannot be selected by comparing the optimal value of a single objective. In order to obtain a set of optimization objectives, the following important concepts about the Pareto system need to be used as follows:

1. Feasible region Ω is defined as: $x \in R_m | g_i x = 0, i = 1, 2, \cdots, u; h_j x \leq 0, j = 1, 2, \cdots, v$. Optimal solution $x^* \in \Omega$, A solution in the feasible region could be a feasible solution
2. Pareto domination: For example, an optimization problem has two feasible solutions marked as x_1 and x_2, x_1, if $f_1(x_1) \leq f_2(x_2)$ set up, that means x1 dominate x_2.
3. Pareto Front: If a feasible solution $x^* \in \Omega$, without any $x \in \Omega$ dominate x^*, call x^* as for optimal solution, the set of all optimal solutions is called the optimal solution set ρ^*. Pareto Front is the (hyper)surface formed by ρ^* in objective space Y_m.

The degree which a decision variable of constraint violation for a constraint is defined as follows:

$$CV_i = \begin{cases} \max\left(0, g_i(x)\right), i = 1, 2, \cdots, u \\ \max\left(0, |h_j(x)| - \delta\right), j = 1, 2, \cdots, v \end{cases} \tag{2}$$

where δ is positive tolerance value, which be used to convert equality constraints to inequality constraints. The feasibility of the solution can be judged by incorporating the constraint violation degree of each constraint variable, if $CV = 0$, The corresponding solution is a feasible solution.

2.2   CCMO Algorithm

CCMO [16] (Cooperative Coevolutionary Multi-objective Optimization) is a renowned framework for solving multi-objective optimization problems. It employs a bi-population structure and assigns different tasks to each population. The main population focuses on solving the original CMOPs (Constrained Multi-Objective Optimization Problems), while the auxiliary population tackles auxiliary problems derived from the original CMOPs. By exchanging solution information, the populations collaboratively converge towards the Pareto frontier. The architecture of CCMO is simple yet highly scalable, allowing for the integration of various constraint handling mechanisms. This enables CCMO to effectively handle a wide range of complex constrained multi-objective optimization problems.

3   The Proposed Algorithm

3.1   Procedure of the Proposed DDCNDS

This section delineates a dynamic evolutionary collaborative strategy involving two distinct populations, as illustrated in Fig. 1. Within this algorithmic framework, the two populations are tasked with addressing distinct optimization problems. The primary population is entrusted with resolving the unconstrained optimization problem, while the second population is dedicated to tackling the constrained optimization problem, thus facilitating an expedited convergence process. To harmonize both feasibility and convergence aspects, the two populations maintain different population sizes at the beginning and end of the evolutionary phase. Specifically, the first population maintains a size:

$$n_1 = \left| \frac{N}{2} \left(1 + \frac{G_{max} - G}{G_{max}} \right) \right| \tag{3}$$

The second population maintains a size:

$$n_2 = \left| \frac{N}{2} \left(1 + \frac{G}{G_{max}} \right) \right| \tag{4}$$

where N is the size of the input population, G_{max} is the maximum number of evolutionary iterations, and G is the current number of evolutionary iterations. In the early stage of evolution, the primary population maintains a larger size, when the algorithm is mainly concerned with convergence. In the late stage of evolution, the size of the second population is enlarged, at which time the algorithm's convergence is enhanced and more attention is paid to obtaining the Pareto optimal solution. Meanwhile, as a two-population co-evolutionary algorithm, sharing offspring is an important way to enhance the information exchange. Therefore, the offspring form other populations are often incorporated to the environment selection of populations to obtain a new generation of populations.

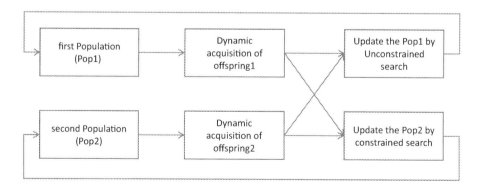

Fig. 1. Workflow of DDCNDS

3.2   Main Framework of DDCNDS

The main steps of DDCNDS are shown in Algorithm 1.

Algorithm 1. DDCNDS

Require: Population size: N, Maximum number of evolutions: G_{max}
Ensure: Main population $Pop2$
/*Initialization*/
$Pop1 \leftarrow$ Initialize()
$Pop2 \leftarrow$ Initialize()
G=0;
while $g \neq G$ **do**
    $Parent1 \leftarrow$ Select N_1 individual by Tournament selection from $Pop1$;
    $Parent2 \leftarrow$ Select N_2 individual by Tournament selection from $Pop2$;
    $Off_1 \leftarrow$ Evolving $Pop1$ by evolution operator GA;
    $Off_2 \leftarrow$ Evolving $Pop2$ by evolution operator DE;
    $Pop_1 \leftarrow Pop_1 \cup Off_1 \cup Off_2$
    $Pop_2 \leftarrow Pop_2 \cup Off_1 \cup Off_2$
    $Pop_1 \leftarrow N_1$ individuals are selected from Pop_1 by using SPEA-II environment
selection strategy
    $Pop_2 \leftarrow$ Select N_2 individuals from Pop_2 by using non-dominated sorting.
end while

In steps 1–4, the algorithm is initiated, initialized, and the number of evaluations is set. Before entering the loop, two populations, Pop1 and Pop2, are randomly generated, and their objective values are determined based on the objective functions of the two populations. Meanwhile, their constraint violations are evaluated. Steps 5–15 encompass the primary cyclic flow of the algorithm. In steps 6 and 7, the number of individuals to be maintained in each iteration for the two populations is determined according to Eqs. 3 and 4, respectively. From these calculations, N1 and N2 individuals are selected as the parental generations of Pop1 and Pop2. In steps 8–9, the two parents select different operators, Parent1 uses the GA operator to generate N1 offspring, and Parent2 uses the DE operator to obtain Off2 as offspring. In steps 10 and 11, the parents of the two populations combine with each other's offspring to form a new population. This strategic combination allows the two populations to share evolutionarily acquired solutions, promoting a faster convergence towards the global optimal solution during the environmental selection session. Subsequently, in step 12, Pop1 uses an archive truncation strategy to maintain the population distribution, while Pop2 utilizes an improved non-dominated solution ordering strategy to select updated individuals from the original population for the new generation. Finally, DDCNDS designates the second population as the output, completing the entire algorithmic process.

To summarize, DDCNDS employs a dynamic strategy to synergize evolution between two populations. Unlike other algorithms, DDCNDS dynamically allocates computation to each population based on the evolutionary stage, thereby significantly reducing the computational cost of the algorithm. In the pre-evolutionary stage, the primary and secondary populations are evenly distributed in the solution space. The secondary population is explored using constraints, and the primary population is searched using unconstraints. At this point the arithmetic is focused on the major population. Through evolutionary coordination, solutions from the primary population help the secondary population in approaching the PF across the infeasible region. In the later stages of evolution, the arithmetic emphasis shifts primarily population to the second population. The secondary population receives support from the primary population to deepen the exploration of the solution space, facilitating the search for Pareto Fronts within locally optimal feasible regions. Towards the end of the evolutionary process, the second population delves extensively into the feasible domain, yielding a set of solutions located on the PF.

3.3   Environmental Selection

Algorithm 2. Environment selection

Require: Mixed population:Pop_1, Population size: N
Ensure: Pop
  $Pf \leftarrow$ Calculate the proportion of feasible solutions for Pop;
  $Fit \leftarrow$ Calculate the population feasibility of Pop;
  $Crowd \leftarrow$ Calculate the crowding degree of the population;
  if $Pf > 0.9$ **then**
      $Crowd{=}Crowd \times Pf - Fitness \times (1 - Pf)$
  else
      $Crowd{=}Crowd \times Pf - Fitness \times (1 - Pf)$
  end if
  $Pop \leftarrow$ Sorting Pop using Congestion Distance Probability;

$$Crowd = P_f \times Crowd + (1 - P_f) \times Fitness \, if \, Pf > 0.9$$
$$Crowd = P_f \times Crowd - (1 - P_f) \times Fitness \, if \, Pf < 0.9 \tag{5}$$

The Algorithm 2 outlines the process of environmental selection. At the outset of the algorithm, the proportion of feasible solutions for the population as well as the degree of crowding is calculated. The fitness of the input population is then determined using the SPEA-2 [17] population fitness calculation strategy. Subsequently, the crowding distance of the new generation is obtained according to the following equation.

In the early stages of evolution, the proportion of feasible solutions in the population is low. Increasing the crowding proportion by decreasing the fitness proportion is conducive to retaining more excellent solutions and helping the population approach the Pareto frontier quickly. In the later stages of evolution, the population's proportion of feasible solutions is higher. Elevating the fitness proportion is beneficial for balancing both feasibility and convergence within the population, thereby promoting local search.

3.4   Analysis

Importantly, compared to the common two-population multi-objective evolutionary algorithms CCMO and FACE [18], the DDCNDS algorithm incorporates several innovations:

1. A dynamic evolutionary collaborative strategy is introduced, allowing the adjustment of the size of two populations according to the stage of evolution. During the early stages of evolution, the primary population maintains a larger size to enhance the global convergence of the algorithm. In the later stages, the search for feasible domains is optimized by increasing the size of the second population to obtain more feasible solutions. In contrast, CCMO and FACE do not alter the population size during evolution.
2. An improved non-dominated sorting strategy is proposed, capable of filtering out solutions with higher non-dominated ranks based on the proportion of feasible solutions within the input population. The adaptation weight of the population is heightened in the early stage to retain more excellent solutions. Increasing the adaptation weight in the later stages of evolution helps balance both feasibility and convergence within the population. In contrast, CCMO and FACE do not dynamically change the solution screening criteria during evolution.

To assess the effectiveness of DDCNDS in addressing complex constrained problems and leveraging information about feasible and infeasible solutions, two problems, MW11 [19], and DAS-CMOP5 [20], were used to observe its performance. In the figures, the distribution of individuals in the early, middle, and final stages of the population for the MW11 and DAS-CMOP5 problems is plotted.

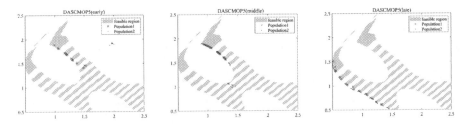

Fig. 2. Shows the performance of the dual-population on the DAS-CMOP5 problem in the early, middle, and last stages.

Figure 2 and Fig. 3 showcase the performance of DDCNDS on DASCMOP5 and MW11 problems. The dynamic population co-evolution strategy allocates computational resources primarily to Population1 in the initial stages. As a result, Pop1 can quickly explore the feasible domain and facilitate the convergence of Pop2 by sharing solutions. In the middle stages of evolution, as computational resources are shifted, Pop2 gradually approaches the Pareto Front (PF).

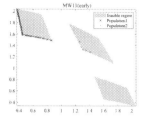 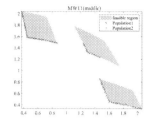 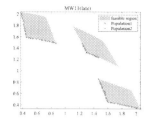

Fig. 3. Shows the performance of the dual-population on the MW11 problem in the early, middle, and last

By the end of the evolutionary process, Pop2 is widely and evenly distributed across the PF, while Pop1, with the support of Pop2, slowly approaches the PF but struggles to maintain a well-distributed population. This phenomenon is also evident in MW11. During the early and middle stages of MW11, Pop1 efficiently spreads across the feasible domain and assists Pop2 in approaching the PF. However, in the later stages of evolution, Pop1 faces the challenge of insufficient population diversity, whereas Pop2 can thoroughly explore the feasible domain and possess a set of solutions situated on the PF. The performance of DDCNDS on MW11 and DAS-CMOP5 demonstrated the feasibility of algorithmic evolution collaboration.

4 Experimental Study

4.1 Algorithm Evaluation Metrics

When evaluating algorithm effectiveness, it is crucial to consider two key aspects: convergence and distribution. In general, the Inverted Generational Distance (IGD) [21] and Hypervolume (HV) [22]metrics are used to quantify the diversity and convergence of the obtained non-dominated reference solution sets. In this paper, we have chosen IGD as a performance metric to evaluate the algorithm.

The IGD metric calculates the average distance between the non-dominated solutions obtained by the algorithm and the Pareto Front (PF). The formula for calculating IGD is as follows:

$$IGD = \sum_{x \in P^*} d(x, PF_{approx}) / |P^*| \tag{6}$$

In this case, P^* represents the solutions uniformly distributed on the Pareto front, and $d(x, PF_{approx})$ represents the minimum Euclidean distance from $x \in P^*$ to PF_{approx}. A lower IGD value indicates better performance of the algorithm in terms of problem representation.

4.2 Experimental Design

To evaluate the effectiveness of the DDCNDS algorithm on benchmark problems, we compared it with five other state-of-the-art algorithms: ToP [23], PPS [12],

Table 1. IGD for the five state-of-the-art CMOEAs and DDCNDS on the MW reference suite.

Problem	ToP	PPS	NSGAIIARSBX	NSGAIII	ANSGAIII	DDCNDS
MW1	NaN	2.78e−3	**1.98e−3**	2.84e−3	3.66e−3	2.1085e−3
	(NaN)	(1.37e−4) −	**(9.17e−5) +**	(1.61e−3) −	(8.31e−3) =	(8.37e−5)
MW2	1.59e−1	1.59e−1	4.82e−2	2.50e−2	**2.36e−2**	2.63e−2
	(1.38e−1) −	(9.98e−2) −	(2.41e−2) −	(8.62e−3) =	**(9.55e−3) =**	(8.52e−3)
MW3	5.75e−1	**6.31e−3**	6.77e−2	1.35e−2	4.12e−2	4.11e−2
	(4.29e−1) −	**(4.76e−4) +**	(2.32e−1) −	(4.92e−2) +	(1.67e−1) −	(4.99e−3)
MW4	4.43e−1	5.57e−2	5.28e−2	**4.11e−2**	4.14e−2	5.40e−2
	(4.45e−1) −	(2.49e−3) −	(1.87e−3) +	**(4.60e−5) +**	(3.44e−4) +	(1.49e−3)
MW5	NaN	3.54e−1	2.48e−1	1.95e−1	2.83e−1	**1.01e−1**
	(NaN)	(3.69e−1) −	(3.55e−1) =	(2.94e−1) −	(3.38e−1) −	**2.56e−1**
MW6	6.19e−1	5.35e−1	1.71e−1	**2.44e−2**	2.71e−2	6.45e−2
	(4.10e−1) −	(3.70e−1) −	(1.90e−1) −	**(1.52e−2) =**	(2.18e−2) =	(1.21e−1)
MW7	3.28e−2	5.56e−3	**5.27e−3**	3.60e−2	3.74e−2	1.10e−1
	(7.94e−2) +	(6.57e−4) +	**(2.55e−4) +**	(1.13e−1) +	(1.13e−1) +	(6.28e−3)
MW8	6.22e−1	1.67e−1	8.56e−2	6.30e−2	1.08e−1	**6.05e−2**
	(3.50e−1) −	(9.79e−2) −	(3.19e−2) −	(5.15e−2) −	(1.48e−1) −	**(8.86e−3)**
MW9	7.46e−1	6.16e−2	**8.09e−3**	9.36e−3	9.23e−3	2.86e−2
	(3.91e−2) −	(1.78e−1) −	**(2.54e−3) +**	(3.99e−3) +	(3.86e−3) +	(9.17e−3)
MW10	5.45e−1	4.53e−1	2.45e−1	**1.45e−1**	1.98e−1	1.80e−1
	(0.00e+0) =	(2.40e−1) −	(1.80e−1) −	**(1.70e−1) +**	(1.62e−1) −	(1.40e−1)
MW11	4.89e−1	7.44e−3	**6.73e−3**	4.00e−1	4.48e−1	2.64e−2
	(2.59e−1) −	(3.74e−4) +	**(2.42e−4) +**	(3.33e−1) =	(3.21e−1) −	(4.48e−3)
MW12	7.14e−1	3.70e−2	5.43e−3	**4.76e−3**	1.21e−2	1.97e−2
	(1.74e−1) −	(1.23e−1) −	(2.12e−4) +	**(2.24e−4) +**	(2.74e−2) +	(8.62e−3)
MW13	6.94e−1	4.98e−1	**1.53e−1**	2.12e−1	2.30e−1	1.94e−1
	(6.39e−1) −	(4.06e−1) −	**(6.19e−2) +**	(3.46e−1) −	(3.51e−1) −	(4.72e−2)
MW14	3.11e−1	1.42e−1	1.89e−1	1.29e−1	**1.11e−1**	1.29e−1
	(2.39e−1) −	(4.36e−2) −	(2.90e−2) −	(2.67e−3) =	**(2.54e−3) +**	(6.41e−3)
+/−/=	1/10/1	3/11/0	7/5/2	6/4/4	5/5/4	

NSGA2ARSBF [24], NSGA3 [25], and ANSGA3 [26]. The experimental settings and parameters were as follows:

Population size: $N_p = 100$. Termination condition: The algorithm will stop after reaching 150,000 function evaluations.

Number of runs: For each case, each algorithm is independently run 30 times.

Crossover operator settings: DE operator with F set to 0.5 and CR set to 1.0. For the SBX operator, the crossover probability (Pc) is set to 0.9, and the distribution index is set to 20. Polynomial mutation operator settings: Mutation probability is set to 1/n and the distribution index is set to 20.

Polynomial mutation operator settings: Mutation probability is set to 1/n and the distribution index is set to 20.

For the selection of the benchmark problems, we chose three specific problems: MW, DCDTLZ [27] and DASCMOP. These benchmark problems allow us to evaluate the diversity, convergence and environmental adaptability of the studied algorithms. The IGDs of these algorithms are listed in Tables 1, 2 and 3. As shown in the table, DDCNDS performs best only in the DCDTLZ problem

Table 2. IGD for the five state-of-the-art CMOEAs and DDCNDS on the DASCMOP reference suite.

Problem	ToP	PPS	NSGAIIARSBX	NSGAIII	ANSGAIII	DDCNDS
DASCMOP1	7.7962e−1	**7.6107e−2**	2.6423e−1	7.3718e−1	7.3251e−1	1.2657e−1
	(4.07e−2) −	**(1.74e−1)** +	(3.13e−1) =	(4.75e−2) −	(3.08e−2) −	(2.49e−1)
DASCMOP2	6.4452e−1	**5.1616e−3**	1.3743e−1	2.6284e−1	2.6470e−1	3.4596e−2
	(2.27e−1) −	**(1.89e−4)** +	(7.87e−2) −	(8.62e−3) −	(1.79e−2) −	(5.76e−2)
DASCMOP3	7.3509e−1	**2.7901e−1**	3.3713e−1	3.5801e−1	3.5891e−1	3.5753e−1
	(8.73e−2) −	**(1.16e−1)** +	(4.23e−2) +	(3.48e−2) −	(7.97e−2) =	(3.50e−2)
DASCMOP4	NaN	1.4721e−1	2.4963e−1	2.1432e−1	1.6060e−1	**1.5313e−2**
	(NaN)	(9.90e−2) −	(1.29e−1) −	(1.39e−1) −	(1.47e−1) −	**(5.45e−2)**
DASCMOP5	NaN	**8.1035e−3**	3.3138e−1	3.5179e−1	4.2222e−1	2.1092e−1
	(NaN)	**(1.32e−2)** +	(1.41e−1) =	(2.41e−1) =	(2.11e−1) −	(2.02e−1)
DASCMOP6	NaN	**2.8883e−1**	5.7797e−1	4.3483e−1	4.6697e−1	4.8398e−1
	(NaN)	**(3.27e−1)** +	(6.90e−2) −	(1.02e−1) −	(8.49e−2) =	(9.12e−2)
DASCMOP7	NaN	8.4331e−2	4.8117e−2	**3.5450e−2**	3.9747e−2	4.5500e−2
	(NaN)	(7.05e−2) −	(2.87e−3) −	**(2.44e−4)** +	(2.35e−3) +	(2.33e−3)
DASCMOP8	NaN	1.0581e−1	6.0320e−2	7.4021e−2	6.5799e−2	**5.8274e−2**
	(NaN)	(8.26e−2) −	(4.27e−3) −	(1.06e−1) −	(8.70e−2) −	**(4.13e−3)**
DASCMOP9	6.2700e−1	**1.8271e−1**	2.4526e−1	4.3090e−1	4.3432e−1	2.4148e−1
	(1.82e−1) −	**(1.26e−1)** =	(1.39e−1) =	(7.31e−2) −	(7.09e−2) −	(1.55e−1)
+/−/=	0/4/0	5/3/1	1/5/3	1/5/3	1/6/2	

Table 3. IGD for the five state-of-the-art CMOEAs and DDCNDS on the DCDTLZ reference suite.

Problem	ToP	PPS	NSGAIIARSBX	NSGAIII	ANSGAIII	DDCNDS
DC1DTLZ1	3.0804e−2	5.0857e−2	1.4329e−2	1.4178e−2	**1.3727e−2**	1.4051e−2
	(5.97e−2) −	(8.05e−2) −	(6.20e−4) −	(1.66e−4) −	**(1.18e−4)** +	(3.74e−4)
DC1DTLZ3	9.9250e−1	3.6097e−1	4.3416e−2	4.5893e−2	4.2101e−2	**4.1104e−2**
	(1.71e+0) −	(1.43e−1) −	(2.12e−3) −	(8.05e−4) −	(1.18e−3) − ‚	**(1.21e−3)**
DC2DTLZ1	NaN	1.0987e−1	1.7273e−1	1.3603e−1	1.1641e−1	**5.1679e−2)**
	(NaN)	(7.12e−2) −	(4.38e−4) −	(5.73e−2) −	(6.87e−2) −	**(5.43e−2)**
DC2DTLZ3	NaN	**3.7758e−1**	5.7144e−1	5.6337e−1	5.6180e−1	6.6535e−1
	(NaN)	**(2.49e−1)** =	(0.00e+0) =	(3.82e−3) =	(0.00e+0) =	(1.33e−1)
DC3DTLZ1	2.4715e+0	4.2894e−1	**1.0763e−1**	1.2370e−1	1.2229e−1	1.2916e−1
	(2.42e+0) −	(4.64e−1) −	**(9.45e−2)** =	(8.93e−2) =	(7.46e−2) =	(8.72e−2)
DC3DTLZ3	8.3571e+0	2.4717e+0	2.0048e+0	1.4308e+0	1.4694e+0	**1.2682e+0**
	(3.60e+0) −	(1.94e+0) −	(4.97e−1) −	(5.19e−1) −	(5.39e−1) −	**(5.52e−1)**
+/−/=	0/4/0	0/1/5	0/3/3	1/2/3	1/3/2	

family and does not perform well in the other problems. In the DCDTLZ problem family, it outperforms in all three problems. On the other hand, PPS excelled in the DASCMOP problem series, achieving the best results in all six problems. More algorithms outperform DDCNDS in the MW series. Specifically, NSGAI-IARSBX and NSGAIII stand out in this regard. NSGAIIARSBX achieves the best performance in five problems, while NSGAIII outperforms in four problems. However, ToP performs relatively poorly among the five compared algorithms, showing no outstanding performance on any of the three problem families.

Although DDCNDS does not dominate in all three problem families, it still has the best overall performance considering all the aspects mentioned above, and it exhibits the fastest convergence in the most problems. This suggests that DDCNDS is a very promising algorithm with great potential and competitiveness in solving CMOPs in general.

5   Conclusion

This study presents a novel approach called the dynamic coordinated dual-population constrained multi-objective evolutionary algorithm. This algorithm effectively manages two dynamic populations and adjusts their sizes during different stages of evolution to achieve a balance between feasibility and convergence. In the initial stage, the algorithm focuses on convergence, while in the later stages, it prioritizes feasibility. Through a comparative analysis of 29 benchmark problems using five state-of-the-art algorithms, the proposed algorithm has demonstrated competitive performance. However, it is important to note that there are several limitations associated with this algorithm. For example, its excessive emphasis on conserving computational resources may lead to compromised convergence. To overcome these limitations, future research will focus on integrating advanced constraint handling techniques to enhance the algorithm's competitiveness in CMOPs.

References

1. Basu, M.: Economic environmental dispatch using multi-objective differential evolution. Appl. Soft Comput. **11**(2), 2845–2853 (2011)
2. Cui, Y., Geng, Z., Zhu, Q., Han, Y.: Multi-objective optimization methods and application in energy saving. Energy **125**, 681–704 (2017)
3. Saravanan, R., Ramabalan, S., Ebenezer, N.G.R., Dharmaraja, C.: Evolutionary multi criteria design optimization of robot grippers. Appl. Soft Comput. **9**(1), 159–172 (2009)
4. Zuo, X., Chen, C., Tan, W., Zhou, M.: Vehicle scheduling of an urban bus line via an improved multiobjective genetic algorithm. IEEE Trans. Intell. Transp. Syst. **16**(2), 1030–1041 (2014)
5. Eiben, A.E., Smith, J.: From evolutionary computation to the evolution of things. Nature **521**(7553), 476–482 (2015)
6. Deb, K., Pratap, A., Agarwal, S., Meyarivan, T.M.: A fast and elitist multiobjective genetic algorithm: NSGA-II. IEEE Trans. Evol. Comput. **6**(2), 182–197 (2002)
7. Jan, M.A., Khanum, R.A.: A study of two penalty-parameterless constraint handling techniques in the framework of MOEA/D. Appl. Soft Comput. **13**(1), 128–148 (2013)
8. Farmani, R., Wright, J.A.: Self-adaptive fitness formulation for constrained optimization. IEEE Trans. Evol. Comput. **7**(5), 445–455 (2003)
9. Liu, J., Teo, K.L., Wang, X., Wu, C.: An exact penalty function-based differential search algorithm for constrained global optimization. Soft. Comput. **20**, 1305–1313 (2016)

10. Peng, C., Liu, H., Gu, F.: An evolutionary algorithm with directed weights for constrained multi-objective optimization. Appl. Soft Comput. **60**, 613–622 (2017)
11. Jiao, R., Zeng, S., Li, C., Ong, Y.-S.: Two-type weight adjustments in MOEA/D for highly constrained many-objective optimization. Inf. Sci. **578**, 592–614 (2021)
12. Fan, Z., et al.: Push and pull search for solving constrained multi-objective optimization problems. Swarm Evol. Comput. **44**, 665–679 (2019)
13. Li, X., et al.: A novel two-stage constraints handling framework for real-world multi-constrained multi-objective optimization problem based on evolutionary algorithm. Appl. Intell. **51**, 8212–8229 (2021)
14. Li, K., Deb, K., Zhang, Q., Kwong, S.: An evolutionary many-objective optimization algorithm based on dominance and decomposition. IEEE Trans. Evol. Comput. **19**(5), 694–716 (2014)
15. Ma, X., et al.: A multiobjective evolutionary algorithm based on decision variable analyses for multiobjective optimization problems with large-scale variables. IEEE Trans. Evol. Comput. **20**(2), 275–298 (2015)
16. Tian, Y., Zhang, T., Xiao, J., Zhang, X., Jin, Y.: A coevolutionary framework for constrained multiobjective optimization problems. IEEE Trans. Evol. Comput. **25**(1), 102–116 (2020)
17. Zitzler, E., Laumanns, M., Thiele, L.: SPEA2: improving the strength pareto evolutionary algorithm. TIK Report, 103 (2001)
18. Yang, Y., Liu, J., Tan, S.: A multi-objective evolutionary algorithm for steady-state constrained multi-objective optimization problems. Appl. Soft Comput. **101**, 107042 (2021)
19. Ma, Z., Wang, Y.: Evolutionary constrained multiobjective optimization: test suite construction and performance comparisons. IEEE Trans. Evol. Comput. **23**(6), 972–986 (2019)
20. Fan, Z., et al.: Difficulty adjustable and scalable constrained multiobjective test problem toolkit. Evol. Comput. **28**(3), 339–378 (2020)
21. Bosman, P.A.N., Thierens, D.: The balance between proximity and diversity in multiobjective evolutionary algorithms. IEEE Trans. Evol. Comput. **7**(2), 174–188 (2003)
22. Zitzler, E., Thiele, L.: Multiobjective evolutionary algorithms: a comparative case study and the strength pareto approach. IEEE Trans. Evol. Comput. **3**(4), 257–271 (1999)
23. Liu, Z., Wang, Y.: Handling constrained multiobjective optimization problems with constraints in both the decision and objective spaces. IEEE Trans. Evol. Comput. **23**(5), 870–884 (2019)
24. Pan, L., Xu, W., Li, L., He, C., Cheng, R.: Adaptive simulated binary crossover for rotated multi-objective optimization. Swarm Evol. Comput. **60**, 100759 (2021)
25. Deb, K., Jain, H.: An evolutionary many-objective optimization algorithm using reference-point-based nondominated sorting approach, part I: solving problems with box constraints. IEEE Trans. Evol. Comput. **18**(4), 577–601 (2013)
26. Jain, H., Deb, K.: An evolutionary many-objective optimization algorithm using reference-point based nondominated sorting approach, part II: handling constraints and extending to an adaptive approach. IEEE Trans. Evol. Comput. **18**(4), 602–622 (2013)
27. Li, K., Chen, R., Fu, G., Yao, X.: Two-archive evolutionary algorithm for constrained multiobjective optimization. IEEE Trans. Evol. Comput. **23**(2), 303–315 (2018)

Dynamic Multi-objective Operation Optimization of Blast Furnace Based on Evolutionary Algorithm

Yumeng Zhao[1,2], Jingchuan Zhang[1,2], Meng Jiang[1,2], Kai Fu[1,2], Qiyuan Deng[1,2], and Xianpeng Wang[3,4(✉)]

[1] Liaoning Engineering Laboratory of Data Analytics and Optimization for Smart Industry, Shenyang 110819, China
[2] Liaoning Key Laboratory of Manufacturing System and Logistics Optimization, Shenyang 110819, China
[3] National Frontiers Science Center for Industrial Intelligence and Systems Optimization, Shenyang, China
[4] Key Laboratory of Data Analytics and Optimization for Smart Industry (Northeastern University), Ministry of Education, Shenyang 110819, China
wangxianpeng@ise.neu.edu.cn

Abstract. Blast furnaces play a critical role in the steel industry, and their operational optimization is crucial for energy conservation and emissions reduction. This paper examines the impact of changes in operational conditions on blast furnace performance. We propose a dynamic multi-objective optimization algorithm based on multiple short time series (MT-DC-RVEA) to solve the constructed dynamic multi-objective operational optimization model for blast furnaces. Experimental results validate the effectiveness of the proposed algorithm in solving the operational optimization model for blast furnace operations.

Keywords: Blast furnace ironmaking · Operation optimisation · Dynamic multiobjective optimisation · Evolutionary algorithm

1 Introduction

The blast furnace, a pivotal production facility at the heart of the steel industry, assumes a critical role in modern industrial operations. The furnace's operational processes are significant energy consumers, concurrently exerting a profound impact on the economic performance of steel enterprises [1]. The blast furnace stands as a colossal cylindrical metal vessel. During furnace operation, iron ore and coke are proportionately layered and introduced into the furnace from the top. Following this, hot air, coal powder, and other auxiliary fuels are introduced into the lower part of the furnace along its circumference. Under high-temperature conditions, a series of intricate physical-chemical reactions occur within the furnace, reducing iron oxides in the iron ore to liquid iron. Eventually, liquid iron is separated from the slag [11]. The liquid iron flows out from

© The Author(s), under exclusive license to Springer Nature Singapore Pte Ltd. 2024
L. Pan et al. (Eds.): BIC-TA 2023, CCIS 2061, pp. 254–261, 2024.
https://doi.org/10.1007/978-981-97-2272-3_19

the tapping hole into the iron ladle, while the slag is discharged from the slag hole, and blast furnace gas is expelled from the furnace top [8].

Given the complexity of the blast furnace ironmaking process, the precision of modeling is particularly pivotal when optimizing blast furnace operating parameters. Zhou et al. employed the NSGA II algorithm to solve a mathematical model for blast furnace charging and operation [12]. In a separate investigation, Zhou et al. utilized the NSGA II algorithm to solve a data-driven model for blast furnace ironmaking [10]. Li et al. used the differential evolution algorithm to solve a data-driven model for the burden surface [4]. Bashista et al. applied the RVEA algorithm to solve a data-driven model for blast furnace ironmaking [5]. Nevertheless, these models disregard the influence of production conditions. Therefore, this paper aims to explore the use of evolutionary algorithms to solve the dynamic optimization problem of blast furnace operations, aiming to enhance industrial production efficiency and sustainability.

This paper has devised an optimization model for blast furnace ironmaking operations based on two paramount metrics that steel enterprises prioritize. One objective is the cost of blast furnace production, reflecting the raw materials, fuels, and economic efficiency of the blast furnace. The second objective is carbon emissions per ton of iron, indicative of the energy consumption and technological proficiency of the blast furnace. The final objective is the production capacity of steel. The predominant component of ironmaking costs lies in raw material expenses, wherein a significant reduction in raw materials tends to decrease the iron output. Such reductions often lead to incomplete smelting, consequently elevating carbon emissions. Based on these considerations, incorporating the environmental impact on blast furnace operating parameters, we have developed a dynamic multi-objective optimization model. To address this model, we propose a Multi-Objective Dynamic Optimization Algorithm based on multi-short time series forecasting (MT-DC-RVEA).

The subsequent sections of this paper are structured as follows: Sect. 2 outlines the methodology solving the dynamic operation optimization model of blast furnace model. In Sect. 3, substantiates the efficacy of solving the blast furnace operation optimization model framework through empirical experiments. The paper concludes with a summary in Sect. 4.

2   Solving Dynamic Multi-objective Operation Optimization of Blast Furnace Model

2.1   Dynamic Multi-objective Operation Optimization of Blast Furnace Model

During the iron smelting process in blast furnaces, there exist variables that are beyond human control but undergo changes with environmental fluctuations, serving as conditional variables that impact optimization objectives. This paper primarily addresses the influence of ore and fuel on optimization goals. Steel mills, aiming to reduce raw material costs and regulate the composition of molten iron, incorporate various types of ore sources. The types and grades of ore delivered cannot be fixed, resulting in significant fluctuations in blast furnace ore feedstock. Similarly, the quality of fuel experiences continuous fluctuations. Within the blast furnace operation, there are several operational variables. This paper focuses on optimizing significant variables based on the operational context of a specific blast furnace in a steel mill. The optimization of these operational variables is intended to decrease costs and carbon emissions while enhancing production. When constructing the model, certain unvarying parameters are established as constants. In the model presented in this paper, fluctuations in ore and fuel prices have been omitted and treated as constants to facilitate computation. In order to smoothly calculate and accurately find the objective function, this paper uses random forest [13] to predict some variables that cannot be directly measured.

This paper constructs a three-objective data-driven optimization model for blast furnace operation. The constructed model is as follows:

$$\begin{aligned} f_1 &= iron_{cost} \\ f_2 &= CO_{2emission} \\ f_3 &= iron_{output} \end{aligned} \tag{1}$$

f_1 is the iron cost. Ignoring the cost of electricity and labor, the cost of making iron in this blast furnace in the model of this paper is mainly in the purchase of fuel and ore. f_2 is the CO_2 of emission. There are many factors that contribute to increased carbon emissions from the blast furnace, including process energy consumption, fuel composition, and resource efficiency. The recycling of carbon within the process is not considered, only the input and output from the outside is considered. f_3 is the hourly iron production. In calculating the objective function, the reciprocal is taken.

A dynamic constrained multi-objective optimization algorithm (MT-DC-RVEA) based on multi-time series prediction is proposed for the blast furnace ironmaking model constructed in this paper. This algorithm, built upon the static algorithm RVEA [2], utilizes an initial population modified by the environment changes predicted through the multi-time series, including the BHT-ARIMA model. This chapter focuses on introducing the approach of the MT-DC-RVEA algorithm in solving the blast furnace operation optimization model.

2.2   MT-DC-RVEA Algorithm for Solving Blast Furnace Operation Optimization Model

Through multiple experimental iterations, we identified a suboptimal population distribution when solving the blast furnace operation optimization model in our research. Consequently, we opted for the RVEA algorithm, known for its superior distribution, as the baseline algorithm. The evolution of the MT-DC-RVEA algorithm is predominantly anchored in the RVEA algorithm. Unlike the majority of existing reference point-based multi-objective evolutionary algorithms [6], RVEA sets itself apart by employing center vectors and radii to define preferred regions.

It introduces a novel angle-based selection criterion. These methodologies confer upon the RVEA algorithm a substantial advantage in distribution compared to other algorithms, thereby offering decision-makers a more comprehensive set of feasible choices.

In the absence of changes in the blast furnace environment, the Pareto frontier of operating variables under the current conditions is determined using the RVEA algorithm. Decision-makers select appropriate operating variables for application in the blast furnace. When the environment undergoes changes, the optimal operating variables also change. To expedite the identification of optimal operating variables, this paper introduces a novel generator for initial operating variables in a new environment. This generator integrates operating variables with Block Hankel Tensor ARIMA (BHT-ARIMA). BHT-ARIMA combines Block Hankel Tensor with ARIMA through low-rank Tucker decomposition. This method exhibits robustness, rapid computational speed, and significant advantages for multiple time series. Boundary detection is applied to the generated operating variables to ensure their distribution within the range of feasible values. In dynamic optimization problems, there are often similarities between adjacent environments, making the optimal solution from the previous environment sometimes applicable to the new environment. To obtain higher-quality operating variables, this paper merges the optimal operating variables from the previous environment with the predicted operating variables for the new environment. The Reference Vector-Guided Selection strategy in the RVEA algorithm is employed to select the optimal operating variables from the merged population. These are then utilized as the initial operating variables in the new environment before further optimization using the C-RVEA algorithm.

3   Experimental Results and Discussion

3.1   Experimental Preparation

In this section, we empirically demonstrate the effectiveness of the proposed approach for solving the optimization model of dynamic in blast furnace operations. All experiments used the same performance metrics and comparative algorithms were applied. In order to compare the results obtained by the algorithms, we use MIGD [7] and MHV [9] as performance metrics for comparison. The optimal values in all experimental outcomes are highlighted in deep gray. A Wilcoxon rank-sum test [3] with a significance level of 0.05 was conducted to compare the algorithm proposed in this paper with other competing algorithms. The test results are represented by "+," "=", and "−", indicating statistical superiority, no significant difference, and inferiority of the proposed algorithm compared to the competitive algorithms, respectively. To facilitate comparison, the contrasting algorithms chosen for this study employ the same baseline algorithm, RVEA. Two dynamic algorithms, IGP-DC-RVEA and ISVM-DC-RVEA, are selected to tackle the optimization model of dynamic constraint operations for blast furnaces proposed in this paper.

The experimental data in this paper are derived from the actual ironmaking process of a large-scale iron and steel enterprise in China, from one of its blast furnaces. Samples were organized on an hourly basis, with the exclusion of samples with missing values and those from periods of furnace instability. Stability was determined by iron yield exceeding the product of the furnace utilization coefficient and furnace volume. This process resulted in 1525 samples. To better emphasize the impact of environmental changes on operational variables and objectives, similar adjacent samples were merged, resulting in a final set of 90 samples. Each sample comprises 49 parameters, including 3 objectives, 19 decision variables, 21 conditional variables, and 6 intermediate variables. These samples were segmented into three parts, constructing three practical problems: Instance 1, Instance 2, and Instance 3. Due to practical considerations, Instance1 exhibits frequent fluctuations in fuel quality, Instance 2 experiences fluctuations in both fuel and ore, and Instance 3 undergoes frequent changes in ore types.

3.2   Parameter Settings

The model tackles three objectives optimization problems, setting the population size for MT-DC-RVEA at 300 with 19 decision variables. Termination occurs for Instance 1, Instance 2, and Instance 3 after 34, 26, and 25 environmental changes, respectively. Testing involves three different change frequencies denoted as τ_t. The evolutionary process spans 100 generations before the initial environmental shift. Each algorithm is run independently 20 times on each problem to compute the mean and variance of their evaluation metrics.

The calculation of MIGD and MHV is conducted using normalized values. The reference point z in MHV is set at $(1.0, 1.0, 1.0)$. Comparative algorithms, IGP-DC-REVA and ISVM-DC-REVA, maintain the same basic parameter settings as those in this study.

3.3   Experimental Results and Analysis

In this section, three different algorithms are applied to test each instance problem individually. The calculated MIGD and MHV values are presented in Table 1 and 2. The table visually highlights that MT-DC-RVEA consistently outperforms across all problems. The convergence of IGP-DC-RVEA is superior to that of ISVM-DC-RVEA, whereas the distribution of ISVM-DC-RVEA excels over IGP-DC-RVEA. These algorithms exhibit low variance after multiple computations, indicating stable and reliable performance.

To demonstrate the practicality of our algorithm, we compare actual data with the values computed by the algorithm. We assess the disparity in carbon emissions and yield calculated by the algorithm under the same cost compared to the actual values. The Pareto fronts for each environment, computed by the algorithm, are averaged, and a set of target values closest to the actual cost data is selected for comparison. The comparative results are depicted in Figs. 1.

Table 1. Mean and variance of MIGD for blast furnace operation optimisation problems

Prob.	τ_t	IGP-DC-RVEA	ISVM-DC-RVEA	MT-DC-RVEA
	5	0.1089(0.0639)+	0.1689 (0.0939)+	0.0554 (0.0018)
Instance1	10	0.0921(0.0525)+	0.1310(0.0852)+	0.0543 (0.0041)
	20	0.0851(0.0241)+	0.1082(0.0326)+	0.0542 (0.0032)
	5	0.1035(0.0462)+	0.1401(0.0841)+	0.0626 (0.0112)
Instance2	10	0.0894(0.0465)+	0.1278(0.0789)+	0.0615 (0.0086)
	20	0.0765(0.0263)+	0.1026(0.0678)+	0.0544 (0.0025)
	5	0.1295(0.0563)+	0.1569(0.0795)+	0.0752(0.0097)
Instance3	10	0.1245(0.0644)+	0.1382(0.0482)+	0.0685(0.0075)
	20	0.1103(0.0458)+	0.1243(0.0486)+	0.0677(0.0082)

Table 2. Mean and variance of MHV for blast furnace operation optimisation problems

Prob.	τ_t	IGP-DC-RVEA	ISVM-DC-RVEA	MT-DC-RVEA
	5	0.3834(0.0443)+	0.4152(0.0226)+	0.4344 (0.0054)
Instance1	10	0.3914(0.0154)+	0.4265(0.0116)+	0.4386 (0.0132)
	20	0.4098(0.0252)+	0.4301(0.0145)+	0.4446 (0.0032)
	5	0.3929(0.0484)+	0.4235(0.0144)+	0.4852(0.0289)
Instance2	10	0.3998(0.0352)+	0.4298(0.0175)+	0.4945 (0.0221)
	20	0.4065(0.0232)+	0.4385(0.0142)+	0.5091 (0.0079)
	5	0.4018(0.0152)+	0.4314(0.0147)+	0.4923(0.0090)
Instance3	10	0.4114(0.0245)+	0.4424(0.0102)+	0.4966(0.0150)
	20	0.4275(0.0123)+	0.4578(0.0098)+	0.5182(0.0069)

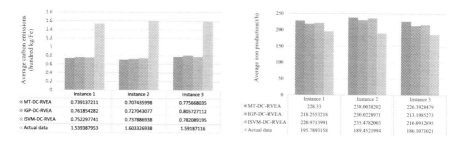

Fig. 1. Comparison of carbon emissions (left) and iron production (right) at the same cost

Figure 1 (left) depicts a comparison of carbon emissions when costs are equal. It is evident that the algorithmically assigned operational variables result in significantly lower carbon emissions than the actual values. However, there is little disparity in carbon emissions among the three algorithms, with a slight advantage observed in the proposed algorithm. Figure 1 (right) compares yields when costs are equal. The molten iron production obtained through the algorithm is noticeably superior to the actual production, promising a substantial enhancement in the economic returns of the steel plant. The algorithm proposed in this paper outperforms the yields obtained through the other two comparison methods.

In conclusion, the MT-DC-RVEA algorithm proposed in this paper is applicable to solving the blast furnace operation optimization model constructed herein. Under equivalent conditions, it has the potential to generate greater profits for steel plants.

4   Conclusion

This paper employs a data-driven approach to construct a blast furnace operation optimization model aimed at minimizing costs, carbon emissions, and yield. Three practical problem sets were formulated using real data from a specific steel plant. Considering a multitude of real-world led to the creation of a complex dynamic multi-objective optimization problem. To tackle this, the paper introduces the MT-DC-RVEA algorithm for resolution. The MT-DC-RVEA algorithm utilizes a multi-time-series model to predict the initial population in new environments, leveraging the RVEA algorithm as the baseline. Two contrasting algorithms were selected for experimental comparison. The experimental findings underscore the significant effectiveness of the MT-DC-RVEA algorithm in resolving the dynamic blast furnace operation optimization model constructed in this study.

Acknowledgments. This research was supported by the Fund for the National Natural Science Foundation of China (62073067, 62303102).

References

1. Agrawal, A., Agarwal, M.K., Kothari, A.K., Mallick, S.: A mathematical model to control thermal stability of blast furnace using proactive thermal indicator. Ironmaking & Steelmaking **46**(2), 133–140 (2019). https://doi.org/10.1080/03019233. 2017.1353765
2. Cheng, R., Jin, Y., Olhofer, M., Sendhoff, B.: A reference vector guided evolutionary algorithm for many-objective optimization. IEEE Trans. Evol. Comput. **20**(5), 773–791 (2016). https://doi.org/10.1109/TEVC.2016.2519378
3. Derrac, J., García, S., Molina, D., Herrera, F.: A practical tutorial on the use of nonparametric statistical tests as a methodology for comparing evolutionary and swarm intelligence algorithms. Swarm Evol. Comput. **1**(1), 3–18 (2011). https://doi.org/10.1016/j.swevo.2011.02.002

4. Li, Y., Li, H., Zhang, J., Zhang, S., Yin, Y.: Burden surface decision using mode with topsis in blast furnace ironmkaing. IEEE Access **8**, 35712–35725 (2020). https://doi.org/10.1109/ACCESS.2020.2974882

5. Mahanta, B.K., Chakraborti, N.: Tri-objective optimization of noisy dataset in blast furnace iron-making process using evolutionary algorithms. Mater. Manuf. Processes **35**(6), 677–686 (2020). https://doi.org/10.1080/10426914.2019.1643472

6. Molina, J., Santana, L.V., Hernández-Díaz, A.G., Coello, C.A.C., Caballero, R.: g-dominance: reference point based dominance for multiobjective metaheuristics. Eur. J. Oper. Res. **197**(2), 685–692 (2009). https://doi.org/10.1016/j.ejor.2008.07.015

7. Muruganantham, A., Tan, K.C., Vadakkepat, P.: Evolutionary dynamic multiobjective optimization via Kalman filter prediction. IEEE Trans. Cybern. **46**(12), 2862–2873 (2016). https://doi.org/10.1109/TCYB.2015.2490738

8. Naito, M., Takeda, K., Matsui, Y.: Ironmaking technology for the last 100 years: deployment to advanced technologies from introduction of technological know-how, and evolution to next-generation process. ISIJ Int. **55**(1), 7–35 (2015). https://doi.org/10.2355/isijinternational.55.7

9. Tian, Y., Cheng, R., Zhang, X., Cheng, F., Jin, Y.: An indicator-based multiobjective evolutionary algorithm with reference point adaptation for better versatility. IEEE Trans. Evol. Comput. **22**(4), 609–622 (2018). https://doi.org/10.1109/TEVC.2017.2749619

10. Zhou, P., Guo, D., Chai, T.: Data-driven predictive control of molten iron quality in blast furnace ironmaking using multi-output LS-SVR based inverse system identification. Neurocomputing **308**, 101–110 (2018). https://doi.org/10.1016/j.neucom.2018.04.060

11. Zhou, P., Guo, D., Wang, H., Chai, T.: Data-driven robust M-LS-SVR-based NARX modeling for estimation and control of molten iron quality indices in blast furnace ironmaking. IEEE Trans. Neural Netw. Learn. Syst. **29**(9), 4007–4021 (2017). https://doi.org/10.1109/TNNLS.2017.2749412

12. Zhou, Q., et al.: Multi-objective optimization of blast furnace dosing and operation based on NSGA-II. In: 2022 4th International Conference on Electrical Engineering and Control Technologies (CEECT), pp. 165–169. IEEE (2022). https://doi.org/10.1109/CEECT55960.2022.10030395

13. Zhou, Z.H.: Machine Learning. Springer, Heidelberg (2021)

HGADC: Hierarchical Genetic Algorithm with Density-Based Clustering for TSP

Zhenghan Song, Yunyi Li, and Wenjun Wang[✉] [iD]

Jinan University - University of Birmingham Joint Institute, Jinan University,
Guangzhou 511400, China
wjwang@jnu.edu.cn

Abstract. Genetic algorithms (GA) are widely used to solve complex combinatorial optimization problems, such as the Traveling Salesman Problem (TSP). However, as the problem being scale-up, the feasible solution space expands extremely large, resulting in a significant decrease in the efficiency of GA. This paper is dedicated to address this challenge on large TSP by proposing an improved Hierarchical Genetic Algorithm with Density-Based Clustering (HGADC). First the Hierarchical Density-Based Clustering (HDBSCAN) algorithm was employed to cluster all cities into clusters. Each cluster will be considered as a sub-problem, by this way we have a hierarchical structure, i.e. low-level intra-cluster sub-problems and high-level inter-clusters problems. For low-level, a novel Elite Ranking strategy GA was proposed. While for high-level, we designed two approaches, one is the Barycenter Method (BM), which condensed the cluster to be an individual city, the other is called Gene Segment Method (GSM), which kept both head and tail of gene segment as a representation of the cluster. Finally, the results of two levels were integrated to form the final solution. HGADC facilitated the preservation of high-quality genes in GA via the clustering method combined with GA, dramatically reduced the problem's size and computational costs. The experiments showed that the HGADC algorithm exhibits promising performance in solving large TSPs in TSPLIB compared with notable methods such as MMAS, PACO-3OPT, and KIG.

Keywords: TSP · Density-Based cluster · Hierarchical GA · Barycenter Method · Gene Segment Method

1 Introduction

The Traveling Salesman Problem (TSP) is a classical NP-hard combinatorial problem that seeks an optimal path traversing a set of cities with precisely once visit. GA has prominently risen as an approach to address TSP as well as analogous combinatorial optimization challenges. Nonetheless, the efficiency

Supported by the project "Study on Robustness and Explainability of Evolutionary Ensemble Learning", 2022ZDZX1006.

© The Author(s), under exclusive license to Springer Nature Singapore Pte Ltd. 2024
L. Pan et al. (Eds.): BIC-TA 2023, CCIS 2061, pp. 262–275, 2024.
https://doi.org/10.1007/978-981-97-2272-3_20

of conventional GA diminishes as the TSP scale grows up. The expansive solution space poses a formidable obstacle, resulting in a degradation in algorithmic performance.

Additionally, due to lack of preserving gene segments, when tackling large TSP, GA often ruin high-quality gene segments to approach optimal solutions. As in Fig. 1, this is a TSP with 8 cities, each data point is named from A to H respectively, and they belong to two local regions, A to D locate in region 1 and E to H are in region 2, which are represented by blue and orange respectively. As demonstrated in Fig. 1(a), this is an optimal solution formed by two optimal gene segments: [A, B, C, D] and [E, F, G, H] according to regions 1 and 2, respectively. Figure 1(b) shows the phenomenon that often occurs in the large-scale TSP solving by GA, i.e., the good gene segments such as [A, B, C, D] and [A, E, B, H] being broken and damaged that obviously reduces the performance of GA.

To address the above problem, this paper introduced a novel GA assisted by clustering techniques. The proposed methodology, named as Hierarchical Genetic Algorithm with Density-Based Clustering (HGADC), strategically integrates the Hierarchical Density-Based Clustering (HDBSCAN) algorithm [1] with Hierarchical Elite Ranking strategy GA (HERGA), thus drastically reduces the problem scale, thus significantly alleviating the computational costs.

When facing large-scale TSP, traditional methods did not take enough care to preserve good gene segments. As demonstrated in Fig. 1(a), good gene segments are often present in local regions, so this paper attempts to first recognize the local regions, i.e., the potential good gene segments by clustering method, in order to preserve the good gene segments. Therefore this integration has the advantage of both preserving high-quality genes and simplifying the original TSP.

This paper considers each region to be an intra-cluster sub-problem, and then a higher level of TSP among clusters, namely inter-clusters problem. HGADC operates at two levels of optimization. For the low-level, ERGA was employed within each cluster to solve individual TSP problems. The high-level concentrates on optimizing paths between these clusters. This paper introduces two distinct approaches to high-level optimization, namely the Barycenter Method (BM) and the Gene Segment Method (GSM), abstracting clusters into a point and a gene segment with a head and tail, respectively. The fusion of these two levels yields a comprehensive TSP solution.

Furthermore, this paper incorporates an elite-rank strategy, augmenting the GA's efficiency in identifying optimal solutions and promoting population diversity. Experimental results demonstrate the promising performance of the HGADC algorithm in solving TSP compared to notable methods. The contributions of this paper are listed as follows:

1. Introduced an approach to preserve high-quality gene segments through Density-Based clustering algorithm for solving TSP.
2. The BM and GSM are proposed for high-level TSP.

As a reminder, Sect. 2 provides a short overview of existing algorithms for the TSP. Moving forward, Sect. 3 delves into the novel approaches proposed in this paper, while Sect. 4 outlines the experimental setup and subsequent analyses. At last, the conclusions are summarized in Sect. 5.

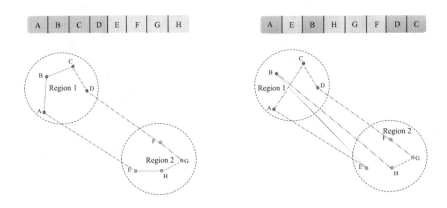

(a) Optimized Solution with good gene segments

(b) Unoptimized Solution with broken gene segments

Fig. 1. Demonstration of gene segments in local regions

2   Existed Approaches for TSP

In the realm of academic exploration, continuous efforts are being made to enhance GA for improved solutions to the TSP. Davis et al. introduced a GA operation named "Order Crossover", endeavoring to preserve the sequence information in parental chromosomes to better adapt to problems requiring consideration of sequence structure [2]. Nevertheless, the performance of "Order Crossover" may be constrained by problem-specific characteristics and may not outperform other operations in certain scenarios. H. D. Nguyen et al. designed a hybrid GA incorporating the Lin-Kernighan heuristic (LK), compensating for the GA's limited local search capabilities [9]. Y. Deng et al. introduced KIG, employing a novel initial population strategy to enhance GA. Built upon the k-means algorithm, it reorganizes travel routes by reconnecting each cluster, forming an initial population from randomly generated travel routes [3]. However, this method is heavily influenced by the initial parameter selection in K-means, leading to reduced accuracy due to excessive randomness in connecting neighboring clusters. F. Yu et al. attempted to solve TSP using an improved Roulette Wheel Selection-Based Genetic Algorithm (RWGA) [11], exhibiting performance gains only for small-scale TSP problems and struggling with larger instances.

In the broader landscape of evolutionary computation algorithms, Dorigo et al. introduced the Ant System [4]. T. Stutzle et al. introduced the Min-Max Ant System (MMAS) [10], setting maximum and minimum thresholds for pheromones to prevent the algorithm from getting stuck in local optima due to excessively high local pheromones. These Ant Systems demonstrate strong global search capabilities in small-scale TSP but face challenges in terms of slow convergence and poor accuracy for medium to large-scale instances. MahiÖ.K et al. optimized the Ant Colony Optimization (ACO) parameters using Particle Swarm Optimization (PSO) and increased algorithm diversity through 3-opt, proposing the PSO-ACO-3OPT algorithm [7]. Although enhancing diversity, the algorithm still faces challenges in convergence speed. Subsequently, Saban et al. proposed the PACO-3OPT algorithm as an improved version [6], demonstrating superior performance in various aspects over the original algorithm, achieving high precision. However, the enhancement of precision with the 3-OPT algorithm poses a significant challenge in terms of computational cost. As a note here, some of the algorithms mentioned above will be discussed in Sect. 4.6 as comparative algorithms.

3 Methodology

In the following sections, this paper denotes the ith city in the TSP by c_i, and n is the number of cities. The flowchart or HGADC is shown in Fig. 2

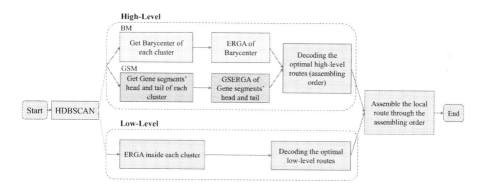

Fig. 2. Flowchart of HGADC

3.1 HDBSCAN

HDBSCAN, a versatile clustering algorithm, dynamically determines optimal density thresholds for individual data points. This adaptability enables the identification of clustering structures across diverse regions and density levels. By employing maximum connected subgraphs at different density thresholds, HDBSCAN ensures stable cluster identification with consistent labels, as detailed in

[1,8]. When faced with clusters of varying sizes, shapes, and densities, HDB-SCAN excels in accommodating this diversity. Its robust performance is particularly evident in handling large-scale datasets, showcasing resilience against noise and density variations in real-world applications. In essence, HDBSCAN cleverly combines elements of Hierarchical Clustering and DBSCAN [5], performing smart DBSCAN clustering within a range of ϵ values. This combination inherits DBSCAN's flexibility in density-based clustering while overcoming its challenge in identifying clusters with different densities. In the context of the local structure of the TSP problem, H-GADC effectively leverages HDBSCAN, proving highly adept at uncovering the intricate local structures inherent in TSP problems.

3.2   Hierarchy: High-Level (Inter-cluster) and Low-Level (Intra-cluster)

In the Hierarchy practice, this paper applies ERGA to optimize two levels and ultimately assembles the optimization results of these two levels into the final solution, thus balancing both local and global optimization.

Intra-cluster. This paper treats each cluster generated by HDBSCAN clustering as an independent TSP. Within each cluster, HERGA is employed to obtain the optimal solution for that specific cluster.

Inter-cluster. The paper considers each cluster as a whole and performs inter-cluster optimization, which is further divided into two approaches: the Barycenter Method(BM) and the Gene Segment Method(GSM). The former is relatively coarse but simple to operate, while the latter is more complex yet more precise. These two optimization methods will be discussed in detail in Sect. 3.4 and Sect. 3.5.

3.3   ERGA: Elite Ranking Strategy GA

The general GA consists of three components: chromosome representation, fitness function, and genetic operators, including selection, crossover, and mutation. ERGA designed in this paper introduces clever improvements to the selection operator. This modification enables the algorithm to exhibit a strong tendency toward searching for the optimal solution while maintaining high diversity. Experimental validation demonstrates its superiority over traditional elite selection, rank selection, and roulette wheel selection. The specific steps are outlined below.

Chromosome Representation. In the context of TSP, a solution is represented as a permutation of the cities, known as a chromosome. Let $C = (c_1, c_2, \ldots, c_n)$ represent a chromosome, where n is the number of cities, and c_i represents the ith city in the order of visitation.

Fitness Function. The fitness of a chromosome is determined by the total distance of the route it represents. For a given permutation C, the total distance $D(C)$ is calculated as the sum of the distances between consecutive cities, including the distance from the last city back to the starting city:

$$D(C) = \sum_{i=1}^{n-1} \text{Dist}(c_i, c_{i+1}) + \text{Dist}(c_n, c_1) \tag{1}$$

where $\text{Dist}(A, B)$ is the Euclidian distance between A and B.

Selection. In the selection operator, this paper adopts the Elite Ranking Selection strategy, should be emphasized which combines elite selection, rank selection, and random selection. Elite-rank strategy initially places the individual with the highest fitness directly into the new population. Subsequently, it calculates the rankings (r_i) for all individuals based on the fitness function. The selection probability (P_i) for each individual is determined using the formula:

$$P_i = \frac{e^{(r_i - 1)}}{\sum_{j=1}^{n} e^{(r_j - 1)}} \tag{2}$$

where n is the total number of individuals in the population, e is the natural exponent

After calculating the probabilities, the Elite Ranking Strategy then selects n_m individuals as parents from the entire population based on these probabilities. For the individuals not selected, n_r individuals are randomly chosen as parents, where $n_m = rate\ of\ retain\ \times n$, $n_r = rate\ of\ regenerate\ \times n$

Algorithm 1. Elite Ranking Strategy

1: Select the individual with the highest fitness and place it directly into the new population.
2: Get the rankings (r_i) for all individuals according to their fitnesses.
3: Calculate selection probabilities (P_i) by eq. (2).
4: Randomly select n_m individuals as parents based on the calculated probabilities.
5: Randomly select n_r individuals as parents from the remaining individuals.

The genetic operator, i.e. crossover and mutation used in this paper are the same as those used in RWGA [11].

3.4 Barycenter Method (BM)

BM seeks to find the optimal connecting paths between different clusters. To achieve this, we temporarily disregard the internal road structures within each cluster and abstract the entire cluster into an individual. In practice, we abstract the cluster into its barycenter, denoted as BC_i, where the formula for the barycenter is as follows:

$$\mathrm{BC}_i = \frac{1}{n_i} \sum_{j=1}^{n_i} c_{ij}, \tag{3}$$

where n_i represents the number of cities in cluster i, and c_{ij} denotes the coordinates of the j-th city in cluster i.

Subsequently, we treat each barycenter as a new input for the TSP and utilize ERGA to find its optimal solution. We believe this optimal solution represents the best order of connecting various cluster barycenters. Finally, according to this optimal order, we connect the enter-points and out-points of each internal path within the clusters to obtain the final TSP solution.

As shown in Fig. 3, the black ×'s are the barycenters of the clusters, and the inter-cluster paths of these three cluster barycenters after optimization with ERGA are marked with a green solid line, which means that the pathways within the three clusters are connected in the order of the connection of the green solid lines.

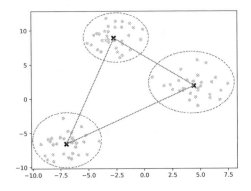

Fig. 3. Demonstration of BM

3.5   Gene Segment Method (GSM)

In the practice of the BM, we use barycenters to substitute these clusters. However, in the step of assembling all intra-cluster paths, however, each predefined head and tail in each cluster are connected without optimization, i.e. the tail of one cluster is connected to the other cluster according to the result of the high-level optimization. We are not sure this is the optimal path as shown in Fig. 4(a), and Fig. 5(a) highlight the heads and tails, which assembled less sagacious than the GSM shown in Fig. 4(b).

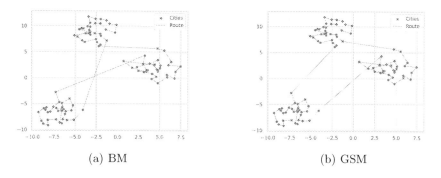

(a) BM   (b) GSM

Fig. 4. Demonstration of BM and GSM on sample data

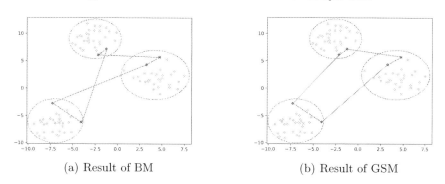

(a) Result of BM   (b) Result of GSM

Fig. 5. Comparison of GM and GSM without considering the inner route

The GSM skips the calculation of the Barycenter and intra-cluster paths and directly uses the head and tail of the intra-cluster paths as the head and tail of the genes, shown in Fig. 5(b). This optimization of the head and tail of genes can be considered as a novel TSP, referred to in this paper as the Gene Segment Traveling Salesman Problem (GSTSP).

In the GSTSP, we extend the traditional TSP to consider pairs of cities. Each city pair (c_{i1}, c_{i2}) represents a single entity, which is indivisible, but the internal order can be exchanged. For example, (c_{i1}, c_{i2}) can be transformed by exchanging the internal order to (c_{i2}, c_{i1}). However, situations like (c_{i1}, c_{j2}) are not allowed (if $i \neq j$). The ultimate optimization goal of GSTSP is to find an optimal path that visits each city pair exactly once and returns to the starting city pair. The set of city pairs is denoted as

$$CP = \{CP_1, CP_2, ..., CP_n\} \tag{4}$$

$$CP_i = Z(c_{i1}, c_{i2}) + (1 - Z)(c_{i2}, c_{i1}), \quad i \in \{1, 2, ..., n\}, \tag{5}$$

where

$$Z = \begin{cases} 1 & if \ (c_{i1}, c_{i2}) \ is \ choosen \\ 0 & if \ (c_{i2}, c_{i1}) \ is \ choosen \end{cases} \tag{6}$$

and n is the number of city pairs.

GSERGA. Based on the above definition, this paper introduces a novel GA designed for the GSTSP called Gene Segment Elite Ranking Strategy GA (GSERGA).

Initialization. In the *Population initialization* part, GSERGA accepts inputs in the form of city pairs $CP_{Z=1} = \{(c_{11}, c_{12}), (c_{21}, c_{22}), ..., (c_{n1}, c_{n2})\}$. It then generates a reversed list $CP_{Z=0} = \{(c_{12}, c_{11}), ..., (c_{n2}, c_{n1})\}$ and combines $CP_{Z=1}$ and $CP_{Z=0}$ into a list $CP_{Total} = \{(c_{11}, c_{12}), ..., (c_{n1}, c_{n2})\} \cup \{(c_{12}, c_{11}), ..., (c_{n2}, c_{n1})\}$. It keeps randomly selecting n city pairs in this new list, and if there is no duplicate city pair in this new list, then it is added to the initial population, performing*population size* a total of times.

Check for Duplicates. In the processing steps of the genetic operator, GSERGA is similar to regular ERGA, but with the addition of a duplicate removal step at each step to ensure that there are no duplicate city pairs. In the *Check for duplicates* step, this paper systematically conducts internal sorting of pairs within the set of individuals (comprising city pairs generated in the *Initialization of Population* step) to eliminate duplicates of city pairs with different internal orders. Subsequently, the sorted city pairs are transcribed into a *tuplepair* variable, which is then inserted into a list named *seenpairs*. This process is iteratively repeated, with each *tuplepair* variable being examined for duplications. If a duplication is detected, the program returns *True*, signifying the presence of duplicates. Conversely, if no duplications are identified, the program returns *False*, indicating the absence of duplicates.

4   Experiments and Analyses

4.1   Experimental Settings and Environment

Software Settings

- Python 3.10
- Google Colab with Intel Xeon CPU @2.20 GHz, 13 GB RAM

GA Parameter Settings

- *number of generations* = 1000
- *population size* = 800
- *mutation rate* = 0.05
- *retain rate* = 0.3
- *random select rate* = 0.6

4.2   Dataset Selection

Sample Data. The Sample data of the example diagrams (Fig. 3, Fig. 5) in this article come from the make_blobs function of the sci-kit learn package. The number of sample points is set to 100 and the random seed is 42, denoted by mb100.

Experimental Data. This article uses data from TSPLIB, a TSP benchmark data set widely used by researchers to evaluate the algorithm effect. In TSPLIB, seven datasets were selected for this paper, which are berlin52, st70, pr107, pr226, lin318, fl417, zi929. And mb100, mb300, mb500, mb1200 are from sci-kit learn package. Ten runs were made for each dataset. For example, a result of a pr226 dataset optimized using HGADC is shown in Fig. 6. The results of a pr226 dataset optimized using HGADC are shown in Fig. 6, with different colored data points representing different clusters generated by HDBSCAN.

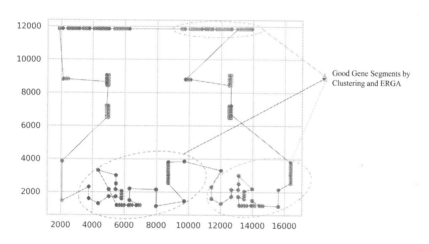

Fig. 6. Visualization of r226 route and local regions

4.3   ERGA vs Existed GA

In this section, this paper compared the performance of ERGA, ordinary GA, and Roulette Wheel Selection-Based Genetic Algorithm (RWGA) on the st70 dataset. Ten experiments were carried out in this paper, and it can be seen that the performance of ERGA is much higher than that of the compared GA. We summarise the result in Table 1.

Table 1. ERGA vs Existed GA.

Method	Time1	Time2	Time3	Time4	Time5	Time6	Time7	Time8	Time9	Time10
ERGA	**734**	**770**	**797**	**762**	**883**	**834**	**762**	**795**	**892**	**947**
GA	2890	2977	2786	2563	2477	2598	2809	2791	2544	2903
RWGA	1479	1426	1698	1541	1655	1829	1502	1493	1479	1732

4.4   HGADC vs ERGA

In this section, this paper compares the performance of ERGA and HGADC. Ten experiments are conducted in this paper, and Table 2 lists the optimal results, average results, worst results, and time of HGADC and ERGA, respectively, which show that the results of HGADC are much higher than those of ERGA, especially in large-sample TSP instances. The advantage of HGADC is undeniable in the large sample of TSP instances. It is worth mentioning that the running time of HGADC is greatly reduced compared to ERGA. For ERGA, experiments with larger samples are no longer performed because they consume too many computational resources. In summary, this paper argues that performing HDBSCAN clustering and performing hierarchical GA will help to retain good gene segments and ultimately give better results and that this operation actually splits the large problem into a series of smaller problems, which is a huge boost to algorithm efficiency.

Table 2. HGADC vs ERGA.

TSP	HGADC				ERGA			
	Best	Ave.	Worst	time/s	Best	Ave.	Worst	time/s
st70	**726**	807	978	56	734	818	947	57
berlin52	**8036**	8589	9093	45	8077	8672	9067	48
mb100	**112**	123	136	42	313	336	348	145
pr107	**49853**	53204	56137	48	112367	126284	141297	177
pr226	**90389**	110582	124536	71	151321	162187	176588	428
mb300	**241**	259	283	89	1503	1859	2032	697

4.5   BM vs GSM

In this section, the comparison of BM and GSM is displayed. Of the two high-level methods proposed in this paper, GSM can achieve higher accuracy due to the more delicate design in the head and tail of gene segments, but BM also has the probability to achieve the same results as GSM because they share a low-level solver, as shown in Table 3.

Table 3. Comparison of GSM and BM.

		st70	mb100	pr107	pr226	mb300	lin318	fl417	mb500	zi929	mb1200
GSM	Best	**726**	**112**	**49853**	**90389**	**241**	**58799**	**14327**	**347**	**178031**	**1431**
	Ave.	807	123	53204	110582	259	64387	15290	368	190282	1689
BM	Best	755	112	52772	96921	241	62842	14879	359	181129	1524
	Ave.	832	137	56096	111209	270	65923	16348	381	193202	1732

4.6   Comprehensive Comparison

In this part, this paper compares HGADC (GSM) with MMAS [10], PACO-3OPT [6], KIG [3] methods. The comparison data are shown in Table 4. From the data, it can be seen that HGADC has a big advantage, its performance is much higher than KIG, slightly higher than MMAS, and higher than PACO-3OPT on a large sample dataset. However, the shortcoming of HGADC is that it is difficult to converge to the optimal accuracy in small samples, for example, in the small-sample instances selected in this paper, the results of HGADC will be a little worse than those of PACO-3OPT, which is considered to be one of the minor disadvantages of prior clustering in this paper, and even if this paper tries its best to overcome the over-emphasis on the local problem brought by the clustering, it is still difficult to be completely solved at a lower complexity solution. However, as this paper has been emphasizing, clustering can help to retain good local features for algorithms that do not converge to high accuracy. And even though clustering may not be optimal, HGADC is guaranteed to be very close to the optimal solution, even in very large samples, which is difficult for other algorithms (e.g., PACO-3OPT) to achieve, and it is hard to run the zi929 and mb1200 samples with PACO-3OPT successfully due to the large amount of consumed time. In summary, HGADC has a good performance in small-sample TSP instances and has an absolute advantage in large-sample TSP instances, which can be closer to the optimal solution while alleviating computational costs.

Table 4. Comparison of HGADC and other advanced algorithms in TSP.

TSP	HGADC(GSM)		MMAS		PACO-3OPT		KIG
	Best	Ave.	Best	Ave.	Best	Ave.	Best
berlin52	**8036**	8589	8029	8211	7891	8038	13098
st70	**726**	807	730	746	689	725	1331
mb100	**112**	123	189	201	151	178	157
pr107	**49853**	53204	53190	55434	52265	53082	58923
pr226	**90389**	110582	98375	120762	99821	110923	123096
mb300	**241**	259	302	331	276	305	298
lin318	**58799**	64387	62176	63211	60328	61762	119847
fl417	**14327**	15290	15398	16277	15099	16082	29231
mb500	**347**	368	401	433	399	421	407
zi929	**178031**	190282	183214	209399	–	–	212652
mb1200	**1431**	1689	1822	2124	–	–	2987

5   Conclusion

In summary, this paper proposes the Hierarchical Genetic Algorithm with Density-Based Clustering (HGADC) to enhance the efficiency of Genetic Algorithms (GA) in handling the Traveling Salesman Problem (TSP). The HGADC simplifies the problem complexity through hierarchy optimization and finding a path to preserve the high-quality genes through searching local regions via clustering, thus alleviating computational costs and improving the quality of results. In the hierarchy optimization, two levels were designed, an Elite Ranking strategy GA is implemented at the low-level, while the Barycenter Method (BM) and Gene Segment Method (GSM) are proposed at the high-level. Hierarchical design makes this algorithm take into account both local and global. In addition, the elite-rank strategy improves the efficiency of GA in finding the optimal solution and maintaining population diversity. The conclusions are summarized as follows:

1. HGADC and ERGA are compared in the full dataset, HERGA is far superior to ERGA, justifying the adoption of clustering to retain superior genes in this paper.
2. We compare HGADC with MMAS, PACO-3OPT, and KIG, HGADC outperforms the other three algorithms on large-scale problems, illustrating the advantages of the layered algorithm proposed in this paper to simplify the problem to solve large-scale TSPs.

Moreover, it must be noted that the heads and tails of each of our intra-clusters are predefined. In future research, we will continue to refine high-level optimization which should be optimized further to achieve more optimal solutions.

References

1. Campello, R.J.G.B., Moulavi, D., Sander, J.: Density-based clustering based on hierarchical density estimates. In: Pei, J., Tseng, V.S., Cao, L., Motoda, H., Xu, G. (eds.) PAKDD 2013. LNCS, vol. 7819, pp. 160–172. Springer, Heidelberg (2013). https://doi.org/10.1007/978-3-642-37456-2_14

2. Davis, L., et al.: Applying adaptive algorithms to epistatic domains. In: IJCAI, vol. 85, pp. 162–164. Citeseer (1985)

3. Deng, Y., Liu, Y., Zhou, D., et al.: An improved genetic algorithm with initial population strategy for symmetric tsp. Math. Probl. Eng. **2015** (2015)

4. Dorigo, M.: Optimization, learning and natural algorithms. Ph.D. thesis, Politecnico di Milano (1992)

5. Ester, M., Kriegel, H.P., Sander, J., Xu, X., et al.: A density-based algorithm for discovering clusters in large spatial databases with noise. In: KDD, vol. 96, pp. 226–231 (1996)

6. Gülcü, Ş, Mahi, M., Baykan, Ö.K., Kodaz, H.: A parallel cooperative hybrid method based on ant colony optimization and 3-opt algorithm for solving traveling salesman problem. Soft. Comput. **22**, 1669–1685 (2018)

7. Mahi, M., Baykan, Ö.K., Kodaz, H.: A new hybrid method based on particle swarm optimization, ant colony optimization and 3-opt algorithms for traveling salesman problem. Appl. Soft Comput. **30**, 484–490 (2015)

8. McInnes, L., Healy, J.: Accelerated hierarchical density based clustering. In: 2017 IEEE International Conference on Data Mining Workshops (ICDMW), pp. 33–42. IEEE (2017)

9. Nguyen, H.D., Yoshihara, I., Yamamori, K., Yasunaga, M.: Implementation of an effective hybrid GA for large-scale traveling salesman problems. IEEE Trans. Syst. Man Cybern. Part B (Cybern.) **37**(1), 92–99 (2007)

10. Stützle, T., Hoos, H.H.: Max-min ant system. Futur. Gener. Comput. Syst. **16**(8), 889–914 (2000)

11. Yu, F., Fu, X., Li, H., Dong, G.: Improved roulette wheel selection-based genetic algorithm for TSP. In: 2016 International Conference on Network and Information Systems for Computers (ICNISC), pp. 151–154. IEEE (2016)

Transformer Surrogate Genetic Programming for Dynamic Container Port Truck Dispatching

Xinan Chen[1]([envelope])[iD], Jing Dong[2,3]([envelope])[iD], Rong Qu[3][iD], and Ruibin Bai[3][iD]

[1] School of Computer Science, University of Nottingham Ningbo China,
Ningbo 315199, China
`xinan.chen@nottingham.edu.cn`
[2] Department of Engineering, University of Cambridge, Cambridge CB2 1TN, UK
`jd704@cam.ac.uk`
[3] School of Computer Science, University of Nottingham, Nottingham NG7 2RD, UK
`{rong.qu,ruibin.bai}@nottingham.edu.cn`

Abstract. In the wake of burgeoning demands on port logistics, optimizing the operational efficiency of container ports has become a compelling necessity. A critical facet of this efficiency lies in practical truck dispatching systems. Although effective, traditional Genetic Programming (GP) techniques suffer from computational inefficiencies, particularly during the fitness evaluation stage. This inefficiency arises from the need to simulate each new individual in the population, a process that neither fully leverages the computational resources nor utilizes the acquired knowledge about the evolving GP structures and their corresponding fitness values. This paper introduces a novel Transformer-Surrogate Genetic Programming (TSGP) approach to address these limitations. The methodology harnesses the accumulated knowledge during fitness calculations to train a transformer model as a surrogate evaluator. This surrogate model obviates the need for individual simulations, thereby substantially reducing the algorithmic training time. Furthermore, the trained transformer model can be repurposed to generate superior initial populations for GPs, leading to enhanced performance. Our approach synergizes the computational advantages of transformer models with the search capabilities of GPs, presenting a significant advance in the quest for optimized truck dispatching in dynamic container port settings. This work improves the efficiency of Genetic Programming and opens new avenues for leveraging GP in scenarios with substantial computational constraints.

Keywords: Evolutionary Algorithm · Truck Dispatching · Dynamic Optimization · Deep Neural Network · Machine Learning

1 Introduction

In the context of the modern global economy, container ports serve as critical nodes in the logistical chain, linking various modes of transportation and

© The Author(s), under exclusive license to Springer Nature Singapore Pte Ltd. 2024
L. Pan et al. (Eds.): BIC-TA 2023, CCIS 2061, pp. 276–290, 2024.
https://doi.org/10.1007/978-981-97-2272-3_21

facilitating the smooth transfer of goods. As international trade volumes continue to rise, the throughput capacity of container ports faces an unprecedented strain, necessitating the development of innovative optimization techniques to bolster operational efficiency [13].

One crucial area of operational efficiency pertains to dispatching trucks within the port. Efficient truck dispatching mechanisms can significantly reduce lead times, minimize fuel consumption, and mitigate the risk of congestion, thereby elevating the overall performance of the port [6]. Many optimization techniques, ranging from linear programming to metaheuristic algorithms, have been proposed to address this complex task.

Genetic Programming (GP), an evolutionary algorithm, has shown promising results in tackling the truck dispatching problem. However, a prominent bottleneck in the practical applicability of GP is the computational overhead incurred during the fitness evaluation stage. Traditional GP algorithms require each newly generated individual in the population to undergo a simulation to calculate its fitness. Not only does this process incur a high computational cost, but it also fails to exploit valuable knowledge acquired during previous algorithm iterations.

This paper aims to mitigate the limitations above by unveiling a novel Transformer Surrogate Genetic Programming (TSGP) framework. The cornerstone of this approach lies in integrating a transformer model trained to act as a surrogate evaluator for the fitness function. This innovative methodology significantly ameliorates the computational time complexity associated with the training phase by obviating the need for resource-intensive simulations. Consequently, our proposed TSGP framework demonstrates faster convergence and superior performance outcomes within the same computational time frame. Detailed experimental comparisons substantiating these claims are presented in Sect. 5.

In addition, the transformer model can be concurrently trained with GP algorithms, utilizing the data generated during the fitness evaluation of GP individuals. This concurrent training process eliminates additional time investment for training the transformer. Notably, the transformer leverages insights from GP individuals' structural attributes and corresponding performance metrics during their fitness evaluation. This enhances the algorithm's efficiency and contributes to a more effective fitness evaluation process. Furthermore, the trained transformer model can be strategically employed to generate superior initial populations for subsequent GP algorithms, thereby augmenting their overall performance.

The primary contributions of this paper can be summarized as follows:

- Introduction of a novel TSGP framework designed to optimize truck dispatching in dynamic container port settings.
- Deployment of a transformer model trained as a surrogate evaluator for the fitness function, significantly reducing the computational overhead associated with traditional Genetic Programming methods.
- Empirical validation demonstrating that the TSGP methodology converges faster and yields superior results within the same computational time frame.

– Illustration of the reusability of the trained transformer model for seeding improved initial populations in Genetic Programming, leading to enhanced algorithmic performance.

The remainder of this paper is organized as follows: Sect. 2 introduces the background of the container port truck dispatching problem and reviews the related work in port logistics optimization. Section 3 details the problem and the proposed methodology. Section 4 provides an empirical evaluation of the proposed technique, including comparisons with traditional GP and other state-of-the-art methods. Finally, Sect. 5 concludes the paper and outlines directions for future research.

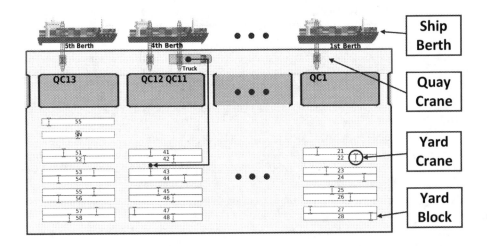

Fig. 1. Typical Layout of a Container Port

2   Background and Lecture Review

Container ports serve as essential nodes within the intricate web of international transportation networks, functioning within a globally interconnected infrastructure to facilitate the smooth and efficient transference of goods [2]. A prototypical container port shown in Fig. 1 is characterized by several critical operational elements, including ship berths, quay cranes (QCs), yard cranes (YCs), yard blocks, and a fleet of internal trucks. These disparate yet interconnected components collaborate synergistically to manage various logistical tasks, from loading and unloading containers to their subsequent intra-port transfer [24].

Truck dispatching - allocating trucks for internal container transport - plays a pivotal role in the varied operations. Efficient truck dispatching is essential for a myriad of reasons. Firstly, trucks play a vital role in the last-mile delivery of goods, linking the port to inland distribution points. Therefore, any inefficiencies in truck dispatching can have a cascading effect on the broader logistics network, incurring considerable delays and increased operational costs [3]. Secondly, truck dispatching in container ports is inherently a complex problem due to various constraints, including but not limited to time windows, truck capacities, and multiple loading and unloading locations within the port [10]. Traditional methods like heuristics and rule-based systems have been employed to address this problem; however, they often fall short in optimizing dispatching routes and schedules, particularly in dynamically changing conditions [5].

The utilization of GP for truck dispatching has emerged as a significant innovation, particularly in light of its ability to tackle intricate optimization tasks [8]. A variety of strategies have been developed to enhance GP performance, including specialized approaches such as double-layer GP [7], neural-network-assisted GP [9], cooperative coevolutionary GP [14], and reinforcement learning assisted GP [20]. These methodologies have collectively demonstrated commendable performance on comparable optimization problems. Despite these advancements, a salient limitation across the spectrum of GP-based methods is their computational demand, primarily owing to the necessity for comprehensive simulations to evaluate the fitness of individuals within the GP population [11].

Consequently, the integration of GP and surrogate models have been explored in previous research [15,21,22]. These studies capitalize on the computational efficiency of surrogate models to approximate fitness functions, thereby streamlining the optimization process [1]. Although these surrogate methods commonly employ simplistic mathematical models that suffice for relatively straightforward problems, they falter when applied to complex tasks such as container port truck dispatching. Empirical evidence from our experiments indicates that surrogate approaches based on simplistic mathematical models are ineffective in accurately approximating the complex dynamics of the port vehicle assignment problem. This inadequacy tends to obfuscate the evolutionary process in GP, complicating its convergence and subsequently diminishing the algorithm's overall performance.

Based on empirical findings, we contend that a more nuanced modeling technique is warranted for constructing a surrogate model capable of adequately simulating the complexities inherent in dry-port dispatching scenarios. The transformative impact of the transformer model [16] across various domains—including but not limited to natural language processing [18], image processing [4], and data prediction [17]—underscores its robust data-fitting and handling capabilities. Consequently, we propose the transformer model as the most suitable candidate for developing a surrogate model tailored to the specific challenges of container port truck dispatching.

Therefore, the present paper focuses on the innovative utilization of the transformer model to address the complex task of truck dispatching in container ports. Specifically, we introduce a surrogate model as an acceleration mechanism for the evolutionary process in Genetic Programming, culminating in a TSGP framework. This paper endeavors to fill existing gaps in the literature by integrating the distinct advantages of both GP and transformer models to achieve enhanced operational efficiency in the dynamic truck dispatching context at container ports.

3   Methodology

This section begins by detailing the complex dynamics and objectives inherent in the problem of truck dispatching within container ports. Building upon existing methodologies, which primarily employ manual heuristics and GP for formulating dynamic dispatching algorithms, we offer a nuanced discussion of these conventional approaches. To advance beyond the limitations of current frameworks, we introduce an innovative transformer-surrogate model specifically engineered to simulate truck dispatching processes within container port environments. This model is then integrated with GP, resulting in the formulation of the TSGP approach. This groundbreaking methodology aims to synergistically leverage the strengths of both GP and transformer models to tackle the multifaceted challenges embedded in this complex optimization problem.

3.1   Problem Description and Objective Function

This study focuses on optimizing dynamic truck dispatching in container ports to minimize ship docking time. Specifically, we address how allocating trucks to QCs and YCs can streamline container loading and unloading. Container sizes are standardized to twenty-foot equivalent units (TEUs), with trucks carrying a maximum of 2 TEUs. Dispatch tasks are bifurcated into single large container and dual small container assignments, with the latter requiring double the loading and unloading time.

QCs can handle either two smaller or one larger container per cycle, while YCs can only manage one. The sequential processing of containers is mandated by load-balancing needs and specific stacking location constraints. Typically, truck congestion and delays arise from front-positioned trucks dominating crane activities or lagging preceding tasks.

Given the port's strict routing regulations and the need for precise calculations related to queuing and waiting times, the internal truck dispatching challenge can be framed as a multi-constrained vehicle routing problem. It involves intricate computations within a directed graph model, primarily centered around minimizing idle times and expediting container throughput.

Let n designate the total number of dispatch tasks, $size_i$ signify the size of task i, E represent the ensemble of task end times, and S indicate the set of task start times. The objective function governing this dynamic truck dispatching problem can be formally expressed as:

$$\max \left(\frac{\sum_{i=1}^{n} size_i}{\max E - \min S} \right) \tag{1}$$

The objective function aims to maximize the task completion rate, measured in TEUs per hour, while minimizing total task duration. Due to its NP-hard complexity, as detailed in our prior work [7], straightforward mathematical modeling is unsuitable. To tackle this, we employ a dynamic dispatching framework that efficiently prioritizes tasks for idle trucks, thereby enhancing overall operational efficiency for both QCs and YCs.

3.2 Manual Heuristics

In most ports, dynamic truck scheduling predominantly relies on a hybrid system that integrates human oversight with carefully crafted manual heuristics shown in Algorithm 1, as discussed in a previous study [5]. This method, serving as the baseline in this paper, has shown effectiveness in contexts like Ningbo Port. However, it is limited by its simplistic architecture and narrow parameter set, often requiring domain-specific expertise for adjustments to evolving operational landscapes. We propose the GP-based dynamic dispatching method to address these limitations and serve as a point of comparison.

Algorithm 1. Manual Heuristic Truck Dispatching Algorithm

Require: Parameters *parameter*, Travel Time t
  function $heuristic(QC, truck)$
    if $crane\_truck\_num < desired\_trucks$ **then**
      $score \leftarrow travel\_time * (truck\_num - prority)$
    else
      $score \leftarrow travel\_time * desired\_trucks$
    end if
    if $truck\_num \geq truck\_limit$ **then**
      $score \leftarrow score + 200000$
    end if
    return $score$
  end function

3.3 Genetic Programming

GP is a specialized machine learning technique inspired by biological evolution employed in this study to optimize computer programs for specific tasks. It uses a predefined fitness function to evaluate a program's efficacy, following the principle of iterative improvement by continuously updating a population of solution candidates [12]. We have chosen the tree structure representation within the GP framework for this study. This choice is based on its intuitive nature and proven efficacy in port dispatch simulations, where the fitness of each solution is assessed through a specialized objective function as defined in Eq. (1). Evolutionary mechanisms like mutation and replication refine these solutions, eventually converging on the one with the highest fitness score as the optimal output.

Building on our previous work [7], this study adopts the categorized Logical Genetic Programming (LGP) shown in Fig. 2 as the GP framework. LGP employs an extended set of operators, including arithmetic, relational, and logical functions, to offer enhanced performance and flexibility. These configurations are consistent with our previous research, where LGP demonstrated superior results.

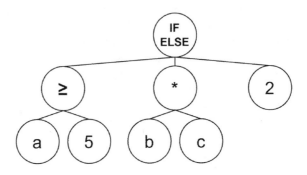

Fig. 2. LGP Tree Structure

In the specific domain of dynamic truck dispatching, GP is used to formulate robust and efficient heuristics. Like manual heuristics, environmental parameters are incorporated into the GP model to rank and optimize tasks. However, conventional GP often suffers from limitations, such as insufficient evolutionary generations within a fixed time frame, leading to poor performance and lack of convergence. To address this, we introduce a transformer surrogate model as an alternative to traditional simulation-based fitness evaluation, aiming for more efficient and effective algorithmic training.

3.4   Transformer Surrogate Model

The limitations inherent in conventional GP implementations, specifically regarding the restricted number of evolutionary generations in unit time and frequent converging failures, necessitate a more efficient approach to evaluating individual fitness. To address this challenge, we introduce a transformer surrogate model as a viable alternative to the traditional simulation-based fitness evaluator.

A transformer model, initially proposed for natural language processing tasks [16], has demonstrated its versatility and efficacy across various domains. In this study, the transformer model is a surrogate for the conventional simulation, providing rapid and accurate evaluations of GP individual candidates. As shown in Fig. 3, in the process of GP evolving, all newly generated individuals have to go through the port truck dispatching simulator to compute their fitness. Even though this part of the work can be calculated in parallel using multi-threaded computation, it still consumes many computational resources. Therefore, we propose to use a transformer surrogate model instead of the port truck dispatching

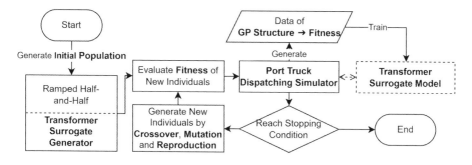

Fig. 3. GP Evolving Process

simulator to accelerate the GP training. The transformer model in the paper follows the default model of the Transformers [19] except for changing the expected and actual output to one double float. According to our experimental results, using the transformer surrogate model can reduce the fitness computation time by more than 65% and is not affected by the increase in the number of tasks.

For the training of the transformer surrogate model, we propose to use the fitness data corresponding to the GP individual structure generated during the GP evolution process to train the transformer directly so that it does not need to consume time to produce data to prepare the transformer. The training process of the transformer can be carried out simultaneously with the GP evolution process, so in this paper, training the transformer surrogate model does not increase the training time of the algorithm. The transformer training process can be carried out simultaneously with the GP evolution process, so in this paper, training the transformer surrogate model does not increase the training time of the algorithm.

In addition, for the trained transformer surrogate model to perform well on datasets with different operating environments and to eliminate the interference of operating environments, the transformer surrogate model will not directly output the same TEU/h data as the simulator but will instead generate a number between 0 and 1, which represents the corresponding size of the GP individual's fitness. Once trained, the model estimates the fitness of new GP individuals with a substantially reduced computational cost.

3.5   Transformer-Surrogate Genetic Programming

After constructing the transformer surrogate model, we integrate it with GP to formulate the TSGP algorithm. As delineated in Algorithm 2, the evolutionary process of TSGP is carried out in multiple stages. Initially, up to generation x, fitness values for GP individuals are computed using a traditional simulator, while simultaneously training the transformer surrogate model with the newly generated data. Subsequently, for y generations, the fitness computation shifts to the transformer surrogate model. After this, the algorithm reverts to the simulator for z generations, updating the surrogate model in tandem. This loop continues unless the termination criteria are met.

Algorithm 2. Transformer-Surrogate Genetic Programming Evolving

---

1: **Initialize:** GP Population, Transformer Surrogate Model
2: $t \leftarrow 0$ ▷ Initialize training generation counter
3: **while** Training time or number of generations not reached **do**
4:     **if** $t < x$ **then**   ▷ First x generations
5:         Compute Fitness of GP individuals using Simulator
6:         Update Transformer Surrogate Model with Individual - Fitness Data
7:     **else if** $t \mod (x + y + z) < x + y$ **then**   ▷ Next y generations
8:         Compute Fitness of GP individuals using Transformer Surrogate Model
9:     **else**   ▷ Next z generations
10:        Compute Fitness of GP individuals using Simulator
11:        Update Transformer Surrogate Model with Individual - Fitness Data
12:     **end if**
13:     Perform GP operations (Selection, Crossover, Mutation)
14:     $t \leftarrow t + 1$ ▷ Increment generation counter
15: **end while**

The hyperparameters x, y, and z can be tuned according to the problem's inherent complexity and optimization objectives. This study empirically sets these parameters to 100, 100, and 10, respectively.

Building upon TSGP, we extend the algorithm to TSGP*, an enhanced version that leverages the trained transformer surrogate model for generating a more effective initial population. Unlike the traditional ramped half-and-half initialization method in standard GP, TSGP* employs a transformer surrogate generator. As elaborated in Algorithm 3, this generator also uses the ramped half-and-half technique to produce initial GP individuals. However, following their generation, the trained transformer surrogate model is invoked to estimate their fitness values. Only those individuals surpassing a fitness threshold f are retained. The value of f is a tunable hyperparameter set to 0.5 in this study.

In the subsequent section, we shall empirically assess the efficacy of both TSGP and TSGP* algorithms by applying them to a real-world case study involving dynamic truck dispatching in container ports. This performance evaluation will juxtapose multiple state-of-the-art methodologies, facilitating a comprehensive comparative analysis.

4   Experiment and Discussion

This section is devoted to a comprehensive evaluation of a diverse array of algorithms, specifically Data-Driven Genetic Programming (GP) [8], Deep Reinforcement Learning Hyper-Heuristic (DRL-HH) [23], Neural Network Assisted Genetic Programming (NN-GP) [9], TSGP, and TSGP*. The primary focus rests

---

Algorithm 3. Transformer Surrogate Generator

1: **Initialize:** Empty GP Population, Transformer Surrogate Model
2: $m_{count} \leftarrow 0$ ▷ Initialize individual counter
3: $f \leftarrow$ Fitness threshold
4: **while** $m_{count} < m$ **do**   ▷ Until m individuals are generated
5:     individual $\leftarrow$ Generate GP tree using ramped half-and-half
6:     fitness $\leftarrow$ Evaluate fitness using Transformer Surrogate Model
7:     **if** fitness $> f$ **then**
8:         Add individual to GP Population
9:         $m_{count} \leftarrow m_{count} + 1$
10:    **end if**
11: **end while**

---

on elucidating the performance merits of TSGP and TSGP*, emphasizing the collaborative efficacy of amalgamating GP and transformer surrogate models. Given its robust and empirically validated performance, the manual heuristic [5] is the chosen benchmark for this comparative study.

The feature set used in this study, often termed GP terminals, closely follows the specifications outlined in our prior research [7]. It encompasses 14 distinctive features, such as truck travel time, the current quantity of QC trucks, the number of trucks in waiting, and the remaining task count, among other salient variables, to accurately portray the operational state of the container port at any given moment. The GP algorithm is configured with a population size of 1024 and employs crossover, mutation, and reproduction rates of 60%, 30%, and 10%, respectively. All the algorithms in this study were subjected to a *10-h training regimen.* . The datasets leveraged in this experiment are consistent with those used in our previous work [7], originating from historical operations at Ningbo-Zhoushan Port. The experimental setup mimics a real-world single ship berth featuring six QCs, with the number of trucks being variable and reflecting the actual operational conditions, ranging between 24 and 48. Uncertain variables like truck travel time and container loading and unloading times are modeled based on empirical distributions derived from genuine operational data.

This study categorized historical task records into ten sets, each representing a distinct operational scenario encountered at different temporal intervals. Five sets were used for training, while the remainder were for testing. Each training or testing set consists of ten instances from a similar time frame, incorporating 200 tasks that include a blend of loading and unloading operations. Each training instance was executed 100 times, employing different random seeds for each run. A comprehensive summary of the average training and testing results is provided in Table 1.

In light of varying benchmark performances observed in real-world data, we introduce the Improvement over the Manual Heuristic (Imp.) as a reference baseline. All subsequent performance metrics are framed as improvements over this established baseline. A t-test conducted on the experimental outcomes ($\alpha = 0.001, p = 0.00$) substantiates that the fully realized TSGP and TSGP* variant

Table 1. GP, DRL-HH, NN-GP, TSGP and TSGP* Experiment Results in 10 Training Hours (TEU/h)

Set	No.	Manual	GP	DRL-HH	NN-GP	TSGP	TSGP*
Train	1	106.87	118.76	117.17	117.84	122.05	130.84
	2	126.85	133.84	139.85	145.21	147.99	149.01
	3	114.27	120.76	127.00	126.14	132.30	134.36
	4	106.31	113.78	112.91	117.14	123.55	122.31
	5	116.88	129.17	125.75	132.60	133.90	136.91
	Avg.	114.236	123.26	124.53	127.79	131.96	134.68
	Imp.	**0.00%**	**7.90%**	**9.02%**	**11.86%**	**15.51%**	**17.90%**
Test	1	105.87	113.05	115.01	111.19	118.70	119.46
	2	118.91	125.93	126.59	134.01	135.50	133.47
	3	115.72	122.58	124.48	126.06	133.81	133.78
	4	121.48	130.10	132.59	134.92	135.93	139.37
	5	117.25	120.63	125.28	127.15	135.72	132.23
	Avg.	115.846	122.46	124.79	126.67	131.93	131.66
	Imp.	**0.00%**	**5.71%**	**7.73%**	**9.34%**	**13.89%**	**13.65%**

exhibits notable performance enhancements over GP, DRL-HH, and NN-GP counterparts—registering an improvement of approximately 7%, 5%, and 3% on the training and test sets, respectively. Notably, the TSGP model demonstrates negligible performance degradation when transitioning from training to test sets. This underscores TSGP's robustness and effective generalization capabilities, rendering it highly applicable to previously unseen datasets.

To probe further into TSGP's potential, we also evaluated TSGP*, an alternative algorithmic formulation incorporating a transformer-based surrogate generator. Empirical evidence attests to substantial performance gains for TSGP* on both training and test sets. This highlights the efficacy of employing a transformer surrogate model in the initial population generation phase, enhancing the overall quality of the GP-based solutions. However, TSGP* does reveal a drawback—increased performance degradation in the test phase, possibly attributable to the lack of diversity in the initial population. Our future work will address this shortcoming to further optimize TSGP*'s efficacy across training and test environments.

In this comprehensive study, we have undertaken a dual-faceted evaluation approach to analyze the efficacy of various algorithms. The first facet involved a time-based comparison where each algorithm was subjected to a fixed training period of 10 h. The second, more granular facet, involved an extended analysis over 300 evolutionary generations for each algorithm, as detailed in Table 2. This two-pronged approach was designed to provide a more holistic understanding of each algorithm's capabilities and limitations.

Our findings, as elucidated by the experimental data, indicate a nuanced spectrum of performance across the algorithms tested. TSGP, when assessed over a congruent number of evolutionary generations, exhibited performance that did not meet the optimal benchmarks. This was largely due to the variation between the real-world operational conditions and those within the simulated environment—a factor that often plagues algorithmic simulations.

However, an intriguing pattern was observed with TSGP*, which integrates the Transformer architecture for initial population generation. This integration appears to confer an advantage, allowing TSGP* to withstand the discrepancies between simulation and reality, thus not significantly affecting its performance.

Moreover, an examination of the time metrics reveals a compelling insight. While heuristic and GP-based methods demonstrate only marginal variations in time consumption across the test set, the training efficiency of both TSGP and TSGP* emerges as distinctly superior. This efficiency is not merely a matter of reduced time expenditure; it is indicative of the depth and adaptability of the algorithms themselves.

The results highlight the proposed surrogate method's profound impact, particularly its ability to accelerate the training process significantly. It is a compelling argument for the surrogate method's integration into the evolutionary computation framework. The method streamlines the computational workflow and ensures that the algorithms remain robust in the face of environmental variability. This is particularly relevant in real-world applications where such variability can often be unpredictable and challenging to simulate.

Table 2. GP, DRL-HH, NN-GP, TSGP and TSGP* Experiment Results in 300 Training Generations (TEU/h)

Set	No.	Manual	GP	DRL-HH	NN-GP	TSGP	TSGP*
Train	1	106.87	115.2474	118.2288	128.6010	120.9443	128.0169
	2	126.85	132.9704	132.4145	135.4286	132.0230	138.4152
	3	114.27	121.0681	132.0767	127.8843	121.7983	133.1774
	4	106.31	118.7571	119.2456	125.5672	117.5625	125.2360
	5	116.88	127.9977	122.5157	132.5091	128.0955	129.9707
	Avg.	114.236	123.2081	124.8962	129.9980	124.0847	130.9632
	Imp.	0.00%	7.85%	9.33%	13.80%	8.62%	14.64%
	Time (hour)	0.00	7.71	22.21	13.58	5.63	5.85
Test	1	105.87	115.8348	118.4098	118.7419	112.3532	124.4371
	2	118.91	127.3668	128.1347	132.6995	126.7159	130.8406
	3	115.72	116.4109	122.1811	125.3569	120.2386	127.3589
	4	121.48	127.4641	125.8459	137.2236	127.5141	132.1725
	5	117.25	125.4904	125.8270	131.5146	129.7641	128.6010
	Avg.	115.846	122.5134	124.0797	129.1073	123.3172	128.6820
	Imp.	0.00%	5.76%	7.11%	11.45%	6.45%	11.08%
	Time(min)	0.31	0.35	0.49	0.51	0.42	0.46

In conclusion, the TSGP framework adeptly amalgamates the strengths of both the transformer and GP paradigms, showcasing superior performance in varied training and testing conditions. The model addresses the high simulation cost inherent in dynamic truck dispatching across multi-scenario ports, even amidst uncertainties. Given its successful implementation, TSGP holds substantial promise for broader applications in dynamic transportation optimization problems, opening up a wide avenue for future research endeavors.

5   Conclusion

This study introduces the TSGP approach, a groundbreaking methodology that addresses the computational inefficiencies in traditional GP applied to truck dispatching in container ports. By employing a transformer model as a surrogate evaluator during the fitness calculation stage, TSGP not only drastically reduces the algorithmic training time but also enhances the performance of GP. The transformer model further serves a dual role by generating optimized initial populations for GPs, demonstrating a synergistic integration of the computational strengths of transformer models with the heuristic search capabilities of GPs.

Our empirical results confirm the efficacy of TSGP, especially its ability to significantly accelerate the training process while maintaining or improving operational efficiency. The transformer model effectively learns to approximate the fitness landscape of the GP, thereby streamlining the evolutionary process. However, while TSGP and TSGP* showed robust performance in both training and testing scenarios, some challenges related to initial population diversity in our extended model, TSGP*, indicate directions for future research.

In summary, the TSGP approach marks a substantial advance in container port optimization, particularly in truck dispatching systems. By marrying the capabilities of transformer models and GPs, we open new avenues for applying machine learning techniques in operational logistics and beyond. The demonstrated success of TSGP and TSGP* sets the stage for their broader application across various domains where computational efficiency and effective heuristic search are critical.

References

1. Alizadeh, R., Allen, J.K., Mistree, F.: Managing computational complexity using surrogate models: a critical review. Res. Eng. Design **31**, 275–298 (2020)
2. Bank, W.: The container port performance index 2020: a comparable assessment of container port performance. World Bank (2021)
3. Bartolacci, M.R., LeBlanc, L.J., Kayikci, Y., Grossman, T.A.: Optimization modeling for logistics: options and implementations. J. Bus. Logist. **33**(2), 118–127 (2012)
4. Chen, H., et al.: Pre-trained image processing transformer. In: Proceedings of the IEEE/CVF Conference on Computer Vision and Pattern Recognition, pp. 12299–12310 (2021)

5. Chen, J., Bai, R., Dong, H., Qu, R., Kendall, G.: A dynamic truck dispatching problem in marine container terminal. In: 2016 IEEE Symposium Series on Computational Intelligence (SSCI), pp. 1–8. IEEE (2016)
6. Chen, X., Bai, R., Dong, H.: A multi-layer GP hyper-heuristic for real-time truck dispatching at a marine container terminal. In: MISTA 2019 (2019)
7. Chen, X., Bai, R., Qu, R., Dong, H.: Cooperative double-layer genetic programming hyper-heuristic for online container terminal truck dispatching. IEEE Trans. Evol. Comput. (2022)
8. Chen, X., Bai, R., Qu, R., Dong, H., Chen, J.: A data-driven genetic programming heuristic for real-world dynamic seaport container terminal truck dispatching. In: 2020 IEEE Congress on Evolutionary Computation (CEC), pp. 1–8. IEEE (2020)
9. Chen, X., Feiyang, B., Qu, R., Jing, D., Bai, R.: Neural network assisted genetic programming in dynamic container port truck dispatching. In: 2023 IEEE International Conference on Intelligent Transportation Systems (ITSC), pp. 1–6. IEEE (2023)
10. Dantzig, G.B., Ramser, J.H.: The truck dispatching problem. Manage. Sci. **6**(1), 80–91 (1959)
11. Hildebrandt, T., Branke, J.: On using surrogates with genetic programming. Evol. Comput. **23**(3), 343–367 (2015)
12. Koza, J.R.: Genetic programming as a means for programming computers by natural selection. Stat. Comput. **4**, 87–112 (1994)
13. Macharis, C., Caris, A., Jourquin, B., Pekin, E.: A decision support framework for intermodal transport policy. Eur. Transp. Res. Rev. **3**, 167–178 (2011)
14. Nguyen, S., Zhang, M., Johnston, M., Tan, K.C.: Automatic design of scheduling policies for dynamic multi-objective job shop scheduling via cooperative coevolution genetic programming. IEEE Trans. Evol. Comput. **18**(2), 193–208 (2013)
15. Nguyen, S., Zhang, M., Tan, K.C.: Surrogate-assisted genetic programming with simplified models for automated design of dispatching rules. IEEE Trans. Cybern. **47**(9), 2951–2965 (2016)
16. Vaswani, A., et al.: Attention is all you need. In: Advances in Neural Information Processing Systems, vol. 30 (2017)
17. Wang, T., et al.: Synchronous spatiotemporal graph transformer: a new framework for traffic data prediction. IEEE Trans. Neural Netw. Learn. Syst. (2022)
18. Wolf, T., et al.: Transformers: state-of-the-art natural language processing. In: Proceedings of the 2020 Conference on Empirical Methods in Natural Language Processing: System Demonstrations, pp. 38–45 (2020)
19. Wolf, T., et al.: Transformers: state-of-the-art natural language processing. In: Proceedings of the 2020 Conference on Empirical Methods in Natural Language Processing: System Demonstrations, pp. 38–45. Association for Computational Linguistics (2020). https://www.aclweb.org/anthology/2020.emnlp-demos.6
20. Yi, W., Qu, R., Jiao, L., Niu, B.: Automated design of metaheuristics using reinforcement learning within a novel general search framework. IEEE Trans. Evol. Comput. (2022)
21. Zeiträg, Y., Figueira, J.R., Horta, N., Neves, R.: Surrogate-assisted automatic evolving of dispatching rules for multi-objective dynamic job shop scheduling using genetic programming. Expert Syst. Appl. **209**, 118194 (2022)
22. Zhang, F., Mei, Y., Nguyen, S., Zhang, M., Tan, K.C.: Surrogate-assisted evolutionary multitask genetic programming for dynamic flexible job shop scheduling. IEEE Trans. Evol. Comput. **25**(4), 651–665 (2021)

23. Zhang, Y., Bai, R., Qu, R., Tu, C., Jin, J.: A deep reinforcement learning based hyper-heuristic for combinatorial optimisation with uncertainties. Eur. J. Oper. Res. **300**(2), 418–427 (2022)
24. Zheng, F., Qiao, L., Liu, M.: An online model of berth and quay crane integrated allocation in container terminals. In: Lu, Z., Kim, D., Wu, W., Li, W., Du, D.-Z. (eds.) COCOA 2015. LNCS, vol. 9486, pp. 721–730. Springer, Cham (2015). https://doi.org/10.1007/978-3-319-26626-8_53

Membrane Computing and DNA Computing

Design and Realization of Encoders Based on Switching Circuit

Zigeng Liu, Yanjun Liu, Yuefei Yang, and Yingxin Hu[✉]

College of Information Science and Technology, Shijiazhuang Tiedao University,
Shijiazhuang 050043, China
huyingxin@mail.stdu.edu.cn

Abstract. With the rapid development of DNA nanotechnology, a variety of flexible molecular computing models and logic computing systems have been proposed and constructed. Among them, the switching circuit model based on DNA strand displacement technology performs well in terms of the decreased complexity in constructing multi-input and multi-output circuits. However, in the previous established circuits based on the switching computing model, the same switching elements are designed differently to generate diverse target strands for different outputs, increasing the complexity and difficulty of the design. Here, a top-down design strategy is proposed in which only once design is required for each switching element. The presented strategy decreased the amount of switching elements and simplifies the design complexity to a certain extent. To verify the strategy, the 4-2 priority and 10-4 priority encoder is realized by this design and validated by Visual DSD. The proposed method has possibilities in design of large-scale DNA logic computing systems with more inputs and outputs.

Keywords: DNA strand displacement · switching circuit · encoder

1 Introduction

Since Adleman first published the theory of DNA computation and solved the Hamiltonian path-related independent set problem [1], DNA computation has become a research hotspot by virtue of its excellent computational ability [2]. DNA computing can not only solve complex mathematical problems, but also accomplish the tasks that are difficult by electronic computers. Thus, this new computing mode is a useful supplement to traditional electronic computers.

Taking advantage of properties such as the programmability and self-assembly of DNA molecules [3], researchers have used DNA to realize a variety of logic computing circuits [4–9]. Seelig et al. designed and realized the digital logic circuits based on DNA strand displacement, including the basic logic AND-gate, OR-gate, NOT-gate [10]. Half-adders [11, 12], half-subtractors [13], full adder [14, 15], encoders [16], parity checkers [17] and other molecular computing devices have been realized. Among them, Qian and Winfree designed a seesaw gate model based on the DNA reversible strand displacement technology and built a digital logic circuit that computed the floor of the square root

© The Author(s), under exclusive license to Springer Nature Singapore Pte Ltd. 2024
L. Pan et al. (Eds.): BIC-TA 2023, CCIS 2061, pp. 293–304, 2024.
https://doi.org/10.1007/978-981-97-2272-3_22

of four-bit binary numbers [18]. Based on the seesaw gate model [19], their groups designed a more complicated artificial neural network computing system called "four-neuron Hopfield associative memory". Since then, researchers have constructed complex logic circuits based on the seesaw model, such as determination of prime numbers [20] and factorial addition [21]. The emergence of seesaw gate model provides a powerful tool for the realization of logic computing system and building the large DNA molecular cascade circuit.

The large number of DNA strands and gates required for the construction of logic circuits by the seesaw gate model increases the design complexity. Fan and his research team proposed the switching circuit model [22], which theoretically requires only two switching elements to realize arbitrary logic circuits. The model has a high degree of modularity, scalability, and robustness. Since then, researches have constructed DNA adder, subtractor, multiplier [23] and multi-input and multi-output encoder [24], which have the advantages of faster operation speed, lower complexity and stronger programmability than the traditional molecular operators. Compared with the circuits built by the seesaw gate model, fewer DNA strands are required and the complexity of circuit design is lower.

However, in the previously built circuits based on the switching computational model, multiple switching elements needs to be designed to obtain different target strands produced by different outputs. This results in the same switching elements in different module circuits to be designed multiple times, which increases the complexity and difficulty of the design to some extent.

In this study, we propose a top-down strategy to design logical computing device. In this strategy, each switch appears only once in the circuit, so each switching element needs to be designed only once. In order to generate different target strands for multiple outputs, we only need to fan out the output strand at the end of the switching circuit. This strategy reduces the number of switching elements, which decreases the workload of specific DNA strand design and simplifies the design complexity to some extent. To validate the strategy, we designed and implemented 4-2 priority encoders and 10-4 priority encoders and simulated by Visual DSD. The results demonstrated the advantages of this strategy compared to see-saw gate model. The proposed strategy will provide support for the future design of large-scale DNA logic computational systems with more inputs and outputs.

2   Encoder Design

In this part, a top-down design strategy was used to construct encoders based on the switching circuit model. First, analyze the system function of the nanodevice and obtain the logic expression based on the truth table. Then, each input signal in the logic expression corresponds to a switching element. The AND and OR relationship between the input signals are accomplished by series connections and parallel connections of the switching elements respectively. Finally, the switching elements are cascaded in series and parallel to fulfill the function of the unified logic circuit. Compared to the previously built circuits based on the switching computational model, only once design of the switching element is required for those signals with multiple occurrences. Meanwhile, fan-out reporting element were used to replace some secondary switch, i.e., the different target strands for multiple outputs were obtained by fan out the output strand at the end of the switching elements.

2.1 The 4-2 Priority Encoder Design

The 4-2 priority encoder is a common electronic component in conventional computers used to encode one of multiple input data bits into a two-bit output. The main purpose of this encoder is to select the input with the highest priority and encode it as an output based on the priority of the input data bits. The 4-2 priority encoder has four inputs and two outputs, allowing multiple signals to be input simultaneously, with the signal priority increasing in order from I_0 to I_3, with the following logical expression.

$$Y_0 = I_3 + \bar{I}_3 \bar{I}_2 I_1 \tag{1}$$

$$Y_1 = I_3 + \bar{I}_3 I_2 \tag{2}$$

I_3 has the highest priority among the four inputs, when I_3 has a signal input and the input is 1, no matter whether I_2, I_1, I_0 have inputs or not, the last output of $Y_1 Y_0$ is 11; when I_2 has inputs and I_3 has no inputs, no matter whether the other inputs have inputs or not, the last output of $Y_1 Y_0$ is 10; when I_1 has inputs and I_3 and I_2 have no inputs, no matter whether the other inputs have inputs or not, the last output of $Y_1 Y_0$ is 01, and when $I_3, I_2,$ and I_1 have no inputs, no matter whether I_0 has inputs or not, the last output of $Y_1 Y_0$ is 00, and its corresponding truth table is shown in Table 1.

Table 1. 4-2 priority encoder truth table

Input variable				Output	
I_3	I_2	I_1	I_0	Y_1	Y_0
1	X	X	X	1	1
0	1	X	X	1	0
0	0	1	X	0	1
0	0	0	1	0	0

According to the logical expressions (1) and (2) and combined with the switching circuit model, the DNA molecule switching circuit with the output Y_0 can be composed of two primary switches (I_3 and I_3') and two secondary switches (I_2' and I_1); and the DNA molecule switching circuit with the output Y_1 can be composed of two primary switches (I_3 and I_3') and one secondary switch (I_2). The I_3 and I_3' switches are repeated in both output module circuits, where logically consistent but repeated switches in different output modules have been designed differently many times in previous studies, and a total of four primary switches and three secondary switches are required for the two output module circuits if they are designed many times. In this study, two different output circuits producing Y_0 and Y_1 are cascaded together, and each switch is designed only once, so the whole circuit can be composed of two primary switches and three secondary switches, and its switching circuit is shown in Fig. 1. In order to produce different target strands from different outputs, it is only necessary to fan out the outputs correspondingly.

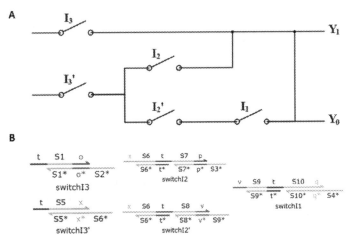

Fig. 1. (A) 4-2 priority encoder switching circuit diagram. (B) Specific DNA strands structure design for switching elements.

This logic circuit uses a three-level cascade. When the input strand for I_3 is fed into the circuit, the corresponding first level switch of the first level is opened directly, and the correct target strand will combine with the corresponding reporting gate through the fan-out to obtain the signal strand of $Y_1 Y_0$. In the absence of the input strand for I'_3, the second level secondary switches corresponding to I_2 and I'_2 are in a blocking state and cannot receive the corresponding input signals. Only when the input strand for I'_3 is present, the first-level switch I'_3 is turned on, and the resulting strand changes the switching state of the second-level I_2 and I'_2 from a blocking state to an activated state, after which the input signals of the corresponding I_2 and I'_2 can be received. When the input strand for I'_3 and I_2 are present, I'_3 causes the corresponding first level switches to open and the circuit is connected. The generated signal strand reacts with the DNA complex in I_2 and activates the switch. The following reaction produces a new signal strand that will combine with the corresponding reporting gates to obtain the signal strand for Y_1. Similarly, when the input strands for I'_3, I'_2 and I_1 are present, two signal strands of $Y_1 Y_0$ will be obtained.

The 4-2 priority encoder designed using the top-down strategy consisted of two primary switches and three secondary switches while four primary switches and three secondary switches were required in the case of original switching circuit model. Compared to the previous method, the amount of the switching elements is decreased, thus reducing the design complexity of the nanodevice.

2.2   The 10-4 Priority Encoder Design

The 10-4 priority encoder is a larger scale electronic component with 10 inputs and 4 outputs. It allows the input of multiple signals simultaneously, with the signal priority increasing in order from I_0 to I_9, with the following logic expression.

$$Y_0 = I_9 + \bar{I}_8 I_7 + \bar{I}_8 I_6 (I_5 + \bar{I}_4 I_3 + \bar{I}_4 \bar{I}_2 I_1) \tag{3}$$

$$Y_1 = \bar{I}_9 \bar{I}_8 (I_7 + I_6 + \bar{I}_5 I_4 I_3 + \bar{I}_5 \bar{I}_4 I_2) \tag{4}$$

$$Y_2 = \bar{I}_9 \bar{I}_8 (I_7 + I_6 + I_5 + I_4) \tag{5}$$

$$Y_3 = I_9 + \bar{I}_9 I_8 \tag{6}$$

I_9 has the highest priority among the ten inputs, when I_9 has a signal input and the input is 1, no matter whether I_8-I_0 have inputs or not, the last output of $Y_3 Y_2 Y_1 Y_0$ is 1001; when I_8 has inputs and I_9 has no inputs, no matter whether the other inputs have inputs or not, the last output of $Y_3 Y_2 Y_1 Y_0$ is 1000. The specific truth table is shown in Table 2.

Table 2. 10-4 priority encoder truth table

Input variable										Output			
I_9	I_8	I_7	I_6	I_5	I_4	I_3	I_2	I_1	I_0	Y_3	Y_2	Y_1	Y_0
1	x	x	x	x	x	x	x	x	x	1	0	0	1
0	1	x	x	x	x	x	x	x	x	1	0	0	0
0	0	1	x	x	x	x	x	x	x	0	1	1	1
0	0	0	1	x	x	x	x	x	x	0	1	1	0
0	0	0	0	1	x	x	x	x	x	0	1	0	1
0	0	0	0	0	1	x	x	x	x	0	1	0	0
0	0	0	0	0	0	1	x	x	x	0	0	1	1
0	0	0	0	0	0	0	1	x	x	0	0	1	0
0	0	0	0	0	0	0	0	1	x	0	0	0	1
0	0	0	0	0	0	0	0	0	1	0	0	0	0

According to the logic expressions (3) to (6) and combined with the switching circuit model, the DNA molecule switching circuit with the output Y_3 can be composed of two primary switches and one secondary switch; the DNA molecule switching circuit with the output Y_2 can be composed of one primary switch and five secondary switches; the DNA molecule switching circuit with the output Y_1 can be composed of one primary switch and seven secondary switches; and the DNA molecule switching circuit with the output

Y_0 can be composed of two primary switches and seven secondary switches. The I_9, I'_9, I_8, etc. switches are repeated in four output module circuits, where logically consistent but repeated switches in different output modules have been designed differently many times in previous studies, and a total of six primary switches and twenty secondary switches are required for the four output module circuits if they are designed many times. In this study, four different output circuits producing Y_0, Y_1, Y_2 and Y_3 are cascaded together, and each switch is designed only once, so the whole circuit can be composed of two primary switches and fifteen secondary switches, and its switching circuit is shown in Fig. 2. In order to produce different target strands from different outputs, it is only necessary to fan out the outputs correspondingly.

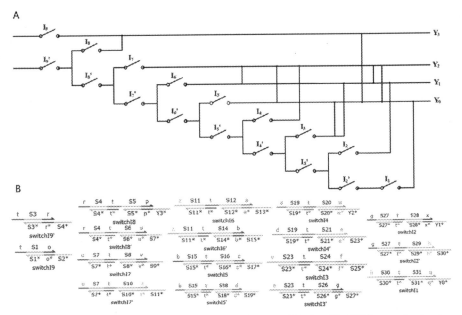

Fig. 2. (A) 10-4 priority encoder switch circuit diagram. (B) Specific DNA strands structure design for switching elements.

This logic circuit uses a nine-level cascade. When the input strand for I_9 is fed into the circuit, the corresponding first level switch of the first level is opened directly, and the correct target strand will combine with the corresponding reporting gate through the fan-out to obtain the signal strand of Y_3Y_0. In the absence of the input strand for I'_9, the second level secondary switches corresponding to I_8 and I'_8 are in a blocking state and cannot receive the corresponding input signals. Only when the input strand for I'_9 is present, the first-level switch I'_9 is turned on, and the resulting strand changes the switching state of the second-level I_8 and I'_8 from a blocking state to an activated state, after which the input signals of the corresponding I_8 and I'_8 can be received. When the input strand for I'_9 and I_8 are present, I'_9 causes the corresponding first level switches to open and the circuit is connected. The generated signal strand reacts with the DNA complex in I_8 and activates the switch. The following reaction produces a new signal strand that will combine with the corresponding reporting gates to obtain the signal strand for Y_3. Similarly, when input strands are present and causes the corresponding switches to open and the circuit is connected, signal strands of $Y_2 Y_1 Y_0$ will be obtained.

The 10-4 priority encoder designed using the top-down strategy consisted of two primary switches and fifteen secondary switches while six primary switches and twenty secondary switches were required in the case of original switching circuit model. Compared to the previous method, the amount of the switching elements was decreased, thus reducing the design complexity of the nanodevice.

3   Simulation Results and Analysis

All the priority encoders were designed and simulated by Visual DSD software. The simulation experiment simulates the priority encoder input signals as 0 or 1 by adjusting the concentration values of the input signal strand, and the output of the encoder is determined by the concentration of the target strand. Finally, the correctness of the encoder is verified by comparing the output with the truth table of the priority encoder. In this experiment, the concentration of the DNA strand in the range of 0–20nM is logic 0, and the concentration in the range of 80–100nM is logic 1.

3.1   The 4-2 Priority Encoder

Experiments are tested in order of priority from highest to lowest. When I_3 has inputs, there are eight possible input situations which are $I_3 I_2 I_1 I_0 = 1000$, $I_3 I_2 I_1 I_0 = 1001$, $I_3 I_2 I_1 I_0 = 1010$, $I_3 I_2 I_1 I_0 = 1011$, $I_3 I_2 I_1 I_0 = 1100$, $I_3 I_2 I_1 I_0 = 1101$, $I_3 I_2 I_1 I_0 = 1110$ and $I_3 I_2 I_1 I_0 = 1111$. Through our experiments we found that the concentration results for each of these situations are very close, so these situations can be represented by $I_3 I_2 I_1 I_0 = 1xxx$. 'x' stands for both 0 and 1, and this expression is used in subsequent narratives. For the 4-2 priority encoder, the four cases of $I_3 I_2 I_1 I_0 = 1xxx$, $I_3 I_2 I_1 I_0 = 01xx$, $I_3 I_2 I_1 I_0 = 001x$, $I_3 I_2 I_1 I_0 = 000x$ were tested in order of priority from highest to lowest, and the simulation results are shown in Fig. 3.

It can be seen from Fig. 3A, the concentration of Y_1 and Y_0 strand increases rapidly to about 100nM and tends to stabilize. The concentration result indicates that $Y_1 Y_0 = 11$. From Fig. 3B, it can be seen that the concentration of Y_1 strand rapidly increases to about 100nM and tends to stabilize, and the concentration of Y_0 strand stabilizes at 0. The concentration result indicates that $Y_1 Y_0 = 10$. From of Fig. 3C, it can be seen that the concentration of Y_0 strand rapidly increases to about 100nM and tends to stabilize, and the concentration of Y_1 strand stabilizes at 0. The concentration results indicates that $Y_1 Y_0 = 01$. From Fig. 3D, it can be seen that the concentrations of Y_0 and Y_1 strand are stabilized at 0. The concentration results indicates that $Y_1 Y_0 = 00$. The above simulation results is in agreement with the expected output of the design. Taking $I_3 I_2 I_1 I_0 = 1xxx$ as an example, we compare two experimental results of the seesaw gate model and the switching circuit model in Fig. 4.

It can be seen the concentration of the Y_1 and Y_0 strand exceeds 80nM within 1000 s and stabilizes around 2000 s from Fig. 4A. And it can be seen the concentration of the Y_1 and Y_0 strand exceeds 80nM around 7000 s and stabilizes around 10000 s from

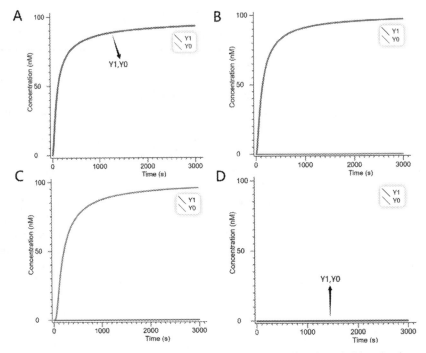

Fig. 3. Simulation results of the 4-2 priority encoder designed using the switching circuit model. Overlapping curves are illustrated by arrows pointing to them. (A) Simulation result when $I_3I_2I_1I_0$ = 1xxx. (B) Simulation result when $I_3I_2I_1I_0$ = 01xx. (C) Simulation result when $I_3I_2I_1I_0$ = 001x. (D) Simulation result when $I_3I_2I_1I_0$ = 000x.

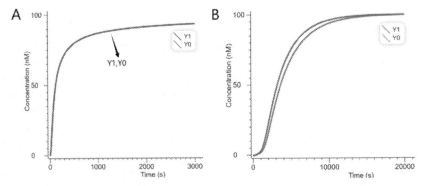

Fig. 4. (A) Simulation results of a 4-2 priority encoder designed using the switching circuit model when $I_3I_2I_1I_0$ = 1xxx. (B) Simulation results of a 4-2 priority encoder designed using the seesaw gate model when $I_3I_2I_1I_0$ = 1xxx.

Fig. 4B. This shows that the 4-2 priority encoder designed using the switching circuit model is faster than using the seesaw gate model in terms of reaction time.

3.2   The 10-4 Priority Encoder

Experiments are tested in order of priority from highest to lowest. For the 10-4 priority encoder, there are ten test cases which are $I_9I_8I_7I_6I_5I_4I_3I_2I_1I_0 = 1xxxxxxxxx$, $I_9I_8I_7I_6I_5I_4I_3I_2I_1I_0 = 01xxxxxxxx$, $I_9I_8I_7I_6I_5I_4I_3I_2I_1I_0 = 001xxxxxxx$, $I_9I_8I_7I_6I_5I_4I_3I_2I_1I_0 = 0001xxxxxx$, $I_9I_8I_7I_6I_5I_4I_3I_2I_1I_0 = 00001xxxxx$, $I_9I_8I_7I_6I_5I_4I_3I_2I_1I_0 = 000001xxxx$, $I_9I_8I_7I_6I_5I_4I_3I_2I_1I_0 = 0000001xxx$, $I_9I_8I_7I_6I_5I_4I_3I_2I_1I_0 = 00000001xx$, $I_9I_8I_7I_6I_5I_4I_3I_2I_1I_0 = 000000001x$, $I_9I_8I_7I_6I_5I_4I_3I_2I_1I_0 = 000000000x$. The concentration of all four target strands in the tenth case was 0 so it was not displayed. The first nine simulation results are shown in Fig. 5.

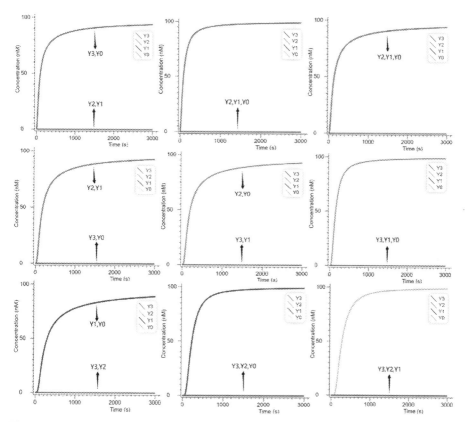

Fig. 5. Simulation results of the 10-4 priority encoder designed using the switching circuit model. The nine concentration plots from left to right and from top to bottom are the results of the simulation in order of priority from highest to lowest.

Due to the large number of cases and the similarity in the analysis of results, we will analyze here the case corresponding to the first picture in Fig. 5 as an example. It can be seen from the picture that the concentration of Y_3 and Y_0 strand is rapidly increased to about 100nM tends to stabilize, and the concentration of Y_2 and Y_1 strand is stable at 0.

The concentration results indicates that $Y_3 Y_2 Y_1 Y_0 = 1001$. All the simulation results are consistent with the design expected results in Fig. 5.

When $I_9 I_8 I_7 I_6 I_5 I_4 I_3 I_2 I_1 I_0 = 1xxxxxxxxx$, we compare two experimental results of the switching circuit model and the seesaw gate model. It can be seen the concentration of the Y_3 and Y_0 strand exceeds 80nM within 500 s and stabilizes around 2000 s from Fig. 6A. And it can be seen the concentration of the Y_3 and Y_0 strand exceeds 80nM around 7000 s and stabilizes around 10000 s from Fig. 6B. This shows that the 10-4 priority encoder designed using the switching circuit model is faster than using the seesaw gate model in terms of reaction time.

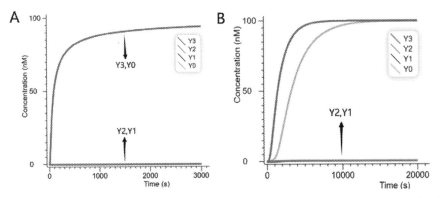

Fig. 6. (A) Simulation result of the 10-4 priority encoder designed using the switching circuit model when $I_9 I_8 I_7 I_6 I_5 I_4 I_3 I_2 I_1 I_0 = 1xxxxxxxxx$. (B) Simulation result of the 10-4 priority encoder designed using the seesaw gate model when $I_9 I_8 I_7 I_6 I_5 I_4 I_3 I_2 I_1 I_0 = 1xxxxxxxxx$.

From the above simulations, it can be seen that the simulation reaction of the two priority encoders designed based on the switching circuit model is relatively faster, which can quickly replace the expected target strands. The reaction takes relatively short time. After the reaction starts, the concentration of the target strand increases rapidly. it can reach 80nM within 1000 s and tend to stabilize around 2000 s. We can determine whether the output is 1 or 0 as early as possible. On the other hand, the simulation time of the priority encoder based on the seesaw gate model takes longer. The concentration of the target strand grows slower. It can only reach 80nM around 7000 s in most cases and takes about 10000 s to stabilize. The reaction time gets longer as the size of the encoder expands. From the reaction time point of view, it is more ideal to use the switching circuit model to construct the encoder. Meanwhile, in our design of the encoder using the switching circuit model, the top-down circuit design strategy play a key role. The number of switching elements required for the circuit is reduced, which reduces the workload of the specific DNA strand design and simplifies the complexity of the logic circuit. The function of the encoder can be represented by an overall switching circuit diagram, which is a great help for the subsequent design of the DNA strand.

4 Conclusion

In this study, we proposed a top-down design strategy for constructing logical computing system based on the switching circuit model. We designed the overall switching circuit based on the function of the encoder, and each switch appears only once in the circuit, so each switching element needs to be designed only once. The output strand is fanned out at the end of the switching circuit, thus allowing multiple outputs to produce different target strands.

Previously established circuits based on switching computational models have the same switching elements repeated and designed differently in multiple output module circuits. For encoders, with the number of inputs and outputs increasing, the more repeated switching elements occurred among which secondary switches account for a large portion. Since the DNA structure corresponding to the secondary switches is more complex than that of the primary switches, this increases the complexity and difficulty of the design. The design strategy in this study reduced the number of switching elements and used fan-out reporting element to replace some secondary switch, decreasing the design complexity in some extent.

On the whole, the proposed strategy reduces the number of switching elements and the design complexity of the DNA strand. In the future, it can be used to design large-scale DNA logic computing systems with more inputs and outputs to promote the development of biological computing systems.

Acknowledgments. This work was funded by Science Research Project of Hebei Education Department (ZD2022098).

References

1. Adleman, L.M.: Molecular computation of solutions to combinatorial problems. Science **266**(5187), 1021–1024 (1994)
2. Zhao, Y., Zhou, S.: Research status and prospects of DNA logic computational modeling. Comput. Appl. Res. **36**(11), 3201–3209 (2019)
3. Zhou, Y., Luo, Y., Jiang, X.: DNA data storage: preservation strategies and data encryption. Synth. Biol. **2**(03), 371–383 (2021)
4. Weng, Z., Yu, H., Luo, W., et al.: Cooperative branch migration: a mechanism for flexible control of DNA strand displacement. ACS Nano **16**(2), 3135–3144 (2022)
5. Massey, M., Ancona, M.G., Medintz, I.L., et al.: Time-gated DNA photonic wires with forster resonance energy transfer cascades initiated by a luminescent terbium donor. ACS Photonics **2**(5), 639–652 (2015)
6. Hu, Y., Xie, C., Xu, F., Pan, L.: A strategy for programming the regulation of in vitro transcription with application in molecular circuits. Nanoscale 5429–5434 (2021)
7. Pan, L., Hu, Y., Ding, T., et al.: Aptamer-based regulation of transcription circuits. Chem. Commun. 7378–7381 (2019)
8. Pan, L., Wang, Z., Li, Y., et al.: Nicking enzyme-controlled toehold regulation for DNA logic circuits. Nanoscale 18223–18228 (2017)
9. Yang, J., Wu, R., Li, Y., et al.: Entropy-driven DNA logic circuits regulated by DNAzyme. Nucleic Acids Res. **16**, 8532–8541 (2018)

10. Seelig, G., Soloveichik, D., Zhang, D.Y., et al.: Enzyme-free nucleic acid logic circuits. Science **314**(5805), 1585–1588 (2006)
11. Chen, X., Liu, X., Wang, F., et al.: Massively parallel DNA computing based on domino DNA strand displacement logic gates. ACS Synth. Biol. **11**(7), 2504–2512 (2022)
12. Li, W., Zhang, F., Yan, H., et al.: DNA based arithmetic function: a half adder based on DNA strand displacement. Nanoscale **8**(6), 3775–3784 (2016)
13. Zhang, S., Wang, K., Huang, C., et al.: Reconfigurable and resettable arithmetic logic units based on magnetic beads and DNA. Nanoscale **7**(48), 20749–20756 (2015)
14. Xie, N., Li, M., Wang, Y., et al.: Scaling up multi-bit DNA full adder circuits with minimal strand displacement reactions. Am. Chem. Soc. **144**(21), 9479–9488 (2022)
15. He, S., Cui, R., Zhang, Y., et al.: Design and realization of triple dsDNA nanocomputing circuits in microfluidic chips. ACS Appl. Mater **14**(8), 10721–10728 (2022)
16. Yao, G., Li, J., Li, Q., et al.: Programming nanoparticle valence bonds with single-stranded DNA encoders. Nat. Mater. **19**(7), 781–788 (2022)
17. Fan, D., Wang, E., Dong, S.: A DNA-based parity generator/checker for error detection through data transmission with visual readout and an output-correction function. Chem. Sci. **8**(3), 1888–1895 (2017)
18. Qian, L., Winfree, E.: Scaling up digital circuit computation with DNA strand displacement cascades. Science **332**(6034), 1196–1201 (2011)
19. Qian, L., Winfree, E., Bruck, J.: Neural network computation with DNA strand displacement cascades. Nature **475**(7356), 368–372 (2011)
20. Ye, M.: A study of molecular logic operations based on DNA strand substitution reactions. Zhengzhou Light Industry Institute (2014)
21. Yuan, G.: Design and implementation of multi-input complex logic circuits based on DNA strand substitution. Zhengzhou Light Industry Institute (2020)
22. Wang, F., Lv, H., Li, Q., et al.: Implementing digital computing with DNA-based switching circuits. Nat. Commun. **11**(1), 1–8 (2020)
23. Xiao, W.: An integer decomposition algorithm based on DNA switching circuits. Guangzhou University (2021)
24. Zhang, X.: Design and implementation of 8-3 priority encoder based on molecular circuits. Guangzhou University (2022)

GSA-Inspired Computational Nanobiosensing for Cancer Detection

Shaolong Shi[1,2], Zhaoyang Jiang[2], and Yifan Chen[2(✉)]

[1] Yangtze Delta Region Institute (Quzhou), University of Electronic Science and Technology of China, Quzhou 324003, China
[2] School of Life Science and Technology, University of Electronic Science and Technology of China, Chengdu 611731, China
`yifan.chen@uestc.edu.cn`

Abstract. We consider the cancer detection procedure (CDP) as the process of the Gravitational Search Algorithm (GSA). Nanorobots coated with protease-sensitive nanomaterials can react with protease which causes unquenching of fluorogenic probes for imaging application. As the protease concentration can induce the different degree of fluorogenic, the strength of fluorogenic can be seen as the tumor-induced biological objective function whose maximum is regarded as the tumor location. The process that employing the externally controllable and trackable nanorobots to detect the tumor in high-risk tissue is treated as using agents to complete the optimization process in the search space. To accomplish the cancer detection procedure using nanorobots with the shortest possible physiological routes and with minimal systemic exposure, it is feasible to adjust the movement of nanorobots according to the information shared by other nanorobots. Following this outline, we consider the mechanism of Gravitational Search Algorithm (GSA) which is more concise and efficient compared with PSO or GA in swarm intelligence field. Every nanorobot in the CDP can be seen as a particle in GSA and the fitness of every nanorobot is associated with the strength of its fluorogenic. To make the reality of the CDP, we should take into account the realistic in vivo propagation scenario of nanorobots swimming in the high-risk tissue to find the tumor location. Finally, we present numerical examples to demonstrate the effective of the CDP in view of GSA.

Keywords: cancer detection · biosensing · swarm optimization · gravitational search algorithm · nanorobots

1 Introduction

Since emergence in the early 1970s, controlled drug delivery systems (DDS) with the purpose to maximize therapeutic activity and to minimize undesirable side-effects have attracted increasing attention [11,20]. Localized delivery refers to direct targeting of anticancer agents using delivery vehicles such as nanorobots, which can be used in the treatment of local tissue malignancies such as prostate,

© The Author(s), under exclusive license to Springer Nature Singapore Pte Ltd. 2024
L. Pan et al. (Eds.): BIC-TA 2023, CCIS 2061, pp. 305–316, 2024.
https://doi.org/10.1007/978-981-97-2272-3_23

head and neck cancers [14]. As a key step of localized delivery, noninvasive cancer detection plays a critical role in the controlled DDS [8,17].

As for noninvasive cancer detection, the traditional ways are medical imaging techniques such as ultrasound, computed tomography, and magnetic resonance imaging (MRI). In the past decades, very little research has been done on early cancer detection, where cancer cells cannot be seen by using traditional medical imaging devices due to their low-resolution [22]. To solve this problem, the traditional way is using contrast agents such as magnetic nanoparticles (MNPs) to provide better delineation between healthy and diseased tissues [21]. However, the current way of contrast agents delivery is relying on the circulatory system which leads to poor targeting efficiency for precise localization of the lesion [2,12]. Enhancing the diagnostic efficacy of nanoparticles necessitates the use of propelling force, and the sensory-based displacement capability [8].

Nanoswimmers assembled by MNPs have been proposed for direct targeting, which use magnetic self-assembly of 50–100 nm iron oxide nanoparticles [6]. The nanoswimmers can swim with a certain direction under time-varying magnetic fields *via* generated by the Helmholtz coil system [5]. Here, we can use nanoswimmers as contrast agents in medical imaging of early stage cancer to provide a higher quality image of the cancer foci. Individual nanoswimmers are limited in their operational ranges and sensing capabilities because of the constrain of their small size. Therefore, nanoswimmers should work together to cover large areas and perform complex functionalities, which can be achieved through swarm intelligence [1,9].

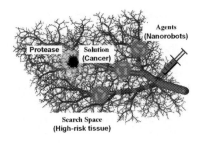

Fig. 1. Pictorial illustration of the externally controllable cancer detection.

There is an intriguing analogy between swarm intelligence algorithm and externally controllable cancer detection as depicted in Fig. 1. The global maximum of a biological objective function associated with a tumor-induced *in vivo* phenomenon with tumor being the epicenter is the cancer to be detected; the domain of the objective function is the tissue region at high risk of malignancy; the agents of the optimization process are externally manoeuvrable nanorobots (i.e., nanoswimmers). An agent locates the optimal solution (i.e., cancer) by moving through the search space (i.e., high-risk tissue) under the guidance of a specified vector (i.e., steering field) produced by the fitness (i.e., sensing information) of other agents (i.e., nanorobots).

Different from the classical mathematical computing by using ideal non-interacting agents, the landscape of the solution space may be altered by agents due to the natural computing feature of the externally controllable cancer detection (i.e., employing natural materials such as nanorobots to compute) where agents interact with the solution space (i.e., nanorobots undergo physical, chemical, and biological reactions in the *in vivo* environment). An external observer can then infer the space by monitoring the movement of agents, which is called "seeing-is-sensing" in [7].

Each cancer cell synthesizes a certain amount of protease and the sum of secreted creates a protease concentration field around the cancer [3,4]. Therefore, we define protease concentration field as a final result of all cancer cells. Nanorobots used in the CDP are modified with nanomaterials with protease-sensitive domains whose cleavage causes unquenching of fluorogenic probes for imaging applications [16,19]. Hence, it is plausible to reflect the protease concentration with the fluorescence intensity, which also represents the cancer detection performance of the corresponding nanorobot. Subsequently, we consider a representative artificial landscape to imitate the protease concentration field and to evaluate the cancer detection performance as a reward function from the perspective of natural computing. Provided with this analogy, Gravitational Search Algorithm (GSA) is considered and applied to the CDP as every particle's mass can be seen as positively correlated with its fitness. Each nanorobot in the CDP can be seen as a particle in the GSA, so the nanorobots can move harmoniously to a better fitness location, which performs a stronger fluorescence intensity.

The paper is organized as follows. In Sect. 2, we introduce the general smart biosensing framework. In Sect. 3, we propose the GSA-inspired CDP. In Sect. 4, we provide numerical examples to demonstrate the effectiveness of the proposed framework. Finally, some concluding remarks are drawn in Sect. 5.

2   Smart Biosensing

In the CDP, we consider a general parameter space $\mathcal{P}$ with the following agent-dependent landscape [15,18]:

$$
\begin{aligned}
f(\vec{x}; A) &= f_{ms}(\vec{x}; A) + f_{act}(\vec{x}; A) \\
&= f_{in}(\vec{x}) + f_{ex}(\vec{x}; A) + f_{act}(\vec{x}; A), \vec{x} \in \mathcal{P},
\end{aligned}
\tag{1}
$$

where $f_{ms}(\vec{x}; A)$ is the externally measurable objective function at location $\vec{x}$ through the observation of agent A, $f_{in}(\vec{x})$ is the intrinsic objective function at $\vec{x}$ independent of the existence of A, which is true objective function of the CDP. $f_{ex}(\vec{x}; A)$ is the disturbance because of the interaction between A and the parameter space $\mathcal{P}$, and $f_{act}(\vec{x}; A)$ is the correction factor accounting for the *activeness* of A. It is assumed that the location of the global maximum remains unchanged regardless of any variation caused by the nanorobots to the landscape, which means the location of cancer foci remains unchanged in the high-risk tissue.

The intrinsic function $f_{in}(\vec{x})$ should reflect the concentration of protease around the cancer location. Protease can be used as biomarker for early diagnostics because of its differential expression in cancer area. Its local production rate is coupled to the state of tumor cells. The protease gradient offers a direction for the angiogenic sprouting of solid tumors. It is assumed that $f_{in}(\vec{x})$ is a nonlinear decreasing function for $|\vec{x}|$. The extrinsic disturbance $f_{ex}(\vec{x}; A)$ is disturbance caused to the protease concentration resulting from the interaction between nanorobot A and human vasculature at $\vec{x}$ (i.e., degradation of nanorobots). The correction factor $f_{act}(\vec{x}; A)$ compensates for the degradation. In this investigation, we assume that the effect of degradation has been roughly counterweighted, i.e., $f_{act}(\vec{x}; A) = -f_{ex}(\vec{x}; A) + \varepsilon(\vec{x}; A)$ with $\varepsilon(\vec{x}; A)$ being the random compensation error. So Eq. (1) can be rewritten as

$$f(\vec{x}; A) = f_{in}(\vec{x}) + \varepsilon(\vec{x}; A), \vec{x} \in \mathcal{P}, \qquad (2)$$

The natural computing problem associated with CDP can be expressed as follows: given a search space (tissue region under surveillance) $\mathcal{P}$ and an objective function f that maps elements of $\mathcal{P}$ into the protease concentration field of agents around the tumor, the agents move through the search space to locate the target according to the concentration of protease in the tissue region under surveillance. The agents will attach to the tumor and stop moving when they meet the tumor, the process of which can be denoted by the expression below:

$$\max_{\vec{x} \in \mathcal{P}} f(\vec{x}; A) \qquad (3)$$

where $\vec{x}$ is a two-dimensional vector.

In the CDP, we consider two different scenarios of the objective function to represent the protease concentration field in the ideal case and in the case that it sustains some disturbance because of some chemical reactions of protease in the high-risk tissue as shown in Fig. 2. As for the objective function, we normalize the maximum value to 1 denoting the situation that there are nanorobots targeting the cancer location. The search space is $-5\,\text{mm} \leq x,y \leq 5\,\text{mm}$.

– *Modified Ackley function (ideal situation)*:

$$f(x,y) = \begin{cases} 1, & if \ \sqrt{x^2 + y^2} \leq 0.25 \\ \dfrac{4}{3}\exp\left(-0.2 \times \sqrt{\dfrac{1}{2}(x^2 + y^2)}\right) \\ +\dfrac{1}{15}\exp\left(\dfrac{1}{2}\right) - \dfrac{1}{15}e - \dfrac{1}{3}, & \text{Otherwise} \end{cases} \qquad (4)$$

– *Modified Ackley function (nonideal situation)*:

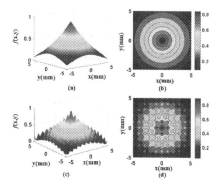

Fig. 2. Illustration of $f(x, y)$, which corresponds to the protease concentration field: (a) Objective function of the ideal case and (b) its contour plot; (c) Objective function of the nonideal case and (d) its contour plot.

$$f(x, y) = \begin{cases} 1, & if \ \sqrt{x^2 + y^2} \le 0.25 \\ \dfrac{4}{3}\exp\left(-0.2 \times \sqrt{\dfrac{1}{2}(x^2 + y^2)}\right) \\ \quad + \dfrac{1}{15}\exp\left[\dfrac{1}{2}\cos(2\pi x) + \dfrac{1}{2}\cos(2\pi y)\right] - \dfrac{1}{15}e - \dfrac{1}{3}, & \text{Otherwise} \end{cases}$$

(5)

As shown in Fig. 2, the two functions represent the situation that the core of a cancer, denoted by a circle of radius 0.5 mm centered at the origin, exhibits protease tropism and pronounced homing of nanorobots. The fist function is convex and quadratic without any local maximum except the global one. The second function is nonconvex with many local maximums. For simplicity, we imprint the reward function in (4) and (5) on the vascular network where the intercapillary distances are $100\,\mu$m, $50\,\mu$m, and $200\,\mu$m for normal ($0.5\,\text{mm} \le |x|, |y| \le 5\,\text{mm}$), peritumoral ($0.25\,\text{mm} \le |x|, |y| \le 0.5\,\text{mm}$), and tumoral ($|x|, |y| \le 0.25\,\text{mm}$) tissues, respectively [15]. The blood inflow and outflow are assumed to be in the bottom left and top right, respectively, where prescribed pressures are set.

3 GSA-Inspired CDP

Gravitational Search Algorithm (GSA) is one of the recent nature inspired algorithm, which is developed by E.Rashedi et al. in 2009 [10, 13]. It is based on the Newton's law of gravity, as shown in equation (6) and the Newton's second law of motion, as shown in Eq. (7).

$$F = G\frac{M_i M_j}{R^2},$$

(6)

$$a = \frac{F}{M}.$$

(7)

where F is the magnitude of the gradvitational force, G is gravitational constant, M_i and M_j are the mass of two different particles, R is the distance between the two particles, and a is the acceleration of the particle, whose mass is M.

Based on (6) and (7), there is an attracting gravity force among all particles of the universe where the effect of bigger and the closer particle is higher. An increase in the distance between two particles means decreasing the gravity force between them as it is illustrated in Fig. 3. In Fig. 3, the sizes of the circles represent their masses, which means the larger the area, the greater the mass; F_{1j} is the force that acting on M_1 from $M_j (j = 2, 3, 4)$ and F is the overall force that acts on M_1 and causes the acceleration in the same direction. In GSA, agents are considered as objects and their performance is measured by their masses. All these objects attract each other by the gravity force, and this force causes a global movement of all objects towards the objects with heavier masses. Hence, masses cooperate using a direct form of communication, through gravitational force. The heavy masses, which correspond to good solutions, move more slowly than lighter ones, this guarantees the exploitation step of the algorithm. As each mass presents a solution, the algorithm is navigated by properly adjusting the gravitational and inertia masses. By lapse of time, we expect that masses be attracted by the heaviest mass. This mass will present an optimum solution in the search space. The GSA could be considered as an isolated system of masses. It is like a small artificial world of masses obeying the Newtonian laws of gravitation and motion.

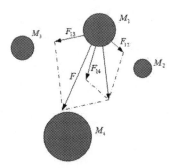

Fig. 3. Gravitational phenomenon.

Considering a system with N agents, we define the position of the ith agent by $\vec{x}_i$, $i = 1, 2, \ldots, N$. The values of the corresponding masses are calculated by the following equations:

$$\begin{cases} m_i(t) = \dfrac{fit_i(t) - worst(t)}{best(t) - worst(t)} \\ M_i(t) = \dfrac{m_i(t)}{\displaystyle\sum_{j=1}^{N} m_j(t)} \end{cases} \tag{8}$$

where $fit_i(t)$ represent the fitness value of the agent i at time t, and, $worst(t)$ and $best(t)$ are defined as follows (for a maximization problem):

$$\begin{cases} best(t) = \max_{i \in 1,2,...,N} fit(t) \\ worst(t) = \min_{i \in 1,2,...,N} fit(t) \end{cases} \tag{9}$$

The force acting on agent 'i' from agent 'j' is defined as follows:

$$\vec{F}_{ij} = G(t) \frac{M_{pi}(t) \times M_{aj}(t)}{R_{ij}(t) + \varepsilon} \tag{10}$$

where M_{aj} is the active gravitational mass related to agent j, M_{pi} is the passive gravitational mass related to agent i, $G(t)$ is gravitational constant at time t, ε is a small constant, and $R_{ij}(t)$ is the Euclidian distance between two agents i and j:

$$R_{ij}(t) = \|\vec{x}_i(t), \vec{x}_j(t)\|_2 \tag{11}$$

$G(t)$ is the gravitational constant whose value reduces with time. It is used to control the search accuracy and defined as follows:

$$G(t) = G_0 \times e^{-\alpha t/T} \tag{12}$$

where G_0 is set to 100, α is set to 20, and T is the total number of iterations (the total age of system). To get a stochastic characteristic of the algorithm, the total force that acts on agent i is supposed to be a randomly weighted sum of the forces exerted from other agents:

$$\vec{F}_i(t) = \sum_{j=1, j \neq i}^{N} rand_j \vec{F}_{ij}(t) \tag{13}$$

where $rand_j$ is a random number in the interval $[0,1]$. According to the law of motion, the acceleration of the agent i at time t is given as follows:

$$\vec{a}_i(t) = \frac{\vec{F}_i(t)}{M_{ii}(t)} \tag{14}$$

where M_{ii} is the inertial mass of ith agent. Furthermore, the next velocity of an agent is considered as a fraction of its current velocity added to its acceleration. Therefore, its position and its velocity could be calculated as follows:

$$\begin{cases} \vec{v}_i(t+1) = rand_i \times \vec{v}_i(t) + \vec{a}_i(t) \\ \vec{x}_i(t+1) = \vec{x}_i(t) + \vec{v}_i(t+1) \end{cases} \tag{15}$$

where $rand_i$ is a uniform random variable in the interval $[0,1]$. This random number is used to give a randomized characteristic to the search.

In the CDP, nanorobots loaded with contrast agent molecules are injected into tissue region at high risk of malignancy. The initial position of the agents is assumed in a one square millimeter area at the lower left corner of the search space (i.e., $-5\,\mathrm{mm} \leq x, y \leq -4\,\mathrm{mm}$). Subsequently, each nanorobot adjust its own location according to (14). Note that the vessel network used in the simulation procedure is a discontinuous two-dimensional grid; therefore all the agents' locations are mapped to the locations associated with the nearest blood vessel. Furthermore, a new agent will be deployed in the injection area if an old agent move outside the search space and it will dissolve in the human body, then the overall number of agents will remain constant to ensure the smooth running of the CDP.

4   Numerical Example

We use several numerical examples to elaborate on the aforementioned GSA-inspired CDP. To demonstrate the effectiveness of it, we also demonstrate the result of normal search method, which means every agent moves in the search space from bottom left to top right without any interaction. The parameters of the CDP are as follows: the population size is $N = 12$, the maximum number of iteration is $t = 100$, the minimum step size is 0.1 mm. Figure 4 shows curves of the best fitness values and mean fitness values over multiple iterations by the two different search methods in the ideal situation described in (3). It can be seen that the agents can reach the best state and keep it after 30 iterations by using the GSA-inspired CDP while the agents can only reach a good state at the 90th iteration and can not keep it by using the normal search method.

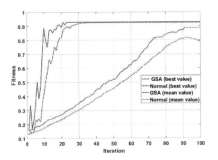

Fig. 4. Fitness curves of the agents by two different search methods in ideal case.

Figure 5 demonstrates the final positions of agents using method of GSA-inspired CDP and normal search method. It can be seen that the cancer is successfully detected with all of the agents locating at the cancer area by using GSA-inspired method while the contrast result is not so good with only a few agents swarming around the cancer area and the others going away. To provide statistical analysis of the robustness and precision of the GSA-inspired CDP, we have carried out 500 independent simulation runs. Figure 6(a) shows the

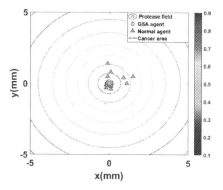

Fig. 5. Final positions of agents using GSA and normal search method in ideal case.

histogram of the quantity of agents locating at the cancer with the mean of 11.966 by using the GSA-inspired CDP. Figure 6(b) shows the histogram of the quantity of agents locating at the cancer with the mean of 4.398 by using the normal search method.

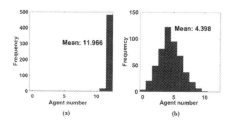

Fig. 6. (a) Histogram of the quantity of agents locating at the target using GSA. (b) Histogram of the quantity of agents locating at the target using normal search method.

The simulation results of CDP in the nonideal situation described in (4) are shown in Fig. 7, Fig. 8, and Fig. 9. It can be seen that the results are similar with those in ideal situation. It is obvious that the proposed GSA-inspired CDP significantly outperforms the normal search method.

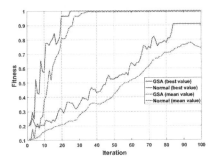

Fig. 7. Fitness curves of the agents by two different search methods in nonideal case.

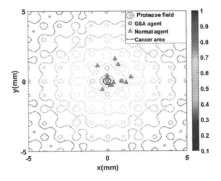

Fig. 8. Final positions of agents using GSA and normal search method in nonideal case.

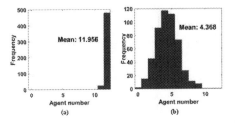

Fig. 9. (a) Histogram of the quantity of agents locating at the target using GSA in the ideal case. (b) Histogram of the quantity of agents locating at the target using normal search method in the nonideal case.

5   Conclusion

We have proposed the GSA-inspired CDP under the rubric of computing inspired cancer detection by taking into account realistic *in vivo* conditions of nanorobots and characteristics of vascular network around the cancer location. Numerical example has demonstrated the effectiveness of the proposed methodology for the protease concentration profile induced by tumor angiogenesis. Future work may include improve the performance of the cancer detection by considering more biological constraints in high-risk tissue. It's also important to examine further the impact of nanorobot nonidealities, such as finite lifespan, imprecise steering, and inaccurate tracking on the performance of cancer detection.

Acknowledgment. This work is supported by the National Natural Science Foundation of China (Grant No. 62102071), the Natural Science Foundation of Sichuan Province (Grant No. 2022NSFSC0873), and the Municipal Government of Quzhou (Grant No. 2023D031).

References

1. Akyildiz, I.F., Brunetti, F., Blázquez, C.: Nanonetworks: a new communication paradigm. Comput. Netw. **52**, 2260–2279 (2008)
2. Bamrungsap, S., et al.: Nanotechnology in therapeutics: a focus on nanoparticles as a drug delivery system. Nanomed.: Nanotechnol. Biol. Med. **7**(8), 1253–1271 (2012)
3. Bartha, K., Rieger, H.: Vascular network remodeling via vessel cooption, regression and growth in tumors. J. Theor. Biol. **241**(4), 903–18 (2006)
4. Carmeliet, P., Jain, R.K.: Molecular mechanisms and clinical applications of angiogenesis. Nature **473**, 298–307 (2011)
5. Cheang, U.K., Kim, H., Milutinovic, D., Choi, J., Rogowski, L., Kim, M.J.: Feedback control of three-bead achiral robotic microswimmers, pp. 518–523 (2015)
6. Cheang, U.K., Kim, M.J.: Self-assembly of robotic micro- and nanoswimmers using magnetic nanoparticles. J. Nanopart. Res. **17**(3), 145 (2015)
7. Chen, Y., Nakano, T., Kosmas, P., Yuen, C., Vasilakos, A.V., Asvial, M.: Green touchable nanorobotic sensor networks. IEEE Commun. Mag. **54**, 136–142 (2016)
8. Felfoul, O., et al.: Magneto-aerotactic bacteria deliver drug-containing nanoliposomes to tumour hypoxic regions. Nat. Nanotechnol. **11**(11), 941–947 (2016)
9. Garnier, S., Gautrais, J., Theraulaz, G.: The biological principles of swarm intelligence. Swarm Intell. **1**, 3–31 (2007)
10. Khajehzadeh, M., Eslami, M., Khajehzadeh, M.: Gravitational search algorithm for optimization of retaining structures (2011)
11. Kwona, E.J., Loa, J.H., Bhatiaa, S.N.: Smart nanosystems: bio-inspired technologies that interact with the host environment (2015)
12. Lee, D.E., Koo, H., Sun, I.C., Ryu, J.H., Kim, K., Kwon, I.C.: Multifunctional nanoparticles for multimodal imaging and theragnosis. Chem. Soc. Rev. **41**(7), 2656–2672 (2012)
13. Rashedi, E., Nezamabadi-pour, H., Saryazdi, S.: GSA: a gravitational search algorithm. Inf. Sci. **179**, 2232–2248 (2009)
14. Shi, S., Chen, Y., Ding, J., Liu, Q., Zhang, Q.: Dynamic in vivo computation for learning-based nanobiosensing in time-varying biological landscapes. IEEE Trans. Evol. Comput. **27**(4), 1100–1114 (2023). https://doi.org/10.1109/TEVC.2022.3198086
15. Shi, S., Chen, Y., Yao, X.: Computing-inspired detection of multiple cancers. In: 2018 IEEE International Conference on Communications (ICC), pp. 1–6. IEEE (2018)

16. Shi, S., Chen, Y., Yao, X., Liu, Q.: Exponential evolution mechanism for in vivo computation. Swarm Evol. Comput. **65**, 100931 (2021). https://doi.org/10.1016/j.swevo.2021.100931, https://www.sciencedirect.com/science/article/pii/S2210650221000924
17. Shi, S., Chen, Y., Yao, X., Zhang, M.: Lightweight evolution strategies for nanoswimmers-oriented in vivo computation. In: 2019 IEEE Congress on Evolutionary Computation (CEC), pp. 866–872 (2019). https://doi.org/10.1109/CEC.2019.8790356
18. Shi, S., Sharifi, N., Chen, Y., Yao, X.: Tension-relaxation in vivo computing principle for tumor sensitization and targeting. IEEE Trans. Cybern. **52**(9), 9145–9156 (2022). https://doi.org/10.1109/TCYB.2021.3052731
19. Shi, S., et al.: Microrobots based in vivo evolutionary computation in two-dimensional microchannel network. IEEE Trans. Nanotechnol. **19**, 71–75 (2020). https://doi.org/10.1109/TNANO.2019.2960126
20. Sun, C., Lee, J.S.H., Zhang, M.: Magnetic nanoparticles in MR imaging and drug delivery. Adv. Drug Deliv. Rev. **60**(11), 1252–65 (2008)
21. Thoidingjam, S., Tiku, A.B.: New developments in breast cancer therapy: role of iron oxide nanoparticles. Adv. Nat. Sci.: Nanosci. Nanotechnol. **8**(2), 023002 (2017)
22. Wintermark, M., et al.: Comparative overview of brain perfusion imaging techniques. J. Neuroradiol. **32**(5), 294–314 (2005)

Text Encryption Scheme Based on Chaotic Map and DNA Strand Displacement

Jing Yang[1]($\boxtimes$), Congcong Liu[1], Zhixiang Yin[2], Shiji Yang[1], Yue Zhang[1], and Guoqing He[1]

[1] School of Mathematics and Big Data, Anhui University of Science and Technology, Huainan 232001, China
jyangh82@163.com
[2] School of Mathematics, Physics and Statistics, Shanghai University of Engineering and Technology, Shanghai 201620, China

Abstract. A text encryption scheme is proposed by combining chaotic map, DNA encoding and DNA strand displacement (DSD). The scheme firstly scrambles the plaintext information and the sequence generated by the two-dimensional logistic chaotic map through Arnold map, and then performs XOR operation with the sequence generated by the one-dimensional logistic chaotic map, and then converts it into a DNA sequence matrix. Secondly, two encryption rules based on DNA strand displacement are proposed according to the diversity of DNA strand displacement, and the new DNA sequence is obtained by randomly selecting these two encryption rules to perform DNA strand displacement reaction on the DNA sequence. Finally, the new DNA sequence is decoded to obtain the encrypted cipher-text. The example analysis demonstrates that the encryption scheme provides effective encryption and security.

Keywords: chaotic map · text encryption · DNA encoding · DNA strand displacement

1 Introduction

With the rapid development of computer and internet technology, people are transmitting information via the internet more and more frequently. In order to prevent the transmitted information from being intercepted and cracked, people are transmitting information via the internet more and more frequently. Information encryption is an important means to ensure information security, but the traditional encryption methods have been difficult to meet the needs of information encryption. Chaos is a kind of complex dynamic behavior similar to random generated by certain nonlinear system. Chaotic systems have such characteristics as long-term unpredictability, good pseudo-randomness, topological transitivity, sensitivity to initial conditions and system parameters [1–4], which enable them to meet the chaos and diffusion principles in cryptographic design systems [5, 6]. Its emergence breaks the limitations of traditional encryption schemes [7–11], and opens up a new path for the technical research of cryptography. Matthews was the first to explicitly

© The Author(s), under exclusive license to Springer Nature Singapore Pte Ltd. 2024
L. Pan et al. (Eds.): BIC-TA 2023, CCIS 2061, pp. 317–330, 2024.
https://doi.org/10.1007/978-981-97-2272-3_24

mention "chaotic ciphers" in 1989 when he designed a sequential cipher algorithm using a one-dimensional chaotic map logistic [12], which has since been developed to some extent.

At present, information encryption algorithms based on chaos are usually based on a simple scrambling-diffusion model, which theoretically has good security, computational efficiency and complexity. However, in the actual encryption process, if only the chaotic system is used for scrambling-diffusion encryption, the random sequence generated by the low-dimensional chaotic system has a poor discrete type and a small chaotic space, resulting in a small key space and easy to decipher [13–16]. The high-dimensional chaotic system has complex structure and many parameters. Although it improves the security of the cryptographic system, it increases the cost of software or hardware implementation and computational complexity [17–19]. Therefore, the application of chaotic system combined with other technologies in information encryption has become a research hotspot in the field of information encryption.

Due to the inherent characteristics of the DNA molecule such as ultra-massive parallelism, large storage capacity and low energy consumption, the encryption algorithms based on DNA computing have been well developed in the field of encryption [20–22]. However, using DNA encryption technology alone for information encryption requires complex biological experiments, which increases the cost of information encryption. Therefore, many scholars began to apply DNA and chaotic systems to the field of information encryption after combining them, and achieved good encryption results. In 2017, Hu *et al.* constructed a new spatiotemporal chaotic system through the Logistic-Sine system, and used DNA deletion and DNA insertion pseudo-operations to confuse DNA-encoded diffusion images, proposing an image cryptosystem [23]. In 2018, Ran *et al.* proposed an image encryption algorithm that fused Logistic-sine mapping, Arnold map and DNA encoding, which improved the algorithm's sensitivity to plaintext and keys, and enhanced the ability to resist statistical attacks [24]. In 2019, Liu *et al.* proposed a color image encryption algorithm based on DNA coding by combining logic diagram, spatial diagram and DNA coding, providing an effective and secure method for real-time image encryption and transmission [25]. In 2020, Li and others proposed an image encryption algorithm based on Chen hyper-chaos and DNA coding [26]. In 2021, Wang *et al.* proposed a new text encryption scheme by combining polymerase chain reaction with 2D-LASM chaotic map [27]. In 2022, Tang *et al.* combined chaotic mapping with DNA strand displacement reaction and proposed a two-layer image encryption scheme. The simulation results and security analysis showed that the proposed encryption scheme can effectively protect image information and resist conventional information attack [28].

In this paper, a text encryption algorithm is proposed by combining DNA strand displacement with chaotic map. After the plaintext information is converted into an one-dimensional decimal matrix through ASCII code, it is firstly scrambled and diffused through one-dimensional logistic chaotic map, two-dimensional logistic chaotic map and Arnold map to complete preliminary encryption. Secondly, according to the diversity of DNA strand displacement, two encryption rules based on DNA strand displacement are proposed, and then two encryption rules based on DNA encoding and DNA strand displacement are used for secondary encryption, and finally the encrypted cipher text is obtained by decoding. Through example analysis, it is found that the encryption scheme has good encryption feasibility and anti-attack capability.

2 Preliminary Knowledge

2.1 Logistic Chaotic Map

2.1.1 One-Dimensional Logistic Chaotic Map

The one-dimensional logistic chaotic map has its origins in the worm-mouth model and is therefore also known as a worm-mouth mapping. Its structure is simple in terms of mathematical form, but it has extremely complex dynamic behavior. It is widely used in the field of secure communication. Its mathematical expression formula is as follows:

$$z_{n+1} = \mu z_n(1 - z_n), z_n \in (0, 1), \mu \in [0, 4]. \tag{1}$$

where μ is the system parameter. According to the bifurcation diagram of the logistic map in Fig. 1(a) and the Lyapunov exponent diagram of the logistic map in Fig. 1(b), when $3.56945972\cdots < \mu < 4$, the Lyapunov exponent of the logistic map is greater than 0, and the chaotic system is in a chaotic state.

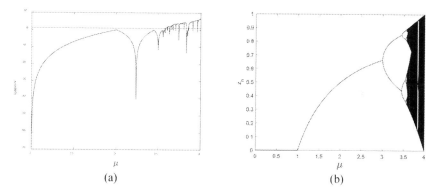

(a)   (b)

Fig. 1. (a) Bifurcation diagram of logistic map; (b) Lyapunov exponents of logistic map.

2.1.2 Two-Dimensional Logistic Chaotic Map

The definition of a two-dimensional logistic chaotic map is as follows:

$$\begin{cases} x_{n+1} = r_1 x_n(1 - x_n) + s_1 y_n^2 \\ y_{n+1} = r_2 y_n(1 - y_n) + s_2(x_n^2 + x_n y_n) \end{cases}. \tag{2}$$

where r_1, r_2, s_1, s_2 are system parameters. When $2.75 \le r_1 \le 3.4$, $2.75 \le r_2 \le 3.45$, $0.15 \le s_1 \le 0.21$, $0.13 \le s_2 \le 0.15$, the two-dimensional logistic chaotic system is in a chaotic state.

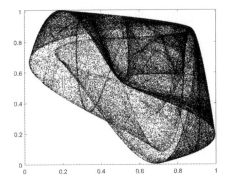

Fig. 2. Chaotic attractor diagram for a two-dimensional logistic system.

When the initial values of the two-dimensional logistic chaotic system $x_0 = 0.1546$, $y_0 = 0.7697$ and the control parameters $r_1 = 3.3$, $r_2 = 3.25$, $s_1 = 0.18$, $s_2 = 0.14$, its chaotic attractor diagram is shown in Fig. 2.

The partial bifurcation diagram of two-dimensional logistic chaotic map is shown in Fig. 3. It can be seen that there are few periodic bifurcations in the bifurcation diagram, which is suitable for encryption systems.

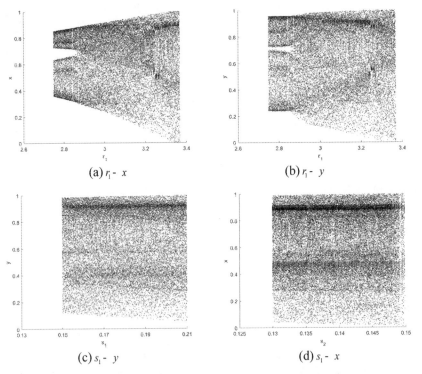

(a) r_1 - x (b) r_1 - y

(c) s_1 - y (d) s_1 - x

Fig. 3. Bifurcation diagram of two-dimensional logistic chaotic map.

2.2 Arnold Map

Arnold map, also known as cat map, were first introduced by Arnold. The definition of the two-dimensional Arnold map on the unit square is

$$\begin{bmatrix} x' \\ y' \end{bmatrix} = \begin{bmatrix} 1 & 1 \\ 1 & 2 \end{bmatrix} \begin{bmatrix} x \\ y \end{bmatrix} \bmod 1, x, y \in (0, 1). \tag{3}$$

where mod represents modular operation.

In order to apply the Arnold map to matrices of size $M \times M$, Eq. (3) is extended. Generalizing its phase space to $\{0, 1, \cdots, M - 1\}$, the generalized Arnold map can be obtained, as shown in Eq. (4):

$$\begin{bmatrix} x' \\ y' \end{bmatrix} = \begin{bmatrix} 1 & a \\ b & ab + 1 \end{bmatrix} \begin{bmatrix} x \\ y \end{bmatrix} \bmod M, x, y \in (0, 1, \cdots, M - 1). \tag{4}$$

where a, b are positive integers.

At this point, the point in the matrix of $M \times M$ at position (x, y) can be transformed to position (x', y') by Eq. (4). When the matrix is a one-dimensional vector, the Arnold transform is applied to any coordinate position $(1, j)$ of the vector to obtain the new coordinate position (p, q), as in Eq. (5):

$$\begin{bmatrix} p \\ q \end{bmatrix} = \begin{bmatrix} 1 & a \\ b & ab + 1 \end{bmatrix} \begin{bmatrix} 1 \\ j \end{bmatrix}. \tag{5}$$

where $p = aj + 1, q = b + (ab + 1)j$.

Considering only q, the transformation from a point at position $(1, j)$ to position $(1, q)$ can be achieved through pseudo-random variables a and b. At the same time, if $ab + 1$ is regarded as a new random number, which is still recorded as a, then q becomes $q = b + aj$.

2.3 DNA Coding Rules

DNA is a macro-molecular compound with a double-stranded structure, consisting of four deoxynucleotides: adenine (A), guanine (G), cytosine (C) and thymine (T). In DNA sequences, they are strictly required to follow the principle of base complementary pairing design, that is, A and T complement, G and C complement. In binary, because 0 and 1 are complementary, 00 and 11 and 10 and 01 are also complementary respectively. Therefore, four deoxynucleotides, A, T, C and G, can be used to code 00, 01, 10 and 11. The binary is combined and the principle of base complementary pairing of DNA, eight DNA coding rules can be obtained, as shown in Table 1.

The characters of plaintext information can be converted into decimal integers with numeric codes ranging from 0–255 through ASCII code, which is within the range of 8-bit binary values. Therefore, each character can be converted into a DNA sequence of 4nt (nt: the length unit of a DNA base). For example, the number code of the character "s" in ASCII code is 115, and its binary sequence is 01110011. If cod rule 3 is used, the resulting DNA cod sequence is AGCG, while cod rule 7 is CATA. If we use the corresponding cod rules to decode the obtained DNA sequence, we can get the original binary sequence 01110011, but when decoding with other cod rules, we cannot get the original binary sequence. So it is impossible to get the original character "s".

Table 1. DNA coding rules

Rule	1	2	3	4	5	6	7	8
A	00	00	01	01	10	10	11	11
C	01	10	00	11	00	11	01	10
G	10	01	11	00	11	00	10	01
T	11	11	10	10	01	01	00	00

2.4 DNA Strand Displacement

DNA strand displacement (DSD) is the process by which a single strand of DNA reacts structurally with a partially complementary double strand, thereby releasing another single strand of DNA. DNA strand displacement relies on its own energy level prim, specific recognition, sensitivity and accuracy to achieve the release and binding of DNA strands without enzyme catalysis. The initial DNA strand is usually a long single strand of DNA that is completely complementary to the substrate. It can specifically recognize and connect to the foothold region of the partially complementary DNA substrate, and then triggers the strand displacement reaction. The target strand is completely replaced, and the basic principle of the reaction is shown in Fig. 4(a). Different from biochemical reactions that require enzyme catalysis, the direction of DNA strand displacement reaction can start either from the 5'end foothold region of the substrate or from the 3'end foothold region of the substrate, as shown in Fig. 4(a) and Fig. 4(b). These two types of DNA strand displacement are denoted as DSD-a and DSD-b, respectively.

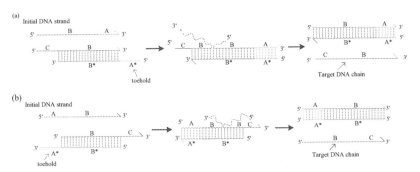

Fig. 4. Basic principles of DNA strand displacement. (a) DSD-a, DNA strand displacement starting from the 5' end foothold region; (b) DSD-b, DNA strand displacement starting from the 3' end foothold region.

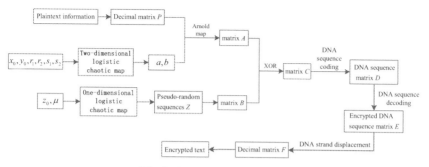

Fig. 5. Encryption Flow Chart.

3 Encryption Algorithms

The text encryption scheme based on chaotic map and DNA strand displacement is mainly divided into two parts: encryption at chaotic level and encryption at DNA strand displacement level. At the chaotic level, the two-dimensional logistic chaotic map and Arnold map are used for position scrambling, and then the logistic chaotic map is used for XOR diffusion. At the DNA strand displacement level, sequence encryption is performed using DNA strand displacement. The flow chart of the encryption algorithm is shown in Fig. 5. The specific encryption steps are as follows:

Step 1: Convert the plaintext message to be encrypted into a numeric code between 0 and 255 to obtain a decimal matrix P of $1 \times m$.

Step 2: The initial values x_0, y_0 and parameters r_1, r_2, s_1, s_2 of the two-dimensional logistic chaotic map are used as the chaotic key, and $2m + t$ times are iterated, where $t = \frac{m^2}{2} t = \frac{m^2}{2}$ to obtain two chaotic sequences x and y with a length of $2m+t$. According to Eq. (6), the sequence x is slightly perturbed by the sequence y, and a pseudo-random matrix E with a length of $2m + t$ is obtained. And through $mod(floor((x + 100) * 10^{10}), 10 * m) + 1$ it is converted into a decimal integer matrix on $[1, 10 * m]$, and the pseudo-random matrix X of size $1 \times 2m$ is obtained after removing the first t item. Here, mod means rounding, and *floor* means rounding down.

$$E = x + h * \sin(y). \tag{6}$$

where $h = (x(1, m/2) + y(1, m/2))/2$.

Step 3: The pseudorandom matrix X and Arnold map are used to scramble the position of matrix P. The values of the integer matrix q are given in Eq. (7), and the specific position permutation formula is given in Eq. (8). The dislocated matrix is denoted as A.

$$q = \mod (b + a. * (1 : m), m) + 1. \tag{7}$$

where $a = X (1 : m)$, $b = X (m + 1 : 2 * m)$.

$$t = A1(j); A1(j) = A1(q(j)); A1(q(j)) = t. \tag{8}$$

where $A1 = P(:)$.

Step 4: The initial value z_0 and parameter μ of the one-dimensional logistic chaotic map are used as the chaotic key, and the first 1000 items are removed after $1000 + m$ iterations to obtain the pseudo random sequence Z. The pseudo-random matrix B is obtained by normalizing the pseudo-random sequence Z to the integer interval [0,255].

Step 5: XOR the scrambled matrix A in step 3 and the pseudo-random matrix B in step 4 to obtain matrix C.

Step 6: First convert matrix C to 8-bit binary matrix, and then convert it to DNA sequence matrix D of size $1 \times 4m$ through randomly selected DNA coding rules. Then, the matrix D is evenly divided into $\frac{m}{4}$ DNA sequence sub-matrices $D_1, D_2, \cdots, D_{m/4}$ of size 1×16. When $\frac{m}{4}$ is not an integer, you need to add a space at the end of the plaintext information in step 1 (that is, the space key on the computer keyboard does not display any content in the ASCII code), so that the length of the plaintext can be divided by 4. Each sub-matrix $D_1, D_2, \cdots, D_{m/4}$ contains a DNA sequence $D'_1, D'_2, \cdots, D'_{m/4}$ with a length of 16nt. Set the left end of each DNA sequence as 5'end and the right end as 3' end.

Step 7: Use the numbers "1" and "2" to represent the DNA strand displacement of DSD-a and DSD-b respectively. Randomly select $\frac{m}{4}$ numbers in the integer interval [1, 2] and integer interval [4, 12] to determine the type of DNA strand displacement and the length of foothold for encrypting DNA sequences $D'_1, D'_2, \cdots, D'_{m/4}$. The DNA sequences $D'_1, D'_2, \cdots, D'_{m/4}$ are scrambled and encrypted by DNA strand displacement to obtain new DNA sequences $E'_1, E'_2, \cdots, E'_{m/4}$, thereby obtaining DNA sequence sub-matrixes $E_1, E_2, \cdots, E_{m/4}$. The obtained DNA sequence sub-matrices $E_1, E_2, \cdots, E_{m/4}$ are combined into a DNA sequence matrix E of size $1 \times 4m$ by column. The specific process of scrambling and encrypting DNA sequences $D'_1, D'_2, \cdots, D'_{m/4}$ using DNA strand displacement is as follows:

When performing displacement encryption on a DNA sequence by DNA strand displacement, the sequence of the DNA substrate involved in the reaction is selected as the key, but not any sequence on the substrate can be used as the key. The initial DNA sequence determines the DNA sequence of the foothold region and the complementary region of the substrate DNA sequence involved in the reaction, however, the principle of complementary base pairing needs to be satisfied. Meanwhile, in order to make the length of the substituted DNA sequence consistent with the length of the original DNA strand, the length of the encryption region should be the same as that of the foothold. The number of displaced bases of the initial DNA sequence is the same as the length of the foothold as the chaotic key. Assuming the length of the foothold is 6 nt, the chaotic key should contain 6 bases. The length of the encryption region should also be 6 nt, and the required 6 bases can be randomly selected, and there are 4^6 possibilities in total.

Assuming that $D_i(i = 1, 2, \cdots, \frac{m}{4})$ is ATCCGCTTACGGTCAC, and D'_i is 5'-ATCCGCTTACGGTCAC-3', the sequence is replaced and encrypted with two types of DNA strand displacement, respectively. When choosing to encrypt it with DNA strand displacement of type DSD-a, assuming that the integer randomly selected on the integer interval [4, 12] is 7, the length of the foothold is 7 nt, and the length of the encrypted area should also be 7 nt. As shown in Fig. 6, the sequence of its foothold area (yellow area in the figure) is GCCAGTG, the sequence of the upper part of the complementary area (blue area in the figure) is ATCCGCTTA, the sequence of the lower part is TAGGC-GAAT, and the sequence assigned to the encrypted area (red area in the figure) in Fig. 6 is CGGATAC. The DNA sequence D_i': 5'-ATCCGCTTACGGTCAC-3' is encrypted into the DNA sequence E_i': 5'-CGGATACATCCGCTTA-3' through the DNA strand displacement of type DSD-a. The chaotic key is that the base replaced with the initial DNA sequence is CGGTCAC.

When choosing to encrypt it with DNA strand displacement of type DSD-b, if the integer randomly selected in the integer interval [4, 12] is also 7, it is similar to the encryption of type DSD-a. As shown in Fig. 7, the sequence of its foothold area (yellow area in the figure) is TAGGCGA, the sequence of the upper part of the complementary area (blue area in the figure) is TACGGTCAC, the sequence of the lower part is ATGCCAGTG, and the sequence assigned to the encryption area (red area in the figure) in Fig. 7 is CGGATAC. The DNA sequence D_i': 5'- ATCCGCTTACGGTCAC-3' was successfully encrypted into the DNA sequence E_i':5'-TACGGTCACGGGATAC-3' through the DNA strand displacement of type DSD-b. The chaotic key is that the base replaced with the initial DNA sequence is ATCCGCT.

Step 8: Randomly select an encoding rule to decode the matrix E, and then convert the decoded matrix into a decimal matrix F.

Step 9: Matrix F is translated into encrypted text by ASCII code.

Since all of the above encryption processes are reversible, the process of decryption is the inverse of the encryption process.

(a)

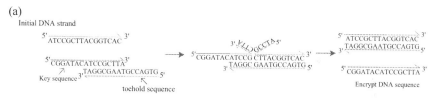

Fig. 6. Encryption of sequence D_i' based on DNA strand displacement of type DSD-a.

(b)

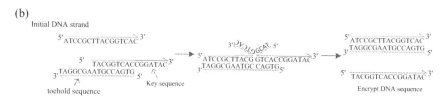

Fig. 7. Encryption of sequence D_i' based on DNA strand displacement of type DSD-b.

4   Instance Validation

To better demonstrate the encryption scheme, the plaintext message "Industry is the parent of success." is used as an example and encrypted based on MATLAB R2017b simulation software.

Step 1: Since the length of the decimal matrix converted from the plaintext message is 34, which is not divisible by 4. It is necessary to add 2 spaces after the plaintext information to obtain a matrix P of size 1×36: [73, 110, 100, 117, 115, 116, 114, 121, 32, 105, 115, 32, 116, 104, 101, 32, 112, 97, 114, 101, 110, 116, 32, 111, 102, 32, 115, 117, 99, 99, 101, 115, 115, 46, 32, 32], which gives $m = 36$ and $t = 648$.

Step 2: The initial values of the two-dimensional logistic chaotic map x_0 and y_0 are set to 0.1546 and 0.7697 respectively, and the parameters r_1, r_2, s_1 and s_2 are set to 3.3, 3.25, 0.18 and 0.14 respectively, with 720 iterations to obtain two chaotic sequences x and y. According to Eq. (6), the sequence x is slightly perturbed by sequence y, and a pseudo-random matrix E with a length of 720 is obtained, and transformed by $\mod(floor((x + 100) * 10^{10}), 10 * m) + 1$ into a matrix of decimal integers on [1,360], and a pseudo-random matrix X of size 1×72 is obtained after removing the first 648 entries: [167, 189, 360, 182, 211, 154, 331, 346, 328, 9, 265, 342, 273, 74, 326, 52, 79, 30, 47, 296, 69, 134, 228, 185, 21, 352, 353, 265, 336, 50, 168, 3, 101, 121, 355, 183, 33, 238, 120, 40, 110, 186, 16, 35, 154, 353, 352, 136, 301, 290, 360, 84, 329, 132, 83, 330, 45, 311, 142, 92, 358, 237, 136, 174, 335, 226, 33, 264, 328, 306, 264, 50].

Step 3: According to Eq. (7) and Eq. (8), the matrix P is positional permuted using the pseudo-random matrix X and the Arnold map to obtain the permuted matrix A: [121, 115, 116, 100, 104, 101, 99, 115, 115, 32, 117, 99, 32, 116, 105, 112, 32, 32, 73, 115, 115, 101, 101, 114, 116, 111, 97, 117, 46, 32, 32, 110, 114, 32, 102, 110].

Step 4: The initial value z_0 and parameter μ of the one-dimensional logistic chaotic map are set to 0.1 and 3.9746 respectively, and the first 1000 items are removed after 1036 iterations to obtain the pseudo random sequence Z. Normalize pseudo-random sequence Z to integer interval [0, 255] to obtain pseudo-random matrix B:[249, 225, 25, 99, 98, 141, 69, 104, 185, 211, 77, 29, 43, 97, 178, 201, 128, 147, 52, 79, 1, 224, 27, 107, 124, 169, 233, 199, 47, 86, 187, 169, 47, 211, 80, 37].

Step 5: XOR the scrambled matrix A in step 3 and the pseudo-random matrix B in step 4 to obtain matrix C:[128, 146, 109, 7, 10, 232, 38, 27, 202, 243, 56, 126, 11, 21, 219, 185, 160, 179, 125, 60, 114, 133, 126, 25, 8, 198, 136, 178, 1, 118, 155, 199, 93, 243, 54, 75].

Step 6: Convert matrix C into an 8-bit binary matrix. The randomly selected DNA coding rule is the fifth type, and the encoded DNA sequence matrix D of size 1×144 is obtained: [AATCCGCCGGCTCCGAAATCTATCCGAACTAGTGCTC-TACCAATCGGGCTTGAGGGGCGTCCCGCGTCTGGCCCGTAATAACAGATAAC CGGCTGAATACCTATGCTACATGACAGAGCAGTGTCGTCATATCACATAGGTG AG].

The matrix D is evenly divided into nine DNA sequence sub-matrices $D_1, D_2, \cdots, D_9$ of size 1×16. Each sub-matrix $D_i(i = 1, 2, \cdots, 9)$ contains a DNA sequence $D'_1, D'_2, \cdots, D'_9$ of length 16nt, setting the left end of each DNA sequence to the 5' end and the right end to the 3' end, the corresponding DNA sequences are shown in Fig. 8.

Step 7: Randomly select 9 numbers from the integer interval [1, 2] as 2, 1, 1, 2, 2, 1, 2, 1 and 2, and randomly select 9 numbers from the integer interval [4, 12] as 8, 5, 11, 4, 10, 12, 9, 7 and 6.

The nine random keys used for DNA strand displacement scramble encryption for $D_i'(i = 1, 2, \cdots, 9)$ by DNA strand displacement technology are TCTATACT, CGAGA, GTAGACTTCAT, GCAG, AATCTACAGA, TAGAGCCAGTAC, GTTAGCAGG, ACGTGCG and TGCTTA. The new DNA sequence $E_1', E_2', \cdots, E_9'$ obtained by performing DNA strand displacement encryption on $E_i'(i = 1, 2, \cdots, 9)$ is shown in Fig. 9, thereby obtaining DNA sequence sub-matrixes $E_1, E_2, \cdots, E_9$.

Step 8: Decode the matrix E. The selected random rule is the second. Then convert the decoded matrix into decimal matrix F: [209, 79, 248, 44, 67, 66, 237, 185, 208, 115, 38, 104, 164, 116, 237, 229, 219, 239, 38, 34, 31, 80, 49, 227, 227, 195, 98, 22, 85, 114, 115, 135, 186, 128, 171, 231].

Step 9: The matrix F is translated into encrypted text by ASCII code.

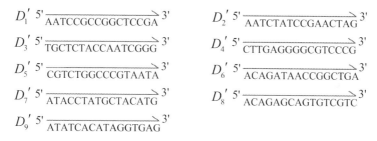

D_1' 5'—————————→3' AATCCGCCGGCTCCGA

D_2' 5'—————————→3' AATCTATCCGAACTAG

D_3' 5'—————————→3' TGCTCTACCAATCGGG

D_4' 5'—————————→3' CTTGAGGGGCGTCCCG

D_5' 5'—————————→3' CGTCTGGCCCGTAATA

D_6' 5'—————————→3' ACAGATAACCGGCTGA

D_7' 5'—————————→3' ATACCTATGCTACATG

D_8' 5'—————————→3' ACAGAGCAGTGTCGTC

D_9' 5'—————————→3' ATATCACATAGGTGAG

Fig. 8. DNA sequence sub-matrices $D_1', D_2', \cdots, D_9'$.

E_1' 5'—————————→3' AATCCGCCTCTATACT

E_2' 5'—————————→3' CGAGAGTAGACACTAG

E_3' 5'—————————→3' TTCATTACCAATCGGG

E_4' 5'—————————→3' CTTGAGGGGCGTGCAG

E_5' 5'—————————→3' CGTCTGAATCTACAGA

E_6' 5'—————————→3' TAGAGCCAGTACCTGA

E_7' 5'—————————→3' ATACCTAGTTAGCAGG

E_8' 5'—————————→3' ACGTGCGAGTGTCGTC

E_9' 5'—————————→3' ATATCACATATGCTTA

Fig. 9. DNA sequence sub-matrices $E_1', E_2', \cdots, E_9'$.

5 Algorithm Security Analysis

5.1 Key Space Analysis

The key space refers to the value space of all keys in the encryption algorithm. The key of the above encryption scheme is composed of two parts: chaotic level and DNA level. The key of chaotic level includes the initial value z_0 and its parameter μ of one-dimensional logistic chaotic map and the initial values x_0, y_0 and its parameter r_1, r_2, s_1 and s_2 of two-dimensional logistic chaotic map. If the floating-point precision of the computer is 10^{-13}, the size of the chaotic key space is $(10^{13})^8 = 10^{104}$. The keys at the DNA level include:

1) In step 6, the matrix C randomly selects a DNA encoding rule and converts it to the DNA sequence matrix D. In step 8, when the DNA sequence matrix E is converted to the decimal matrix F, the DNA coding rules are also randomly selected once. Therefore, DNA encoding rules were randomly selected twice in the whole encryption process, and the size of the key space was $8 \times 8 = 64$;

2) In step 7, it is necessary to randomly select $\frac{m}{4}$ numbers in the integer interval $[1, 2]$ and the integer interval $[4, 12]$ to determine the DNA strand displacement type and the length of the foothold for encryption of $D'_i (i = 1, 2, \cdots, \frac{m}{4})$, and generate 4–12 DNA key sequences according to the length of the foothold. The key space is at least $2^{\frac{m}{4}} \times 9^{\frac{m}{4}} \times 4^m$. So the key space at the DNA level is $64 \times 2^{\frac{m}{4}} \times 9^{\frac{m}{4}} \times 4^m$.

Therefore, the key space of the entire encryption algorithm is $64 \times 2^{\frac{m}{4}} \times 9^{\frac{m}{4}} \times 4^m \times 10^{104}$, which is much larger than 2^{100}, and the more plaintext information characters, the larger the key space. This key space is large enough to withstand an exhaustive attack against the key.

5.2 Sensitivity Analysis of Keys

Key sensitivity refers to the fact that a small change in a key during decryption, with other keys unchanged, can lead to a huge change in the decryption result. Chaotic systems, on the other hand, are very sensitive to initial values, and a small change in the initial value or parameters can lead to a huge change in the result. For example, if only the parameter r_2 of the two-dimensional logistic chaotic map is changed from 3.25 to 3.250000000000001, the decryption result will become "atsryssfp.hcndItt e re ou iens scu". It can be seen that a very slight change in the key will give a decryption result that is completely different from the plaintext message. The key of this encryption scheme is therefore sensitive enough to provide good confidentiality security and can effectively resist brute force attacks.

6 Conclusion

According to the diversity of DNA strand displacement, two encryption rules based on DNA strand displacement are proposed, and a text encryption scheme based on chaotic map and DNA strand displacement is proposed. When using these two DNA strands to replace the encryption rules, the key space of the encryption algorithm is increased due to the difference in the length of the foothold region. The example analysis shows that the encryption scheme has better encryption effect and security. This encryption scheme can also be extended to encrypt images.

Acknowledgment. This project is supported by National Natural Science Foundation of China (No. 62272005, No. 61672001).

References

1. Liu, H., Zhao, B., Huang, L.: A novel quantum image encryption algorithm based on crossover operation and mutation operation. Multimed. Tools Appl. **78**(14), 20465–20483 (2019)

2. Chen, S.X., Tang, Y.R.: A triple dislocation algorithm for RGB color images based on chaotic systems. J. Chongqing Univ. Posts Telecommun. Nat. Sci. Edition **30**(6), 812–818 (2018)
3. Chai, X., Chen, Y., Broyde, L.: A novel chaos-based image encryption algorithm using DNA sequence operations. Opt. Lasers Eng. **88**, 197–213 (2017)
4. Zhu, S.Q., Li, J.Q., Ge, G.Y.: A new algorithm for image encryption based on a new four-dimensional discrete chaotic mapping. Comput. Sci. **44**(1), 188–193 (2017)
5. Shannon, C.E.: Communication theory of secrecy systems. Bell Syst. Tech. J. **28**(4), 656–715 (1949)
6. Liu, Y.: A study on the characteristics of chaotic motion and its application in cryptography. Electron. Test. **9**, 35–40 (2016)
7. Kocarev, L., Jakimoski, G., Stojanovski, T., et al.: From chaotic maps to encryption schemes. In: IEEE International Symposium on Circuits & Systems. IEEE (1998)
8. Wheeler, D.D.: Problems with chaotic cryptosystems. Cryptologia **13**, 243–250 (1989)
9. Alvarez, G., Li, S.: Some basic cryptographic requirements for chaos-based cryptosystems. Int. J. Bifurc. Chaos **16**(08), 2129–2151 (2006)
10. Pak, C., Huang, L.: A new color image encryption using combination of the 1D chaotic map. Signal Process. **138**, 129–137 (2017)
11. Wang, X.Y., Zhu, X.Q., Zhang, Y.Q.: An image encryption algorithm based on Josephus traversing and mixed chaotic map. IEEE Access **6**, 23733–23746 (2018)
12. Matthews, R.: On the derivation of a chaotic encryption algorithm. Cryptologia **13**(1), 29–42 (1989)
13. Cao, C., Sun, K., Liu, W.: A novel bit-level image encryption algorithm based on 2D-LICM hyperchaotic map. Signal Process. **143**, 122–133 (2018)
14. Zhu, H., Zhao, Y., Song, Y.: 2D logistic-modulated-sine-coupling-logistic chaotic map for image encryption. IEEE Access **7**, 14081–14098 (2019)
15. Zhang, Y., Tang, Y.: A plaintext-related image encryption algorithm based on chaos. Multimed. Tools Appl. (2018)
16. Akhavan, A., Samsudin, A., Akhshani, A.: Cryptanalysis of an image encryption algorithm based on DNA encoding. Opt. Laser Technol. **77**(6), 6647–6669 (2017)
17. Han, R.: Image encryption algorithm based on high-dimensional cat mapping. J. Longdong Coll. **29**(05), 9–12 (2018)
18. Xu, B., Sun, Y.W., Li, Y., Wang, Y.S.: Improved image encryption algorithm based on high-dimensional chaotic system. J. Jilin Univ. (Inf. Sci. Edition) **30**(01), 12–17 (2012)
19. He, J.N.: A new algorithm for image grouping encryption based on high-dimensional chaotic systems. Comput. Eng. Des. **31**(21), 4546–4549 (2010)
20. Alruily, M., Shahin, O.R., Al-Mahdi, H., Taloba, A.I.: Asymmetric DNA encryption and decryption technique for Arabic plaintext. J. Ambient Intell. Humaniz. Comput. 1–17 (2021)
21. Sreeja, C.S., Misbahuddin, M.: DNA cryptography for secure data storage in cloud. Int. J. Netw. Secur. **20**(3), 447–454 (2018)
22. Leier, A., Richter, C., Banzhaf, W., Rauhe, H.: Cryptography with DNA binary strands. BioSystems **57**(1), 13–22 (2000)
23. Hu, T., Liu, Y., Gong, L.H., Guo, S.F., Yuan, H.M.: Chaotic image cryptosystem using DNA deletion and DNA insertion. Signal Process. **134**, 234–243 (2017)
24. Ran, W., Wei, P.C., Duan, A.: Image encryption algorithm based on multi-chaotic mapping and DNA coding. Comput. Eng. Des. (2018)
25. Liu, P., Zhang, T., Li, X.: A new color image encryption algorithm based on DNA and spatial chaotic map. Multimed. Tools Appl. **78**(11), 14823–14835 (2019)
26. Li, Z.Y., Jiang, A.P., Shen, Y.Q.: An image encryption algorithm based on Chen hyperchaos and DNA coding. J. Nat. Sci. Heilongjiang Univ. **37**(05), 602–609 (2020)

27. Wang, X.Y., Yin, Z.X., Tang, Z., Yang, J., Cui, J.Z., Xu, R.J.: A 2D-LASM chaotic text encryption algorithm based on polymerase chain substitution reaction. J. Guangzhou Univ. (Nat. Sci. Edition) **20**(02), 23–27+34 (2021)
28. Tang, Z., Yin, Z.X., Wang, R.S., Wang, X.Y., Yang, J., Cui, J.Z.: A double-layer image encryption scheme based on chaotic maps and DNA strand displacement. J. Chem. **2022**, 1–10 (2022)

Numerical P Systems with Thresholds and Petri Nets

Luping Zhang[✉] and Zhimeng Zhang

Jiangxi Engineering Technology Research Center of Nuclear Geoscience Data Science
and System, Jiangxi Engineering Laboratory on Radioactive Geoscience and Big
Data Technology, School of Information Engineering, East China University
of Technology, Nanchang 330013, China
zhanglp@ecut.edu.cn

Abstract. Membrane computing provides efficient computing devices
for broad applications due to its distributed storage and the parallel
processing. As computing devices in membrane computing, numerical P
systems with thresholds (NPT systems) are proven to be Turing universal, and their computational and operational semantics need to be further investigated. In this work, the intrinsic relationship between NPT
systems and Petri nets is concerned. The ingredients of Petri nets are
associated with the elements of numerical variables in NPT systems,
and the operations of Petri nets are associated with the evolutions of
NPT systems. The results on the boundedness and reachability of NPT
systems are obtained by using the relationship between NPT systems
and Petri nets.

Keywords: Membrane computing · Numerical P systems ·
Threshold · Petri net

1 Introduction

Membrane computing proposed by Păun is a new branch of bio-inspired computing which interprets the information processing within a single cell or among a
group of cells from a computational perspective [11]. The membrane systems in
membrane computing are known as P systems [15]. In a P system, information
is encoded as objects distributed in delimited membranes, and it is processed
in a parallel manner [17]. Distributed and parallel P systems have been successfully applied in modeling biological systems [4], processing digital images (eg.,
the segmentation and skeletonization aspects of the digital images) [3], designing
robot controllers based on P systems [2], and so on.

Numerical P system is defined as a numerical computing device where a
piece of information is encoded by numerical variables distributed in the cell-
like structure, and the information is processed by the use of production-
repartition programs associated with numerical variables [12]. In the research
of the theories of numerical P systems, the systems are usually used as number generating/accepting devices [23], function computing devices, and gener-
ating/accepting devices [20,21]. A number of variants of numerical P systems

© The Author(s), under exclusive license to Springer Nature Singapore Pte Ltd. 2024
L. Pan et al. (Eds.): BIC-TA 2023, CCIS 2061, pp. 331–338, 2024.
https://doi.org/10.1007/978-981-97-2272-3_25

have been developed as the computing devices mentioned above by introducing the restrictions on the use of programs, such as the thresholds of the production functions [10], the enzymatic conditions [19], and the migrating variables [22].

Numerical P systems with thresholds (NPT systems) are a variant of numerical P systems where a program is enabled if and only if each of the variables in the production function meets the threshold condition [18]. The computational power of NPT systems has been proven to be equivalent to that of Turing machines when they are used as number generating/accepting and function computing devices [9]. More operational and computational semantics of NPT systems need to be investigated. Petri nets are a class of computing models with strong mathematical support for describing and analyzing distributed and concurrent computation [16]. Petri net was first introduced for the modeling and analyzing of membrane systems in [13] where a systematic and structural link between a Petri nets and a membrane system was established. Then, the behaviors of membrane systems were employed to propose new Petri nets [8]. The usefulness of using both Petri nets and membrane systems in simulating generally biological processes is demonstrated in [1]. The aforementioned research indicates that there will be many interesting results if the Petri nets were employed to analyze NPT systems.

In this work, the intrinsic relationship between NPT systems and Petri nets is concerned. There are 3 ingredients in an NPT system: the hierarchical structure, the variables with values, and the programs. The structure of NPT systems is described by the connections between places and transitions in the related Petri nets; the numerical variables are associated with the places where the value of a variable is associated with the number of tokens in a place; the programs in NPT systems are associated with the transitions in the related Petri nets. In this way, the Petri net can simulate NP systems with thresholds where each production function is a linear function with positive coefficients.

This work is organized as follows. In the next section, the symbols and notations NPT systems and Petri nets are briefly introduced. In Sect. 3, the evolution of an NPT system are related with that of a Petri net, and the results on the boundedness and reachability of NPT systems are obtained by designing corresponding Petri nets. In the last section, the conclusions are provided.

2   Notations and Definitions

In this section, the definitions of the NPT systems and Petri nets are briefly recalled.

Definition 1. *An NPT system of degree* m *($m \geq 1$) is a construct*

$$\Pi = (m, H, \mu, T_h, (Var_1, Pr_1, Var_1(0)), \ldots, (Var_m, Pr_m, Var_m(0)), V_{out})$$

where

- m *($m \geq 1$) is the number of membranes in the system;*

- H is the label set for the m membranes;
- μ is the hierarchical membrane structure represented by nested brackets with labels in H;
- T_h $(T_h \in \mathbb{Q})$ is the threshold;
- Var_i $(1 \leq i \leq m)$ is the set of variables in i-th membrane;
- Pr_i $(1 \leq i \leq m)$ is the finite set of programs in i-th membrane, where each program consisting of a production function $F_{i,j}(v_{i_1}, \ldots, v_{i_k})|_{T_h}$ and a repartition protocol $c_{j_1}|v_{j_1} + \ldots + c_{j_{n_j}}|v_{j_{n_j}}$ $(c_{j_l} \in N - \{0\}, 1 \leq l \leq n_j)$ has the form

$$F_{i,j}(v_{i_1}, \ldots, v_{i_k})|_{T_h} \to c_{j_1}|v_{j_1} + \ldots + c_{j_{n_j}}|v_{j_{n_j}};$$

- $Var_i(0)$ is the set of initial values of the variables in membrane i;
- Var_{out} is the set of output variables.

An NPT system consists of the hierarchical structure where information is encoded by distributed variables, and the information is processed by the use of programs. The NPT systems where each production function is a linear function with integer coefficients are proved to be Turing universal.

Definition 2. *A Petri net is a construct*

$$\mathcal{N} = (P, T_r, A, W, P_0, M_0),$$

where

- $P = \{P_0, P_1, \ldots, P_m\}$ is a finite set of places;
- $T_r = \{T_{r_0}, T_{r_1}, \ldots, T_{r_n}\}$ is a finite set of transitions such that $P \cap T_r = \emptyset$;
- $A \subseteq (P \times T_r) \cup (T_r \times P)$ is a finite set of arcs;
- $W : A \to \mathbb{N}$ assigns weights to each of the arcs;
- $P_0 \in P$ is the output place with no outgoing arc;
- $M_0 : P \to \mathbb{N}$ specifies an initial number of tokens in each place.

Petri nets are a type of distributed and concurrent model. Since the similar way of information processing, Petri nets are introduced in the research of NPT systems.

3 Main Results

In this section, the intrinsic relationship between NPT systems and Petri nets is investigated. The results on the dynamics of NPT systems are obtained by using the relationship and the existing results on Petri nets.

Theorem 1. *An NPT system Π where each production function is a linear function with integers is linked with a Petri net $[\![\Pi]\!]$ in which if there is a transition $C \to C'$ in the NPT system Π, then there exists an associated transition $M \to M'$ in the Petri net $[\![\Pi]\!]$; conversely, if there is a firing transition $M \to M'$ in the Petri net $[\![\Pi]\!]$, then there exists a transition $C \to C'$ in the system Π.*

Proof. Consider an NPT system Π where each production function is a linear function with integers. Construct a Petri net $[\![\Pi]\!] = (P, T_r, A, W, P_0, M_0)$ associated with system Π: each place in the net $[\![\Pi]\!]$ corresponds to a variable in system Π where the number of tokens in a place in the net $[\![\Pi]\!]$ equals to the value of the variable; each transition in the net $[\![\Pi]\!]$ corresponds to a program in system Π where a weight on an arc equals to the coefficient of the variable in the production function or the coefficient of the variable in the repartition protocol.

Here, a mapping $G : T_r \to \mathbb{N}$ is introduced to the Petri net where G assigns a threshold to each transition. A transition is enabled if the number of tokens in its input places is more than the threshold given by G. The occurrence of the enabled transition removes the tokens in the input places and distributes the tokens in the transition to the output places. The amount of tokens in the transition received from the input place depends on the number of tokens in the input place and the weight on the arc from the place to the transition. The number of tokens in the output places received from the transition depends on the number of tokens in the transition and the weights on the arc from the transition to the output places.

The working mode of net $[\![\Pi]\!]$ is associated with the maximally parallel manner of system Π where all applicable programs are chosen to be simultaneously applied in the restricted way that a variable participates in only one of the chosen programs in a step. A marking M_c of the net $[\![\Pi]\!]$ refers to the configuration C of the system Π. For example, the initial configuration C_0 of system Π is described by the initial marking M_0 where the i-th element of the configuration C_0 is the initial value of i-th variable, and the value equals to the initial number of tokens in i-th place which is the i-th element of the initial marking M_0. A set of the used Transitions T_{s_c} in the net $[\![\Pi]\!]$ in a step corresponds to the set of used programs s_c in system Π. In each step, all the Transitions in the set T_{s_c} are used corresponding to the use of the programs s_c in system Π. The state $M_c[T_{s_c}\rangle$ denotes that the transitions in the set T_{s_c} are enabled when the marking of the net $[\![\Pi]\!]$ is M_c. If there is a configuration transition $C \to C'$ using a set of programs s_c in the system Π, there is a firing $M_c \to M_{c'}$ using the set of transitions T_{s_c} in the related net $[\![\Pi]\!]$. Conversely, if there is a firing $M_c \to M_{c'}$ using a set of transitions T_{s_c} in the net $[\![\Pi]\!]$, there is a transition $C \to C'$ using a set of programs s_c in system Π. Thus, the theorem holds.

Example 1 is provided to describe the details of the Petri net designed for a given NPT system. Petri nets are a class of graphical models. In the graph of a Petri net, each circle corresponds to a place storing tokens, each rectangle corresponds to a transition, and arcs connect places and transitions. A computation of the NPT system can be described by a sequence of transitions starting from the initial configuration $M_0[T_{r_1}\rangle M_1[T_{r_2}\rangle \ldots M_l[T_{r_l}\rangle M_{r_{l+1}}$, where each configuration in the sequence is reachable.

Example 1. The NP system having a lower threshold

$$\Pi_0 = (1, \{1\}, [\,[\ ]_2]_1, 1, (Var_1, Pr_1, Var_1(0)), (Var_2, Pr_2, Var_2(0)), \{x_0\}\})$$

is a construct, where

- $T_h = 1$ is the lower threshold;
- $Var_1 = \{x_1\}$, $Pr_1 = \{3x_1|_1 \to 1|x_1 + 2|x_0\} \cup \{x_1|_1 \to 1|x_0\}$, $Var_1(0) = (1)$;
- $Var_2 = \{x_0\}$, $Pr_1 = \emptyset$, $Var_2(0) = (0)$.

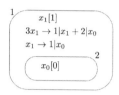

Fig. 1. NPT system Π_0.

The system Π_0 is shown in Fig. 1. At step 1, the program $3x_1|_1 \to 1|x_1 + 2|x_0$ and the program $x_1|_1 \to 1|x_0$ is enabled. If the program $3x_1|_1 \to 1|x_1 + 2|x_0$ is applied, then the variables are updated to $x_1 = 1$ and $x_0 = 2$. The two programs in membrane 1 are enabled again. If the program $x_1|_1 \to 1|x_0$ is applied, then the variables are updated to $x_1 = 0$ and $x_0 = 1$.

Consider the Petri net $\mathcal{N}_1 = (\{P_0, P_1\}, \{T_{r_1}, T_{r_2}\}, A, W, P_0, M_0)$ with the mapping $G : T_r \to 1$. The pictorial representation of $\mathcal{N}_1$ is shown in Fig. 2. The places P_0 and P_1 are associated with the variable x_1 and x_0, the transition T_{r_1} and T_{r_2} are associated with the programs $3x_1|_1 \to 1|x_1 + 2|x_0$ and $x_1|_1 \to 1|x_0$, respectively. The weight on the arc from place P_1 to transition T_{r_1} is 3 and the weight on the arc from transition T_{r_1} to place P_0 is 2. Except that all the weights are 1 according to the programs. The initial number (k) of tokens in each place is represented by the (k) black dots.

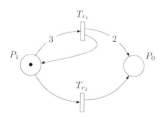

Fig. 2. NPT system Π_0.

The initial configuration $M_0 = (1, 0)$. If the program $3x_1|_1 \to 1|x_1 + 2|x_0$ is applied, then $M_0[T_{r_1}\rangle M_1$ where $M_1 = (1, 2)$. If the program $x_1|_1 \to 1|x_0$ is applied, then $M_0[T_{r_1}\rangle M_2$ where $M_1 = (0, 1)$.

Reachability is a fundamental property of Petri nets. For a given a Petri net $\mathcal{N}$ with initial marking M_0, the marking M is reachable if there is a finite number n and a sequence of firing $M_0 \to M_1 \to \cdots \to M_n = M$ in the net $\mathcal{N}$. On the basis of intrinsic relationship between NPT systems and Petri nets, the configuration C is reachable if there is a finite number n and a sequence of transitions $C_0 \to C_1 \to \cdots C_n = C$ in a given NPT system. The reachability of NPT system is discussed as follows.

Theorem 2. *Let C_0 and C ($C_0 \neq C$) be two configurations of an NPT system Π where each production function is a linear function with integers. It is decidable whether there is a integer n and a sequence transitions $C_0 \to C_1 \to \cdots C_n = C$ in the system Π.*

Proof. The theorem is actually the decidability of the reachability for the NPT where each production function is a linear function with integers. Based on the results of Theorem 1, the reachability problem of the NPT system Π is the reachability problem of the related Petri net $\llbracket \Pi \rrbracket$: whether the marking M can be reached from the initial marking M_0 where the markings M_0 and M which are associated with the given configurations C_0 and C, respectively.

It is proved that the reachability problem of a Petri net is decidable, and the detailed proof can be found in the references [5,6]. Thus, the reachability problem of the NPT system Π is decidable.

Boundness is another fundamental property of Petri nets. For a given a Petri net $\mathcal{N}$, the net $\mathcal{N}$ is bounded if the set of reachable markings is finite. Based on the notion of boundness in Petri nets, an NPT system where each production function is a linear function with integers is finite if the set of reachable markings of the associated Petri net is finite. The following result is on the boundness of NPT systems.

Theorem 3. *It is decidable whether an NPT system where each production function is a linear function with integers is bounded.*

Proof. The boundness problem of a Petri net is proved to be decidable by using the Rackoff algorithm which requires at most space $2^{cn\log n}$ (c is a constant) [14] or the Lipton algorithm which requires at least space $2^{c\sqrt{n}}$ (c is a constant) [7]. Since the intrinsic relationship between NPT systems and Petri nets, the boundness problem of the system Π is decidable.

4   Conclusion

The NPT systems where each production function is a linear function with integer coefficients are proved to be Turing universal. To obtain more operational and computational semantics on NPT systems, Petri nets are introduced to simulate and analyze the computation of NPT systems where each production function is a linear function with integer coefficients. For a given NPT system, the elements of the system are associated with the ingredients of Petri nets: the structure of an NPT system is described by the connections between places and transitions in the related Petri nets; the numerical variables are associated with the places where the value of a variable is associated with the number of tokens in a place; the programs in NPT systems are associated with the transitions in the related Petri nets. The configuration of an NPT system is associated with the marking of the related Petri net, and a computation of an NPT system is described by a sequence of firing in the related Petri net. In this way, the modeling and analysis of the NPT system can be performed by the related Petri net. Further, the results on the boundedness and reachability of NPT systems are obtained by using the existing results in Petri nets and designing the associated Petri nets.

References

1. Assaf, G., Heiner, M., Liu, F.: Coloured fuzzy petri nets for modelling and analysing membrane systems. Biosystems **212**, 104592 (2022)
2. Buiu, C., Florea, A.G.: Membrane computing models and robot controller design, current results and challenges. J. Membr. Comput. **1**(4), 262–269 (2019)
3. Díaz-Pernil, D., Gutiérrez-Naranjo, M.A., Peng, H.: Membrane computing and image processing: a short survey. J. Membr. Comput. **1**, 58–73 (2019)
4. Duan, Y., Rong, H., Qi, D., Valencia-Cabrera, L., Zhang, G., Pérez-Jiménez, M.J.: A review of membrane computing models for complex ecosystems and a case study on a complex giant panda system. Complexity **2020**, 1–26 (2020)
5. Kosaraju, S.R.: Decidability of reachability in vector addition systems. In: 14th Annual ACM Symposium on Theory of Computing, San Francisco, pp. 267–281 (1982)
6. Lambert, J.L.: A structure to decide reachability in petri nets. Theoret. Comput. Sci. **99**, 79–104 (1992)
7. Lipton, R.J.: The reachability problem requires exponential space. Department of Computer Science, Yale University (1976)
8. Liu, F., Heiner, M.: Petri nets for modeling and analyzing biochemical reaction networks. In: Chen, M., Hofestädt, R. (eds.) Approaches in Integrative Bioinformatics, pp. 245–272. Springer, Heidelberg (2014). https://doi.org/10.1007/978-3-642-41281-3_9
9. Liu, L., Yi, W., Yang, Q., Peng, H., Wang, J.: Small universal numerical P systems with thresholds for computing functions. Fund. Inform. **176**(1), 43–59 (2020)
10. Pan, L., Zhang, Z., Wu, T., Xu, J.: Numerical P systems with production thresholds. Theoret. Comput. Sci. **673**, 30–41 (2017)
11. Păun, Gh.: Computing with membranes. J. Comput. Syst. Sci. **61**(1), 108–143 (2000)
12. Păun, Gh., Păun, R.: Membrane computing and economics: numerical P systems. Fund. Inform. **73**(1), 213–227 (2006)
13. Qi, Z., You, J., Mao, H.: P systems and petri nets. In: Martín-Vide, C., Mauri, G., Păun, G., Rozenberg, G., Salomaa, A. (eds.) WMC 2003. LNCS, vol. 2933, pp. 286–303. Springer, Heidelberg (2004). https://doi.org/10.1007/978-3-540-24619-0_21
14. Rackoff, C.: The covering and boundedness problem for vector addition systems. Theoret. Comput. Sci. **6**, 223–231 (1978)
15. Rong, H., Duan, Y., Zhang, G.: A bibliometric analysis of membrane computing (1998–2019). J. Membr. Comput. **4**(2), 177–207 (2022)

16. Rozenberg, G., Thiagarajan, P.S.: Petri nets: basic notions, structure, behaviour. Curr. Trends Concurr.: Overviews Tutor. 585–668 (1986)
17. Song, B., Li, K., Orellana-Martín, D., Pérez-Jiménez, M.J., Pérez-Hurtado, I.: A survey of nature-inspired computing: membrane computing. ACM Comput. Surv. (CSUR) **54**(1), 1–31 (2021)
18. Zhang, Z., Pan, L.: Numerical P systems with thresholds. Int. J. Comput. Commun. Control **11**(2), 292–304 (2016)
19. Zhang, Z., Su, Y., Pan, L.: The computational power of enzymatic numerical P systems working in the sequential mode. Theoret. Comput. Sci. **724**, 3–12 (2018)
20. Zhang, Z., Wu, T., Pan, L.: On string languages generated by sequential numerical P systems. Fund. Inform. **145**(4), 485–509 (2016)
21. Zhang, Z., Wu, T., Pan, L., Păun, Gh.: On string languages generated by numerical P systems. Roman. J. Inf. Sci. Technol. **18**(3), 273–295 (2015)
22. Zhang, Z., Wu, T., Păun, A., Pan, L.: Numerical P systems with migrating variables. Theoret. Comput. Sci. **641**, 85–108 (2016)
23. Zhang, Z., Wu, T., Păun, A., Pan, L.: Universal enzymatic numerical P systems with small number of enzymatic variables. Sci. China Inf. Sci. **61**, 1–12 (2018)

An Efficient Graph Theoretic Algorithm for Channel Routing in VLSI Design with Given Constraint Graph

Ming Liu[1] and Xianya Geng[2(✉)]

[1] Department of Electrical and Electronic Engineering, Anhui Vocational and Technical College of Industry and Trade, Huainan 232001, China
[2] School of Mathematics and Big Data, Anhui University of Science and Technology, Huainan 232001, China
gengxianya@sina.com

Abstract. In VLSI design, the channel routing is one of the most significant detailed routings. Given a channel with length in 2-layer Manhattan model. Given a channel with length in 2-layer Manhattan model, Szeszler proved that the width (number of tracks required for routing) of the channel is at most 7/4, and this upper bound can be achieved by a linear time algorithm. In this paper, the channel with horizontal constraint graph is considered as a path. An efficient graph theoretic algorithm is presented, compared with the latest results, our algorithm yields a better bound on the width of the channel.

Keywords: Channel routing · Manhattan model · VLSI

1 Introduction

In VLSI design, the problem of completing the necessary interconnections between different modules is called the routing problem. In general, due to the problem complexity, the routing problem can be divided into two steps: global routing and detailed routing. One of the most significant detailed routings is channel routing [1–11]. The channel routing problem (CRP) is the problem of interconnecting all the nets in a channel using minimum possible routing area.

We use the expression of graph theory to describe the channel routing problem, a channel is defined by a rectangular grid G of size $(w + 2)n$ consisting of horizontal tracks (numbered from 0 to $w + 1$) and vertical columns (numbered from 1 to n), where w represents the width and n stands for the length of the channel. The top and bottom points of G are called terminals.

A channel routing problem is a set of $N = \{N_1, N_2, ..., N_t\}$ pairwise disjoint nets. A channel routing problem is called bipartite if each net contains exactly two terminals, one on the top, and another on the bottom side. A channel routing problem is dense if each terminal on the top and bottom sides belongs to a net. A net is called trivial if it consists of two terminals which are located in the same column. In this paper, we always suppose that each net contains at least two terminals.

© The Author(s), under exclusive license to Springer Nature Singapore Pte Ltd. 2024
L. Pan et al. (Eds.): BIC-TA 2023, CCIS 2061, pp. 339–344, 2024.
https://doi.org/10.1007/978-981-97-2272-3_26

There are two constraints for the net in a channel: horizontal constraint and vertical constraint.

The horizontal constraint is defined as the constraint that two nets cannot overlap on the horizontal layer. Let l_i be the leftmost and r_i be the rightmost column of net i. A net i is said to span the c-th column if $l_i \leq c \leq r_i$. The set of columns $[l_i, r_i]$ is called the span of net i. If and only if the spans of net i and net j overlap, there is a horizontal constraint between them. The horizontal constraints are often stood for an undirected graph, which is the horizontal constraint graph (HCG) (see Fig. 1), where vertices stand for the nets and edges represent the horizontal constraints.

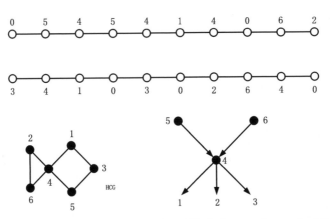

Fig. 1. Horizontal constraint graph and Vertical constraint graph of G

Let Z_i be the set of nets that span the i-th column, $d_{max} = \max\{|Z_i|, \text{i is a column}\}$ is known as the density of the CRP. Obviously, d_{max} is a lower bound on S since nets spanning the same column can't be distributed to the same track.

The vertical constraint is defined as the constraint that two nets cannot overlap on the vertical layer. Note that if net i is linked to the i-th column in the top row and net j is linked to the c-th column in the bottom row i ≠ j, then net i must be assigned to a track higher than net j. In this condition, we denote that net i must precede net j, moreover, there is a vertical constraint from i to j. The Vertical constraints define the order of sections between nets. The vertical constraints are often stood for a directed graph, which is the vertical constraint graph (VCG) (see Fig. 1), where vertices stand for the nets and arcs represent the vertical constraints.

A solution of a channel routing problem is said to belong to the Manhattan model if consecutive layers contain only wire segments of different directions. That is, layers with horizontal (from east to west) and vertical (from north to south) wire segments alternate. In this paper we restrict that the length of the channel cannot be extended by introducing extra columns.

2-layer Manhattan routing has always been one of the most concerned and well-studied issues in VLSI routing. The initial classic result in the topic of VLSI routing is probably Gallai's linear time algorithm that solves the single row routing problem, a special case of channel routing, with optimal width in the 2-layer Manhattan model.

Theorem 1 (Dvid Szeszlr [7]). A channel routing problem is unsolvable in the 2-layer Manhattan model (with an arbitrary width) if and only if it is bipartite, dense and has at least one non-trivial net. Moreover, it can be solved with width at most $3/2n$ in the bipartite, and $7/4n$ in the general case (where n represents the length of the channel), if a specification is solvable.

2 Method

Recall that, if each net contains exactly two terminals, one on the top, and another on the bottom side, then the channel routing problem is called bipartite. Now we consider horizontal constraint graph of the net in this problem is a path.

Theorem 2.1. If a channel with horizontal constraint graph be a path, then the vertical constraint graph has no directed cycle.

Proof. Let the horizontal constraint graph G has n vertices, denoted by $v_1, v_2, \ldots, v_n$ and v_i, v_{i+1} ($1 \leq i \leq n-1$) have an edge joined. By the define of horizontal constraint graph, net $v_i (1 \leq i \leq n-1)$ has horizontal constraint with net v_{i+1} and other two nets have not horizontal constraint.

Now we proof by contradiction. If these is a direct cycle belongs to the vertical constraint graph, denoted by $u_1, u_2, \ldots, u_n, u_1$. Recall that if two nets have vertical constraint, then they also have horizontal constraint. Then the net u_1 and u_{i+1} ($1 \leq i \leq n-1$) have horizontal constraint, and u_s and u_1 have horizontal constraint, then the vertex $u_1, u_2, \ldots, u_n, u_1$ obtain a cycle in horizontal constraint graph, a contradiction.

So if a channel with horizontal constraint graph be a path, then the vertical constraint graph has no directed cycle. Since we consider channel routing problem is bipartite, that is all the net has two terminals, the directed path of vertical constraint graph have length at most n.

An algorithm for 2-layer Manhattan routing problem with horizontal constraint graph be a path, based on the notions of horizontal constraint graph and vertical constraint graph, is presented in this section. The algorithm proceeds in three cases outlined below.

Route all trivial nets straight down in the obvious fashion. Henceforth we do not include these columns and nets below.

Case 1: Vertical Constraint Graph of the Net Has No Edge. In this case, each two nets have not vertical constraint, we only consider horizontal constraint. Since horizontal constraint graph is a path, we only need to route the net in this path one by one. Denote the path by $v_1, v_2, \ldots, v_n$, we assign the first track on the bottom layer to net v_1. By horizontal constraint, we must assign the second track on the bottom layer to net v_2. Since v_3 has no horizontal constraint with v_1, then we assign the first track on the bottom layer to net v_3, using the same method, we can assign the second track on the bottom layer to net v_4, , the rest nets can also using this method. Then we route them in the most straightforward way: in the corresponding track of the bottom layer we introduce a horizontal wire segment connecting the columns of the two terminals, switch to the top layer at both ends of this segment and attach to the two terminals (a specification is shown in Fig. 2, where $n = 6$).

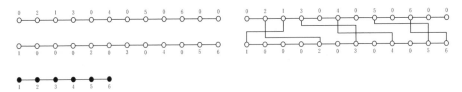

Fig. 2. Vertical constraint graph of the net has no edge of G and $n = 6$

Case 2: Vertical Constraint Graph of the Net Has Directed Paths with Length Less Than n. In this case, suppose that these are k directed paths and we denote these k directed paths by $P_{n_1}, P_{n_2}, \ldots, P_{n_k}$, where n_i denote the number of nets which belongs to directed path P_{n_i}, and we denote these nets by $N_{i1}, N_{i2}, \ldots, N_{in_i}$. So we have $n_1 + n_2 + \ldots + n_k \leq n$. The algorithm proceeds in four phases outlined below.

Phase 1: Route all trivial nets straight down in the obvious fashion. Henceforth we do not include these columns and nets below. We first choose a directed path arbitrary, without loss of generality, say P_{n_1}. Proceeding from North to South, we assign a separate track on the bottom layer to each net of P_{n_1} in the order of directed path P_{n_1} and route them in the most straightforward manner: in the corresponding track of the bottom layer we introduce a horizontal wire segment connecting the columns of the two terminals, and switch to the top layer at both ends of this segment and attach to the two terminals. Therefore, we have consumed the first n_1 continuous tracks to route the nets of P_{n_1}.

Phase 2: In the rest directed paths, we also choose a directed path arbitrary, without loss of generality, say P_{n_2}. Now we route the nets belong to this directed path. For the net N_{21}, if in the horizontal constraint graph H, there is not an edge to connect the vertices N_{21} and N_{11}, then we assign N_{21} a track to be the same as N_{11} on the bottom layer (that is in the first track) and route it in the above straightforward way. This assign is reasonable because N_{21} and N_{11} neither have horizontal constraint nor have vertical constraint. If in the horizontal constraint graph H, there is an edge to connect the vertices N_{21} and N_{11}, then we consider N_{21} and N_{12}, if in the horizontal constraint graph H, there is not an edge to connect the vertices N_{21} and N_{11}, then we distribute N_{21} a track to be the same as N_{12}, on the bottom layer and route it in the way mentioned above, otherwise, we consider N_{21} and N_{1i} until these exit a net N_{1i} ($3 \leq i \leq n_1$) which doesn't have an edge connect N_{21} and N_{1i} ($3 \leq i \leq n_1$) in H. If all of the nets in the directed path P_{n_1} have an edge with N_{21} in the H, then we assign the track $n_1 + 1$ and route it in the same way as above.

Phase 3: In the rest directed paths, we also choose a directed path arbitrary, without loss of generality, say P_{n_3}. Now we route the nets belong to this directed path. For the net N_{31}, we compare it to nets which belong to the first track (there are may be two nets, N_{11} and N_{21}, and at least there is a net N_{11}), if in the horizontal constraint graph H, the net N_{31} and the nets which belongs to the first track doesn't have an edge, then we assign N_{31} a track to be the first track and route it in the above straightforward way. If in the horizontal constraint graph H, the net N_{31} and the nets which belongs to the first track at least have a edge, then we consider N_{31} and the nets belong to the second track with the above method, in the second track, there are may be two nets, N_{12} and N_{21} or N_{22}, and at least there is a net N_{12}. For other nets belong to the directed path P_{n_3}, using the same method, we can assign a track to each of these nets and route these nets in the above straightforward way.

Phase 4: For the rest directed paths, we can use the same method as the step 3. So, we can assign a track to each of these nets and route these nets in the above straightforward way (a specification is shown in Fig. 3, where $n = 6$ and $k = 2$).

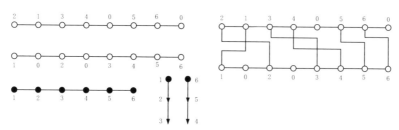

Fig. 3. Vertical constraint graph of the net has directed paths with length less than n and $n = 6$

Case 3: Vertical Constraint Graph of the Net is Directed Path with Length n. In this case, vertical constraint graph of the net is a directed path with length n. Proceeding from north to south, we assign a separate track on the bottom layer to each net in the order of directed path and route them in the above straightforward way.

3 Results

3.1 Running Time Analysis

For case 1 and case 3, we need constant time to route the nets of this path. For case 2, when we choose the first directed path in step 1, we need constant time to route the nets of this path. When we route the second directed path in step 2, we need n_1 compares. When we route the second directed path in step 3, we need at most $n_1 + n_2$ compares. And when we route the kth directed path in step k, we need at most $n_1 + n_2 + \ldots + n_{k-1}$ compares.

Then if we complete routing of a channel problem, we need all of the compares C are:

$$
\begin{aligned}
C &= n_1 + n_1 + n_2 + \ldots + n_1 + n_2 + n_{k-1} \\
&= (k-1)n_1 + (k-2)n_2 + \ldots + n_{k-1} \\
&\leq (k-1)(n_1 + n_2 + \ldots + n_{k-1})
\end{aligned}
\tag{1}
$$

And since $n > n^* = n_1 + n_2 + \ldots + n_{k-1} + n_k$, then $k < n$. Therefore, we have

$$
C < n \cdot n = n^2
\tag{2}
$$

Since every compare need constant time, we give a polynomial time algorithm to solve this routing problems.

3.2   The Upper Bound

In the above algorithm, we give a solution for this particular routing problem. Now we analyse the upper bound of our algorithm and compare to other algorithms.

In [8], the author routes this problem with n tracks, he assign a separate track on the bottom layer to each net. In our algorithm, we consider the horizontal constraint and vertical constraint graph of the nets, we can route this problem with no more than n tracks.

For case 1, we can route this problem with 2 tracks.

For case 2, if the longest directed path is Pm and we assign the first nm tracks to route these nets, for others nets belong to the rest directed paths, we can also route them in the first nm tracks when the algorithm is complete. If the nets belong to the later directed path cannot use the tracks which have been used with the nets belong to the ahead directed paths, we route this problem with n tracks.

For case 3, we can route this problem with n tracks.

So, we give a better polynomial time algorithm to solve the 2-layer Manhattan channel routing problem.

Acknowledgements. This work was supported in part by the Natural Science Foundation of Educational Commission of Anhui Province of China (KJ2021A1341).

References

1. Baker, B.S., Bhatt, S.N., Leighton, F.T.: An approximation algorithm for Manhattan routing. In: Proceedings of the Fifteenth Annual ACM Symposium on Theory of Computing, pp. 477–486 (1983)
2. Gao, S., Kaufmann, M.: Channel routing of multiterminal nets. J. ACM (JACM) **41**(4), 791–818 (1994)
3. Geng, X.: A polynomial time algorithm for 2-layer Manhattan channel. Int. J. Appl. Math. Stat. **29**(5), 76–83 (2012)
4. Johnson, D.S.: The NP-completeness column: an ongoing guide. J. Algorithms **6**(3), 434–451 (1985)
5. Marek-Sadowska, M., Kuh, E.S.: General channel-routing algorithm. IEE Proc. G Circ. Dev. Syst. [See also IEE Proc.-Circ. Dev. Syst.] **3**(130), 83–88 (1983)
6. Recski, A., Strzyzewski, F.: Vertex disjoint channel routing on two layers. In: Proceedings of the 1st Integer Programming and Combinatorial Optimization Conference, pp. 397–405 (1990)
7. Recski, A.: Some polynomially solvable subcases of the detailed routing problem in VLSI design. Discrete Appl. Math. **115**(1–3), 199–208 (2001)
8. Szeszlér, D.: A new algorithm for 2-layer, Manhattan channel routing. In: Proceedings of the 3rd Hungarian-Japanese Symposium on Discrete Mathematics and Its Applications, pp.179–185 (2003)
9. Szkaliczki, T.: Optimal routing on narrow channels. Periodica Polytech. Electr. Eng. (Arch.) **38**(2), 191–196 (1994)
10. Szymanski, T.G.: Dogleg channel routing is NP-complete. IEEE Trans. Comput. Aided Des. Integr. Circ. Syst. **4**(1), 31–41 (1985)
11. Yoeli, U.: A robust channel router. IEEE Trans. Comput. Aided Des. Integr. Circ. Syst. **10**(2), 212–219 (1991)

Extremal Values of Generalized Somber Index in Chemical Graphs

Xingyue Dong and Xianya Geng[(⊠)]

School of Mathematics and Big Data, Anhui University of Science and Technology,
Huainan 232001, China
2022201427@aust.edu.cn

Abstract. A graph G consists of a set of vertices V(G) and a set of edges E(G). A new vertex-degree-based molecular structure descriptor was introduced, named "Generalized Somber index". In this paper we determine the extremal values of generalized Somber index in chemical graphs, chemical trees and hexagonal systems. Especially it has been proved that the Somber index could predict some physicochemical properties and make their chemical applications.

Keywords: Somber index · Generalized Somber index · Chemical graphs · Chemical trees · Hexagonal systems

1 Normalized Introduction

A graph G consists of a set of vertices V(G) and a set of edges E(G). The degree d_u of the vertex $u \in V(G)$ is the number of vertices adjacent to u. If there is an edge from vertex u to vertex v, we indicate this by writing uv(or vu). Recently, a new vertex-degree molecular structure descriptor was put forward in [10], the Somber index, defined as

$$SO(G) = \sum_{uv \in E(G)} \sqrt{(d_u)^2 + (d_v)^2}, \tag{1}$$

$$SO_{red}(G) = \sum_{uv \in E(G)} \sqrt{(d_u - 1)^2 + (d_v - 1)^2}. \tag{2}$$

The geometric interpretation of the degree radius of an edge uv motivated this topological index, which is the distance from the origin to the ordered pair (d_u, d_v), where $d_u \leq d_v$. In this paper, we are concerned with the Somber index of chemical graphs. Recall that G is a chemical graph if $d_u \leq 4$ for all $u \in V(G)$. We characterize the graphs extremal with respect to the Somber index over the following sets: (connected) chemical graphs with n vertices, chemical trees with n vertices and hexagonal systems with h hexagons.

Supported by National Science Foundation of China (Grant No. 12171190) and Natural Science Foundation of Anhui Province (Grant No. 2008085MA01).

© The Author(s), under exclusive license to Springer Nature Singapore Pte Ltd. 2024
L. Pan et al. (Eds.): BIC-TA 2023, CCIS 2061, pp. 345–360, 2024.
https://doi.org/10.1007/978-981-97-2272-3_27

Chemical structures can be conveniently represented by graphs with $d_u \leq 4$ for all $u \geq 4$, which are called chemical graphs or molecular graphs.

In [6], the extremal value of the chemical graphs, the chemical trees and the hexagonal systems are calculated respectively. Subsequently, in [12], the author also studied the classification of non-pendent (chemical) graphs with respect to the Somber Index, and determine the minimum Somber Indices of tetracyclic (chemical) graphs and the reduced Somber Indices of the minimum tricyclic graphs $SO_{red}(G)$. In this paper, we give the generalized definition of Somber index $SO_k(G)$ and the corresponding Somber index about chemical graphs, trees and hexagonal systems.

$$SO_k(G) = \sum_{uv \in E(G)} \sqrt{(d_u - k)^2 + (d_v - k)^2} \ (k \geq 0), \tag{3}$$

when $k = 0$, $SO_k(G) = SO(G)$; when $k = 1$, $SO_k(G) = SO_{red}(G)$.

For recent results in the theory of vertex-degree-based topological indices we refer to [1,3,5,10,16].

2   Generalized Somber Index of Chemical Graphs

Let G be a chemical graph (with no isolated vertices), $n_x = n_x(G)$ the number of vertices of G of degree x, and $m_{x,y} = m_{x,y}(G)$ the number of edges of G joining a vertex of degree x with a vertex of degree y. Clearly

$$n = n_1 + n_2 + n_3 + n_4, \tag{4}$$

$$n = n_1 + n_2 + n_3 + n_4, \tag{5}$$

and it is also well-known that the following relations hold:

$$\begin{cases} 2m_{1,1} + m_{1,2} + m_{1,3} + m_{1,4} = n_1 \\ m_{1,2} + 2m_{2,2} + m_{2,3} + m_{2,4} = 2n_2 \\ m_{1,3} + m_{2,3} + 2m_{3,3} + m_{3,4} = 3n_3 \\ m_{1,4} + m_{2,4} + m_{3,4} + 2m_{4,4} = 4n_4 \end{cases} \tag{6}$$

Let
$$T = \{(x, y) \in N \times N : 1 \leq x \leq y \leq 4\}. \tag{7}$$

It follows easily from the two equations above that

$$n = \sum_{(x,y) \in T} \frac{x + y}{xy} m_{x,y}. \tag{8}$$

This result is a consequence from [2].

Note that the Eq. (3) is equivalent to

$$SO_k(G) = \sum_{(x,y) \in T} \sqrt{(x - k)^2 + (y - k)^2} m_{x,y} \ (k \geq 0). \tag{9}$$

Theorem 1. *Let G be a chemical graph with n vertices. Then*

$$SO_k(G) \leq 2\sqrt{2}|4 - k|n. \tag{10}$$

Equality occurs if and only if G is a 4-regular graph.

Proof. Let $A = \{(x, y) \in T : (x, y) \neq (4, 4)\}$. It is easy to find that

$$\sqrt{(x - k)^2 + (y - k)^2} - 2\sqrt{2}|4 - k|\frac{x + y}{xy} < 0, \tag{11}$$

for all $(x, y) \in A$. By Eq. (9), and (11), we compute that

$$
\begin{aligned}
SO_k(G) &= \sqrt{2}|4 - k|m_{4,4} + \sum_{(x,y)\in A} \sqrt{(x - k)^2 + (y - k)^2}m_{x,y} \\
&= 2\sqrt{2}|4 - k|\left[n - \sum_{(x,y)\in A} \frac{x + y}{xy}m_{x,y}\right] \\
&\quad + \sum_{(x,y)\in A} \sqrt{(x - k)^2 + (y - k)^2}m_{x,y} \\
&\leq 2\sqrt{2}|4 - k|n.
\end{aligned}
\tag{12}
$$

If $SO_k(G) = 2\sqrt{2}|4 - k|n$, then by inequality above,

$$
2\sqrt{2}|4 - k|n + \sum_{(x,y)\in A}\left[\sqrt{(x - k)^2 + (y - k)^2} - 2\sqrt{2}|4 - k|\frac{x + y}{xy}\right]m_{x,y}
$$
$$
= 2\sqrt{2}|4 - k|n,
\tag{13}
$$

and by inequality above, we conclude that $m_{x,y} = 0$ for all $(x, y) \in A$. In other words, G is a 4-regular graph. Conversely, if G is a 4-regular graph , then

$$SO_k(G) = \sqrt{2}|4 - k|m_{4,4} = \sqrt{2}|4 - k|\frac{4n_4}{2} = 2\sqrt{2}|4 - k|n. \tag{14}$$

$\square$

Theorem 2. *Let G be a chemical graph with n vertices.*
Case1. If n is even, then $\frac{1}{2}\sqrt{2}|1 - k|n \leq SO_k(G)$. Equality occurs if and only if $G \cong \frac{n}{2}|1 - k|P_2$.
Case2. If n is odd, then $\frac{n-3}{2}\sqrt{2}|1-k|+2\sqrt{(1 - k)^2 + (2 - k)^2} \leq SO_k(G)$. Equality occurs if and only if $G \cong \frac{n-3}{2}|1 - k|P_2 \oplus P_3$.

Proof. Assume that n is even and let $B = \{(x, y) \in T : (x, y) \neq (1, 1)\}$. Clearly,

$$\sqrt{(x - k)^2 + (y - k)^2} - \frac{\sqrt{2}}{2}|1 - k|\frac{x + y}{xy} > 0, \tag{5}$$

for all $(x, y) \in B$. According to Eq. (8), (9), and (11) we deduce that

$$
\begin{aligned}
SO_k(G) &= \sqrt{2}|1 - k|m_{1,1} + \sum_{(x,y)\in B} \sqrt{(x - k)^2 + (y - k)^2}\, m_{x,y} \\
&= \frac{1}{2}\sqrt{2}|1 - k| \left[n - \sum_{(x,y)\in B} \frac{x + y}{xy} m_{x,y} \right] \\
&\quad + \sum_{(x,y)\in B} \sqrt{(x - k)^2 + (y - k)^2}\, m_{x,y} \\
&\geq \frac{1}{2}\sqrt{2}|1 - k|n.
\end{aligned}
\tag{15}
$$

If $\frac{1}{2}\sqrt{2}|1 - k|n = SO_k(G)$ then, by inequality above,

$$
\frac{1}{2}\sqrt{2}|1-k|n + \sum_{(x,y)\in B} \left[\sqrt{(x - k)^2 + (y - k)^2} - \frac{1}{2}\sqrt{2}|1 - k|\frac{x + y}{xy} \right] m_{x,y} = \frac{1}{2}\sqrt{2}n,
\tag{16}
$$

and by Eq. (11), we deduce that $m_{x,y} = 0$ for all $(x, y) \in B$. Since n is even, then clearly $G \cong \frac{n}{2}|1 - k|P_2$. The converse is clear.

This completes the proof of Case 1.

Assume that n is odd. Since G has no isolated vertices, $m_{1,1} \leq \frac{n-3}{2}$. Let

$$
C = \{(x, y) \in T : (x, y) \neq (1, 1)\ and\ (x, y) \neq (1, 2)\},
\tag{17}
$$

it is easy to check that

$$
\sqrt{(x - k)^2 + (y - k)^2} - \sqrt{(1 - k)^2 + (2 - k)^2}\, \frac{x + y}{xy} > 0,
\tag{6}
$$

for all $(x, y) \in C.$, now by Eq. (8), (9), and (6) it follows that

$$
\begin{aligned}
SO_k(G) &= \sqrt{2}|1 - k|m_{1,1} + \sqrt{(1 - k)^2 + (2 - k)^2}\, m_{1,2} \\
&\quad + \sum_{(x,y)\in C} \sqrt{(x - k)^2 + (y - k)^2}\, m_{x,y} \\
&= \left[\sqrt{2}|1 - k| - \frac{4}{3}\sqrt{(1 - k)^2 + (2 - k)^2} \right] m_{1,1} \\
&\quad + \frac{2}{3}\sqrt{(1 - k)^2 + (2 - k)^2}\, n
\end{aligned}
\tag{18}
$$

$$
\begin{aligned}
&+ \sum_{(x,y)\in C} \left[\sqrt{(x - k)^2 + (y - k)^2} - \sqrt{(1 - k)^2 + (2 - k)^2}\frac{x + y}{xy} \right] m_{x,y} \\
&\geq \left[\sqrt{2}|1 - k| - \frac{4}{3}\sqrt{(1 - k)^2 + (2 - k)^2} \right] \frac{n - 3}{2} + \frac{2}{3}\sqrt{(1 - k)^2 + (2 - k)^2}\, n \\
&= \frac{n - 3}{2}\sqrt{2}|1 - k| + 2\sqrt{(1 - k)^2 + (2 - k)^2}.
\end{aligned}
\tag{19}
$$

If the left and right sides of the equation above are equal then, by equality above, $m_{1,1} = \frac{n-3}{2}$ and $m_{x,y} = 0$ for all $(x, y) \in C$. This clearly implies that $G \cong \frac{n-3}{2}|1 - k|P_2 \oplus P_3$. The converse is clear.

This completes the proof of Case 2.

$\square$

For connected chemical graphs we have the following result.

Theorem 3. *Let G be a connected chemical graph with n vertices. Then*

$$\frac{n-3}{2}\sqrt{2}|1 - k| + 2\sqrt{(1-k)^2 + (2-k)^2} \leq SO_k(G) \leq 2\sqrt{2}|4 - k|n. \quad (20)$$

The equality occurs in the left if and only if $G \cong P_n$, the equality occurs in the right if and only if G is a connected 4-regular graph.

Proof. This result is a consequence of Theorem 1 and Theorem 2 in [10].

$\square$

3   Generalized Somber Index of Chemical Trees

Let D_n be the set of chemical trees with n vertices. As a consequence of Theorem 2 in [10], the minimal tree with respect to the Somber index over the set D_n is the path T_n. Now, we find the maximal values of SO_k over the set D_n. Note that if $P \in D_n, m_{1,1} = m_{1,1}(P) = 0$.

Lemma 1. *Let $P \in D_n$ with $n \geq 3$. Then*

$$SO_k(P) = \frac{2\sqrt{(1-k)^2 + (4-k)^2} - \sqrt{2}|4 - k|}{3} n$$

$$+ \frac{2\sqrt{(1-k)^2 + (4-k)^2} - 5\sqrt{2}|4 - k|}{3} \quad (21)$$

$$+ \sum_{(x,y)\in T} h_k(x, y)m_{x,y},$$

where

$$h_k(x, y) = \sqrt{(x - k)^2 + (y - k)^2} + \frac{4\left[\sqrt{2}|4 - k| - \sqrt{(1-k)^2 + (4-k)^2}\right]}{3} \frac{x + y}{xy}$$

$$- \frac{5\sqrt{2}|4 - k| - 2\sqrt{(1-k)^2 + (4-k)^2}}{3}. \quad (22)$$

Proof. For any $P \in D_n$ with $n \geq 3$, the following relations holds :

$$n - 1 = \sum_{(x,y)\in T} m_{x,y}. \quad (23)$$

The relations Eq. (8) and Eq. (23) can be rewritten as follows:

$$5m_{1,4} + 2m_{4,4} = 4n - \sum_{(x,y)\in D'} 4\frac{x+y}{xy} m_{x,y},$$

$$m_{1,4} + m_{4,4} = n - 1 - \sum_{(x,y)\in D'} m_{x,y}, \tag{24}$$

where $D' = \{(x,y) \neq (1,1), (x,y) \neq (1,4), (x,y) \neq (4,4)\}$. From previous relations we obtain the following expressions for $m_{1,4}$ and $m_{4,4}$:

$$3m_{1,4} = 2n + 2 - \sum_{(x,y)\in d'} \left(4\frac{x+y}{xy} - 2\right) m_{x,y},$$

$$3m_{4,4} = n - 5 + \sum_{(x,y)\in D'} \left(4\frac{x+y}{xy} - 5\right) m_{x,y}. \tag{25}$$

The Somber index of P is

$$\begin{aligned}
SO_k(P) &= \sqrt{(1-k)^2 + (4-k)^2}\, m_{1,4} + \sqrt{2}|4-k|m_{4,4} \\
&\quad + \sum_{(x,y)\in D'} \sqrt{(x-k)^2 + (y-k)^2}\, m_{x,y} \\
&= \frac{2\sqrt{(1-k)^2 + (4-k)^2} + \sqrt{2}|4-k|}{3}\, n \\
&\quad + \frac{2\sqrt{(1-k)^2 + (4-k)^2} - 5\sqrt{2}|4-k|}{3} \\
&\quad + \sum_{(x,y)\in D'} h_k(x,y) m_{x,y}.
\end{aligned} \tag{26}$$

Since f(1, 4) = f(4, 4)= 0 and $m_{1,1} = 0$, we obtain the result. It is easy to deduce that

$$h(2,2) < h(1,2) < h(2,3) < h(2,4) < 0 = h(1,4) = h(4,4), \tag{27}$$

$$h(2,3) < h(3,3) < h(1,3) < h(3,4) < 0 = h(1,4) = h(4,4), \tag{28}$$

$$h(2,2) < 2h(2,4), \tag{29}$$

$$h(3,3) < 2h(3,4). \tag{30}$$

$\square$

Theorem 4. *Let $P \in D_n$ with $n \geq 3$. Then*

$$\begin{aligned}
SO_k(P) &\leq \frac{2\sqrt{(1-k)^2 + (4-k)^2} + \sqrt{2}|4-k|}{3}\, n \\
&\quad + \frac{2\sqrt{(1-k)^2 + (4-k)^2} - 5\sqrt{2}|4-k|}{3}.
\end{aligned} \tag{31}$$

Equality occurs if and only if $n_2(P) = n_3(P) = 0$.

Proof. The inequality is obtained from relation Eq. (23) in Lemma 1 and from inequalities Eq. (27) and Eq. (28).

If

$$SO_k(P) = \frac{2\sqrt{(1-k)^2 + (4-k)^2} + \sqrt{2}|4-k|}{3} n$$
$$+ \frac{2\sqrt{(1-k)^2 + (4-k)^2} - 5\sqrt{2}|4-k|}{3},$$

(32)

then by inequalities above we conclude that $m_{x,y} = 0$ for all $(x,y) \in D'$ and this occurs if and only if P has no vertices of degree 2 or 3. Conversely, if P has no vertices of degree 2 or 3, by relation Lemma 1 and the fact that $f(1,4) = f(4,4) = 0$ and $m_{1,1} = 0$, we obtain

$$SO_k(P) = \frac{2\sqrt{(1-k)^2 + (4-k)^2} + \sqrt{2}|4-k|}{3} n$$
$$+ \frac{2\sqrt{(1-k)^2 + (4-k)^2} - 5\sqrt{2}|4-k|}{3}$$
$$+ \sum_{(x,y)\in D'} f_k(x,y)m_{x,y}$$

(33)

$$= \frac{2\sqrt{(1-k)^2 + (4-k)^2} + \sqrt{2}|4-k|}{3} n$$
$$+ \frac{2\sqrt{(1-k)^2 + (4-k)^2} - 5\sqrt{2}|4-k|}{3}.$$

Let n be a positive integer and consider the following subsets of D_n:

$$D_{00}(n) = \{P \in D_n : n_2(P) = n_3(P) = 0\},$$
$$D_{10}(n) = \{P \in D_n : n_2(p) = 1, n_3(P) = 0\},$$
$$D_{01}(n) = \{P \in D_n : n_2(P) = 0, n_3(P) = 1\}.$$

(34)

Let $n \geq 3$. it was proved that $D_{10}(n) \neq \emptyset$ if and only if $n \equiv 1 \pmod{3}$, $D_{00}n \neq \emptyset$ if and only if $n \equiv 2 \pmod{3}$.

$\square$

Theorem 5. *Let n be a positive integer. Then, among all trees in D_n, the maximal value of SO_k is attained in:*
Case1. $U \in D_{00}(n)$ if $n \equiv 2 \pmod{3}$ and $n \geq 5$;
Case2. $V \in D_{10}(n)$ such that $m_{1,2}(V) = 0$ if $n \equiv 0 \pmod{3}$ and $n \geq 9$;
Case3. $W \in D_{01}(n)$ such that $m_{1,3}(W) = 0$ if $n \equiv 1 \pmod{3}$ and $n \geq 13$.

Proof. Note that SO_k is constant in $D_{00}(n)$ since for any $U \in D_{00}(n)$,

$$SO_k(U) = \frac{2\sqrt{(1-k)^2 + (4-k)^2} + \sqrt{2}|4-k|}{3} n$$
$$+ \frac{2\sqrt{(1-k)^2 + (4-k)^2} - 5\sqrt{2}|4-k|}{3}.$$

(35)

The graphs in $D_{00}(n)$ are depicted in Fig. 1.
This completes the proof of Case 1.

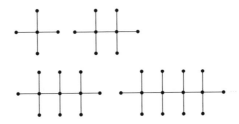

Fig. 1. Trees in D_{00}.

If $n \equiv 0 \pmod 3$, then $D_{00}(n) = \emptyset$ and $D_{01}(n) = \emptyset$. By relations Eq. (9), Eq. (27), and Eq. (28) above, and a relation in Lemma 1 we have

$$
\begin{aligned}
SO_k(P) \leq {} & \frac{2\sqrt{(1-k)^2 + (4-k)^2} + \sqrt{2}|4-k|}{3} n \\
& + \frac{2\sqrt{(1-k)^2 + (4-k)^2} - 5\sqrt{2}|4-k|}{3} \\
& + h(1,2)m_{1,2} + h(2,2)m_{2,2} + h(2,3)m_{2,3} + h(2,4)m_{2,4} \\
\leq {} & \frac{2\sqrt{(1-k)^2 + (4-k)^2} + \sqrt{2}|4-k|}{3} n \\
& + \frac{2\sqrt{(1-k)^2 + (4-k)^2} - 5\sqrt{2}|4-k|}{3} \\
& + h(2,4)(m_{1,2} + 2m_{2,2} + m_{2,3} + m_{2,4}) \\
= {} & \frac{2\sqrt{(1-k)^2 + (4-k)^2} + \sqrt{2}|4-k|}{3} n \\
& + \frac{2\sqrt{(1-k)^2 + (4-k)^2} - 5\sqrt{2}|4-k|}{3} \\
& + 2n_2(V)h(2,4).
\end{aligned}
\tag{36}
$$

If $n_2(P) = 0$, then

$$
\begin{aligned}
SO_k(P) < {} & \frac{2\sqrt{(1-k)^2 + (4-k)^2} + \sqrt{2}|4-k|}{3} n \\
& + \frac{2\sqrt{(1-k)^2 + (4-k)^2} - 5\sqrt{2}|4-k|}{3},
\end{aligned}
\tag{37}
$$

since the equality only occurs if $P \in D_{00}(n)$ and $D_{00} = \emptyset$.

If $n_2(P) \geq 1$, then

$$
\begin{aligned}
SO_k(P) &\leq \frac{2\sqrt{(1-k)^2 + (4-k)^2} + \sqrt{2}|4-k|}{3} n \\
&\quad + \frac{2\sqrt{(1-k)^2 + (4-k)^2} - 5\sqrt{2}|4-k|}{3} \\
&\quad + 2n_2(V)f(2,4) \\
&\leq \frac{2\sqrt{(1-k)^2 + (4-k)^2} + \sqrt{2}|4-k|}{3} n \\
&\quad + \frac{2\sqrt{(1-k)^2 + (4-k)^2} - 5\sqrt{2}|4-k|}{3} \\
&\quad + 2h(2,4).
\end{aligned}
\tag{38}
$$

The inequality in the previous relation occurs if and only if $m_{2,4} = 2$ and $m_{1,2} = m_{1,3} = m_{2,3} = m_{3,3} = m_{3,4} = m_{2,2} = 0$. This implies $n_2(P) = 1$ and $n_3 = 0$. Then, if $n \equiv 0 \pmod 3$ and $n \geq 9$, then the maximal value of SO_k over D_n is attained at $V \in D_{10}(n)$ such that $m_{1,2}(V) = 0$.

The graphs in $D_(10)$ with $m_{1,2} = 0$ are depicted in Fig. 2.

This completes the proof of Case 2.

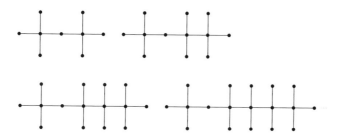

Fig. 2. Trees in D_{10} with $m_{1,2} = 0$.

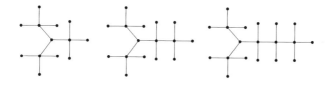

Fig. 3. Trees in D_{01} with $m_{1,3} = 0$.

If $n \equiv 1 \pmod 3$, then $D_{00}(n) = \emptyset$ and $D_{10}(n) = \emptyset$. By relations above we have

$$
\begin{aligned}
SO_k(P) &\leq \frac{2\sqrt{(1-k)^2+(4-k)^2}+\sqrt{2}|4-k|}{3}\,n \\
&\quad + \frac{2\sqrt{(1-k)^2+(4-k)^2}-5\sqrt{2}|4-k|}{3} \\
&\quad + h(1,3)m_{1,3}+h(2,3)m_{2,3}+h(3,3)m_{3.3}+h(3,4)m_{3,4} \\
&\leq \frac{2\sqrt{(1-k)^2+(4-k)^2}+\sqrt{2}|4-k|}{3}\,n \\
&\quad + \frac{2\sqrt{(1-k)^2+(4-k)^2}-5\sqrt{2}|4-k|}{3} \\
&\quad + h(3,4)(m_{1,3}+m_{2,3}+2m_{3,3}+m_{3,4}) \\
&= \frac{2\sqrt{(1-k)^2+(4-k)^2}+\sqrt{2}|4-k|}{3}\,n \\
&\quad + \frac{2\sqrt{(1-k)^2+(4-k)^2}-5\sqrt{2}|4-k|}{3} \\
&\quad + 3n_3(P)h(3,4).
\end{aligned}
\tag{39}
$$

If $n_3(P) = 0$, then

$$
\begin{aligned}
SO_k(P) &< \frac{2\sqrt{(1-k)^2+(4-k)^2}+\sqrt{2}|4-k|}{3}\,n \\
&\quad + \frac{2\sqrt{(1-k)^2+(4-k)^2}-5\sqrt{2}|4-k|}{3},
\end{aligned}
\tag{40}
$$

since the equality only occurs if $T \in D_{00}(n)$ and $D_{00}(n) = \emptyset$.

If $n_3(P) \geq 1$, then

$$
\begin{aligned}
SO_k(P) &\leq \frac{2\sqrt{(1-k)^2+(4-k)^2}+\sqrt{2}|4-k|}{3}\,n \\
&\quad + \frac{2\sqrt{(1-k)^2+(4-k)^2}-5\sqrt{2}|4-k|}{3} \\
&\quad + 3n_3(P)h(3,4) \\
&\leq \frac{2\sqrt{(1-k)^2+(4-k)^2}+\sqrt{2}|4-k|}{3}\,n \\
&\quad + \frac{2\sqrt{(1-k)^2+(4-k)^2}-5\sqrt{2}|4-k|}{3} \\
&\quad + 3h(3,4).
\end{aligned}
\tag{41}
$$

Equality in the previous relation occurs if and only if $m_{3,4} = 3$ and $m_{1,2} = m_{1,3} = m_{2,2} = m_{2,3} = m_{2,4} = m_{3,3} = 0$. This implies $n_2(P) = 0$ and $n_3(P) = 1$.

Then, if $n \equiv 1 \pmod 3$ and $n \geq 13$, then the maximal value of SO_k over D_n is attained at $W \in D_{01}(n)$ such that $m_{1,3}(W) = 0$.

The graphs in D_{01} with $m_{1,3} = 0$ are depicted in Fig. 3.

This completes the proof of Case 3.

$\square$

4   Generalized Somber Index of Hexagonal Systems

An important class of chemical graphs are the hexagonal systems, natural graph representations of benzenoid hydrocarbons. For the definition, notation and details of their theory we refer to [8,14]. The number of fissures, bays, coves, and fjords of a hexagonal system H is denoted by $f = f(H)$, $B = B(H)$, $C = C(H)$ and $F = F(H)$, respectively (see Fig. 4. The number of inlets is

$$r = r(H) = f + B + C + F. \tag{42}$$

We denote by $n = n(H)$ the number of vertices of H, $n_i = n_i(H)$ the number of internal vertices of H, and by $e = e(H)$ the number of hexagonal of H.

Lemma 2. *Let H be a hexagonal system with h hexagons, r inlets and n_i internal vertices. Then*

$$
\begin{aligned}
SO_k(H) = {} & \left[2\sqrt{2}|2 - k| + 3\sqrt{2}|3 - k| \right] h \\
& + \left[2\sqrt{(2 - k)^2 + (3 - k)^2} - \sqrt{2}|2 - k| - \sqrt{2}|3 - k| \right] r \\
& - \sqrt{2}|2 - k|n_i + \sqrt{2} \left[4|2 - k| - 3|3 - k| \right].
\end{aligned}
\tag{43}
$$

Proof. Since H only has vertices of degree 2 and 3,

$$SO_k(H) = \sqrt{2}|2 - k|m_{2,2} + \sqrt{(2 - k)^2 + (3 - k)^2}\,m_{2,3} + \sqrt{2}|3 - k|m_{3,3}. \tag{44}$$

The result follows from the previously known relations [15]:

$$
\begin{aligned}
m_{2,2} &= n - 2h - r + 2, \\
m_{2,3} &= 2r, \\
m_{3,3} &= 3h - r - 3,
\end{aligned}
$$

and [4]

$$n = 4h + 2 - n_i. \tag{45}$$

Let L_h be the linear hexagonal chain (see Fig. 5. Since $r(L_h) = 2(h - 1)$ and $n_i(L_h) = 0$, it follows from Lemma 2 that

$$
\begin{aligned}
SO_k(L_h) = {} & \left[4\sqrt{(2 - k)^2 + (3 - k)^2} + \sqrt{2}|3 - k| \right] h + 6\sqrt{2}|2 - k| \\
& - \sqrt{2}|3 - k| - 4\sqrt{(2 - k)^2 + (3 - k)^2}.
\end{aligned}
\tag{46}
$$

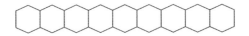

Fig. 4. Linear hexagonal system L_h

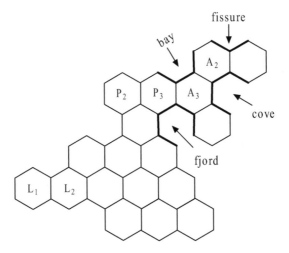

Fig. 5. Types of hexagons in a hexagonal system and some structural features on the perimeter

Now, it was shown in [4] that if H is a hexagonal system with h hexagons then,

$$r(H) \leq 2(h - 1) = r(L_h). \tag{47}$$

Based on this result, among all hexagonal systems with h hexagons, we can show that L_h has the maximal SO_k index.

□

Theorem 6. *Let H be a hexagonal system with h hexagons. Then*

$$SO_k(H) \leq SO_k(L_h). \tag{48}$$

Proof. By relations above, bearing in mind that

$$2\sqrt{(2 - k)^2 + (3 - k)^2} - \sqrt{2}|2 - k| - \sqrt{2}|3 - k| > 0, (k \geq 0), \tag{49}$$

we deduce that

$$SO_k(L_h) - SO_k(H) = \left[2\sqrt{(2 - k)^2 + (3 - k)^2} - \sqrt{2}|2 - k| - \sqrt{2}|3 - k|\right]$$
$$[2(h - 1) - r(H)] + \sqrt{2}|2 - k|n_i \geq 0. \tag{50}$$

Finding the maximal value of SO_k among all hexagonal systems with h hexagons, which this problem is much more complicated. This is due to the fact that if H and H' are hexagonal systems with h hexagons, then by Lemma 2,

$$SO_k(H) - SO_k(H') = \left[2\sqrt{(2-k)^2 + (3-k)^2} - \sqrt{2}|2-k| - \sqrt{2}|3-k|\right]$$
$$\left(r - r'\right) + \sqrt{2}|2-k| \left(n_i' - n_i\right). \tag{51}$$

Unfortunately, the hexagonal systems with minimal number of inlets have $r = \left\lceil\sqrt{3(h-1)}\right\rceil$ inlets [7], meanwhile the hexagonal systems with maximal value of internal vertices have $n_i = 2h - 1 - \left\lceil\sqrt{12h-3}\right\rceil$ internal vertices [11].

$\square$

Problem 1. Among all hexagonal systems with h hexagons, which hexagonal systems have minimal value of SO_k?

Recall that a catacondensed hexagonal system H is a hexagonal system such that $n_i(H) = 0$. One special catacondensed hexagonal system is C_h, shown in Fig. 6.

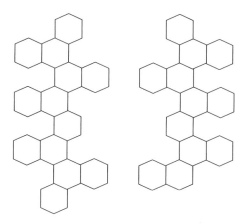

Fig. 6. Catacondensed hexagonal system C_h when h is even (on the left) and when h is odd (on the right)

It was shown in [13] that if H is a catacondensed hexagonal system with h hexagons then,

$$r(C_h) = \left\lceil\frac{h}{2} + 1\right\rceil \leq r(H). \tag{52}$$

It follows from Lemma 2, since $n_i(C_h) = 0$ we deduce that

$$
SO_k(C_h) = \left[2\sqrt{2}|2 - k| + 3\sqrt{2}|3 - k|\right] h
$$
$$
+ \left[2\sqrt{(2 - k)^2 + (3 - k)^2} - \sqrt{2}|2 - k| - \sqrt{2}|3 - k|\right] \left[\frac{h}{2} + 1\right]
$$
$$
+ 4\sqrt{2}|2 - k| - 3\sqrt{2}|3 - k|. \tag{53}
$$

Theorem 7. *Let C_h be a catacondensed hexagonal system with h hexagons. Then*

$$
SO_k(C_h) \leq SO_k(H) \leq SO_k(L_h). \tag{54}
$$

Proof. By Theorem 6, we only have to verify the left inequality. Since $n_i(H) = 0$, bearing in mind that

$$
2\sqrt{(2 - k)^2 + (3 - k)^2} - \sqrt{2}|2 - k| - \sqrt{2}|3 - k| > 0, \tag{55}
$$

where $k \geq 0$, from relations above it follows that

$$
SO_k(H) - SO_k(C_h) = \left[2\sqrt{(2 - k)^2 + (3 - k)^2} - \sqrt{2}|2 - k| - \sqrt{2}|3 - k|\right]
$$
$$
\left(r - \left[\frac{h}{2} + 1\right]\right) \geq 0. \tag{56}
$$

Another interesting class of hexagonal systems are the hexagonal chains. Denote by $L_1 = L_1(H), L_2 = L_2(H), A_2 = A_2(H)$, and $A_3 = A_3(H)$, the number of $L_1 - type, L_2 - type, A_2 - type$, and $A_3 - type$ hexagons, respectively (see Fig. 6). Hexagonal chains are catacondensed hexagonal systems for which $A_3(H) = 0$.

$\square$

Theorem 8. *If H is a hexagonal chain with n hexagons then,*

$$
SO_k(H) = \left[4\sqrt{(2 - k)^2 + (3 - k)^2} + \sqrt{2}|3 - k|\right] h
$$
$$
+ \left[\sqrt{2}|2 - k| - 2\sqrt{(2 - k)^2 + (3 - k)^2} + \sqrt{2}|3 - k|\right] A_2 \tag{57}
$$
$$
+ \left[6\sqrt{2}|2 - k| - 4\sqrt{(2 - k)^2 + (3 - k)^2} - \sqrt{2}|3 - k|\right].
$$

Proof. Theorem 8 follows from Eq. (44) and the relations for hexagonal chains [14]:

$$
m_{22} = A_2 + 6,
$$
$$
m_{23} = 4(h - 1) - 2A_2, \tag{58}
$$
$$
m_{33} = h - 1 + A_2.
$$

Clearly, if H is a hexagonal chain with h hexagons then,

$$A_2(H) \leq h - 2, \tag{59}$$

the equality occurs if and only if H is an hexagonal chain such that $L_2(H) = 0$ (i.e. H is a fibonacene chain [9], see Fig. 7). For any fibonacene chain N, by Theorem 8,

$$SO_k(N) = \left[2\sqrt{(2-k)^2 + (3-k)^2} + 2\sqrt{2}|2-k| \right] h + 4\sqrt{2}|2-k| - 3\sqrt{2}|3-k|. \tag{60}$$

$\square$

Theorem 9. *Let H be a hexagonal chain with h hexagons. Then*

$$SO_k(N) \leq SO_k(H) \leq SO_k(L_h), \tag{61}$$

where N is a fibonacene chain.

Proof. By Theorem 6, we only have to prove the left inequality. Let N be a fibonacene chain. By relations above, bearing in mind that $\sqrt{2}|2-k| + \sqrt{2}|3-k| - 2\sqrt{(2-k)^2 + (3-k)^2} < 0(k \geq 0)$, we deduce that

$$SO_k(H) - SO_k(N) = \left[\sqrt{2}|2-k| + \sqrt{2}|3-k| - 2\sqrt{(2-k)^2 + (3-k)^2} \right] \\ [A_2 - (h-2)] \geq 0. \tag{62}$$

$\square$

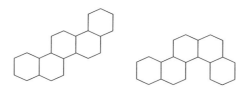

Fig. 7. Two different fibonacene chains with h = 5 hexagons.

5   Normalized Concluding Remarks

In this paper, we obtain the explicit graphs extremal with constant k ($k \geq 0$) for the expected values of generalized Somber indices of chemical graphs, trees with n vertices and the hexagonal systems with n hexagons, discuss the extremal value among them, respectively. Somber index of various graphs of sets have attracted a lot of attention, and their composition and structure are also being studied in the direction of graph theory. Especially, the topological indices and graph invariants based on distances between vertices of a graph are widely used for characterizing molecular graphs, predicting biological activity of chemical compounds, establishing relationships between structure and properties of molecules, and making their chemical applications.

Acknowledgements. This work is partially supported by National Science Foundation of China(Grant No. 12171190), Natural Science Foundation of Anhui Province (Grant No. 2008085MA01).

References

1. Ali, A., Elumalai, S., Mansour, T.: On the symmetric division Deg index of molecular graphs. MATCH Commun. Comput. Chem. **83**, 193–208 (2020)
2. Ali, A., Furtula, B., Gutman, I., Vukicevic, D.: Augmented Zagreb index: extremal results and bounds. MATCH Commun. Comput. Chem. **85**, 211–244 (2020)
3. Aouchiche, M., Ganesan, V.: Adjusting geometric-arithmetic index to estimate boiling point. MATCH Commun. Comput. Chem. **84**(2), 483–497 (2020)
4. Cruz, R., Rada, J., Gutman, I.: Convex hexagonal systems and their topological indices. MATCH Commun. Comput. Chem. **68**, 97–108 (2012)
5. Cruz, R., Monsalve, J., Rada, J.: On chemical trees that maximize atom-bond connectivity index, its exponential version, and minimize exponential geometric-arithmetic index. MATCH Commun. Comput. Chem. **84**, 691–718 (2020)
6. Cruz, R., Monsalve, J., Rada, J.: Sombor index of chemical graphs. Appl. Math. Comput. **399**, 126018 (2021)
7. Cruz, R., Duque, F., Rada, J.: Hexagonal Systems with Minimal Number of Inlets. MATCH Commun. Comput. Chem. **76**, 707–722 (2016)
8. Gutman, I., Cyvin, S.J.: Introduction to the Theory of Benzenoid Hydrocarbons. Springer, Heidelberg (2012)
9. Gutman, I., Klavar, S.: Chemical graph theory of fibonacenes. MATCH Commun. Comput. Chem. **55**, 39–54 (2006)
10. Gutman, I.: Geometric approach to degree-based topological indices: Sombor indices. MATCH Commun. Comput. Chem. **86**(1), 11–16 (2021)
11. Harary, F., Harborth, H.: Extremal animals. J. Comput. Syst. Sci. **1**, 1–8 (1976)
12. Liu, H., You, L., Huang, Y.: Extremal Sombor indices of tetracyclic (chemical) graphs. MATCH Commun. Comput. Chem. **88**, 573–581 (2022)
13. Rada, J.: Bounds for the Randić index of catacondensed systems. Utilitas Math. **62**, 155–162 (2002)
14. Rada, J.: Extremal properties of hexagonal systems. In: Bounds in Chemical Graph Theory-Mainstreams, pp. 239–286 (2017)
15. Rada, J., Araujo, O., Gutman, I.: Randić index of benzenoid systems and phenylenes. Croat. Chem. Acta **74**, 225–235 (2001)
16. Xu, K., Gao, F., Das, K.C., Trinajstić, N.: A formula with its applications on the difference of Zagreb indices of graphs. J. Math. Chem. **57**, 1618–1626 (2019)

Study on the Genetic Links Between Type 2 Diabetes Mellitus and Glioma by Bioinformatics

Yidan Sang, Mengyang Hu, Na Wang, Yangyinchun Bao, Xuemei Yan, Yafei Dong[✉], and Luhui Wang[✉]

College of Life Science, Shaanxi Normal University, Xi'an 710119, China
{dongyf,wangluhui}@snnu.edu.cn

Abstract. Diabetes will increase the risk of most cancers, such as pancreatic, breast and prostate cancer. However, some studies showed patients with diabetes are less susceptible to glioma than the people with normal blood sugar. This inverse relationship between diabetes and glioma remained elusive. Here, we aimed to use bioinformatics methods to reveal the underlying molecule mechanisms between diabetes and glioma. We obtained the gene expression profiles of diabetes (GSE25724) from GEO database. Obtained gene expression profiles of glioma and normal brain tissue from TCGA and GTEX databases. Nine differentially expressed genes (HSPE1-MOB4, BMP5, CD99, MAFB, PLP2, CTSC, NAMPT, VIM, IL13RA2) with opposite expression changes were identified in 2 diseases. These genes were down-regulated in diabetes, but up-regulated in glioma. We then conducted an integrated analysis of these genes, including protein–protein interaction (PPI) network, Gene Ontology (GO) term enrichment and Kyoto Encyclopedia of Genes and Genomes (KEGG) pathway analysis. We validated the expression of hub genes in external datasets (GSE7014 and GSE4290) and studied the correlation between hub genes.

Keywords: Type 2 Diabetes Mellitus · Glioma · Bioinformatics · Genetic Links

1 Introduction

Type II diabetes (T2DM), also known as non-insulin-dependent diabetes, is due to insulin resistance or insulin secretion is relatively insufficient, leading to increased blood sugar. About 90% of diabetes patients have type II diabetes. Obesity is the main cause of type 2 diabetes. Obesity can lead to enlargement of adipocytes and metabolic disorders, leading to impaired insulin signaling and insulin resistance. After insulin resistance, cells are unable to utilize blood sugar. Blood sugar can only be discharged through urine, which is a typical symptom of diabetes. Researchers have found that diabetes will increase the risk of most cancers, such as pancreatic cancer, breast cancer and prostate cancer [1]. However, people with normal blood glucose levels are more likely to suffer from glioma, and diabetes patients are less likely to suffer from glioma. That is, diabetes is inversely related to glioma [2].

Glioma is the most common primary central nervous system tumor. The cells that make up the nervous system mainly include neurons and glial cells, glial cells play a

© The Author(s), under exclusive license to Springer Nature Singapore Pte Ltd. 2024
L. Pan et al. (Eds.): BIC-TA 2023, CCIS 2061, pp. 361–370, 2024.
https://doi.org/10.1007/978-981-97-2272-3_28

supporting, nourishing, and repairing role on neural cells. Glioma originates from the malignant transformation of glial cells. According to the WHO Guidelines for the classification of central nervous system tumors, glioma is classified as grades I to IV with increasing malignancy [3]. Low-grade glioma has the lowest malignancy, the likelihood of recurrence is low, and the prognosis is good. Grade IV glioma, also known as glioblastoma. If left untreated, patients with glioblastoma (GBM) have an average lifespan of only about 4 months. And the average survival time after treatments such as surgery and chemotherapy are only 12–14 months, the three-year survival period is only about 6% [4]. There were no obvious clinical symptoms in the early stage of glioma, so many patients have progressed to terminal phase when glioma is diagnosed. This makes glioma difficult to treat and becomes a significant challenge for neurosurgical oncologists. The pathogenesis of glioma is not yet clear, and the two identified risk factors are radiation and genetic factors.

So far, many studies have found an inverse association between diabetes and glioma. The study by Luqian Zhao et al. showed that diabetes mellitus is significantly associated with reduced risk of glioma [5]. However, this inverse association between diabetes and glioma remained elusive. Therefore, this study explored the genetic links between diabetes and glioma through bioinformatics based on high-throughput data of biopsy samples. We hope this study can help researchers understand the pathogenesis of glioma and contributes to the glioma treatment.

2   Materials and Methods

2.1   Data Source

The gene expression profiles of T2DM (GSE25724) were downloaded from the Gene Expression Omnibus (GEO, http://www.ncbi.nlm.nih.gov/geo/) database. The gene expression profiles of GBM and normal brain tissue from The Cancer Genome Atlas (TCGA, https://portal.gdc.cancer.gov/) and Genotype-Tissue Expression (GTEX, https://www.gtexportal.org/home/) databases. GSE25724 dataset includes 7 T2DM human islets and 6 non-diabetic islet samples. 153 primary tumor samples of GBM were obtained from the TCGA database, and 1153 normal brain tissue samples were obtained from the GTEx database.

2.2   Acquisition of DEGs and DEGs with Opposite Expression Changes in T2DM and GBM

Our study used "limma" package of R software to select DEGs from T2DM (GSE25724) data set, and the criteria for screening out T2DM DEGs were set as adjusted p-value < 0.05 and |log2 Fold Change| > 1. we used "DESeq2" package of R software to perform differential expression analysis for GBM data, and the criteria for screening out GBM DEGs were set as P < 0.05 and |log2 Fold Change| > 2.5. Using the "ggscatter" function of R software to create volcano maps and visualize DEGs. DEGs with opposite expression changes in T2DM and GBM were acquired by Venny2.1.0 (http://www.liu xiaoyuyuan.cn/), then used an Excel spreadsheet to beautify the Venn diagrams.

2.3 Functional Analysis

We performed protein interaction analysis, KEGG and GO enrichment analysis by the STRING online website (https://string-db.org/). Protein interaction networks can help us understand the interactions between proteins, including gene expression regulation, energy and substance metabolism, cell cycle regulation and so on. KEGG is a database that integrates genomic, chemical, and systemic functional information, and it is a popular pathway search database highly used by biologists. The Gene Ontology database divides the function of genes into three parts: cellular component (CC), molecular function (MF), and biological process (BP). By using the GO database, we can obtain the functions of our target genes in CC, MF, and BP. Then imported the protein interaction table into Cytoscape3.9.0 (https://cytoscape.org/) software to beautify the PPI network diagram. Visualized KEGG and GO enrichment results by "ggplot2" and "cowplot" package of R sofware. Use the "analyze network" tool of Cytoscape software to perform topology calculations on the network and identified hub genes. Then we validated the expression of hub genes in external datasets (GSE7014 and GSE4290) and used the ggviolin function of R software to create violin diagrams to visualize the expression levels of hub genes in T2DM and GBM. We evaluated the correlation between hub genes by Spearman correlation analysis.

3 Results

3.1 Identification of DEGs with Opposite Expression Changes in T2DM and GBM

813 DEGs were identified in the T2DM (GSE25724), including 32 up-regulated genes and 781 down-regulated genes. 2050 DEGs were identified in the GBM dataset, including 603 up-regulated genes and 1447 down-regulated genes. Figure 1 showed the DEGs between T2DM human islets and non-T2DM islet samples, and between T2DM primary tumor samples and normal brain tissue samples through volcano plots. DEGs with opposite expression changes in 2 diseases were visualized by Venn diagrams (Fig. 2). We found 9 genes (HSPE1-MOB4, BMP5, CD99, MAFB, PLP2, CTSC, NAMPT, VIM, IL13RA2) were down-regulated in T2DM, but up-regulated in GBM. No genes were up-regulated in diabetes and down-regulated in GBM.

3.2 Protein–Protein Interaction Analysis

We imported the 9 genes obtained in the previous work into STRING database to generate important protein pairs with a minimum interaction score of 0.400 (medium confidence). HSPE1-MOB4 did not find proteins that interact with it. Afterwards, we imported the obtained protein interaction table into Cytoscape 3.9.0 for visualization. As shown in Fig. 3A, we constructed a PPI network that have 49 nodes and 284 edges. The red nodes located in the center represented 8 DEGs obtained from 2 diseases, the yellow nodes represented genes that interact with 8 DEGs, and edges represented the interconnections between different genes. We found that DEGs interact with some inflammatory factors (IL2, IL4, IL13) and tumor related genes (TP53, ATF2, FOXO3). Use the "analyze

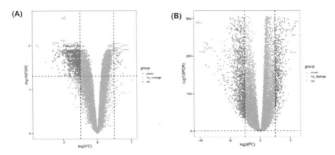

Fig. 1. Volcano plots showed DEGs between T2DM human islets and non-T2DM islet samples (A), and DEGs between primary GBM samples and normal brain tissue samples (B).

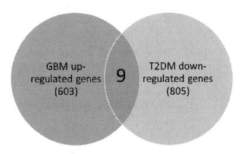

Fig. 2. Venn diagram depicted DEGs with opposite expression changes in 2 diseases.

network" tool of Cytoscape software to perform topology calculations on the network and identified key genes. Based on the degree algorithm, 10 hub genes (TP53, JUN, ESR1, FOS, EP300, FOXO1, MAPK8, SIRT1, BMP2 and JUNB) were identified for further analysis. As shown in Fig. 3B, the core gene of hub gene network was TP53. TP53 is a tumor suppressor gene that can prevent tumor cell division and induce tumor cell apoptosis, and the occurrence and development of glioma are related to exon mutations in the TP53 gene. The color depth and node size represented the degree of genes in the network. Table 1 showed the top ten hub genes calculated based on topological analysis.

Table 1. Topological analysis for top 10 hub genes

Hub gene	Closeness Centrality	Stress	degree	Radiality
TP53	0.701492537	2750	28	0.984802432
JUN	0.671428571	1464	25	0.982522796
ESR1	0.652777778	1186	23	0.98100304
FOS	0.652777778	1020	23	0.98100304

(continued)

Table 1. (*continued*)

Hub gene	Closeness Centrality	Stress	degree	Radiality
EP300	0.626666667	926	21	0.978723404
FOXO1	0.602564103	542	19	0.976443769
MAPK8	0.56626506	314	18	0.972644377
SIRT1	0.546511628	712	18	0.970364742
BMP2	0.546511628	1650	16	0.970364742
JUNB	0.528089888	288	16	0.968085106

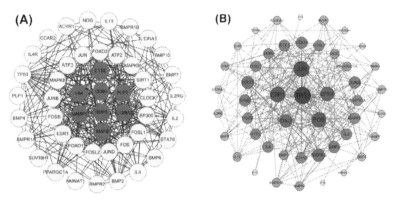

Fig. 3. PPI network: (A) Red nodes were DEGs with opposite expression changes in T2DM and GBM, yellow nodes were genes that interact with 8 DEGs. (B) The screening of hub genes from the PPI network, color depth and node size represented the degree of genes. (Color figure online)

3.3 KEGG and GO Enrichment Analysis

The KEGG and GO enrichment results obtained from STRING database were visualized by R software, displaying the top 10 reliable pathways of KEGG and GO (Fig. 4). As shown in Fig. 4A, KEGG had enriched into some pathways related to inflammatory factors and immune mechanisms. T-helper 1 cell (Th1), Th2 cell, Th17 cell, and Regulatory T cell (Treg) all belong to CD4+ T cells. Th1 cells mainly secrete pro-inflammatory cytokines such as INF-γ, IL-1β, TNF-α, mediating cellular immune response. Th2 cells mainly secrete factors such as IL-4, IL-5, IL-10, and IL-13, promoting antibody production and mediating humoral immune responses. In healthy cells, Th1 and Th2 cell differentiation are in equilibrium. In diseased cells, there is an imbalance in the differentiation of Th1 and Th2 cells [6]. TGF-β stimulates the differentiation of naive CD4+T cells into Treg cells. TGF-β and IL-16 stimulate the differentiation of naive CD4+T cells into Th17 cells. Meanwhile, IL-6 can inhibit Treg cells. Treg cells and Th17 cells have opposite immune regulatory functions. Treg cells mediate immune tolerance, while Th17 mediate immune rejection [7]. Compared to people with normal blood glucose, the proportion of Th1 and Th17 cells increase in diabetes patients, while the proportion of Treg cells significantly decreased, and there was no significant change in Th2 cells. Ultimately, it leads to an increase in the ratio of Th1/Th2 cells and Th17/Treg cells [8]. Prakash Somi Sankaran fed C57bl/6 mice and Balbc mice strains with high-fat-diet to induce diabetes. C57bl/6 strains tended to develop Th1-type immune response, while Balbc mice developed Th2 type immune response. It was found that Th1-biased mice strains were susceptible to obesity induced by high-fat diet, resulting in insulin resistance and diabetes [9]. Similarly, Montgomery et al. studied five different mouse strains and found that Balbc with Th2-type immune response tendency is the only strain that will not suffer from obesity or diabetes [10]. In glioma tissues, the Th2 cytokine genes have a significant expression advantage. Using IFN-γ cytokines to induce the transformation of Th2 to Th1 can significantly inhibit the proliferation of glioma cells. Using IL-4 to enhance the switch of Th2 cells could stimulate the growth of glioma cells [11]. Therefore, we suspected that people who tend to generate Th1-type immune response are susceptible to diabetes, while people who tend to generate Th2-type immune response are susceptible to glioma. This may be one of the reasons why diabetes patients are less to suffer from glioma. However, a tendency towards Th2-type immune response is not unique to glioma, and it is also present in other cancers. So, there are other negative correlation mechanisms between diabetes and glioma that we have not yet found. At the same time, patients with diabetes have decreased Treg cells and increased Th17 cells. Treg cells can help normal cells resist immunity, and the decrease in Treg cells increases the likelihood of pancreatic islet cells being attacked by autoimmunity. According to reports, TGF-β secreted by Treg cells could reduce weight and insulin resistance [12]. Therefore, the decline of Treg cells is one of the causes of diabetes. In glioma tissues, both Treg and Th17 cells increase. Treg cells can help glioma tissues have immune escape ability and promote tumor development [13]. Th17 cells secrete IL-17 factor and induce pathological features such as persistent inflammation.

As shown in Fig. 4B, GO enriches into pathways related to the synthesis, transcription, and metabolism of biological macromolecules. The image showed the top 10 pathways enriched into BP, MF, and CC based on P-value, including positive regulation of transcription by RNA polymerase II, positive regulation of nitrogen compound metabolic process, transcription regulator complex, transcription factor AP-1 complex, chromatin binding, sequence-specific DNA binding and so on. The shape of point represented category of the pathway. The color of point represented the p-value, and the lower the p-value, the higher the reliability. The size of the point represented the number of genes enriched into this pathway.

3.4 Validation of Hub Genes Performance

We used GSE7014 and GSE4290 datasets to study the mRNA expression of identified hub genes in T2DM and GBM. GSE7014 included three groups of samples: T1DM patients, T2DM patients, and healthy samples. We compared the expression levels of hub genes between T2DM patients and healthy samples.

As shown in Fig. 5A, there is no significant change in the expression of 10 hub genes in diabetes patients. GSE4290 had 23 samples from epilepsy patients are used as non-tumor samples. 157 different types of glioma samples include 26 astrocytomas, 50 oligodendrogliomas and 81 glioblastomas. As shown in Fig. 5B, the difference had

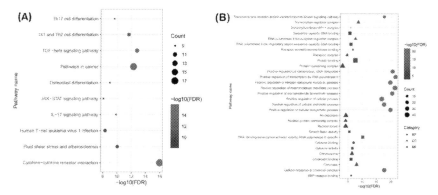

Fig. 4. The KEGG and GO enrichment results were visualized by bubble plots: (A) the Top 10 KEGG pathways, (B) Top 10 GO pathways regarding molecular function, biological process, and cellular.

existed in the expression of BMP2, TP53, MAPK8, and SIRT1 genes between healthy and glioma tissues. BMP2 is one of the bone morphogenetic proteins. Research have shown that the expression of BMP2 was up regulated in glioma, and its expression level was positively correlated with the malignancy of the tumor [14]. TP53 is an important tumor suppressor gene, and its encoded protein p53 can regulate cell cycle and prevent cell carcinogenesis. The mutation of TP53 has a high incidence in glioma issues the incidence of TP53 gene mutations is 50%–60% in low-grade astrocytoma, and it is very low in oligodendroglioma. The incidence of secondary GBM is 70%, and the incidence of primary GBM is 25%–37%. The mutant p53 protein will lose its inhibitory effect on tumor tissue, leading to uncontrolled cell growth and promoting tumor development [15]. Compared with healthy tissues, the MAPK8 gene is downregulated in glioma. MAPK8 is a member of the MAPK family. Kappadakunnel M et al. found MAPK8 is related to survival in glioma patients, increased expression levels of MAPK8 were associated with longer survival [16]. SIRT1 is a highly conserved deacetylase in the nucleus that can deacetylate non-histone proteins such as PARP1 and p53. The role of SIRT1 in tumors has a double-sided nature. It can not only inhibit the occurrence and development of tumors through mechanisms such as inhibiting inflammation and interacting with tumor related genes, but also promote tumor proliferation and development by regulating tumor related genes, epithelial mesenchymal transformation, and other mechanisms [17].

We evaluated the correlation between hub genes based on Spearman correlation analysis (Fig. 6). Blue represented a negative correlation between genes, while orange represented a positive correlation between genes. The larger the size of point, the stronger the correlation between genes. We found a strong positive correlation between FOS and junB genes in two datasets. Both FOS and junB proteins belong to the activator protein-1 and participate in the transcriptional regulation of many growth factors and cytokines, leading to a series of physiological and pathophysiological changes in the body. The signal transduction pathway mediated by activating protein-1 is associated with the occurrence of various malignant tumors and metabolic diseases.

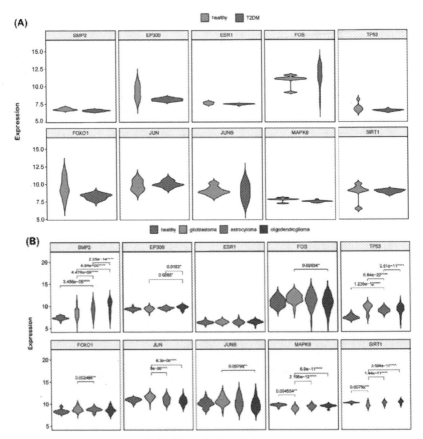

Fig. 5. Validation of hub genes performance. The expression level of 10 hub genes in T2DM (A) and glioma (B).

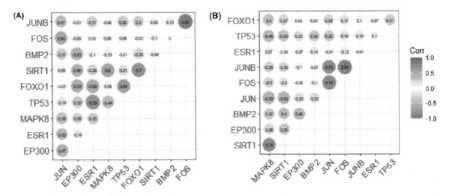

Fig. 6. Spearman correlation analysis for evaluating the correlation between hub genes in T2DM (A) and glioma (B) datasets. (Color figure online)

4   Limitations

We enriched DEGs with opposite expression changes in T2DM and glioma, hoping to find some pathways that can explain the mechanism that diabetes and glioma are not easy to appear at the same time. Then we found T2DM, and glioma patients have opposite regulation in Th1/Th2 cell differentiation pathway, that may be one of the reasons why diabetes patients are not susceptible to glioma. But there are no relevant experiments to confirm this conjecture, so it has uncertainty. We plan to study the causal relationship between Th1-type immune response and diabetes, and the causal relationship between Th2-type immune response and glioma through genome-wide association analysis in our future work to determine whether the conjecture is correct.

5   Conclusions

We identified DEGs with opposite expression changes in T2DM and GBM and enriched these DEGs. We found that T2DM and GBM patients have different changes in immune response related pathways. We conducted PPI analysis on DEGs and found that DEGs interact with some inflammation related genes and tumor related genes. We screened hub genes from the PPI network based on degree algorithm, studied the mRNA expression of hub genes in T2DM and GBM, and calculated the correlation between hubs. Then we found that four genes (BMP2, TP53, MAPK8, and SIRT1) are associated with the development of glioma. These genes may be potential targets for the treatment of glioma.

Acknowledgements. This work is supported by the National Natural Science Foundation of China (No. 62073207), the basic natural science research program of Shaanxi Province (No. 2020JM-298).

References

1. Giovannucci, E., et al.: Diabetes and cancer: a consensus report. Diabetes Care **33**(7), 1674–1685 (2010)
2. Schwartzbaum, J., et al.: Associations between prediagnostic blood glucose levels, diabetes, and glioma. Sci. Rep. **7**(1), 1436 (2017)
3. McNamara, C., et al.: 2021 WHO classification of tumours of the central nervous system: a review for the neuroradiologist. Neuroradiology **64**(10), 1919–1950 (2022)
4. Urbańczyk, H., Strączyńska-Niemiec, A., Głowacki, G., Lange, D., Miszczyk, L.: Case presentation - a five-year survival of the patient with glioblastoma brain tumor. Rep. Pract. Oncol. Radiother. J. Greatpoland Cancer Center Poznan Pol. Soc. Radiat. Oncol. **19**(5), 347–351 (2014)
5. Zhao, L., Zheng, Z., Huang, P.: Diabetes mellitus and the risk of glioma: a meta-analysis. Oncotarget **7**(4), 4483–4489 (2016)
6. Zhu, X., Zhu, J.: CD4 T helper cell subsets and related human immunological disorders. Int. J. Mol. Sci. **21**(21), 8011 (2020)
7. Lee, G.R.: The balance of Th17 versus Treg cells in autoimmunity. Int. J. Mol. Sci. **19**(3), 730 (2018)

8. Zhang, C., et al.: The alteration of Th1/Th2/Th17/Treg paradigm in patients with type 2 diabetes mellitus: relationship with diabetic nephropathy. Hum. Immunol. **75**(4), 289–296 (2014)

9. Sankaran, P.S.: High-fat-diet induced obesity and diabetes mellitus in Th1 and Th2 biased mice strains: a brief overview and hypothesis. Chronic Dis. Transl. Med. **9**(1), 14–19 (2023)

10. Montgomery, M.K., et al.: Mouse strain-dependent variation in obesity and glucose home-ostasis in response to high-fat feeding. Diabetologia **56**(5), 1129–1139 (2013)

11. Li, G., Hu, Y.S., Li, X.G., Zhang, Q.L., Wang, D.H., Gong, S.F.: Expression and switching of TH1/TH2 type cytokines gene in human gliomas. Chin. Med. Sci. J. = Chung-kuo i Hsueh k'o Hsueh tsa Chih **20**(4), 268–272 (2005)

12. Zhang, S., et al.: The alterations in and the role of the Th17/Treg balance in metabolic diseases. Front. Immunol. **12**, 678355 (2021)

13. Amoozgar, Z., et al.: Targeting Treg cells with GITR activation alleviates resistance to immunotherapy in murine glioblastomas. Nat. Commun. **12**(1), 2582 (2021)

14. Liu, C., Tian, G., Tu, Y., Fu, J., Lan, C., Wu, N.: Expression pattern and clinical prognostic relevance of bone morphogenetic protein-2 in human gliomas. Jpn. J. Clin. Oncol. **39**(10), 625–631 (2009)

15. Comprehensive genomic characterization defines human glioblastoma genes and core pathways. Nature **455**(7216), 1061–1068 (2008)

16. Kappadakunnel, M., et al.: Stem cell associated gene expression in glioblastoma multiforme: relationship to survival and the subventricular zone. J. Neurooncol **96**(3), 359–367 (2010)

17. Liang, S.P., et al.: Activated SIRT1 contributes to DPT-induced glioma cell parthanatos by upregulation of NOX2 and NAT10. Acta Pharmacol. Sin. **44**(10), 2125–2138 (2023)

An Investigation into the Use of DNA Strand Displacement Reaction Networks for Subset Sum Problem Solutions

Yawen Zheng, Jing Yang[✉], Tongtong Zhang, and Tianyi Jiang

School of Mathematics and Big Data, Anhui University of Science and Technology,
Huainan 232001, Anhui, China
1157201914@qq.com

Abstract. This paper investigates an innovative application of DNA strand displacement reaction networks in solving the subset sum problem. The subset sum problem is a significant issue in computer science, and this study introduces a method based on DNA strand displacement reaction networks, simulating the specific computational process of solving the subset sum problem through visual DSD. The method involves three cascaded reaction modules—weighted, sum and threshold—ultimately expressed through the output of a single-stranded DNA. Experimental results demonstrate the method's effectiveness in identifying the presence of a target subset in a given set, where the sum equals a specific target value. The computational model presented in this paper lays the groundwork for the applicability of DNA computing in complex problem-solving scenarios, while also providing a foundation for the future development of DNA-based computer systems and biocomputing applications.

Keywords: DNA Strand Displacement · DNA Computation · NP Problem · Subset Sum Problem

1 Introduction

Traditional computing approaches had reached their physical limits, constrained by the rapid advancements in science and technology. This was particularly evident in combinatorial optimization problems like the subset sum problem, frequently applied in work scheduling and resource allocation. The need for a more effective solution intensified as the problem's size increased and its computational complexity surged. DNA computing garnered considerable interest, specifically in DNA strand displacement reaction networks, as a novel molecular self-assembly method. With notable mobility and convenient experimental settings, it demonstrated significant potential for addressing intricate computational challenges. In the late 1990s, Yurke [1] and others pioneered DNA strand displacement, introducing a "tweezer" device based on this process. Its precise control and high programmability, stemming from its cascading nature, substantially enhanced its utility for complex function development. Researchers employed this method to construct diverse computational models, including neural networks [2] and chemical reaction networks [3]. Various logic gates, such as AND gates, OR gates, NOT gates, XOR

© The Author(s), under exclusive license to Springer Nature Singapore Pte Ltd. 2024
L. Pan et al. (Eds.): BIC-TA 2023, CCIS 2061, pp. 371–383, 2024.
https://doi.org/10.1007/978-981-97-2272-3_29

gates, and NAND gates [4–7], were developed. In 2011, Qian and Winfree [8] devised the seesaw gate, one of the most intricate biological digital logic circuits known to date. Academics applied the DNA strand displacement approach to address a wide array of computing problems, including 0–1 integer programming problems [9] and probabilistic reasoning [10]. In 2022, Anna P. *et al.* [11] demonstrated the technique for temporal storage in a DNA-only system. The current method, implementable with a basic two-input AND-gate logic circuit, facilitated temporal storage and decision-making based on specific combinations and their relative timing.

In this paper, we explore the application of DNA strand displacement reaction networks to address the subset sum problem by transforming it into a process of altering DNA chain concentrations. In this intricate procedure, concentrations of different elements undergo transformation through chain displacement reactions, converging into the concentration of a single DNA strand. This manipulation is instrumental in effectively resolving the subset sum problem. Simultaneously, DNA strand displacement can coordinate the convergence of different concentrations into a unified output, serving as a key mechanism to address the complexities of the subset sum problem. This approach not only demonstrates the applicability of DNA strand displacement in computational problem-solving but also emphasizes the potential for innovative solutions to combinatorial optimization challenges through the utilization of molecular self-assembly methodologies.

2   DNA Strand Displacement and Subset Sum Problems

2.1   DNA Strand Displacement

DNA Strands Displacement (DSD) is a molecule-based method that uses molecular recognition qualities and the standard Watson-Crick base pairing principle to initiate a sequence of programmable chemical reactions without the need for extra enzymes or proteins. To put it simply, it is a precisely controlled process of numerous DNA strands exchanging and replacing one another.

A single strand of DNA, known as the invading strand, interacts with a double-stranded DNA whose partial sequence complements the invading strand in a DNA strand replacement reaction. By binding to one strand of the target double-stranded DNA, the invading strand uses the complementary pairing principle to reject (replace) the original other strand. Due to the sequence-specific nature of this mechanism, the reaction can only take place in the presence of adequate pairwise complementarity between the double-stranded DNA and the invasive strand. Figure 1 illustrates the reaction's basic idea.

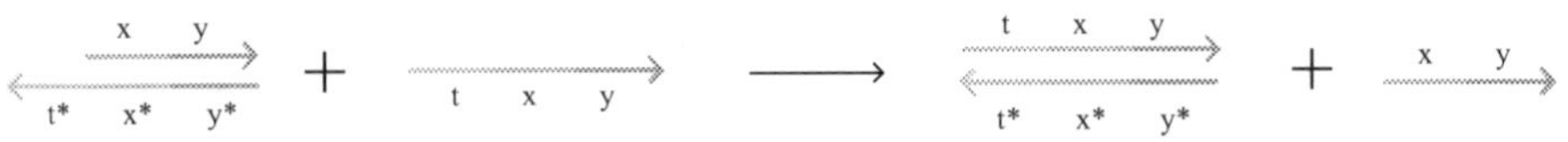

Fig. 1. Principle of DNA strand displacement reaction.

2.2 Subset Sum Problems

The Subset Sum Problems (SSP) is an NP-complete issue and a famous problem in computer theory. It is crucial to investigate this issue in the domains of operations research, computer science, and mathematics. Studying subset sum problems can lead to advancements in a number of fields, including operations research, computational complexity theory, algorithm design, and more.

For the subset sum problem, consider the following explanation: Find out if there is a subset of integers $I \in \{1, 2, \cdots, n\}$ such that $\sum w_i = T, i \in I$ for a given set of integers $W = \{w_1, w_2, \cdots, w_n\}$ and a target number T.

3 A Subset Sum Problems and Computational Models Based on DNA Strand Displacements

3.1 Computational Model

For the subset sum problem, the solution process is divided into three steps. First, for a given set of integers w_i, if they are present in the subset, it is transformed into a single-stranded DNA x_i of concentration w_i, i.e., a weighted reaction is performed. Then, sum over w_i. Finally, the subset that matches is found by means of the target number. In the three steps, three reaction modules, i.e., weighted reaction module, sum reaction module and threshold reaction module are designed to achieve the goal. The reactions are cascaded in the order of the three reaction modules and finally the conforming subset is obtained through the output. The three modules mentioned above are specifically depicted in Fig. 2:

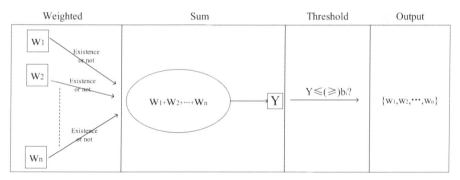

Fig. 2. Reaction Modules. It includes weighted, sum and threshold three modules, finally the output module.

Weighted Reaction Module. In the set of integers, we refer to the members as w_i as weights. A weighted reaction module was created in this instance. In case the element is included in the subset, the definition of single-stranded DNA Xi is $x_i = 1$. In contrast, the definition of single-stranded DNA Xi is $x_i = 0$. Weight w_i is assigned to x_i when

1nM single-stranded DNA Xi is input, and w_i nM single-stranded DNA Xi is output. It is important to note that the weighted reaction module can be omitted in order to get directly to the sum reaction module when the weights $w_i = 1$. For the weighted reaction module, the chemical reaction equation is

$$Xi \rightarrow w_i Xi,$$
$$\downarrow implement,$$
$$Xi + Mi \rightarrow Wi + waste1,$$
$$Wi + Ni \rightarrow w_i Xi + waste2. (i = 1 : n)$$

(1)

As seen in Fig. 3, the chemical steps were carried out by two DSD reactions with auxiliary double-stranded DNA Mi and Ni. The number of $5' - x_i - a - 3'$ in the auxiliary double-stranded DNA Mi and the number of $x_i - a - b - c$ and $x_i^* - a$ in the auxiliary double-stranded DNA Ni are determined by the weight of x_i, w_i. The intermediate product Wi was created by combining supplemental double-stranded DNA Mi with single-stranded DNA Xi. Wi gives w_i weights by binding to auxiliary double-stranded DNA Ni, which releases w_i nM Xi. The starting Xi and Mi concentrations in both DSD reactions meet the conditions $[Xi]_0 = [Mi]_0 = 1nM$, $[Ni]_0 = 6nM$.

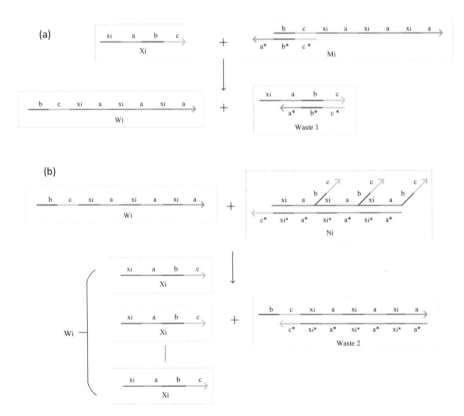

Fig. 3. Weighted reaction module. (a) $Xi + Mi \rightarrow Wi + waste1$; (b) $Wi + Ni \rightarrow w_i Xi + waste2$.

Sum Reaction Module. We were able to successfully weight w_i in the weighted reaction module. Subsequently, summing over w_i, in which a sum reaction module is intended to superimpose the concentrations of the various input signals to the concentration of Y and transform various input signals $(x_1, x_2, \cdots, x_i)$ to the same output (Y). Ultimately, $w_1 x_1 + w_2 x_2 + \cdots + w_i x_i \xrightarrow{DSD} (w_1 + w_2 + \cdots + w_i)Y$ is obtained by using the sum reaction module. The chemical reaction equation of the sum reaction module is

$$Xi \to Y,$$
$$\downarrow implement,$$
$$Xi + Gi \to Y + waste3. (i = 1 : n)$$

(2)

As seen in Fig. 4, the chemical reactions were carried out by a DSD reaction between auxiliary double-stranded DNA Gi and single-stranded DNA Xi. Following the weighted reaction module, w_i nM is the Xi concentration. Subsequently, 1nM Xi generates the output signal 1nM Y in the sum reaction module, converting Xi of w_i nM to Y of w_i nM, or $w_1 x_1 + w_2 x_2 + \cdots + w_i x_i \xrightarrow{DSD} (w_1 + w_2 + \cdots + w_i)Y$. The starting concentrations of Gi and Xi in the DSD reaction meet the condition $[Xi]_0 = [Gi]_0$.

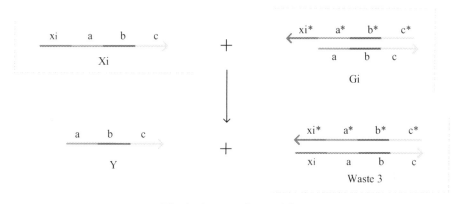

Fig. 4. Sum reaction module.

Threshold Reaction Module. $w_1 x_1 + w_2 x_2 + \cdots + w_i x_i \xrightarrow{DSD} (w_1 + w_2 + \cdots + w_i)Y$ Is produced in the sum reaction module. To choose a subset that meets the target number, analysis is performed by comparing the concentration of the output signal Y with the target number. However, the threshold reaction module was created to replace concentration detection with fluorescence detection in order to improve the accuracy of the results due to the concentration detection's restricted accuracy. For the threshold reaction module, the chemical reaction equation is

$$Y \xrightarrow{k1} Q,$$
$$Y \xrightarrow{k2} F,$$
$$\downarrow implement,$$

$$[Y]_0 \le b_i? \begin{cases} Yes, \ Y_{[Y]_0} + P_{[P]_0=b_i} \xrightarrow{k1} Q_{[Q]_\infty=[Y]_0} + waste4, \\ No, \begin{cases} Y_{[Y]_0} + P_{[P]_0=b_i} \xrightarrow{k2} Q_{[Q]_\infty=b_i} + waste4, \ (i = 1:n) \\ Y_{[Y]_0-b_i} + T_{[T]_0=b_i} \xrightarrow{k2} F + waste5. \end{cases} \end{cases} \tag{3}$$

Figure 5 illustrates how the chemical processes were carried out using DSD reactions of auxiliary double-stranded DNA P and T. The DSD reaction was found to rely exponentially on foothold length in earlier studies of DSD kinetics. The footing domain b^* (3nt) for the threshold reaction module in this paper is a shortened form of $a^* - b^*$ (6nt), which results in a significantly higher reaction rate, k1, than k2.The equivalent strand Y prefers to bind on the lengthy fixed point domain $a^* - b^*$ when the concentration $[Y]_0 \le [P]_0$.The first strand Y is bound to the long footing point domain $a^* - b^*$ if concentration $[Y]_0 > [P]_0$, while the remaining strands Y can only be bound to the short footing point domain b^*.Consequently, only when the concentration $[Y]_0 > [P]_0$, where $[P]_0$ is the threshold concentration, does fluorescence release. The sum reaction module supplies the initial concentration of the input signal Y in the DSD reaction. The constraints, that is, $[P]_0 = b_i$, supply the starting concentration of P, and $[T]_0 = b_i$ sets the initial concentration of T. Crucially, if the number of targets is equal to b_i, only then is a subset that satisfies the requirements obtained.

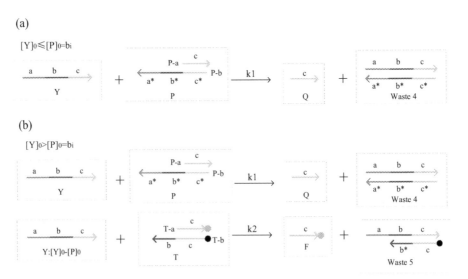

Fig. 5. Threshold reaction module. (a) Initial concentration $[Y]_0 \le [P]_0$; (b) Initial concentration $[Y]_0 > [P]_0$.

3.2 Analysis of Examples

Let's put this computational model into practice with a real-world example. We have the following set of integers: $S = \{3, 4, 12, 5, 2, 1\}$, and $T = 9$. The set of integers in this issue consists of 6 elements, and there are $2^6 = 64$ subsets of those elements. Finding the subset that satisfies the requirement is the next three phases.

Step 1: The weights of the six items in the set of integers are 3, 4, 12, 5, 2, and 1. Therefore, the weighted reaction module is built initially.

For element 3, the chemical reaction equation for the weighted reaction module is

$$
\begin{aligned}
X1 &\to 3X1, \\
&\downarrow implement, \\
X1 + M1 &\to W1 + waste1, \\
W1 + N1 &\to 3X1 + waste2.
\end{aligned}
\tag{4}
$$

As illustrated in Fig. 6, the chemical process achieved by the DSD reaction network goes through two substitution reactions.

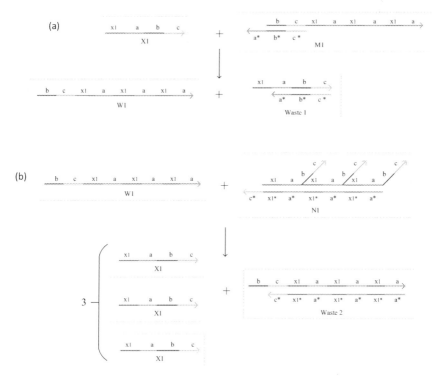

Fig. 6. Element 3 weighted reaction module. (a) $X1 + M1 \to W1 + waste1$; (b) $W1 + N1 \to 3X1 + waste2$.

Next, we used Visual DSD to simulate the weighted reaction module. The concentration of the auxiliary double-stranded DNA $M1$ and $N1$ were set to 1nM and 6nM, respectively, while the single-stranded DNA *X1* was utilized as the reactant in the simulation at a concentration of 1nM. The reaction process was timed to take 10,000 s in total. The weighted reaction module's whole simulation time series plot is shown in Fig. 7. The starting concentration of *X1* decreases and the concentration of the produced intermediates (*sp5, sp6*) increases during the simulation when single-stranded DNA *X1* enters the reaction module and binds to the auxiliary double-stranded DNA *M1*. Following their peak concentrations, these intermediates start to react with auxiliary double-stranded DNA *N1*, which results in a decrease in *N1* concentration and an increase in the concentrations of the ultimate product (*sp4*) and single-stranded DNA *X1*. *M1* and *N1* consumed 1nM each during this time, and the intermediate *sp5* was produced and then consumed. The labels "generated waste products" refer to *sp4* and *sp6*. Significantly, the process was able to raise the single-stranded DNA *X1* concentration to 3nM, confirming the weighted reaction module's theoretical viability. The accuracy of the experiment was ensured by the fact that no unexpected products were seen, just the expected intermediate products (*sp5, sp6*) and ultimate product (*sp4*), as proven by complete simulation time series plots.

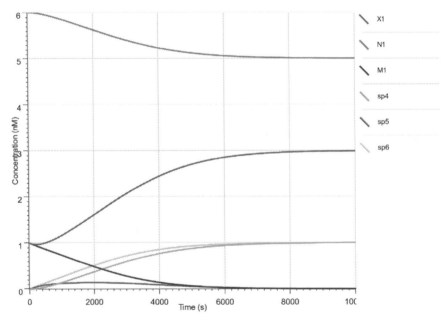

Fig. 7. Element 3 weighted reaction module. The complete simulation time series graph of the element 3 weighted reaction module.

They are exposed to a weighted reaction module for the remaining items, which are 4, 12, 5, 2, and 1, respectively. Element 1 can proceed directly to step 2, the sum reaction module, as it is not required to go through a weighted reaction. The weighted reaction module's simulation sequence diagram for the remaining elements is displayed in Fig. 8.

Step 2: The sum reaction module, which supplies the initial concentration of the input signals in step 1, transforms the various input signals into the same output, *Y*.

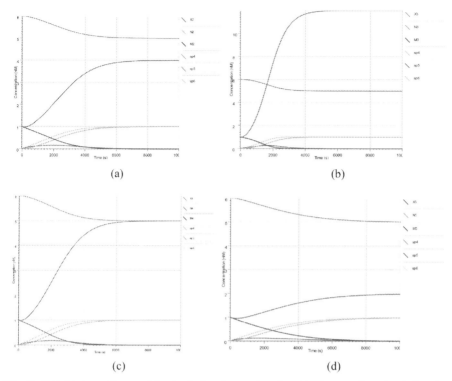

Fig. 8. Weighted reaction module simulation time series plot. (a) The complete simulation time series graph of the element 4 weighted reaction module; (b) The complete simulation time series graph of the element 12 weighted reaction module; (c) The complete simulation time series graph of the element 5 weighted reaction module; (d) The complete simulation time series graph of the element 2 weighted reaction module.

The chemical reaction equation for the cumulative reaction module for 3, 4, 12, 5, 2, 1 is

$$Xi \rightarrow Y,$$
$$\downarrow implement, \quad (5)$$
$$Xi + Gi \rightarrow Y + waste3.$$

As seen in Fig. 9, the DSD reaction carries out the chemical processes. As a result, step 2 involves adding up each of the 64 subsets.

Visual DSD is used as an example to simulate the aforementioned sum reaction module for the subset {3, 4, 12, 5, 2, 1}. Step 1 supplies the initial concentration of Xi, and the concentration profile of the single-stranded DNA Y is displayed in Fig. 10. Simulation results of {3, 4, 12, 5, 2, 1}. Summation reaction. Step 1 supplies the initial concentration of Xi, and Fig. 10 displays the concentration profile of the single-stranded DNA Y. We discover that the subset {3, 4, 12, 5, 2, 1} adds up to 27, and the simulation result confirms that the sum reaction module's conclusion is accurate.

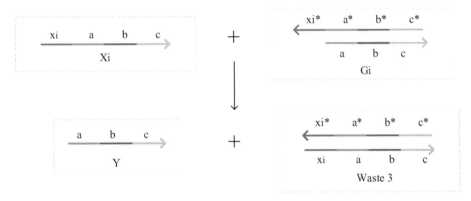

Fig. 9. Subset sum reaction module.

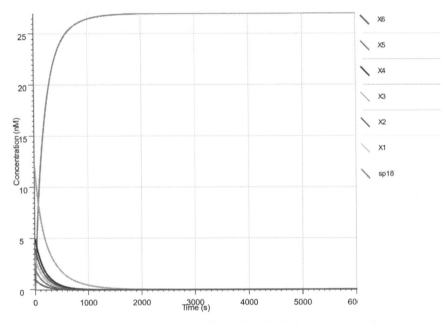

Fig. 10. Simulation results of {3, 4, 12, 5, 2, 1}. Summation reaction.

Step 3: This stage's goal is to pinpoint a certain subset of the target number that the threshold reaction module can meet. The findings of subsets {4, 5} and {3, 4, 5, 2, 1}, which are shown in Fig. 11 and Fig. 12, were simulated using Visual DSD software. In the experiment, a target number was used to calculate the beginning concentration of P, and the input signal Y's initial concentration was set based on the sum reaction module. Additionally, we started with a T concentration of 9nM.

Figure 11's simulation results show that while the concentration of P is constant, there is a noticeable increase in the concentration of the end product (output) and a considerable drop in the concentrations of Y and T. These modifications suggest that

T and Y are involved in the DSD reaction, are consumed, and that Y's consumption prevents it from reacting with P in the future. As a result, the P concentration is stable. The veracity of this reaction is supported by the rise in concentration of the finished good (output).

We verified the subset {4, 5} as the target subset since the final concentration matched the predetermined target number of 9 and no fluorescence signal was observed. On the other hand, Fig. 12 depicts an alternative situation. The goal thresholds were exceeded by the concentrations of Y, P, and T due to their considerable changes in concentration. Additionally, the results are validated by the measured fluorescence signal. All of these measures combined show that the subset {3, 4, 5, 2, 1} does not meet the requirements to be considered a target subset.

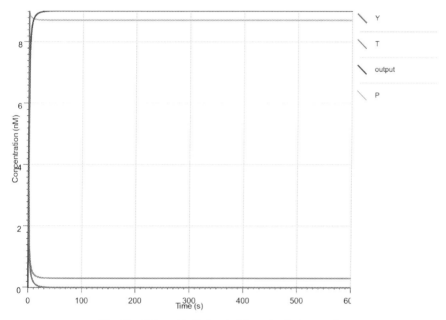

Fig. 11. Subset {4, 5} thresholding simulation results.

Lastly, this analysis finds the subsets {4, 5}, {3, 4, 2}, and {3, 5, 1}, which correspond to the number of objectives and the best solution to the problem, respectively. The usefulness of the approach is illustrated in this work by an analysis of a six-element subset sum issue, highlighting the high practicality of subset sum problems. It is important to note that this approach functions just as well with additional elements. In order to solve the subset sum problem, the DNA strand displacement (DSD) reaction network was used due to its high computational construction logicalness, accuracy, and sensitivity. However, in real biochemical investigations, the approach might also encounter some difficulties. For instance, the number of DNA strands needed to solve the problem grows with its size, which may result in an increase in experimental complexity; the accuracy of the results must be confirmed by additional biochemical experiments; and the results

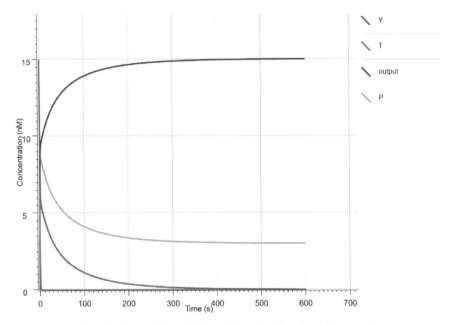

Fig. 12. Subset {3, 4, 12, 5, 2, 1} thresholding simulation results.

must be read using a specific biological method. It is crucial to remember that issues like leakage reactions could arise during the tests because of time constraints, the nature of the experimental setting, etc., all of which could get more difficult as the number of components rises.

4 Summary

In tackling the subset sum problem, this paper introduces a computer model grounded in DNA strand displacement reactions. By employing DNA strand displacement techniques, the correlation between variations in DNA chain inputs and outputs and the target sum is elucidated. The method successfully addresses a subset sum problem involving six elements, with the simulation of all pertinent DNA strand displacement reaction processes and changes in DNA chain concentrations carried out using Visual DSD software. While theoretical validation has been achieved for the subset sum problem with six elements, experimental data verification is still pending. Due to temporal and experimental constraints, an increase in the number of elements may lead to leakage responses, characterized by mismatch occurrences in the chain branching migration process, thereby reducing the generation rate. Although leakage responses do not substantially alter the research findings, it is crucial to acknowledge that biochemical experiments and potential imperfections may compromise computational accuracy, particularly when confronted with larger or more complex subset sum problems. Therefore, maintaining precision remains paramount for the validity of subsequent research endeavors.

In addition to being a fundamental technology for nanotechnology and molecular computing, DNA strand displacement also creates new opportunities for molecular-level information processing and control. Nevertheless, this methodology is not without its inherent obstacles, including intricate molecular dynamics, issues in designing DNA strands, and intricate interactions during the reaction. Future research in this area will concentrate on resolving these problems and achieving the accurate design and management of large-scale complicated reaction networks. With considerable advantages over conventional computational techniques, the DNA strand displacement computational strategy used in this work exhibits enormous potential for managing large-scale datasets in particular circumstances. This work provides a strong basis for the investigation of potential DNA-based computing systems and biocomputing applications, in addition to demonstrating the usefulness of DNA computing in the resolution of challenging computational issues.

Acknowledgement. This project is supported by National Natural Science Foundation of China (No. 62272005).

References

1. Yurke, B., Turberfield, A.J., Mills, A.P., Jr., et al.: A DNA-fuelled molecular machine made of DNA. Nature **406**(6796), 605–608 (2000)
2. Qian, L., Winfree, E.: A simple DNA gate motif for synthesizing large-scale circuits. J. Roy. Soc. Interface **8**(62), 1281–1297 (2011)
3. Qian, L., Winfree, E., Bruck, J.: Neural network computation with DNA strand displacement cascades. Nature **475**(7356), 368–372 (2011)
4. Seelig, G., Soloveichik, D., Zhang, D., et al.: Enzyme-free nucleic acid logic circuits. Science **314**(5805), 1585–1588 (2006)
5. Carell, T.: Molecular computing: DNA as a logic operator. Nature **469**(7328), 45–46 (2011)
6. Song, T., Eshra, A., Shah, S., et al.: Fast and compact DNA logic circuits based on single-stranded gates using strand-displacing polymerase. Nat. Nanotechnol. **14**(11), 1075–1081 (2019)
7. Qian, L., Winfree, E.: Scaling up digital circuit computation with DNA strand displacement cascades. Science **332**(6034), 1196–1201 (2011)
8. Tang, Z., Yin, Z., Wang, L., et al.: Solving 0–1 integer programming problem based on DNA strand displacement reaction network. ACS Synth. Biol. **10**(9), 2318–2330 (2021)
9. Zhang, Q., Wang, X., Wang, X., et al.: Solving probability reasoning based on DNA strand displacement and probability modules. Comput. Biol. Chem. **71**, 274–279 (2017)
10. Lapteva, A.P., Sarraf, N., Qian, L.: DNA strand-displacement temporal logic circuits. J. Am. Chem. Soc. **144**(27), 12443–12449 (2022)
11. Adleman, L.M.: Molecular computation of solutions to combinatorial problems. Science **266**(5187), 1021–1024 (1994)
12. Head, T., Rozenberg, G., Bladergroen, R.S., et al.: Computing with DNA by operating on plasmids. Biosystems **57**(2), 87–93 (2000)
13. Yang, J., Zhang, C., Xu, J., et al.: A novel computing model of the maximum clique problem based on circular DNA. Sci. China Inf. Sci. **53**, 1409–1416 (2010)
14. Zhao, S., Yu, L., Yang, S., et al.: Boolean logic gate based on DNA strand displacement for biosensing: current and emerging strategies. Nanoscale Horiz. **6**(4), 298–310 (2021)
15. Xu, J., Qiang, X., Zhang, K., et al.: A DNA computing model for the graph vertex coloring problem based on a probe graph. Engineering **4**(1), 61–77 (2018)
16. Wang, Y.F., Yuan, G., Sun, J.: Four-input multi-layer majority logic circuit based on DNA strand displacement computing. IEEE Access **8**, 3076–3086 (2019)

Convolutional Codes Based Index-Free Coding Strategy for High-Density DNA Storage

Wanqing Chen, Zixiao Zhang, Zuqi Liu, and Fei Xu[✉]

Key Laboratory of Image Information Processing and Intelligent Control
of Education Ministry of China, School of Artificial Intelligence and Automation,
Huazhong University of Science and Technology, Wuhan 430074, China
fxu@hust.edu.cn

Abstract. DNA data storage has become a promising solution for large-scale data storage as the cost of high-throughput DNA synthesis and DNA sequencing drops rapidly. In previous DNA data storage methods, the overall storage density is compromised as a specific portion of DNA bases are used as index segments to facilitate data recovery. In this work, a novel DNA storage strategy termed as Index-Free High-Density Convolutional Codes (IFHDCC) is introduced. Due to the serial coding nature of the convolutional code in the IFHDCC strategy, the correlation of data segment information between groups is fully exploited, the same organisational efficiency as index segment method is achived without sacrificing storage density. IFHDCC strategy generates DNA sequences compatible with current synthesis and sequencing technologies by superimposing pseudo-random sequences and employing row and column interleaving methods. To evaluate IFHDCC, we have constructed a dataset containing videos, images, audios, and documents. The computer simulation experiments conducted on the dataset show that the information density of our strategy is 1.75 bits/nt, which is a great improvement over the DNA Fountain strategy (1.57 bits/nt). Furthermore, the convolution matrix used in IFHDCC adds an additional layer of security, DNA sequences can only be decoded accurately with the correct convolution matrix, enhancing the security of information transmission.

Keywords: DNA storage · Coding strategy · Inter-group correlation · Convolution codes

1 Introduction

With the rapid development of information technology, the discrepancy between the exponential growth of data and the limited capacity of current mainstream

This work is supported by the National Key Research and Development Program of China (2021YFF1200200), the National Natural Science Foundation of China (62072201) and Provincial Key R&D Program of Hubei (2021BAA168).

© The Author(s), under exclusive license to Springer Nature Singapore Pte Ltd. 2024
L. Pan et al. (Eds.): BIC-TA 2023, CCIS 2061, pp. 384–395, 2024.
https://doi.org/10.1007/978-981-97-2272-3_30

storage media has become increasingly conspicuous. The pressure on data storage has led to an urgent requirement for more efficient storage mediums. DNA, with its attractive advantages of high information density, remarkable durability, robust replicability, and low maintenance cost, emerges as a medium with great potential for data storage [1,2]. A series of validation studies has demonstrated the immense value of DNA in the field of information storage [3–10].

DNA data storage consists of six main components: encoding, synthesis, storage, retrieval, sequencing, and decoding [11]. Specifically, binary files are typically mapped into corresponding base sequences consisting of Adenine (A), Thymine (T), Cytosine (C), and Guanine (G) using specific transcoding rules. Due to the synthesis limitations [12], long sequences are divided into shorter fragments of equal length. Each fragment contains an index segment to record address information [8]. The ends of these fragments are tagged with primers for synthesis into gene fragments or an oligonucleotide pool, which are then stored either in vivo or in vitro [13–15]. During retrieval, Polymerase Chain Reaction (PCR) amplification identifies specific files, allowing the splicing of short fragments into longer sequences for decoding and original file recovery [16].

Previous studies have explored the maximum Shannon information capacity in DNA storage [6,17]. Ideally, the information capacity could reach 2 bits per nucleotide, considering the four possible options: A, T, C, and G. However, DNA encoding faces physical limitations, such as difficulties in synthesizing sequences with high GC content and long homopolymer runs (e.g., AAAAAAA, TTTTTT...) [18,19], which are prone to sequencing errors. The processes of oligonucleotide synthesis, PCR amplification, and DNA storage can introduce deletion, insertion, and substitution errors [3,12,18,20]. To improve the decoding correctness, it is necessary to consider DNA biochemical constraints and introduce error correction codes while designing DNA coding algorithms. Nevertheless, the redundancy, along with the space occupied by index segments in short fragments, reduces actual DNA storage density compared to the theoretical maximum [3–5,7,8,21].

Various coding schemes have been proposed to address these challenges [21–25]. In 2013, Goldman et al. [7] applied Huffman coding and trinary codec conversion to cut the source file into separate blocks of data fixed at 117-bit in length, including a 17-bit base-length index segment and a "check digit" DNA fragment to detect base errors, effectively increasing the coding potential to 1.58 bits/nt. In 2016, Bornholt et al. [4] improved Goldman's coding scheme using the XOR coding method, resulting in information density of 0.88 bits/nt. In 2016, Blawat et al. [3] proposed forward error correction coding methods with information density up to 0.92 bits/nt. However, these encoding schemes with index segments limit further increases in storage density. To tackle this issue, the DNA fountain algorithm [6] proposed by Y. Erlich's research team increased the information potential to 1.98 bits/nt. Nevertheless, the main drawback of the algorithm was the risk of decoding failure due to the Luby transform when dealing with specific binary features. Thus, the algorithm introduced enough logical redundancy to guarantee the correct decoding rate, resulting in an actual information density of 1.57 bits/nt.

In this work, we propose an innovative DNA storage strategy called Index-Free High-Density Convolutional Codes (IFHDCC) to enhance information storage density. Pseudo-random overlay coding and row interleaved coding are performed before convolutional coding to achieve GC content between 45–55% and homopolymer length ≤ 5. In the process of convolutional coding, long binary information is separated into short segments, corresponding to long DNA sequences divided into 160-nt short segments. The encoding result of each short segment is not only related to its own information bits but also to the information bits of the previous short segment. The IFHDCC leverages the information correlation of each short clip to remove the index segments and increase the storage density. We designed a computer simulation experiment in which files of different data types were compressed into a compressed package and coded and decoded using the IFHDCC proposed in this work to achieve an information density of 1.75 bits/nt. The contributions of this work are summarized as follows:

1. Pseudo-random overlay coding and row interleaved coding, converting binary source code to quadratic code to maintain GC content around 50%. The approach, along with row interleaved coding before and after convolutional coding, effectively reduces homopolymer length, aligning with DNA biochemical constraints.
2. The design of IFHDCC exploits code group correlations, eliminating index segments and increasing DNA information storage density to 1.75 bits/nt, which is 11.5% higher than the information density of the previous work in simulated conditions.
3. The convolution matrix in IFHDCC enhances data security, as DNA sequences can only be correctly decoded with the correct matrix, thereby securing information transmission.

2   IFHDCC Strategy

In this section, we introduce the principle of IFHDCC strategy and demonstrate the encoding and decoding processes in detail with some examples. The role of encoder is to encode a binary file into DNA sequences, ensuring compatibility with contemporary synthesis and sequencing technologies. Additionally, we delineate the process of decoding data from these DNA sequences and its subsequent restoration to the original file format. The IFHDCC strategy is a perfect replacement for index segments and has error correction properties to some extent. The inspiration comes from convolutional coding [26–28], proposed by Elias in 1955, which has been widely used in various fields of communication.

2.1   Encoders

Traditional storage medium is fundamentally different from DNA. The former is to write and read information in order. While the latter, oligos, are randomly mixed in the solution, meaning that it is quite difficult to find one or more oligos with specific information. Because the file is divided into many blocks, and then is encoded into many oligos. In order to restore this file completely, a common method is to design an index segment and order each oligonucleotide by number, which is also often used in previous studies [3–5,7,8,21], but the index segments will further lead to the decrease of encoding density.

To solve the low density caused by index segments, the IFHDCC strategy proposed in this work utilizes the correlation between groups of information, removes the index segments while having the same effect as indexing, and stores the entire DNA strand as a data segment, thus effectively improving the encoding density.

Assuming the maximum length for DNA synthesis is limited to only 4-nt, we need to store a binary sequence, denoted as A_1, within the DNA. The specific bit sequence to be encoded is as follows:

$$A_1 = \begin{bmatrix} 01\,11\,00\,10\,10\,01 \end{bmatrix}.$$

It converts the sequence A_1 into the following quadratic sequence by translating 00, 01, 10, 11 into 0, 1, 2, 3 respectively:

$$A_2 = \begin{bmatrix} 1\,3\,0\,2\,2\,1 \end{bmatrix}.$$

The sequence A_2 is input into a convolutional encoder to generate a new data sequence B_1. The components of B_1 belong to the finite field of four elements, 0, 1, 2, 3. The parameters of the encoder are set to (2, 1, 2). Here, the first two parameters indicate that two symbols are output for every input symbol, and the last parameter represents the number of shift registers used. Generally, the initial state of the shift register is 0. The encoding process involves matrix multiplication. Specifically, a 1-symbol subsequence of A_2 (represented as a column vector) is multiplied by a matrix G, which has 2 rows and 3 columns, to produce 2 symbols for B_1. The matrix G is configured as follows:

$$G = \begin{bmatrix} 1\,1\,1 \\ 1\,0\,1 \end{bmatrix} \tag{1}$$

Furthermore, to account for the initial state of the register, two additional zeros should be appended to the beginning of the sequence A_2. Consequently, the updated sequence, now denoted as A_3, is modified as follows:

$$A_3 = \begin{bmatrix} 0\,0\,1\,3\,0\,2\,2\,1 \end{bmatrix}.$$

During the encoding process, the matrix G is multiplied with a subsequence of A_3, specifically $\{0, 0, 1\}$, to produce the first two symbols of the sequence B_1, which are $\{1, 1\}$. This calculation is illustrated as follows:

$$\begin{bmatrix} 1\,1\,1 \\ 1\,0\,1 \end{bmatrix} \begin{bmatrix} 0 \\ 0 \\ 1 \end{bmatrix} \ (\text{mod } 4) = \begin{bmatrix} 1 \\ 1 \end{bmatrix}. \tag{2}$$

Subsequently, the subsequence of A_3 shifts one step forward, changing to $\{0, 1, 3\}$, and is then multiplied with the matrix G. This process is repeated until the final symbol of A_3 is reached. As a result of these calculations, the sequence B_1 is obtained as follows:

$$B_1 = \begin{bmatrix} 1\,1\,0\,3\,0\,1\,1\,1\,0\,2\,1\,3 \end{bmatrix}.$$

However, considering our initial assumption that the maximum length of an oligonucleotide is 4-nt, the sequence B_1 must be divided into three blocks to adhere to this constraint. It is important to note that there is no need to add an index to each block. Consequently, the sequence B_1 is updated to B_2, which is structured as follows:

$$B_2 = \begin{bmatrix} 1\,0\,0 \\ 1\,1\,2 \\ 0\,1\,0 \\ 3\,1\,3 \end{bmatrix}.$$

In practice, after thorough analysis, we have discovered that this operation can be further simplified by performing a matrix convolution of G with the matrix composed of the subsequences of A_3. This approach is adopted to facilitate parallel processing by computers, as demonstrated in the following illustration:

$$\begin{bmatrix} 1\,1\,1 \\ 1\,0\,1 \end{bmatrix} \begin{bmatrix} 0\,1\,0 \\ 0\,3\,2 \\ 1\,0\,2 \\ 3\,2\,1 \end{bmatrix} \pmod 4 = \begin{bmatrix} 1\,0\,0 \\ 1\,1\,2 \\ 0\,1\,1 \\ 3\,1\,3 \end{bmatrix}. \tag{3}$$

The resulting calculation is identical to that of B_2. It should be noted that the red part of Eq. (3) corresponds directly with the blue part. This consistency is a characteristic feature of the convolutional encoding process. Additionally, the first two zeros present in the first column of the matrix signify the initial state of the shift register.

Sequences B_2 are then translated into oligonucleotides by converting the elements $\{0,1,2,3\}$ to their respective nucleotide bases $\{A, T, C, G\}$. This marks the completion of the encoding process. Consequently, we obtain the following sequences, which are the actual sequences encoded into the DNA:

$$B_3 = \begin{bmatrix} T\,A\,A \\ T\,T\,C \\ A\,T\,T \\ G\,T\,G \end{bmatrix}.$$

Two critical issues still require attention: the presence of long homopolymer runs and high GC content, both of which are associated with increased dropout rates. To address the issue of long homopolymer runs, the use of interleaved coding can be effective. For instance, consider the scenario where the information in sequence N_1 needs to be stored in DNA:

$$N_1 = \begin{bmatrix} T\,A\,A\,A\,A\,A\,C\,G\,T\,T\,G\,G \end{bmatrix}.$$

The sequence N_1 is divided equally into two segments, resulting in the following:

$$N_2 = \begin{bmatrix} T\,A\,A\,A\,A\,A \\ C\,G\,T\,T\,G\,G \end{bmatrix}.$$

The sequence N_2 is read sequentially by column, yielding the following result:

$$N_3 = \begin{bmatrix} T\,C\,A\,G\,A\,T\,A\,T\,A\,G\,A\,G \end{bmatrix}.$$

To address the issue of high GC content, we can employ the method of integrating a pseudo-random sequence into the information source, thereby reducing the overall GC content. Consider an extreme case where the sequence S exhibits a very high GC content when its contents are mapped to nucleotide bases. This scenario is illustrated as follows:

$$S = \begin{bmatrix} 2\,2\,2\,2\,2\,2\,3\,3\,3\,0\,0\,0 \end{bmatrix}.$$

We apply specific rules to generate a pseudo-random sequence, denoted as P, which is outlined as follows:

$$P = \begin{bmatrix} 3\,1\,2\,0\,3\,1\,0\,2\,1\,3\,1\,0 \end{bmatrix}.$$

We execute a bit-by-bit addition of the sequence S and the pseudo-random sequence P, taking the modulus 4 of the result. This operation yields the sequence H, as shown below:

$$H = \begin{bmatrix} 1\,3\,0\,2\,1\,3\,3\,1\,0\,3\,1\,0 \end{bmatrix}.$$

It is important to note that both of the aforementioned methods are reversible, provided that the parameters used in the process are known.

The encoding process outlined above constitutes the coding component of the IFHDCC strategy, as detailed in Algorithm 1. Key parameters within this strategy are defined as follows: S represents the source file designated for encoding into DNA, M indicates the length of the oligonucleotide, and the tuple (n, k, r) specifies the parameters of the convolutional encoder. Within this context, k refers to the number of input characters, n to the number of output characters, and r to the number of shift registers.

Algorithm 1. Encode

Input: S, G
Output: C
1: reconstructe S by interleaved coding and adding pseudo-random sequence
2: $S = \begin{bmatrix} s_1 & s_2 & \cdots & s_b \end{bmatrix}$
3: $S' = \begin{bmatrix} 0_{r \times 1} & s_{1[-r:]} & \cdots & s_{b-1[-r:]} \\ s_1 & s_2 & \cdots & s_b \end{bmatrix}$
4: $C = (G \cdot S') \pmod 4$
5: recombinate C only by interleaved coding
6: **return** C

The term $s_{b-1[-r:]}$ represents the last r characters of s_{b-1}. This segment plays a crucial role in accurately sequencing the jumbled oligonucleotides present in the solution, as its function is analogous to that of an index. Additionally, it's evident that the coding efficiency of this method is represented by the ratio k/n, while the information capacity is expressed as $2k/n$. Furthermore, G denotes the convolutional matrix, which is presented in the form of a block matrix as shown below:

$$G = \left[A_{n \times r} \middle| \begin{matrix} B_{1 k \times k} \\ B_{2 (n-k) \times k} \end{matrix} \right]_{n \times (r+k)}. \tag{4}$$

Importantly, in our work, each symbol within the matrix G is assigned, with uniform probability, to be any element from the set $\{0, 1\}$. Additionally, for ease of calculation, we have configured B_1 as an identity matrix.

2.2 Decoders

We now describe the decoding process. Assuming that we have accurately sequenced the DNA, resulting in the sequence B_2. Additionally, we have knowledge of both the convolution matrix and the initial state of the register. In this context, x_1 represents the first input character. This implies that at least one of the following equations can be solved:

$$\begin{cases} \begin{bmatrix} 1 & 1 & 1 \\ 1 & 0 & 1 \end{bmatrix} \begin{bmatrix} 0 \\ 0 \\ x_1 \end{bmatrix} \ (\mathrm{mod}\ 4) = \begin{bmatrix} 1 \\ 1 \end{bmatrix} \\[12pt] \begin{bmatrix} 1 & 1 & 1 \\ 1 & 0 & 1 \end{bmatrix} \begin{bmatrix} 0 \\ 0 \\ x_1 \end{bmatrix} \ (\mathrm{mod}\ 4) = \begin{bmatrix} 0 \\ 1 \end{bmatrix} \\[12pt] \begin{bmatrix} 1 & 1 & 1 \\ 1 & 0 & 1 \end{bmatrix} \begin{bmatrix} 0 \\ 0 \\ x_1 \end{bmatrix} \ (\mathrm{mod}\ 4) = \begin{bmatrix} 0 \\ 2 \end{bmatrix} \end{cases}. \tag{5}$$

Equation (5) can be further simplified into a matrix multiplication form, as illustrated below:

$$\begin{bmatrix} 1 & 1 & 1 \\ 1 & 0 & 1 \end{bmatrix} \begin{bmatrix} 0 & 0 & 0 \\ 0 & 0 & 0 \\ x_1^{(1)} & x_1^{(2)} & x_1^{(3)} \end{bmatrix} \ (\mathrm{mod}\ 4) = \begin{bmatrix} 1 & 0 & 0 \\ 1 & 1 & 2 \end{bmatrix}. \tag{6}$$

By dividing the matrix along the dotted line and suitably reformulating the equation, we can derive the following equation:

$$\begin{bmatrix} 1 \\ 1 \end{bmatrix} \begin{bmatrix} x_1^{(1)} & x_1^{(2)} & x_1^{(3)} \end{bmatrix} = \begin{bmatrix} 1 & 0 & 0 \\ 1 & 1 & 2 \end{bmatrix}. \tag{7}$$

This implies that we only need to solve the following equation:

$$1 \begin{bmatrix} x_1^{(1)} & x_1^{(2)} & x_1^{(3)} \end{bmatrix} = \begin{bmatrix} 1 & 0 & 0 \end{bmatrix}, \tag{8}$$

and then use the obtained solution for validation purposes:

$$1 \left[x_1^{(1)} \ x_1^{(2)} \ x_1^{(3)} \right] = \left[1 \ 1 \ 2 \right] . \tag{9}$$

Clearly, the only correct solution is $x_1^{(1)} = 1$. This indicates that the sequence $[1, 1, 0, 3]$ corresponds to sequence number one, and the first decoded character is 1. However, it's important to note that achieving a result may not always be straightforward after just one round of solution and verification. Often, this process needs to be repeated multiple times. Once the first character has been decoded, the equation to decode the second character can be formulated as follows:

$$\begin{bmatrix} 1 \ 1 \ 1 \\ 1 \ 0 \ 1 \end{bmatrix} \begin{bmatrix} 0 \\ 1 \\ x_2 \end{bmatrix} \ (\mathrm{mod} \ 4) = \begin{bmatrix} 0 \\ 3 \end{bmatrix}, \tag{10}$$

and the second decoded character is 3. By following these steps and iteratively applying this process, the entire decoding procedure can be completed without the need for index segments.

Correspondingly, the detailed decoding algorithm can be found in Algorithm 2.

Algorithm 2. Decode
___
Input: $C = \left[c_1 \ c_2 \ \cdots \ c_b \right]$, G, $J = 0_{r \times 1}$
Output: S
1: restore C by deinterleaved coding
2: **for** $i = 1; i \le b; i + + $ **do**
3:     **for** $j = 1; j \le b; j + + $ **do**
4:         $D = (c_j - AJ + 4) \, \%4$
5:         $D \rightarrow \begin{bmatrix} D_{1_{k \times 1}} \\ D_{2_{(n-k) \times 1}} \end{bmatrix}$
6:         $B_1 X = D_1 \ \rightarrow X$
7:         **if** $(B_2 X = D_2)$ **then**
8:             $J = \begin{bmatrix} J_{[k:]} \\ X \end{bmatrix}$
9:         **end if**
10:     **end for**
11: **end for**
12: **return** S
___

In step 6 of the algorithm, we circumvent the need for complex matrix inversion operations by setting B_1 as a unit matrix.

3   Results

In this section, we introduce the results of computer simulation experiments. In order to demonstrate the efficiency of the IFHDCC algorithm and quantify its featured parameters in comparison with other early DNA-based storage coding schemes, we constructed a dataset as the original input data and encoded this dataset by using the IFHDCC algorithm for comparing the encoding performance. The entire computer simulation experiment was conducted in a Windows 10 environment running on an i7 CPU with 16GB of random access memory using Python 3.11.5.

The input dataset was in the form of a compressed package, including images of the South Gate of Huazhong University of Science and Technology (HUST), a HUST promotional video, a piano song, and a PDF document as shown in Table 1.

Table 1. The input data.zip (0.76 Mbyte).

Size	Type	Name
18 Kbyte	Image	hust.jpg
86 Kbyte	Music	Peter Jeremias Coming Home.mp3
598 Kbyte	Video	hust1.5s.mp4
76 Kbyte	PDF	DNAfountain-1.pdf

We encoded the dataset of 799744 bytes using IFHDCC, where parameters were set as: $M = 160$, $(n, k, r) = (16, 14, 28)$. The input file was reconstructed by the row and column interleaving method and superimposed pseudo-random sequences, and the input file was split into 3198550 segments, followed by repeating the encoding steps of the IFHDCC strategy to create valid oligonucleotides. The length of the DNA oligonucleotides was $320/2 = 160$-nt. After using the interleaved coding and addition of pseudo-random sequence methods, the length of the homopolymer in the DNA sequences was less than 5-nt and the GC content was 52%, whereas without these two methods, the length of the homopolymer amounted to 7-nt and the GC content was 60%. Decoding the oligos pool fully recovered the entire input file with zero errors, according to Algorithm 2. The entire experiment took 3.5 s to encode, while decoding took nine minutes. We achieved an information density of 1.75 bits/nt, which is 11.5% higher than the information density of the previous work in simulated conditions as shown in Table 2.

Table 2. Comparison of DNA storage coding schemes and experimental results.

Parameter	Church	Goldman	Grass	Bornholt	Blawat	Erlich	This work
Input data (Mbytes)	0.65	0.75	0.08	0.15	22	2.15	0.76
Robustness to dropouts	No	Repetition	RS	Repetition	RS	Fountain	Convolution
Error correction/detection	No	Yes	Yes	No	Yes	Yes	Yes
Net information density (bits/nt)	0.83	0.33	1.14	0.88	0.92	1.57	1.75
Realized capacity	45%	18%	62%	48%	50%	86%	87.5%
Number of oligos	54,898	153,335	4,991	151,000	1,000,000	72,000	22,847

In light of these results, IFHDCC offers the opportunity to generate DNA sequences that are highly adaptable to synthesis and sequencing processes while maintaining a relatively high information density. This is essential for improving the utility of DNA storage.

4 Conclusion

In this work, we propose an efficient DNA storage strategy called IFHDCC, which has several advantages. Convolutional coding in the IFHDCC strategy ensures that the coding result of each fragment depends not only on its own information bits but also on the information bits of the previous fragment, which exploits the information relevance of each short fragment to remove the indexed fragment and increases the storage density. The pseudo-random superposition and row interleaving methods in the IFHDCC strategy can generate DNA sequences that satisfy biochemical constraints. Due to the key property of convolution matrix in IFHDCC, the original file can be decoded correctly only if the correct convolution matrix is known, which improves the security of information delivery. The IFHDCC strategy can dynamically adjust the parameter values of the convolutional encoder, thus obtaining different coding schemes with different coding densities and robustness for different file types, which provides a certain degree of flexibility. We conducted computer simulation experiments to compress files such as images and videos into compressed packages. The data is then encoded and decoded by the IFHDCC strategy. The experimental results show that our strategy increases the DNA information storage density to 1.75 bits/nt, which is 11.5% higher than the previous strategies.

Despite these achievements, there are still several areas that need to be further explored and improved. For example, it is important to investigate whether the performance of encoded sequences has different robustness depending on the different distributions of the convolution matrix. The IFHDCC strategy in the nature of serial coding is difficult to solve the sequence loss problem and needs to be combined with other external codes to improve the robustness. Our strategy is only simulated on a computer, and experimental validation by DNA synthesis and sequencing will further guide the improvement and refinement of the algorithm if it is implemented in subsequent work.

References

1. Bancroft, C., Bowler, T., Bloom, B., Clelland, C.: Long-term storage of information in DNA. Science **293**(5536), 1763–1765 (2001)
2. Wallace, M.: Molecular cybernetics: the next step? Kybernetes **7**(4), 265–268 (1978)
3. Blawat, M., Gaedke, K., Huetter, I., Chen, X., Turczyk, B., Inverso, S., Pruitt, B., Church, G.: Forward error correction for DNA data storage. Procedia Comput. Sci. **80**, 1011–1022 (2016)

4. Bornholt, J., Lopez, R., Carmean, D., Ceze, L., Seelig, G., Strauss, K.: A DNA-based archival storage system. In: Proceedings of the Twenty-First International Conference on Architectural Support for Programming Languages and Operating Systems, pp. 637–649 (2016)

5. Church, G., Gao, Y., Kosuri, S.: Next-generation digital information storage in DNA. Science **337**(6102), 1628 (2012)

6. Erlich, Y., Zielinski, D.: DNA fountain enables a robust and efficient storage architecture. Science **355**(6328), 950–954 (2017)

7. Goldman, N., et al.: Towards practical, high-capacity, low-maintenance information storage in synthesized DNA. Nature **494**(7435), 77–80 (2013)

8. Grass, R., Heckel, R., Puddu, M., Paunescu, D., Stark, W.: Robust chemical preservation of digital information on DNA in silica with error-correcting codes. Angew. Chem. Int. Ed. **54**(8), 2552–2555 (2015)

9. Tabatabaei Yazdi, S., Yuan, Y., Ma, J., Zhao, H., Milenkovic, O.: A rewritable, random-access DNA-based storage system. Sci. Rep. **5**(1), 1–10 (2015)

10. Yazdi, S., Kiah, H., Garcia-Ruiz, E., Ma, J., Zhao, H., Milenkovic, O.: DNA-based storage: trends and methods. IEEE Trans. Mol. Biol. Multi-Scale Commun. **1**(3), 230–248 (2015)

11. Jordan, B.: DNA for information storage? Med. Sci.: M/S **34**(6–7), 622–625 (2018)

12. Yazdi, S., Gabrys, R., Milenkovic, O.: Portable and error-free DNA-based data storage. Sci. Rep. **7**(1), 1–6 (2017)

13. Nguyen, H., et al.: On-chip fluorescence switching system for constructing a rewritable random access data storage device. Sci. Rep. **8**(1), 1–11 (2018)

14. Shah, S., Dubey, A., Reif, J.: Programming temporal DNA barcodes for single-molecule fingerprinting. Nano Lett. **19**(4), 2668–2673 (2019)

15. Sheth, R., Wang, H.: DNA-based memory devices for recording cellular events. Nat. Rev. Genet. **19**(11), 718–732 (2018)

16. Ceze, L., Nivala, J., Strauss, K.: Molecular digital data storage using DNA. Nat. Rev. Genet. **20**(8), 456–466 (2019)

17. MacKay, D., Mac Kay, D., et al.: Information Theory, Inference and Learning Algorithms. Cambridge University Press, Cambridge (2003)

18. Ross, M., et al.: Characterizing and measuring bias in sequence data. Genome Biol. **14**(5), 1–20 (2013)

19. Schwartz, J., Lee, C., Shendure, J.: Accurate gene synthesis with tag-directed retrieval of sequence-verified DNA molecules. Nat. Methods **9**(9), 913–915 (2012)

20. Erlich, Y., et al.: DNA Sudoku—harnessing high-throughput sequencing for multiplexed specimen analysis. Genome Res. **19**(7), 1243–1253 (2009)

21. Ping, Z., et al.: Towards practical and robust DNA-based data archiving using the yin-yang codec system. Nat. Comput. Sci. **2**(4), 234–242 (2022)

22. Anavy, L., Vaknin, I., Atar, O., Amit, R., Yakhini, Z.: Data storage in DNA with fewer synthesis cycles using composite DNA letters. Nat. Biotechnol. **37**(10), 1229–1236 (2019)

23. Chen, W., et al.: An artificial chromosome for data storage. Natl. Sci. Rev. **8**(5), nwab028 (2021)

24. Choi, Y., et al.: High information capacity DNA-based data storage with augmented encoding characters using degenerate bases. Sci. Rep. **9**(1), 1–7 (2019)

25. Ren, Y., et al.: DNA-based concatenated encoding system for high-reliability and high-density data storage. Small Methods **6**(4), 2101335 (2022)
26. Elias, P.: Coding for noisy channels. IRE Conv. Rec. **3**, 37–46 (1955)
27. Forney, G.: The Viterbi algorithm. Proc. IEEE **61**(3), 268–278 (1973)
28. Viterbi, A.: Error bounds for convolutional codes and an asymptotically optimum decoding algorithm. IEEE Trans. Inf. Theory **13**(2), 260–269 (1967)

Author Index

© The Editor(s) (if applicable) and The Author(s), under exclusive license
to Springer Nature Singapore Pte Ltd. 2024
L. Pan et al. (Eds.): BIC-TA 2023, CCIS 2061, pp. 397–399, 2024.
https://doi.org/10.1007/978-981-97-2272-3

Printed in the United States
by Baker & Taylor Publisher Services